I0055292

New Developments in Industrial Microbiology

New Developments in Industrial Microbiology

Edited by Edward Grace

New York

Published by Syrawood Publishing House,
750 Third Avenue, 9th Floor,
New York, NY 10017, USA
www.syrawoodpublishinghouse.com

New Developments in Industrial Microbiology
Edited by Edward Grace

International Standard Book Number: 978-1-68286-740-2 (Hardback)

Cataloging-in-Publication Data

New developments in industrial microbiology / edited by Edward Grace.
p. cm.
Includes bibliographical references and index.
ISBN 978-1-68286-740-2
1. Industrial microbiology. 2. Microbial products. 3. Microbiology. I. Grace, Edward.
QR53 .N48 2019
660.62--dc23

TABLE OF CONTENTS

PREFACE

Industrial microbiology is a branch of biotechnology, which is concerned with the production of industrial products by the manipulation of microorganisms. It also involves maximizing product yields through gene amplification using plasmids and vectors. The production of antibiotics, enzymes, alcohol, vitamins, etc. is an important application of industrial microbiology. In agriculture, innovative biopesticides and microbial inoculants are applied to avoid the chemical use of pesticides and fertilizers. This book presents the complex subject of industrial microbiology in the most comprehensible and easy to understand language. It is a valuable compilation of topics, ranging from the basic to the most complex advancements in this domain. As this field is emerging at a rapid pace, the contents of this book will help the readers understand the modern concepts and applications of the subject.

This book is the end result of constructive efforts and intensive research done by experts in this field. The aim of this book is to enlighten the readers with recent information in this area of research. The information provided in this profound book would serve as a valuable reference to students and researchers in this field.

At the end, I would like to thank all the authors for devoting their precious time and providing their valuable contribution to this book. I would also like to express my gratitude to my fellow colleagues who encouraged me throughout the process.

Editor

1

Thermus thermophilus as source of thermozymes for biotechnological applications: homologous expression and biochemical characterization of an α-galactosidase

Martina Aulitto[1], Salvatore Fusco[2], Gabriella Fiorentino[1], Danila Limauro[1], Emilia Pedone[1], Simonetta Bartolucci[1] and Patrizia Contursi[1*]

Abstract

Background: The genus *Thermus*, which has been considered for a long time as a fruitful source of biotechnological relevant enzymes, has emerged more recently as suitable host to overproduce thermozymes. Among these, α-galactosidases are widely used in several industrial bioprocesses that require high working temperatures and for which thermostable variants offer considerable advantages over their thermolabile counterparts.

Results: *Thermus thermophilus HB27* strain was used for the homologous expression of the TTP0072 gene encoding for an α-galactosidase (*Tt*GalA). Interestingly, a soluble and active histidine-tagged enzyme was produced in larger amounts (5 mg/L) in this thermophilic host than in *Escherichia coli* (0.5 mg/L). The purified recombinant enzyme showed an optimal activity at 90 °C and retained more than 40% of activity over a broad range of pH (from 5 to 8).

Conclusions: *Tt*GalA is among the most thermoactive and thermostable α-galactosidases discovered so far, thus pointing to *T. thermophilus* as cell factory for the recombinant production of biocatalysts active at temperature values over 90 °C.

Keywords: α-Galactosidase, *Thermus thermophilus*, Thermozymes, Recombinant expression, Themostability

Background

After cellulose, the second most abundant biopolymer on earth is hemicellulose, an heterogeneous polymer of pentoses (xylose and arabinose) and hexoses (glucose, galactose, mannose) [1]. Among these, mannans comprise linear or branched polymers derived from sugars such as D-mannose, D-galactose and D-glucose. Moreover, they represent the major source of secondary cell wall found in conifers (softwood) and leguminosae [2]. Based on their sugar composition, mannans are classified in four subfamilies: i.e. mannans, glucomannans, galactomannans and galactoglucomannans. The concerted action of different hydrolytic enzymes such as β-glucosidases (EC 3.2.1.21), endo-mannanases (EC 3.2.1.78), mannosidases (EC 3.2.1.25) and α-galactosidases (EC 3.2.1.22) is needed to achieve the degradation of galactoglucomannans, the most complex mannans subfamily [3].

α-Galactosidases (α-D-galactoside galactohydrolase EC 3.2.1.22) are exoglycosidases that catalyse the cleavage of the terminal non-reducing α-1,6-linked galactose residues present in different galactose-containing oligo- and polysaccharides. α-Galactosidases are widely distributed in microorganisms, plants, animals and mammalians, including humans [4]. The localization of α-galactosidases can be cytoplasmic (e.g. *Escherichia*

*Correspondence: contursi@unina.it
[1] Dipartimento di Biologia, Università degli Studi di Napoli Federico II, Complesso Universitario Monte S. Angelo, Via Cinthia, 80126 Naples, Italy
Full list of author information is available at the end of the article

coli), lysosomal (e.g. *Homo sapiens*) or extracellular (e.g. yeast) [5, 6].

α-Galactosidases have a great potential in both biotechnological and medical applications. For instance, these enzymes are used for the treatment of Fabry's disease [7], in xenotransplantation [8] and in blood group transformation for safety transfusion [9]. Furthermore, α-galactosidases offer a promising solution for the degradation of raffinose in beet sugar industry [10], pulp and paper manufacturing [11] as well as in soy food [12] and animal feed processing [13].

The interest towards enzymes hydrolysing hemicellulose, such as α-galactosidases, has particularly increased in recent years. Indeed, they are extensively used in synergic combination with cellulases [14] for the production of bioethanol from lignocellulose [15]. In this regard, it is noteworthy that the pretreatment of raw lignocellulosic material requires elevated temperatures; therefore, thermostable and thermoactive enzymes are suitable for this purpose [16], since they can withstand high temperature, extreme pH and pressure, as well as the presence of some inhibitors, including toxic metals. Furthermore, as catalysts, thermostable enzymes might provide additional advantages, since higher temperatures often promote: (1) better enzyme penetration, (2) cell wall disorganisation of the raw materials, (3) increase of the substrate solubility and (4) reduction of contamination. Over all, these features may improve the global yield of the process [17].

In recent years, the thermophilic microorganism *T. thermophilus* has emerged has a rich source of polymer-degrading enzymes (e.g. α-amylases, xylanases, esterases, lipases, proteases and pullulanases). Indeed, this thermophilic bacterium is able to grow on different organic sources such as various proteinaceous and carbohydrates substrates [18].

Although *E. coli* has been successfully employed to produce thermophilic recombinant proteins/enzymes [19–24], in some cases their expression level can be too low to perform functional and structural characterization as well as to exploit their biotechnological potential [25–27]. For this reason, efficient and reliable "hot" expression systems are needed.

Some genetic systems for both archaeal and bacterial thermophilic microorganisms have been designed [28–32]. However, the amount of the recombinant proteins is in general lower than that achieved through conventional mesophilic expression systems [33].

A thermophilic expression system for *T. thermophilus* HB27 was previously developed and proved to be suitable to achieve high expression levels of heterologous proteins [34, 35]. Therefore, we have used this system to produce a novel α-galactosidase (named *Tt*GalA) from *T. thermophilus* HB27. In particular, the coding gene was over-expressed in the native host and the recombinant enzyme was purified and characterized. *Tt*GalA exhibits an optimal hydrolytic activity at high temperature (90 °C and pH 6.0), a good catalytic efficiency ($k_{cat} = 709.7/s$) and a significant thermostability (30 h at 70 °C), which are all interesting features for industrial applications.

Results and discussion

The genus *Thermus* comprises thermophilic and hyperthermophilic bacteria [18], which represent genetic reservoirs of several thermozymes potentially useful for industrial bioprocesses at high temperature [36, 37]. This aspect, along with other intrinsic features such as the natural competence of many strains, high growth rates and biomass yields, make these thermophiles suitable models for the production of thermozymes [27, 38].

In this work we have used *T. thermophilus* HB27::*nar* as host for the homologous expression of a novel α-galactosidase (*Tt*GalA), one of the most thermoactive α-galactosidases identified so far.

Sequence analysis

An in-depth survey of the carbohydrate-active enzymes database CAZy (http://www.cazy.org/) was carried out to identify putative α-galactosidases in the genome of *T. thermophilus* HB27. The only protein sequence retrieved is encoded by TTP0072 (Accession No. AAS82402) and shows high sequence identity with previously characterized α-galactosidases from *Thermus* sp. strain T2 (75%; Accession No. BAA76597), *Thermus brockianus* (73%; Accession No. AF135398), *Thermotoga neapolitana* (36%; Accession No. AF011400), *Thermotoga maritima* (35%; Accession No. AJ001776), *Sulfolobus solfataricus* (39%; Accession No. Q97U94) and *H. sapiens* (23%; Accession No. CAA29232.1). In general, sequence identities among the analysed proteins fall into regions that are "signature" of the catalytic activity and/or of the structural properties of α-galactosidases. A multiple alignment (Fig. 1a) revealed the presence of a consensus motif (FEVFQIDDGW) in the TTP0072 translated sequence characteristic of α-galactosidases. This is generally localised within the central region of bacterial enzymes (CAZy family GH36) or at the amino-terminal region of eukaryotic variants (CAZy family GH27). The presence of such a sequence points out to a common reaction mechanism and/or a similar substrate binding site [39]. Two aspartic residues i.e. D193–194 underlined in the above consensus sequence are followed by a cysteine or a glycine in the GH27 or GH36 members, respectively. Moreover, GH27 and GH36 families share another common feature, which is the presence of two additional aspartic acids involved in the acid–base catalytic mechanism (D301 and D355 in the protein from *T. thermophilus* HB27,

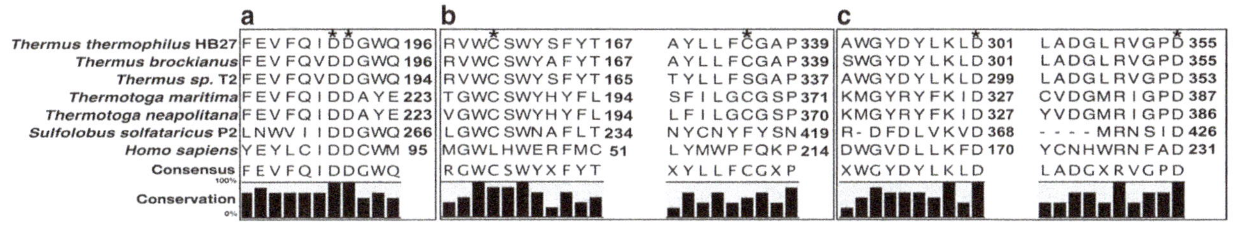

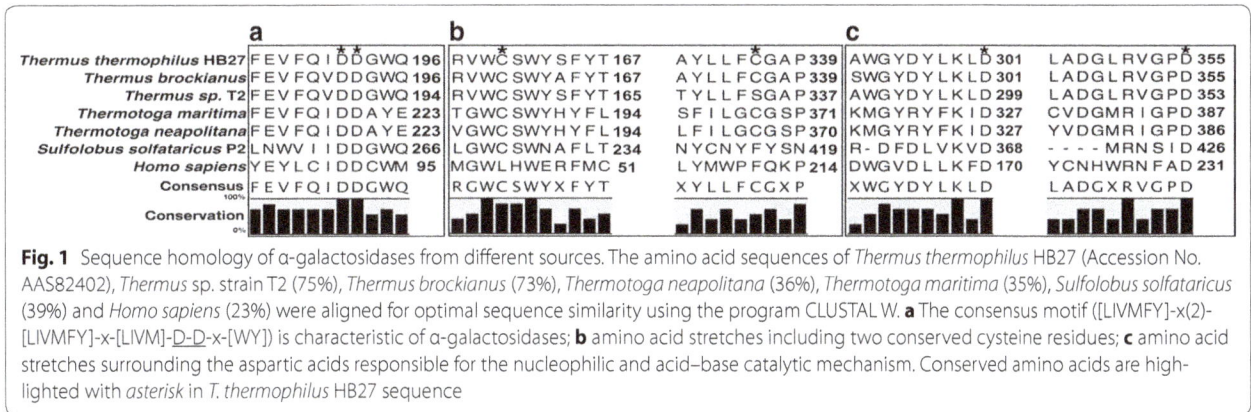

Fig. 1 Sequence homology of α-galactosidases from different sources. The amino acid sequences of *Thermus thermophilus* HB27 (Accession No. AAS82402), *Thermus* sp. strain T2 (75%), *Thermus brockianus* (73%), *Thermotoga neapolitana* (36%), *Thermotoga maritima* (35%), *Sulfolobus solfataricus* (39%) and *Homo sapiens* (23%) were aligned for optimal sequence similarity using the program CLUSTAL W. **a** The consensus motif ([LIVMFY]-x(2)-[LIVMFY]-x-[LIVM]-D-D-x-[WY]) is characteristic of α-galactosidases; **b** amino acid stretches including two conserved cysteine residues; **c** amino acid stretches surrounding the aspartic acids responsible for the nucleophilic and acid–base catalytic mechanism. Conserved amino acids are highlighted with *asterisk* in *T. thermophilus* HB27 sequence

Fig. 1). At consensus level, whereas the motif including the catalytic nucleophile (D301) is fully conserved [K(Y/V/L/W)**D**], the A/B aspartic acid motif (D355) is more variable (RXXX**D**) (Fig. 1c). The co-presence of these two motifs defines the sub-group identity of GH36bt (where "bt" stands for bacterial thermophilic) [40] that also includes the thermophilic *Thermotoga* and *Thermus* α-galactosidases. Another characteristic of the GH36bt enzymes is the presence of two conserved cysteine residues (C161 and C336) (Fig. 1b) whose functional, structural and stabilization role is questioned.

The TTP0072 gene is located on the megaplasmid pTT27 upstream of a gene encoding for the galactose-1-phosphate uridylyltransferase gene (*galT*) and partially overlapping it. This operon-like structure closely resembles that found in other *Thermus* spp. [41], thus suggesting that their functional association might be important for galactose metabolism.

Cloning, expression and purification of *Tt*GalA

The TTP0072 gene was cloned both in pET28(b) and in pMKE2 vectors to compare the expression levels of the N-terminal His-tagged enzymes *Ec*GalA and *Tt*GalA in the heterologous (*E. coli*) as well as in the homologous (*T. thermophilus* HB27::*nar*) hosts. In the case of *E. coli*, after a two-step purification procedure i.e. a thermal precipitation at 70 °C and an affinity chromatography through His-trap column, the final protein amount was very low (0.5 mg/L) (Fig. 2a) (Additional file 1: Table S1).

On the other hand, the enzyme *Tt*GalA was expressed at higher level in *T. thermophilus* HB27::*nar* with a final amount of 5 mg/L (tenfold higher than *Ec*GalA). Moreover, its specific activity (338 U/mg) was twofold higher than that of *Ec*GalA (Additional file 1: Tables S1, S2). As previously observed, the over-production of soluble and active enzymes in mesophilic hosts is in some cases limited by: (1) differences in the codon usage [42]; (2) domain misfolding during protein synthesis at a temperature (37 °C), which is far lower than the optimal growth temperature of the native host; (3) requirement of specific chaperone(s), cofactors, metals as well as genus- or species-specific post-translational modifications [27].

Given the significant higher amount of the soluble active enzyme produced by the thermophilic host, all the subsequent characterizations were performed on the *Tt*GalA enzyme. The *nar* promoter, which drives the expression of the genes cloned in the pMKE2 vector, is induced by the combined action of nitrate and anoxia in the facultative anaerobic derivatives of *T. thermophilus* HB27 [34]. In our tested conditions, the biomass yield of the culture achieved under anaerobic growth was rather low, thus negatively affecting the amount of enzyme produced by the cells. To overcome

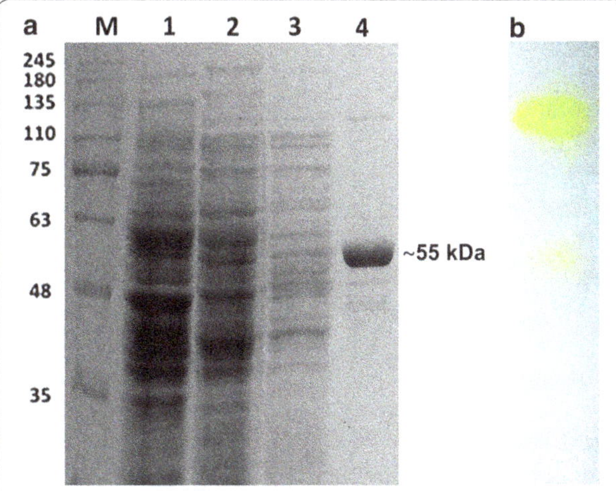

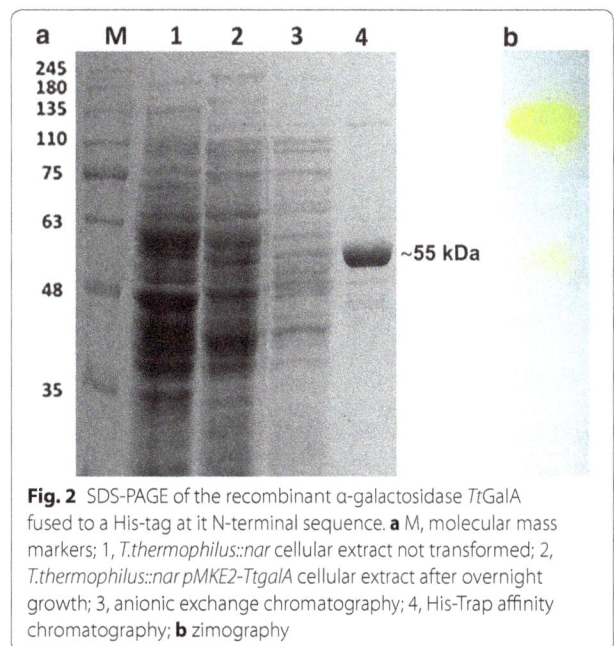

Fig. 2 SDS-PAGE of the recombinant α-galactosidase *Tt*GalA fused to a His-tag at it N-terminal sequence. **a** M, molecular mass markers; 1, *T.thermophilus::nar* cellular extract not transformed; 2, *T.thermophilus::nar pMKE2-TtgalA* cellular extract after overnight growth; 3, anionic exchange chromatography; 4, His-Trap affinity chromatography; **b** zimography

this limitation, the expression of *Tt*GalA was carried out by growing *T. thermophilus* HB27::*nar* cells aerobically to reach an higher biomass yields. A total protein amount, similar to that of the cells cultured anaerobically, was achieved.

We resolved to set up a purification protocol from crude extracts of aerobically cultured cells based on anionic exchange followed by affinity His-trap chromatography. The recombinant enzyme purified to homogeneity displays a single band on SDS-PAGE with an estimated molecular weight (MW) of 55 kDa (Fig. 2a), which is consistent with the predicted MW of a his-tagged monomer (55.8 kDa). The identity of the recombinant *Tt*GalA was verified by mass spectrometry (data not shown). It is important to note that the presence of the His-tag at the N-terminus of the recombinant enzyme ensures that the reported experiments are not affected by the presence of the endogenous enzyme. Furthermore, zymography revealed the presence of two bands with hydrolytic activity indicating that: (1) *Tt*GalA adopts an oligomeric structure; (2) the oligomer is far more active than the monomer and (3) it is partially resistant to the denaturing conditions employed in the SDS-PAGE. These results prompted us to further investigate on the quaternary structure of *Tt*GalA (Fig. 2b).

Characteristics of the recombinant *Tt*GalA

Determination of the molecular weight

To assess the quaternary structure, size-exclusion chromatography coupled with a triple-angle light scattering-QELS, was performed. This analysis showed a molecular weight of about 320 kDa ± 0.2% (RH = 8.1 nm ± 3%), thus indicating that *Tt*GalA is a hexamer in solution. This oligomeric structure is in agreement with that of some previously characterised α-galactosidases, which adopt dimeric (*Thermotoga maritima*) [43], trimeric (*Sulfolobus solfataricus*) [40], tetrameric (*T. brockianus*) [39], octameric (*Thermus* sp. *strain T2*) structure [44]. This complex oligomeric state might correlate with the stability at higher temperature of *Tt*GalA, as it was showed for *B. stearothermophilus* α-galactosidases [45].

To understand if C161 and C336 had a structural role (Fig. 1b), the purified *Tt*GalA was analysed on SDS-PAGE with and without a reducing agent. In the sample containing β-mercaptoethanol *Tt*GalA is present mainly in monomeric form, while in the absence of this reducing agent the protein forms essentially high MW oligomers (data not shown). Indeed, the widespread stabilizing role of intracellular disulphide bonds in thermophiles and hyperthermophiles, has been already established as a strategy for protein stabilization [46].

Catalytic and stability properties

*Tt*GalA is able to hydrolyze *p*NP-α-D-galactopyranoside, but shows negligible activity on both *p*NP-α-substituted (D-glucose, D-mannose, L-rhamnose) and *p*NP-β-substituted (D-galactose, D-glucose, and D-mannose) (Table 1). In particular, the enzyme has a barely detectable activity for β anomer of galactose (6.75 U/mg), which is 50-fold lower than activity over *p*NP-α-galactose (338.0 U/mg). Therefore, the kinetic enzymatic properties were determined, using *p*NP-α-D-galactopyranoside as substrate, at optimal pH and temperature (Table 2). *Tt*GalA shows higher affinity towards its substrate (K_M = 0.69 mM) compared to α-galactosidase from *Thermus* sp. strain T2 (K_M = 4.7 mM) [44] and *T. brockianus* (K_M = 2.5 mM) [39]. However, comparing the kinetic parameters between the most known thermoactive (T_{opt} 105 °C) α-galactosidase of *Thermotoga neapolitana* (*Tn*GalA) and the *Tt*GalA, the K_M value is very similar. Interestingly, the different catalytic constant of 152.5/s towards 709.7/s, for *Tn*GalA and *Tt*GalA respectively [47] (Table 2) reflects a great efficiency of *Tt*GalA on *p*NP-α-D-galactopyranoside substrate. This aspect constitutes an important criterion for employing different enzyme variants for industrial purposes. For istance, the high catalytic efficiency of *Tt*GalA makes it a suitable candidate to enhance pulp bleachability in combination with other hemicellulases exhibiting different catalytic specificity [11].

*Tt*GalA is among the most thermoactive and pH tolerant α-galactosidases known so far. Indeed, when assayed at different pH and temperatures, it exhibited an optimal hydrolytic activity at 90 °C and pH 6.0 (Figs. 3, 4). As reported for an α-galactosidase from

Table 1 Substate specificity of *Tt*GalA

Substrate	Specific activity (U/mg)
*p*NP-α-D-mannose	0
*p*NP-β-glucose	0.65
*p*NP-α-L-rhamnose	1.5
*p*NP-α-D-glucose	2.16
*p*NP-β-galactose	6.75
*p*NP-β-mannose	7.13
*p*NP-α-D-galactose	338.0

Table 2 Kinetic parameters for the hydrolysis *p*NPG hydrolysis at 90 °C by the *Tt*GalA

K_M (mM)	V_{max} (U/mg)	k_{cat} (/s)	k_{cat}/K_M (/s M)
0.69 ± 0.017	338.0 ± 7.9	709.7 ± 17.7	$1.03 \times 10^4 \pm 0.025 \times 10^4$

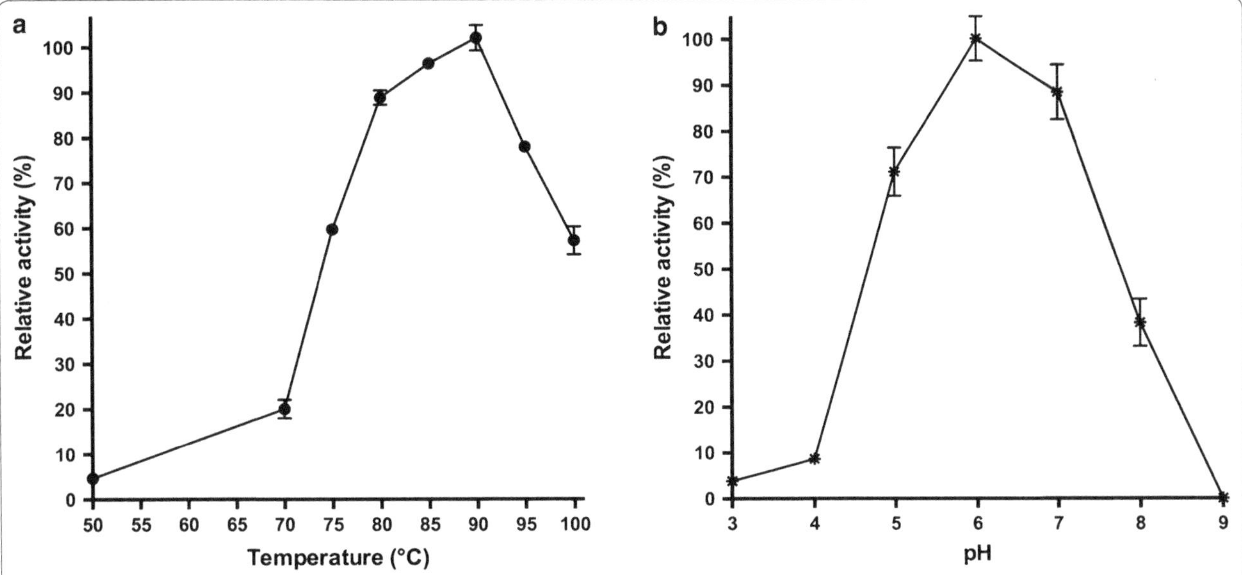

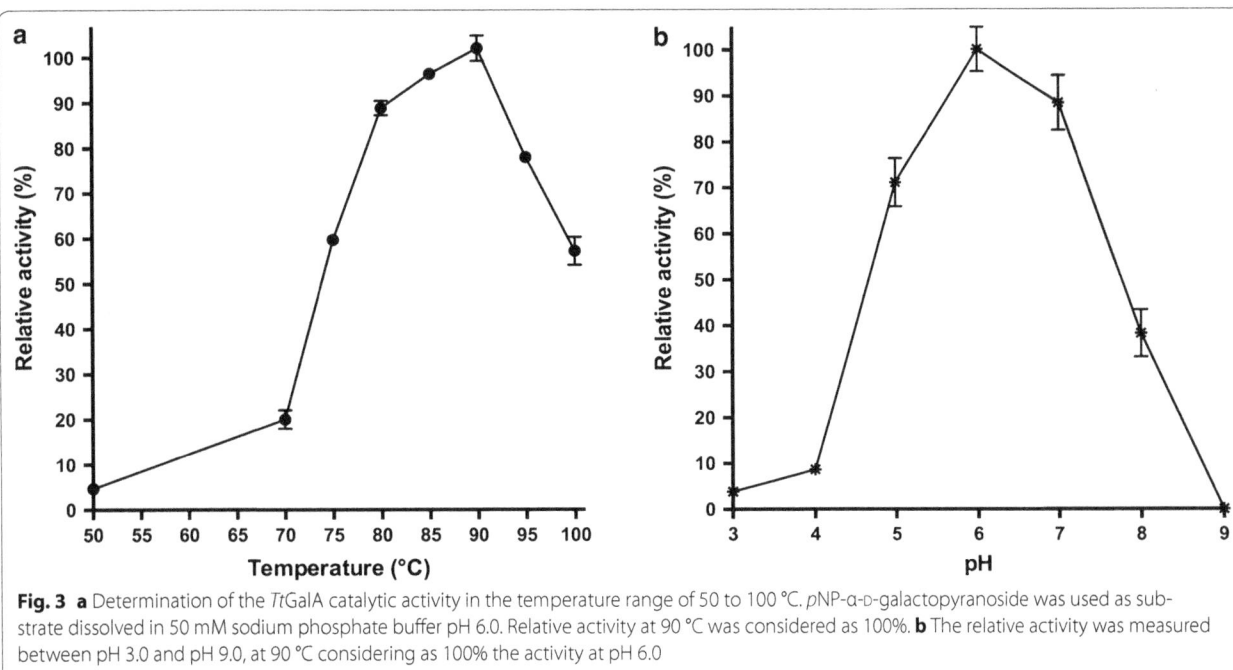

Fig. 3 **a** Determination of the *Tt*GalA catalytic activity in the temperature range of 50 to 100 °C. *p*NP-α-D-galactopyranoside was used as substrate dissolved in 50 mM sodium phosphate buffer pH 6.0. Relative activity at 90 °C was considered as 100%. **b** The relative activity was measured between pH 3.0 and pH 9.0, at 90 °C considering as 100% the activity at pH 6.0

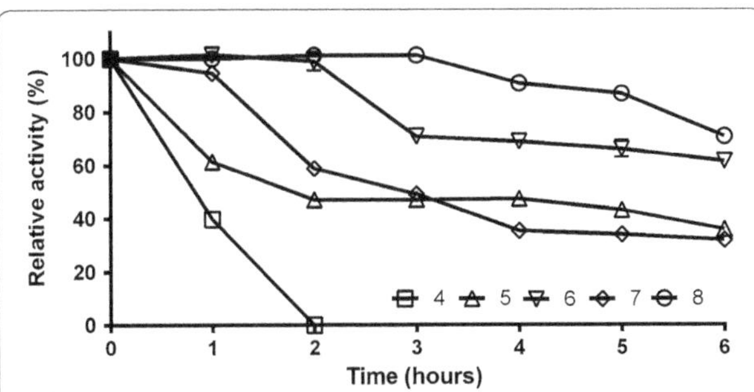

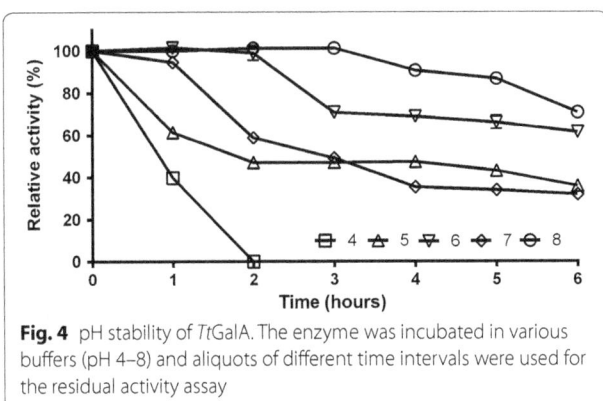

Fig. 4 pH stability of *Tt*GalA. The enzyme was incubated in various buffers (pH 4–8) and aliquots of different time intervals were used for the residual activity assay

Bacillus megaterium VHM1 (optimal pH 7–7.5) [48], it is possible to foresee the employment of *Tt*GalA for the removal of oligosaccharides from soya based foods, thus improving their nutritive value. Noteworthy, *Tt*GalA might be a better catalyst for this process, since its pH optimum (pH 6.0) is even closer to that of the soymilk hydrolysis process (pH 6.2–6.4).

Interestingly, the retained activity was greater than 40% within the pH range from 5.0 to 8.0, which is a quite wide tolerance range compared to other characterised thermostable α-galactosidases [40]. Therefore, pH shifts during on-going enzymatic reaction in industrial processes could have a minor impact on its activity (Fig. 3). The recombinant enzyme did not lose activity after 2 h of incubation at pH 6.0 and 8.0, and it retained up to 60% of residual activity after 6 h (Fig. 4). Approximately 50–60% loss in activity was recorded on either side of the pH optimum after 6 h incubation, while at pH 4.0, the activity rapidly dropped, as expected from the pH-dependence data (Fig. 4). Because of its considerable tolerance towards neutral/slightly alkaline pH values, *Tt*GalA could be employed in hydrothermal processing in which water in liquid phase or in vapour phase is used to pretreat lignocellulosic biomasses [49]. In this process, controlling the pH around neutral values minimizes the formation of fermentation inhibitors [50]. Typically, the optimal temperature for catalysis of thermophilic enzymes mirrors the growing temperature of the native host, like the α-galactosidase from *Thermus* sp. strain T2 (75 °C) [44], whereas their activity is limited at lower temperatures. *Tt*GalA is unusual since its optimum was set at 90 °C and its activity was lower at temperature ≤70 °C, which is the optimal growth temperature for most of *Thermus* species. Moreover, *Tt*GalA is among the most thermophilic α-galactosidases so far characterised, such as that of *T. neapolitana* (105 °C) [47], *T. brockianus* (94 °C) and *T. maritima* (95 °C) [39, 43].

The thermal inactivation data indicate that the *Tt*GalA has a half-life of 60 min (Fig. 5) at its optimal temperature (90 °C). Moreover, residual activity higher than 90% was detected up to 6 h of incubation at 70 °C (Fig. 5), retaining 50% of its activity after 30 h (not shown).

It has already been reported that previously characterised enzymes from *Thermus* species display in vitro an optimal catalytic temperature higher than the growing

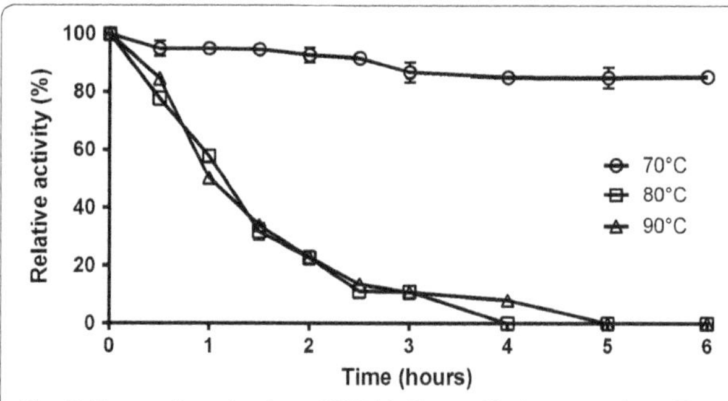

Fig. 5 Thermal inactivation of *Tt*GalA. The purified enzyme from *T. thermophilus* was incubated in 20 mM sodium-phosphate buffer pH 6.0 at 90, 80 and 70 °C for different period of time and then assayed for residual activity at 90 °C

temperature of the native host [51], thus making this microorganism a fruitful source of enzyme catalytically active at temperature above 70 °C. Due to the elevated temperatures used during some industrial applications (such as sugar manufacturing processes and/or raw material pretreatments in bioethanol production), stability at high temperatures is an important feature for the utilization of α-galactosidases, since it prevents the loss of ternary/quaternary structures that leads to enzyme inactivation as for the mesophilic counterparts [36, 37].

Effect of metal ions

Metal ions can be released during processing of biomass as consequence of corrosion of pretreatment equipment, resulting in the liberation of heavy metal ions, which can be inhibitory to biocatalysts [52]. Moreover, other cations can derive from chemicals used to adjustment the pH. Noteworthy, several divalent cations are potent inhibitors of α-galactosidases; therefore, we tested their effect on the enzyme activity over a range of concentration from 0.5 to 5 mM (Table 3). Interestingly, the enzyme turned to be slightly activated in the presence of Co^{+2}, Mn^{+2}, Zn^{+2} at 1 mM concentration suggesting that it is a metalloenzyme or requires divalent cations as cofactors. On the other hand, this effect is negligible with monovalent cations (Table 3). The inhibition effect occurred only at the highest concentration of metal ions tested (5 mM). To further investigate on the possible role of metal ions, the enzyme activity was assayed in the presence of EDTA (5 mM) and a 20% reduction of its catalytic activity was observed. The inhibitory effect of EDTA was studied up to 40 mM concentration (not shown) resulting in a linear decrease of the enzymatic activity, thus indirectly confirming the role as cofactor of the metal ions (Co^{+2}, Mn^{+2}, Zn^{+2}) (Table 3). Similar to these findings, EDTA slightly inhibited also the α-galactosidase activity from *Lenzites elegans* [53], whereas it has no effect on the activity of α-galactosidases from *B. megaterium* and *Ganoderma lucidum* [48, 54].

Table 3 Influence of metal ions on the relative activity of *Tt*GalA

Metal ions	0.5 mM	1 mM	2.5 mM	5 mM
Mg^{+2}	99.3 ± 0.7	91.7 ± 0.3	89.9 ± 2.5	55.6 ± 0.5
Ca^{+2}	95.9 ± 0.7	95.8 ± 0.1	90.1 ± 2.4	27.8 ± 0.4
Cu^{+2}	87.7 ± 3.1	77.2 ± 1.6	63.9 ± 0.97	22.2 ± 0.3
Li^{+}	106.3 ± 0.5	113.9 ± 3.6	114.8 ± 3.1	93.6 ± 2.5
Zn^{+2}	100.6 ± 2.2	110.3 ± 0.6	70.4 ± 0.6	15.1 ± 0.3
Mn^{+2}	103.4 ± 1.0	128.5 ± 1.9	114.8 ± 3.0	20.2 ± 0.6
Co^{+2}	101.8 ± 2.6	136.2 ± 0.4	92.8 ± 0.9	30.1 ± 0.7

Inhibition of activity

Since detergents are reported to strongly inhibit the α-galactosidase activity in *Glycine max* and *Pencillium griseoroseum* [55, 56], we assayed *Tt*GalA in the presence of 5 mM detergents. The enzyme turned out to be very sensitive to the common anionic detergent SDS, which leads to a complete loss of function possibly due to the disruption of enzyme native structure. By contrast, the non-ionic detergents Tween 20 and Triton X-100 had a less marked effect with reduction of the enzyme activity to ~40 and ~19%, respectively (Table 4).

During pretreatment and saccharification of lignocellulosic biomasses, several sugars are released. These can inhibit the activity of glycoside hydrolases during the saccharification phase [49, 50]. Moreover, various sugars were also reported to inhibit the α-galactosidase activity, for instance, an α-galactosidase from *Aspergillus nidulans* is competitively inhibited by D-galactose and D-glucose [57]. Accordingly, the activity of *Tt*GalA turned out to be inhibited by the presence of several saccharides, such as D-galactose, D-saccarose and D-arabinose (Table 4). Nevertheless, TtGalA might have a potential use in the sugar

Table 4 Influence of additives on the activity of *Tt*GalA

Compound (5 mM)	Relative activity (%)
EDTA	79.9
D-Galactose	27.9
D-Saccarose	44.6
D-Arabinose	54.7
Urea	60.0
Guanidine chloride	36.8
SDS	1.6
Tween 20	39.8
Triton X100	18.8

beet industry for raffinose hydrolysis, because it retains 44.6% of its activity in presence of sucrose [10].

Finally, *Tt*GalA was assayed in presence of caotropic agents such as urea and guanidine chloride. The partial reduction of the catalytic activity is in agreement with its intrinsic stability as thermozyme [20].

Conclusions

In this work, we report the biochemical characterization of a thermoactive and thermostable α-galactosidase from *T. thermophilus HB27* (*Tt*GalA). Moreover, the drawbacks of using a heterologous mesophilic host (*E. coli*) for the production of this thermozyme have been highlighted. Indeed, "hot" expression systems are in some cases indispensable to get functional thermozymes in reasonable amounts.

Interestingly, the long-term retained activity of *Tt*GalA (30 h at 70 °C) might pave the way to its utilization after thermal pretreatment of lignocellulosic biomass (pre-saccharification), when the temperature is still too high for the fungal enzymes currently used for the hydrolysis of the biomass.

Despite its already interesting catalytic features, a fine-tuning of *Tt*GalA enzymatic properties, through genetic engineering, will be attempted to make it even more suitable for industrial applications.

Methods

Bacterial strains and growth conditions

Thermus thermophilus HB27 strain was purchased from DSMZ. A frozen (−80 °C) stock culture was streaked on a *Thermus* Medium (TM) solidified by the addition of 0.8% Gelrite® (Sigma) and incubated at 70 °C overnight [58]. *T. thermophilus* HB27::*nar* strain, kindly provided by Prof. J. Berenguer (Universidad Autónoma de Madrid) was used for the homologous expression of *Tt*GalA. *E. coli* strains were grown in Luria–Bertani (LB) medium at 37 °C with 50 μg/ml kanamycin, 33 μg/ml chloramphenicol as required. *E. coli* DH5α and BL21-CodonPlus (DE3)-RIL (Stratagene, La Jolla, CA, USA) strains were used for DNA manipulations and for the heterologous expression of the recombinant α-galactosidase, respectively.

Cloning and sequencing of TTP0072 gene

A single colony of *T. thermophilus* HB27 was inoculated into TM liquid medium and genomic DNA was isolated using DNeasy® 124 Tissue kit, (Qiagen), according to the instruction manufacturer. TTP0072 gene, encoding for a putative α-galactosidase, was amplified by PCR from *T. thermophilus* HB27 genomic DNA using the primers 5′GGAGGGCATATGAGGCTGAA3′ (*Nde*I site underlined) and 5′CGGTGGAAGCTTTTATAGAAGG3′ (*Hin*dIII site underlined) and Phusion Taq Polymerase (NEB). The amplification was carried out at 94 °C for 1 min, 58 °C for 1 min and 72 °C for 1 min, for 35 cycles. The PCR product was purified with QIAquick PCR purification kit (Qiagen Spa, Milan, Italy) and cloned in pCR4-TOPO_vector using the TOPO TA CLONING Kit (Invitrogen). The identity of the cloned DNA fragment was confirmed by DNA sequencing (BMR Genomics). Then, the insert was sub-cloned into the *Nde*I/*Hin*dIII digested pET28(b) and for pMKE2 [34] vectors for *E. coli* and *T. thermophilus* HB27::*nar* expression, respectively.

Expression and purification of recombinant *Ec*GalA and *Tt*GalA

The recombinant α-galactosidases expressed in *E. coli* (*Ec*GalA) and *T. thermophilus* HB27::*nar* (*Tt*GalA) bear an His-tag at their N-terminus. To express *Ec*GalA, *E. coli* BL21-CodonPlus (DE3)-RIL was transformed with the recombinant plasmid pET28(b)-*EcgalA*. Protein expression was induced by adding 0.5 mM of isopropil-β-D-1-thiogalactopyranoside (IPTG) to exponentially growing cells (0.5 OD_{600}) and culturing them for 12 h. Since *Ec*GalA was poorly expressed, different approaches were attempted to achieve a sufficient amount of soluble recombinant protein: (1) varying the induction time (2, 4, 6, and 8 h and overnight induction); (2) lowering the temperature during induction (down to 20 °C); (3) decreasing the IPTG concentration (0.01–0.1 mM). At every conditions, the expression levels were monitored by SDS-PAGE and enzymatic assays. However, none of these strategies turned out to reasonably increase the final amount of the recombinant protein.

For the homologous expression 200 ng of pMKE2-*TtgalA* plasmid were added to exponentially growing (0.4 OD_{600nm}) *T. thermophilus* HB27::*nar* cells and transformants were selected on TM plates at 70 °C containing 50 μg/ml kanamycin [59]. The induction of *Tt*GalA expression was performed as previously described [34].

Crude extracts from both *E. coli* BL21-CodonPlus (DE3)-RIL and *T. thermophilus* HB27::*nar* cells were prepared following a similar procedure. Cell pellets were collected from 1-L cultures by centrifugation at 4000×*g* for 15 min at 4 °C and resuspended in buffer A (50 mM Tris–HCl, pH 7.5 and 500 mM NaCl) for *Ec*GalA and in buffer B (50 mM Tris–HCl, pH 7.5) for *Tt*GalA purification, respectively. Protease inhibitor cocktail tablets (Roche) were added in both cases. The cells were homogenized by sonication (Sonicator:Heat System Ultrasonic, Inc.) for 5 min, alternating 30 s of pulse-on and 30 s of pulse-off and clarification of the cell extract was obtained by centrifugation at 40,000×*g* for 20 min at 4 °C. Purification of *Ec*GalA from the soluble fraction was performed through a two-step procedure, i.e. (1) thermal precipitation at

70 °C for 10 min followed by centrifugation at 5000×*g* for 20 min at 4 °C; (2) affinity chromatography on a His-Trap column (1 ml, GE Healthcare) connected to an AKTA Explorer system (GE Healthcare). The affinity chromatography was equilibrated with buffer A and the *Ec*GalA was eluted with the same buffer A supplemented with a linear gradient of imidazole (0–250 mM).

*Tt*GalA was purified through a similar procedure, except that the first thermal precipitation step was substituted with an anionic exchange chromatography on a Hi-trap Q HP column (5 ml, GE Healthcare). The column was equilibrated in buffer B and elution was performed in the same buffer through a linear gradient from 0 to 500 mM NaCl. The affinity chromatography was carried out under the same conditions described above. Protein identity was further verified by Western blot analysis using anti-His antibodies and LC–MS/MS analysis.

Protein concentration was estimated by using bovine serum albumin as standard according to Bradford [60]. Protein fractions displaying α-galactosidase hydrolysing activity toward *p*-nitrophenyl-α-D-galactopyranoside (herein named *p*NP-α-D-galactopyranoside, Sigma) were pooled, dialyzed against 20 mM Tris–HCl pH 7.5 and analysed by 12% SDS-PAGE [61]. The *Tt*GalA activity was detected through zimography in 12% SDS-PAGE under not reducing conditions. After electrophoresis, the gel was soaked in 2.5% Triton X-100 for 30 min at 4 °C and then was incubated with 20 mM of *p*NP-α-D-galactopyranoside solution at 90 °C for 10 min.

Molecular weight determination

The native molecular weight of the purified *Tt*GalA was analysed by gel-filtration chromatography connected to Mini DAWN Treos light-scattering system (Wyatt Technology) equipped with a QELS (quasi-elastic light scattering) module mass value and hydrodynamic radius (*R*h) measurements [62]. Five hundreds micrograms of protein (1 mg/ml) were loaded on a S200 column (16/60, GE Healthcare) with a flow-rate of 0.5 ml/min and equilibrated in buffer A. Data were analysed using Astra 5.3.4.14 software (Wyatt Technology).

Determination of the α-galactosidase activity

The *Tt*GalA standard assay was performed by using *p*NP-α-D-galactopyranoside as substrate (2.0 mM) in 50 mM sodium phosphate buffer (pH 6.0) at 90 °C in 160 µl of the reaction mixture using 50 ng of the enzyme. All assays were performed in triplicate in a 96-well microplate reader (Synergy H4, Biotek), on at least 3 different enzyme preparations. The reaction was stopped, after 10 min, by addition of 1 volume of cold 0.5 M Na_2CO_3 and the concentration of the released *p*-nitrophenol (molar extinction coefficient, 18.5/mM cm) was determined by measuring A_{405nm}. One unit of enzyme activity (U) was defined as the amount of enzyme required to release 1 µmol *p*-nitrophenol per minute, under the above assay conditions.

Catalytic and stability properties

The optimal pH and temperature were determined by performing the *p*NP-α-D-galactopyranoside assay between pH 3.0–9.0 using the following buffers (each 50 mM): citrate phosphate (pH 3.0–5.0), sodium phosphate (pH 6.0–9.0). Thermal inactivation assays were performed by incubating 50 ng of enzyme at different temperatures (70, 80, 90 °C) in buffer sodium phosphate at pH 6.0 and taking aliquots at regular time intervals to measure the residual enzyme activity under standard assay conditions (90 °C for 10 min, pH 6.0). To test enzyme stability to pH, the assays were performed by incubating *Tt*GalA at 70 °C in various buffers (pH 4.0–8.0). The residual activity was measured at different time intervals following the α-galactosidase standard assay.

Inhibition of activity

The effect of $MgCl_2$, $CaCl_2$, $CuCl_2$, LiCl, $ZnSO_4$, $MnCl_2$, $CoSO_4$ on the α-galactosidase activity was studied over a range of concentrations (0.5–5.0 mM) by pre-incubating the enzyme (50 ng) for 5 min at room temperature with the specific metal ions and by measuring the residual activity under standard assay conditions.

The influence of reducing (DTT, β-mercaptoethanol) and chelating (EDTA) agents as well as of sugars (D-galactose, sucrose, L-arabinose, and D-fucose), each tested at 5 mM concentration, was studied under the same pre-incubation and assay conditions as above.

Substrate specificity

*Tt*GalA was tested for the hydrolysis of *p*NP-α-substituted hexoses (D-glucose, D-mannose, L-rhamnose) and of *p*NP-β-substituted hexoses (D-galactose, D-glucose, and D-mannose) at a concentration of 2 mM under standard conditions (90 °C and pH 6.0 for 10 min). The kinetic parameters were determined using different *p*NP-α-D-galactopyranoside concentration (ranging from 0.0125 to 2 mM) with 50 ng of *Tt*GalA for 3 min. The Michaelis constant (K_M) and V_{max} were calculated by non-linear regression analysis using GraphPad 6.0 Prism software.

Abbreviations

bt: bacterial thermophilic; CAZy: carbohydrate-active enzymes; DTT: dithiothreitol; *Ec*GalA: *Thermus thermophilus* α-galactosidase expressed in *E. coli*; EDTA: ethylenediaminetetraacetic acid; GH: glycoside hydrolase; *g*: gravity;

h: hour(s); IPTG: isopropil-β-D-1-tiogalattopiranoside; k_{cat}: catalytic constant; K_M: michaelis constant; LB: Luria-Bertani broth; LC–MS/MS: liquid chromatography-tandem mass spectrometry; *p*NP: P-nitrophenol; QELS: quasi-elastic light scattering; *R*h: hydrodynamic radius; s: second(s); TM: thermus medium; *Tn*GalA: *Thermotoga neapolitana* α-galactosidase; *Tt*GalA: *Thermus thermophilus* α-galactosidase; SDS-PAGE: sodium dodecyl sulphate-polyacrylamide gel electrophoresis.

Authors' contributions

MA and SF performed experiments. SF, GF, DL, EP, SB and PC supervised the project. MA and PC drafted the manuscript. All authors read and approved the final manuscript.

Author details

[1] Dipartimento di Biologia, Università degli Studi di Napoli Federico II, Complesso Universitario Monte S. Angelo, Via Cinthia, 80126 Naples, Italy. [2] Division of Industrial Biotechnology, Department of Biology and Biological Engineering, Chalmers University of Technology, Gothenburg, Sweden.

Acknowledgements

We thank Dr. Andrea Carpentieri for performing the LC–MS/MS analysis and Prof. José Berenguer who kindly provided us with the pMKE2 plasmid.

Competing interests

The authors declare that they have no competing interests.

Funding

This work was supported by BIOPOLIS: PON03PE_00107_1 CUP: E48C14000030005.

References

1. Pauly M, Keegstra K. Plant cell wall polymers as precursors for biofuels. Curr Opin Plant Biol. 2010;13:304–11.
2. Malgas S, van Dyk JS, Pletschke BI. A review of the enzymatic hydrolysis of mannans and synergistic interactions between β-mannanase, β-mannosidase and α-galactosidase. World J Microbiol Biotechnol. 2015;31:1167–75.
3. Chauhan PS, Puri N, Sharma P, Gupta N. Mannanases: microbial sources, production, properties and potential biotechnological applications. Appl Microbiol Biotechnol. 2012;93:1817–30.
4. Marraccini P, Rogers WJ, Caillet V, Deshayes A, Granato D, Lausanne F, Lechat S, Pridmore D, Pétiard V. Biochemical and molecular characterization of alpha-D-galactosidase from coffee beans. Plant Physiol Biochem. 2005;43:909–20.
5. Pourcher T, Bassilana M, Sarkar HK, Kaback HR, Leblanc G. Melibiose permease and alpha-galactosidase of *Escherichia coli*: identification by selective labeling using a T7 RNA polymerase/promoter expression system. Biochemistry. 1990;29:690–6.
6. Yang D, Tian G, Du F, Zhao Y, Zhao L, Wang H, Ng TB. A fungal alpha-galactosidase from *Pseudobalsamia microspora* capable of degrading raffinose family oligosaccharides. Appl Biochem Biotechnol. 2015;176:2157–69.
7. Linthorst GE, Hollak CE, Donker-Koopman WE, Strijland A, Aerts JM. Enzyme therapy for Fabry disease: neutralizing antibodies toward agalsidase alpha and beta. Kidney Int. 2004;66:1589–95.
8. Lim HG, Kim GB, Jeong S, Kim YJ. Development of a next-generation tissue valve using a glutaraldehyde-fixed porcine aortic valve treated with decellularization, α-galactosidase, space filler, organic solvent and detoxification. Eur J Cardiothorac Surg. 2015;48:104–13.
9. Olsson ML, Hill CA, De La Vega H, Liu QP, Stroud MR, Valdinocci J, Moon S, Clausen H, Kruskall MS. Universal red blood cells—enzymatic conversion of blood group A and B antigens. Transfus Clin Biol. 2004;11:33–9.
10. Linden JC. Immobilized α-D-galactosidase in the sugar beet industry. Enzym Microb Technol. 1982;4:130–6.
11. Clarke JH, Davidson K, Rixon JE, Halstead JR, Fransen MP, Gilbert HJ, Hazlewood GP. A comparison of enzyme-aided bleaching of softwood paper pulp using combinations of xylanase, mannanase and alpha-galactosidase. Appl Microbiol Biotechnol. 2000;53:661–7.
12. Pessela BC, Fernández-Lafuente R, Torres R, Mateo C, Fuentes M, Filho M, Vian A, García JL, Guisán JM, Carrascosa AV. Production of a thermoresistant alpha-galactosidase from *Thermus* sp. strain T2 for food processing. Food Biotechnol. 2007;21:91–103.
13. Prashanth SJ, Mulimani V. Soymilk oligosaccharide hydrolysis by *Aspergillus oryzae* α-galactosidase immobilized in calcium alginate. Process Biochem. 2005;40:1199–205.
14. Rosgaard L, Pedersen S, Langston J, Akerhielm D, Cherry JR, Meyer AS. Evaluation of minimal *Trichoderma reesei* cellulase mixtures on differently pretreated Barley straw substrates. Biotechnol Prog. 2007;23:1270–6.
15. Sun Y, Cheng J. Hydrolysis of lignocellulosic materials for ethanol production: a review. Bioresour Technol. 2002;83:1–11.
16. Linares-Pasten JA, Andersson M, Karlsson NE. Thermostable glycoside hydrolases in biorefinery technologies. Current Biotechnol. 2014;3:26–44.
17. Paës G, O'Donohue MJ. Engineering increased thermostability in the thermostable GH-11 xylanase from *Thermobacillus xylanilyticus*. J Biotechnol. 2006;125:338–50.
18. Henne A, Brüggemann H, Raasch C, Wiezer A, Hartsch T, Liesegang H, Johann A, Lienard T, Gohl O, Martinez-Arias R, et al. The genome sequence of the extreme thermophile *Thermus thermophilus*. Nat Biotechnol. 2004;22:547–53.
19. Limauro D, D'Ambrosio K, Langella E, De Simone G, Galdi I, Pedone C, Pedone E, Bartolucci S. Exploring the catalytic mechanism of the first dimeric Bcp: functional, structural and docking analyses of Bcp4 from *Sulfolobus solfataricus*. Biochimie. 2010;92:1435–44.
20. Contursi P, D'Ambrosio K, Pirone L, Pedone E, Aucelli T, She Q, De Simone G, Bartolucci S. C68 from the *Sulfolobus islandicus* plasmid-virus pSSVx is a novel member of the AbrB-like transcription factor family. Biochem J. 2011;435:157–66.
21. Prato S, Vitale RM, Contursi P, Lipps G, Saviano M, Rossi M, Bartolucci S. Molecular modeling and functional characterization of the monomeric primase–polymerase domain from the *Sulfolobus solfataricus* plasmid pIT3. FEBS J. 2008;275:4389–402.
22. Fiorentino G, Del Giudice I, Bartolucci S, Durante L, Martino L, Del Vecchio P. Identification and physicochemical characterization of BldR2 from *Sulfolobus solfataricus*, a novel archaeal member of the MarR transcription factor family. Biochemistry. 2011;50:6607–21.
23. Contursi P, Farina B, Pirone L, Fusco S, Russo L, Bartolucci S, Fattorusso R, Pedone E. Structural and functional studies of Stf76 from the *Sulfolobus islandicus* plasmid-virus pSSVx: a novel peculiar member of the winged helix-turn-helix transcription factor family. Nucleic Acids Res. 2014;42:5993–6011.
24. Contursi P, Fusco S, Limauro D, Fiorentino G. Host and viral transcriptional regulators in Sulfolobus: an overview. Extremophiles. 2013;17:881–95.
25. Francis DM, Page R. Strategies to optimize protein expression in *E. coli*. Curr Protoc Protein Sci. 2010;Chapter 5:Unit 5.24.21–9.
26. Rosano GL, Ceccarelli EA. Recombinant protein expression in *Escherichia coli*: advances and challenges. Front Microbiol. 2014;5:172.
27. Hidalgo A, Betancor L, Moreno R, Zafra O, Cava F, Fernández-Lafuente R, Guisán JM, Berenguer J. *Thermus thermophilus* as a cell factory for the production of a thermophilic Mn-dependent catalase which fails to be synthesized in an active form in *Escherichia coli*. Appl Environ Microbiol. 2004;70:3839–44.
28. Contursi P, Cannio R, She Q. Transcription termination in the plasmid/virus hybrid pSSVx from *Sulfolobus islandicus*. Extremophiles. 2010;14:453–63.
29. Fusco S, She Q, Bartolucci S, Contursi P. T(lys), a newly identified Sulfolobus spindle-shaped virus 1 transcript expressed in the lysogenic state, encodes a DNA-binding protein interacting at the promoters of the early genes. J Virol. 2013;87:5926–36.
30. Contursi P, Fusco S, Cannio R, She Q. Molecular biology of fuselloviruses and their satellites. Extremophiles. 2014;18:473–89.

31. Prato S, Cannio R, Klenk HP, Contursi P, Rossi M, Bartolucci S. pIT3, a cryptic plasmid isolated from the hyperthermophilic crenarchaeon *Sulfolobus solfataricus* IT3. Plasmid. 2006;56(35–45):31.
32. Bartolucci S, Contursi P, Fiorentino G, Limauro D, Pedone E. Responding to toxic compounds: a genomic and functional overview of Archaea. Front Biosci (Landmark Ed). 2013;18:165–89.
33. Suzuki H, Yoshida K-I, Ohshima T. Polysaccharide-degrading thermophiles generated by heterologous gene expression in *Geobacillus kaustophilus* HTA426. Appl Environ Microbiol. 2013;79:5151–8.
34. Moreno R, Haro A, Castellanos A, Berenguer J. High-level overproduction of His-tagged Tth DNA polymerase in *Thermus thermophilus*. Appl Environ Microbiol. 2005;71:591–3.
35. Wu WL, Liao JH, Lin GH, Lin MH, Chang YC, Liang SY, Yang FL, Khoo KH, Wu SH. Phosphoproteomic analysis reveals the effects of PilF phosphorylation on type IV pilus and biofilm formation in *Thermus thermophilus* HB27. Mol Cell Proteomics. 2013;12:2701–13.
36. Viikari L, Alapuranen M, Puranen T, Vehmaanperä J, Siika-Aho M. Thermostable enzymes in lignocellulose hydrolysis. Adv Biochem Eng Biotechnol. 2007;108:121–45.
37. Elleuche S, Schäfers C, Blank S, Schröder C, Antranikian G. Exploration of extremophiles for high temperature biotechnological processes. Curr Opin Microbiol. 2015;25:113–9.
38. Sarmiento F, Peralta R, Blamey JM. Cold and hot extremozymes: industrial relevance and current trends. Front Bioeng Biotechnol. 2015;3:148.
39. Fridjonsson O, Watzlawick H, Gehweiler A, Rohrhirsch T, Mattes R. Cloning of the gene encoding a novel thermostable alpha-galactosidase from *Thermus brockianus* ITI360. Appl Environ Microbiol. 1999;65:3955–63.
40. Brouns SJ, Smits N, Wu H, Snijders AP, Wright PC, de Vos WM, van der Oost J. Identification of a novel alpha-galactosidase from the hyperthermophilic archaeon *Sulfolobus solfataricus*. J Bacteriol. 2006;188:2392–9.
41. Fridjonsson O, Watzlawick H, Mattes R. The structure of the alpha-galactosidase gene loci in *Thermus brockianus* ITI360 and *Thermus thermophilus* TH125. Extremophiles. 2000;4:23–33.
42. Jenney FE, Adams MW. Hydrogenases of the model hyperthermophiles. Ann NY Acad Sci. 2008;1125:252–66.
43. Liebl W, Wagner B, Schellhase J. Properties of an α-galactosidase, and structure of its gene galA, within an α- and β-galactoside utilization gene cluster of the hyperthermophilic bacterium *Thermotoga maritima*. Syst Appl Microbiol. 1998;21:1–11.
44. Ishiguro M, Kaneko S, Kuno A, Koyama Y, Yoshida S, Park GG, Sakakibara Y, Kusakabe I, Kobayashi H. Purification and characterization of the recombinant *Thermus* sp. strain T2 alpha-galactosidase expressed in *Escherichia coli*. Appl Environ Microbiol. 2001;67:1601–6.
45. Gote M, Khan M, Gokhale D, Bastawde K, Khire J. Purification, characterization and substrate specificity of thermostable α-galactosidase from *Bacillus stearothermophilus* (NCIM-5146). Process Biochem. 2006;41:1311–7.
46. Beeby M, O'Connor BD, Ryttersgaard C, Boutz DR, Perry LJ, Yeates TO. The genomics of disulfide bonding and protein stabilization in thermophiles. PLoS Biol. 2005;3:e309.
47. Duffaud GD, McCutchen CM, Leduc P, Parker KN, Kelly RM. Purification and characterization of extremely thermostable beta-mannanase, beta-mannosidase, and alpha-galactosidase from the hyperthermophilic eubacterium *Thermotoga neapolitana* 5068. Appl Environ Microbiol. 1997;63:169–77.
48. Patil A, Praveen Kumar S, Mulimani VH, Veeranagouda Y, Lee K. α-Galactosidase from *Bacillus megaterium* VHM1 and its application in removal of flatulence-causing factors from soymilk. J Microbiol Biotechnol. 2010;20:1546–54.
49. Hu F, Ragauskas A. Pretreatment and lignocellulosic chemistry. Bioenergy Research. 2012;5:1043–66.
50. Jönsson LJ, Martín C. Pretreatment of lignocellulose: formation of inhibitory by-products and strategies for minimizing their effects. Bioresour Technol. 2016;199:103–12.
51. Blank S, Schröder C, Schirrmacher G, Reisinger C, Antranikian G. Biochemical characterization of a recombinant xylanase from *Thermus brockianus*, suitable for biofuel production. JSM Biotechnol Biomed Eng. 1027;2014:2.
52. Turner P, Mamo G, Karlsson EN. Potential and utilization of thermophiles and thermostable enzymes in biorefining. Microb Cell Fact. 2007;6:9.
53. Sampietro D, Quiroga E, Sgariglia M, Soberón J, Vattuone MA. A thermostable α-galactosidase from *Lenzites elegans* (Spreng.) ex Pat. MB445947: purification and properties. Antonie Van Leeuwenhoek. 2012;102:257–67.
54. Sripuan T, Aoki K, Yamamoto K, Tongkao D, Kumagai H. Purification and characterization of thermostable α-galactosidase from *Ganoderma lucidum*. Biosci Biotechnol Biochem. 2003;67:1485–91.
55. Falkoski DL, Guimarães VM, Callegari CM, Reis AP, de Barros EG, de Rezende ST. Processing of soybean products by semipurified plant and microbial alpha-galactosidases. J Agric Food Chem. 2006;54:10184–90.
56. Mi S, Meng K, Wang Y, Bai Y, Yuan T, Luo H, Yao B. Molecular cloning and characterization of a novel α-galactosidase gene from *Penicillium* sp. F63 CGMCC 1669 and expression in *Pichia pastoris*. Enzym Microb Technol. 2007;40:1373–80.
57. Ríos S, Pedregosa AM, Fernández Monistrol I, Laborda F. Purification and molecular properties of an alpha-galactosidase synthesized and secreted by *Aspergillus nidulans*. FEMS Microbiol Lett. 1993;112:35–41.
58. Del Giudice I, Limauro D, Pedone E, Bartolucci S, Fiorentino G. A novel arsenate reductase from the bacterium *Thermus thermophilus* HB27: its role in arsenic detoxification. Biochim Biophys Acta Proteins Proteom. 2013;1834:2071–9.
59. Koyama Y, Hoshino T, Tomizuka N, Furukawa K. Genetic transformation of the extreme thermophile *Thermus thermophilus* and of other *Thermus* spp. J Bacteriol. 1986;166:338–40.
60. Bradford MM. A rapid and sensitive method for the quantitation of microgram quantities of protein utilizing the principle of protein-dye binding. Anal Biochem. 1976;72:248–54.
61. Laemmli UK. Cleavage of structural proteins during the assembly of the head of bacteriophage T4. Nature. 1970;227:680–5.
62. Limauro D, De Simone G, Pirone L, Bartolucci S, D'Ambrosio K, Pedone E. *Sulfolobus solfataricus* thiol redox puzzle: characterization of an atypical protein disulfide oxidoreductase. Extremophiles. 2014;18:219–28.

2

Gas fermentation: cellular engineering possibilities and scale up

Björn D. Heijstra*, Ching Leang and Alex Juminaga

Abstract

Low carbon fuels and chemicals can be sourced from renewable materials such as biomass or from industrial and municipal waste streams. Gasification of these materials allows all of the carbon to become available for product generation, a clear advantage over partial biomass conversion into fermentable sugars. Gasification results into a synthesis stream (syngas) containing carbon monoxide (CO), carbon dioxide (CO_2), hydrogen (H_2) and nitrogen (N_2). Autotrophy–the ability to fix carbon such as CO_2 is present in all domains of life but photosynthesis alone is not keeping up with anthropogenic CO_2 output. One strategy is to curtail the gaseous atmospheric release by developing waste and syngas conversion technologies. Historically microorganisms have contributed to major, albeit slow, atmospheric composition changes. The current status and future potential of anaerobic gas-fermenting bacteria with special focus on acetogens are the focus of this review.

Keywords: Climate change, GHG, Waste gas, Syngas, Fermentation, Gas contaminants, Carbon recycling, Carbon capture and utilization, Scale up

Background

The critical need for technologies to limit greenhouse gas (GHG) outputs and slow down warming of the Earth is rapidly accepted. Essential to this is a further improvement in global awareness of nations and generations, and their demand for sustainable technology development and products. At the same time continued questioning of polluting industry and government enforced further tightening of emission rules is essential. Sadly, 2016 marks the year the global atmospheric CO_2 level, measured at Mauna Loa Observatory, permanently reached values over 400 ppm [1]. This level is thought to have an impact extending far beyond our lifetime and the link to the increasing average global temperature is undeniable. On the 14th of November 2016 the World Meteorological Organization (WMO) reported at the 22nd session of the Conference of the Parties (COP22) United Nations global climate summit in Morocco that 2016 is on track to be the hottest year on record. The vast amount of published research on climate change is unanimous and unequivocal pointing to the carbon footprint of the expanding world population. The urgency to reduce emissions and divest from fossil fuels has been recognized by World leaders from over 190 countries who negotiated the Paris Agreement at the 21st Conference of the Parties to the United Nations Framework Convention on climate change [2]. This agreement was signed by 174 countries on 22 April 2016 in New York and each country that ratifies the agreement will have to set emission reduction or limitation targets, known as "nationally determined contribution," or "NDC," however the targets will be voluntary [2].

Available gaseous feedstocks

A variety of large scale industrial processes generate side streams containing low to medium and high BTU (British Thermal Units) value off gases. Examples are steel mills, ferroalloy industries, refineries and chemical plants producing high CO containing gases with variable compositions of H_2, CO_2, CH_4 and N_2. Many of these gases are flared or preferably burned for internal energy generation within the production facility. Another large gas source, biomass gasification to generate fermentable syngas, is recognized as an alternative to lignocellulosic biomass to

*Correspondence: bjorn.heijstra@lanzatech.com
LanzaTech, Inc., 8045 Lamon Ave, Suite 400, Skokie, IL, USA

fuel conversion. Virtually any waste product can be recycled by turning this into syngas [3–6].

When derived from biomass, syngas can be variable in H_2 (1.2–7.3 mol%) [7, 8] which makes this less suited for catalytic processes such as the Fischer–Tropsch Process (FTP) which require a fixed H_2:CO ratio of 2:1 [9, 10]. In addition, non-lignocellulosic biomass gasification such as municipal solid waste (MSW) is another rapidly growing gas source with limited impact on land usage and a preferred technology in crowded nations. Within petrochemical refineries (syn)gas or natural gas to liquid (GTL) technologies are well developed but require high capital investment to be economically viable and compared to petroleum based fuels have high greenhouse gas (GHG) output [11]. Within petrochemical refineries several streams of 'stranded gas' often remain underutilized due to logistical and economic barriers [12]. To limit carbon emissions into the atmosphere governments are increasingly exploring regulatory incentives while planned CO_2 capping can provide economic benefits [13]. New regulatory opportunities can be expected to arise, further growing the gaseous pool available for conversion by gas fermentation.

According to life cycle analysis (LCA) studies, in many of the feed stock examples mentioned above a microorganism based gas to liquid conversion could be an economically profitable proposition while simultaneously decreasing GHG emissions when compared to fossil gasoline [14, 15].

Gas fermentation process

The advantages of gas fermentation have been made clear in recent reviews [16–19]. The available macro gas composition determines the organisms available for conversion: autotrophic acetogenic, carboxydotrophic, and methanotrophic bacteria can fix the carbon from CO, CO_2 or CH_4 containing gases, respectively. Although chemical processes are generally faster than biological conversions, the high enzymatic specificities of biological reactions result in higher product selectivity with the formation of fewer by-products.

In this review we present data from acetogens which can conserve energy through CO_2 (CO) fixation via the Wood-Ljungdahl pathway (WLP). This is the most efficient known pathway to convert CO_2 to secreted organic products [20, 21]. The key intermediate of the WLP, acetyl-CoA, is a precursor for enzymatic production of various other organic compounds, production of which can be of commercial interest [20, 22–25].

H_2 can provide an additional energy source and certain acetogens are able to grow and produce ethanol from CO_2 and H_2 [26], providing direct CO_2 sequestration into products. Direct input of wind, hydro or solar generated electrons could further improve carbon capture utilization (CCU) in these naturally occurring microbial cell factories. Sakimoto *et all* showed a remarkable biomimetic approach with direct electron input into the WLP of *Moorella thermoacetica* by photosensitizing these nonphotosynthetic microbes using a biological-inorganic hybrid approach. This is a true solar to chemical carbon dioxide reduction with 90% selectivity to acetate and 10% selectivity to biomass [27]. A wide variety of CO_2 reduction technologies remain under development and each could have its own positive impact reducing atmospheric CO_2 levels [28–31].

A critical aspect of any fermentation involving gases as a substrate is the ability of the gas to solubilize in the liquid to a concentration that does not inhibit microbial metabolism. Inhibition can occur by the substrate being too concentrated [32] or by a low volumetric mass transfer coefficient (k_{La}) when substrate availability can become rate-limiting. A variety of reactor configurations attempting to achieve an optimal and controllable k_{La} have been extensively discussed in the literature: Continuous stirred tank reactors (CSTR's), bubble columns, loop reactors, immobilized beds, and hollow fiber membrane columns each have certain process dependent benefits and specific volumetric mass-transfer coefficients [4, 6, 18, 33–36].

Detailed gas composition

The wide variety of industries producing waste gas streams invariably introduce impurities due to process variables and trace elements in process feed stocks. These impurities can affect downstream conversion performance, compounds such as ash, char, tar and aromatics, lipophilic compounds that are known to accumulate into lipid bilayers affecting their functional properties [37]. Halogens and mono nitrogenous species such as hydrogen cyanide (HCN), ammonia (NH_3), nitrogen oxide (NOx) and other known enzyme inhibiting gases such as acetylene (C_2H_2), ethylene (C_2H_4), ethane (C_2H_6) and oxygen (O_2) can be present [3, 5, 6, 38, 39]. Sulfur compounds in the gas such as hydrogen sulfide (H_2S), carbonyl sulfide (COS), carbon disulfide (CS_2) can in turn negatively affect catalyst based scrubbing systems and their atmospheric release is restricted by environmental regulations.

For many of the above compounds commercially available scrubbing systems exist, however microbial gas fermentation as the downstream process is a relatively new addition. Monitoring optimal scrubbing system performance, including peak loads, saturation and regen cycles is critical to effectively maintain a reactive microbial population. A complete understanding of upstream process variability effect on gas contaminants production, together with the effect that accumulating and reactive impurities have, could reduce treatment costs. However, assuming feed gas process stability, at macro and micro composition, is an unrealistic expectation and can cause production delays at scale [40].

Gas contaminant process tolerance

A distinct advantage of the biological conversion route is that a biocatalyst is versatile, constantly renewing due to its growth rate and as a consequence also capable of adapting to its environment. The biocatalyst is therefore less susceptible to poisoning by sulfur, chlorine and tar contaminants than inorganic catalysts which in turn have a much longer residence time, and therefore exposure to, the aforementioned gas contaminants [41, 42]. However tolerance levels to certain compounds is low, C_2H_2, HCN and NO are considered particularly troublesome as they are known to inhibit enzymes responsible for initial harvesting of carbon and energy from syngas in acetogenic organisms [43].

Hydrogen cyanide can be formed in gasifiers fed with nitrogen containing materials, and output concentrations can be influenced by gasifier operation parameters [44, 45]. Enzyme specific tolerance has been reported where cyanide specifically interacted with Fe-hydrogenases but not with di-nuclear metal centers as found in NiFe or FeFe hydrogenases [39]. In another study it was found cyanide acts as a competitive inhibitor acting on the Ni-4Fe-5S center of carbon monoxide dehydrogenase (CODH) [46], a key enzyme of the WLP [39, 44–47].

Besides cyanide, nitric oxide can be cogenerated in gasifiers. NO is a radical gas and used within biological systems as a transcriptional regulator [48]. At high concentrations however this reactive gas interacts within the cell to form toxic nitrogen oxides that inhibit key enzymes and at high concentrations prevent microbial growth [49]. A report on the inhibition of hydrogenase activity within a syngas operating system found tolerance levels to 40 ppm without compromising productivity while 200 ppm levels resulted in complete enzyme inactivation [49]. Biological tolerance can be based on conversion of NO to less reactive compounds such as nitrate (NO_3) nitrous oxide (N_2O) or ammonia (NH_3) [48].

Acetylene dissolves well in aqueous solution, up to 47 mM at standard conditions and is a well-known inhibitor of metalloproteins due to reversible binding to the catalytic site [50]. Acetylene can reversibly inhibit hydrogenases limiting energy generation through H_2 uptake [51, 52]. Due to the high reactivity with metalloenzymes tolerance levels are found to be low. Using 10% (v/v) C_2H_2 fed to *Rhodospirillum rubrum*, it was found that CO-linked hydrogenases had 50% reduced activity [52]. However it was found that only NiFe hydrogenases, not Fe hydrogenases, are inhibited by acetylene binding [51]. Using the rate of methanogenesis in marine sediments to study inhibitory compounds it was found acetylene irreversibly inhibits methane production while ethylene had a reversible inhibitory effect [53]. In the same study ethane was found to have no effect. Ethylene has also been described as a toxic compound to the gas fermentation process [38]. For commercialization of their gas fermentation process LanzaTech has performed extensive laboratory gas contaminant exposure tests on continuously grown *Clostridium autoethanogenum* cultures. Test results indicate ethylene appears to have limited to no effect on gas uptake rates in *C. autoethanogenum* cultures tested at up to a partial pressure of 10 mbar (Fig. 1).

For obligate anaerobic *Clostridium* species in industrial settings, oxygen and reactive oxygen species (ROS) are considered gas contaminants although some species are reported to withstand microoxic conditions [54–56]. In laboratory experiments on *C. autoethanogenum* under a partial pressure of up to 8 mbar oxygen an impact on CO utilization was measurable (Fig. 2). After reducing the oxygen concentration to 2 mbar the carbon monoxide uptake levels increased again indicating the tolerance level and reversible nature of the oxidative effect.

Synthetic biology development

Synthetic biology and metabolic engineering approaches play an essential role in expanding acetogen product spectrum beyond the native products, such as ethanol, acetate and butanediol (BDO) to other fuels and commodity chemicals. These approaches had been applied to classic model microorganisms, such as *E. coli* and yeast which have been successfully engineered to produce non-native products at commercial scale [57–60]. On the other hand, acetogenic clostridia had long been considered challenging hosts for genetic modification. The slow development of reliable molecular biology tools is partly contributed by a strong native restriction-modification

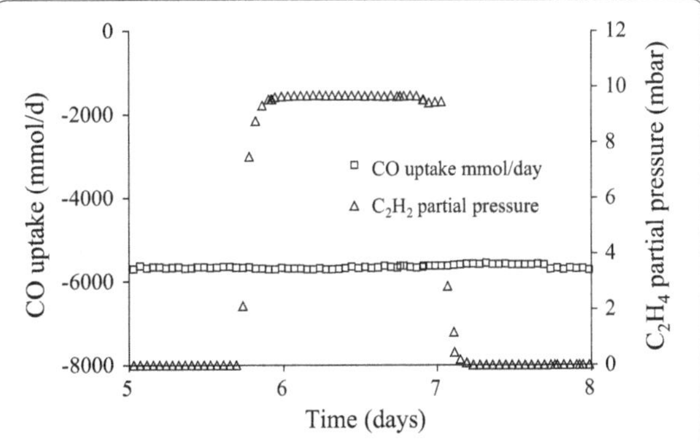

Fig. 1 CO consumption profile of a continuously operating *C. autoethanogenum* gaseous fermentation undergoing addition of ethylene by sparging with ethylene containing Nitrogen. CO consumption remains stable around 5800 mmol CO/day

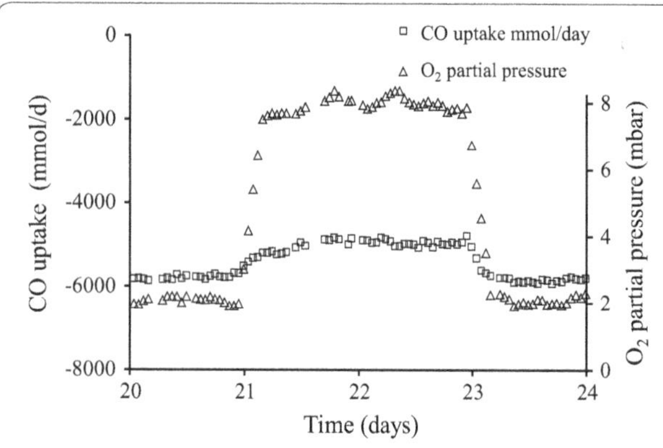

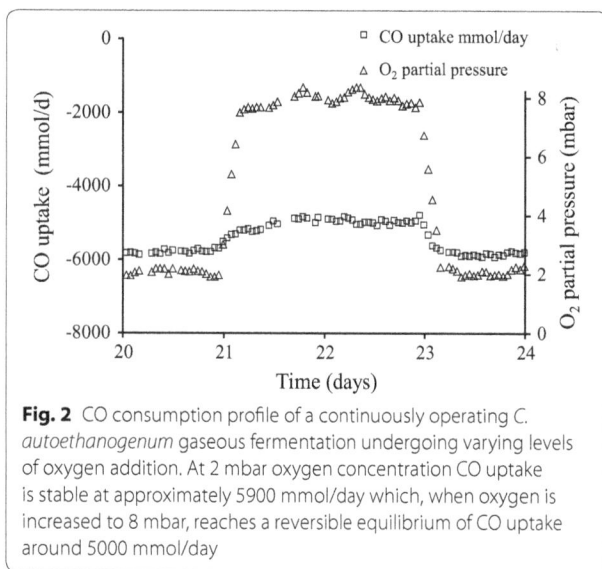

Fig. 2 CO consumption profile of a continuously operating *C. autoethanogenum* gaseous fermentation undergoing varying levels of oxygen addition. At 2 mbar oxygen concentration CO uptake is stable at approximately 5900 mmol/day which, when oxygen is increased to 8 mbar, reaches a reversible equilibrium of CO uptake around 5000 mmol/day

system, non-standard culturing conditions (toxic gas at pressure and obligate anaerobic), and slow doubling times. Since the successful demonstration of gas fermentation at pilot and pre-commercial scale as mentioned below, significant progress had been made in understanding acetogens at both the molecular and system biology levels [61–63]. Most notably, whole genome sequences, genome scale models, transcriptomic, proteomic studies and genetic tools have now been developed for these organisms. [18, 22, 26, 56, 61–71].

DNA transfer

In order to genetically modify a microorganism, whether to delete a competing pathway or to introduce a new product pathway, it is imperative to have a reliable method to introduce foreign DNA into the cell. Electroporation and conjugation are the most frequently used methods for introducing foreign DNA into acetogens [26, 62, 72]. These strategies have been successfully demonstrated in *C. ljungdahlii*, *C. autoethangenum*, *C. aceticum*, *A. woodii* and *M. thermoacetica* [22, 62, 65, 73–76]. The highest transformation efficiency was reported to be around 1.7×10^4 cfu/μg DNA for *C. ljungdahlii* in acetogens and the authors successfully introduced suicide vector with homology arms for chromosomal modification [62, 77]. Although electrocompetent cells preparation is elaborate, the method is donor cell independent, unlike conjugation. Further improvement of electroporation efficiency has been achieved through in vitro methylation or disruption of host's restriction endonuclease, such as those examples in *C. acetobutylicum* [78], *C. pasteurianum* [79] and *C. cellulolyticum* [80], when the methylation/restriction patterns are identified either through restriction digestion pattern identification or PacBio sequencing [79, 81, 82].

In addition, conjugation is used broadly among *Clostridium* species, mainly because during conjugation DNA is transferred from donor to recipient cell as a single strand, not recognizable by the recipient's restriction modification system. This method has been successfully used in *C. autoethanogenum* [26] and *A. woodii* [83]. In combination the two methods provide a robust basis platform for routine and advanced synthetic biology discovery.

Genome modification

Homologous recombination utilizing host's own recombination machinery is widely used for genome engineering. More specifically, a plasmid that carries homologous arms to the upstream and downstream areas of target gene(s), is introduced into the host. In order to select for a double crossover event (gene deletion), a positive selection (such as antibiotic resistance cassettes) or combination with a negative selection (such as *mazF* [84] or *pyrE* [85]) is used. Other variant methods that rely on homologous recombination also include Allele-Coupled Exchange (ACE) [86], Triple crossover [87] and scar-less, marker-less knockout or knock-in using two negative selection markers (*C. thermocellum*), detailed information has recently been reviewed [88]. In some instances, specific DNA sequences which can be recognized by site-specific recombinases, flanking the antibiotic resistance cassettes were introduced into the chromosome at the same time during the double crossover event. The antibiotic resistance cassettes can then be excised out of the chromosome by the site-specific recombinase and produce a marker-less mutant [77].

Other genetic modification tools utilizing RNA machinery, such as the group II intron gene inactivation [89] and CRISPR/Cas9 (Clustered Regularly Interspaced Short Palindromic Repeats/CRISPR-associated protein 9), a RNA-guided prokaryotic immune system which can cleave foreign DNA [90]. The group II intron method had been applied to different *Clostridium* species including acetogens such as *C. autoethanogenum* [26, 61], and others [91]. This method, based on RNA-mediated, retro-homing mechanism [89], provides a quick and easy gene inactivation tool without relying on host recombination machinery, thus bypassing the low occurrence of double crossover events, and resulted in greater success in genome editing in acetogenic *Clostridium*. However, the nature of group II intron mutagenesis is based on insertion of the group II DNA at the target gene, therefore, this method is flawed with the possibility of polar effects on downstream genes.

It was recently reported that the CRISPR/Cas9 system from *Streptococcus pyogenes* was successfully applied to acetogens, many other bacteria, and also yeast and Eukaryotes due to high and reliable efficiency, the simplicity in design and fast turnaround to generate scar-less mutants [90, 92–96]. Moreover, CRISPR/Cas9 system has been reported to target multiple genes at the same time (multiplex gene editing) [92], which allows for engineering bacterial strains with desired phenotypes in a one-step. This system has also reported to be able to edit bacterial strains at the single nucleotide level [97]. The CRISPR/Cas9 system has rapidly become the preferred method for genome editing in most organisms, facilitating rapid functional analysis and strain development for industrial applications.

Genetic parts

In addition to chromosomal editing tools, genetic parts such as promoters, terminators ribosomal binding sites (RBS) [98, 99] are essential for both strain and pathway development. Unlike other model microorganisms for which commercial genetic parts and even software designing tools are available, acetogens' part library is less well-developed, the majority of genetic parts such as the promoters are extracted from close *Clostridium* relatives or from its own genome. Recently inducible promoter systems had been successfully developed in *C. ljungdahlii* and *C. autoethanogenum*, respectively [25, 87]. It is critical to develop an organism specific validated library of genetic parts.

One limiting factor to carry out promoter screening in acetogens is the lack of fluorescent reporter protein that would allow signal to correlate with the amount of translation from a given quantity of mRNAs transcribed. So far, there has only been a flavin-based fluorescent protein derived from *Pseudomonas putida* that works under anaerobic conditions [100]. This has been used to characterize two endogeneous promoters of *C. cellulolyticum* [101]. However, it remains to be determined if this flavin-based fluorescence system will work in acetogens. Thus for most parts, promoters in acetogens are characterized using either the *gusA* or *catP* systems, encoding β-glucuronidase and chloramphenicol acetyltransferase, respectively [25, 87]. Characterizing promoter strength, based on the enzymatic activities, is however less straightforward and time consuming.

Metabolic engineering in gas fermentation

Gas fermentation offers the benefit of not using heterologous feedstocks such as sugars that affect food supply chain. Metabolic engineering of acetogens in an industrial setting has been reviewed at length elsewhere [18]. The central metabolic pathway in acetogens begin with the reduction of CO/CO_2 to acetyl-CoA through the WLP. Depending on the choice of strains and feedstocks used, various native products can be produced, including acetate, ethanol, 2,3-BDO, lactate, butyrate, etc. (Table 1 in [18] and reference therein). The metabolic profiles of acetate, ethanol and 2,3-BDO produced by various industrial strains have recently been summarized [19]. At LanzaTech a proprietary process has been developed that maximizes the conversion of CO to ethanol in *C. autoethanogenum* using steel mill off-gas. Furthermore, it has been demonstrated that deletion of the *budA* gene encoding for an enzyme catalyzing 2,3-BDO production resulted in an increase in ethanol selectivity and titer as a result of diminished production of 2,3-BDO [61, 102]. The ethanol pools currently produced from the demo plants around the world have been converted into the jet fuels by the catalytic process known as alcohol-to-jet, which involves dehydration to alkenes and oligomerization to the targeted C-length [103].

To enhance process viability, the conversion of gas to more valuable products than ethanol have to be developed. There have been several reported successes in expressing heterologous pathways to produce acetone, butanol, butyrate, and isopropanol, in acetogens [22, 25]. Recent publication by the White Dog Lab even employed a co-feeding strategy, producing a mix of acetone, isopropanol, ethanol, at 12.5 g/L in *C. ljungdahlii* with a combination of CO and sugar [104]. In addition to these products, LanzaTech has also developed and owns several patent families exemplifying the synthesis of higher value products such as 3-hydroxypropionic acid, methyl ethyl ketone, and mevalonate, by expressing corresponding biosynthetic pathway genes from photosynthetic bacteria *Chloroflexus aurantiacus, Klebsiella, E. coli* and even plant [105–108]. In most instances, the productions were demonstrated using a plasmid platform under the control of native promoter systems.

Pathway and strain optimization

In order to scale up production, pathway gene expression needs to be optimized to minimize metabolic bottlenecks and un-wanted side products [109–111]. Even though the number of publications on this topic in the field of gas fermentation is limited, many of the approaches developed through the metabolic engineering of *E. coli* and yeast are applicable to the gas fermentation organisms. In general, the strategy involves multilayers of analysis and debugging, both at the biosynthetic pathway level as well as the overall metabolic flux level of the host cells [112, 113]. Due to the inherent complexity of a biological system, however, debugging bottlenecks one gene at a time is tedious and time consuming. Thus, it is more efficient to manipulate the gene expressions systematically, refactoring the biosynthetic pathway via modular design, combinatorial analysis and high-throughput screening, to identify the best combination of genes and promoters, and other transcriptional elements such as ribosomal binding sites (RBS), and terminators. [109, 114, 115]. Additionally, routine targeted proteomics and metabolomics can be performed to rapidly assess gene expressions and key metabolites accumulation [116–119]. With the technologies developed in the field of synthetic biology for the past 10 years, including computer-aided pathway design algorithms [120–122], DNA assembly and sequencing [123–125], it is now routine to screen a large combinatorial libraries. When combined with rational design and effective screening methodologies, the combinatorial library facilitates the search for ideal pathway combinations for highly productive strains [126, 127].

Use of omics based technology to monitor bioprocess performance

Nextgen sequencing has become a powerful tool in process optimization. Routine sequence analysis at genomic and transcriptomic levels are carried out to determine gene expression and mutation rate, which directly relate to process productivity and stability at molecular level. One recent study linked the genomic and metabolic analysis of various acetogens to confirm the involvement of the acetaldehyde oxidoreductase (AOR) in ethanol production and NADPH-dependent alcohol dehydrogenase (ADH) in the hydration of acetone to isopropanol in acetogens [19]. Moreover, *C. autoethanogenum* has been the subject of a multi-omics investigation to compare energy metabolism between autotrophic and heterotrophic growth [61]. The study highlighted the interplay of hydrogenases and the electron-bifurcating Nfn complex in ethanol formation during the autotrophic growth. The study also concluded that the overall energy yield does not change during the autotrophic or heterotrophic growth. The vast data provided by omics analysis from production plants, can be used to further improve pathway and strain design.

Metabolic flux analysis is often used in conjunction to the omics analyses to debug bottlenecks through the metabolic flux of interest [128]. A metabolic flux analysis on the syngas species, *Clostridium tyrobutyricum*, correlated increase in NADH with increase in butanol production [129, 130]. Moreover, genome-scale metabolic flux balance analysis has been used to construct spatiotemporal metabolic models for *Clostridium ljungdahlii* [131]. When combined with the Optknock computation, the models could predict new gene knockout targets relevant to the overproduction of ethanol, lactate and 2,3-BDO in a bubble column reactor [132].

Scale-up

As described above the research output in the gas fermentation field and the synthetic biology capabilities on its subject microorganisms have been rapidly expanding. However, in 2016 two of the three companies that own and operate scaled up gas fermentation facilities suspended operations. This immediately raises the question whether gas fermentation is scalable. Below we briefly summarize what is known about these three companies and for the first time present gas fermentation production data from a LanzaTech demonstration facility located within a steel mill plant in China.

Three companies, Coskata, INEOS Bio, and LanzaTech have operated pilot and demonstration plants for extended periods of time. Coskata's technology reformed methane into syngas with a H_2:CO ratio of between 2:1 and 3:1, followed by fermentation of this syngas to ethanol. This approach seeks to take advantage of the current low price of natural gas in geographies such as the US. While Coskata announced that it was to cease operation in 2015, the technology developed in this company now forms the basis of a new company: Synata Bio [133].

INEOS recently announced it is selling the INEOS Bio facility in Vero Beach, FL, USA [134]. This name-plate 8 million gallon per year (Mgy) semi-commercial facility was built as a joint venture with New Planet Energy Holdings, LLC. Commissioned in 2012, the facility used lignocellulosic biomass and MSW for generating syngas and coproduced 6 MW of electrical power. In July 2013 the company announced successful production of ethanol at its facility [135]. In September 2014 operational changes were imposed to optimize the technology and de-bottleneck the plant to achieve full production capacity [40].

LanzaTech

LanzaTech was founded in 2005 and after extensive piloting at a modest capacity steel mill plant in New Zealand, it partnered with 2 larger Chinese steel mills to build gas fermentation demonstration facilities. The first Demonstration unit was located at one of BaoSteel's mills near Shanghai (operational since 2012) and the second at a Shougang steel mill near Beijing (operational since 2013), both facilities have a 100,000 gpy pre-commercial capacity. Typical production results from the second facility (Fig. 3) are shown below. To our knowledge this is the first time continuous, long term gas fermentation production data has been published from a demonstration facility. It is important to note this facility is running directly off steel mill produced off-gas and operational set-ups are a reality of scaled up operations. The gas fermentation process has proven robust to a wide variety of process upsets such as: macro gas concentration

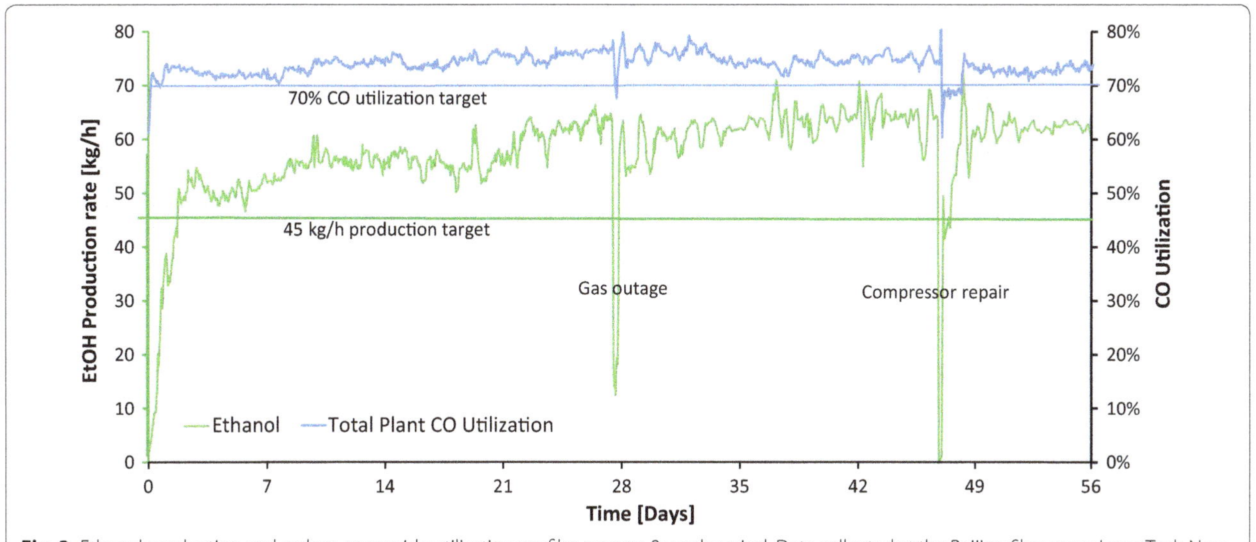

Fig. 3 Ethanol production and carbon monoxide utilization profiles over an 8 week period. Data collected at the Beijing Shougang LanzaTech New Energy Science & Technology Co., Ltd, a 0.1 Mgy ethanol capacity demonstration facility

fluctuations, presence of gas contaminants, intermittent gas supply and equipment failure which can be replaced during the continuous fermentation.

The Shougang facility earned the Roundtable on Sustainable Biomaterials (RSB) certification for sustainability [136]. The RSB is the most robust and credible global sustainability standard and certification system for biofuels and biomaterials production. Here we present production and gas utilization data from a typical run from the RSB certified plant. The resulting ethanol from the LanzaTech Demo facilities has been turned into jet fuel ready for a test flight scheduled for 2017 [137].

In 2015, both China Steel Corporation of Taiwan and ArcelorMittal of Luxembourg approved commercial projects with LanzaTech. The former will be a 17 Mgy facility with the intention to scale up to 34 Mgy [138]. The latter 21 Mgy facility will be built at ArcelorMittal's flagship steel plant in Ghent, Belgium with intention to construct further plants across ArcelorMittal's operations [139]. If scaled up to its full potential at steel mills in Europe alone, the technology could enable the production of around 104 Mgy with the potential to displace 1.6 million barrels of fossil fuel-derived gasoline on a BTU basis.

Summary and outlook

Gas fermentation is rapidly becoming an established platform for the conversion of (waste) gas to valuable liquid chemicals. Clear advantages are process stability and tolerance to inhibitory compounds and therefore flexibility in gas feedstock sourcing. Process upsets, either upstream or downstream can occur with limited warning at scaled up operations. Resilience of the microbial culture to upsets can be enhanced by engineering design to limit their impact. The production of ethanol has been proven robust at scaled up operations, the next stage is now set for expanding the product portfolio utilizing advanced synthetic biology technologies developed for gas fermenting microorganisms. This allows for a profitable carbon recycling operation, producing sustainable chemicals independent of carbon credits, to further limit GHG emission. With an industrially robust strain, efficient genetic toolbox, advanced synthetic biology capabilities, and scalable reactor design, the field of gas fermentation remains on course to reduce global carbon emissions.

Authors' contributions

AJ and CL wrote synbio sections, BH wrote gaseous feedstocks, scale up and prepared figures. All authors edited the complete manuscript. All authors read and approved the final manuscript.

Acknowledgements

We thank the complete and dedicated team involved at the Beijing Shougang LanzaTech New Energy Science & Technology Co., Ltd. for the scaled up fermentation data presented in Fig. 3. We thank Melvin Moore and Steven Glasker for data gathered in Figs. 1 and 2.

Competing interests

The authors declare that the review was written in the absence of any commercial or financial relationships that could be construed as a potential competing interests.

LanzaTech, Inc has commercial interest in gas fermentation.

Funding
We thank the following investors in LanzaTech's technology: Sir Stephen Tindall, Khosla Ventures, Qiming Venture Partners, Softbank China, the Malaysian Life Sciences Capital Fund, Mitsui, Primetals, CICC Growth Capital Fund I, L.P. and the New Zealand Superannuation Fund.

References

1. Kahn B. The world passes 400 PPM threshold permanently. Clim Cent. http://www.climatecentral.org/news/world-passes-400-ppm-threshold-permanently-20738.
2. United Nations. Adoption of the Paris agreement. 2015;21932:32. http://unfccc.int/resource/docs/2015/cop21/eng/l09r01.pdf.
3. Griffin DW, Schultz MA. Fuel and chemical products from biomass syngas: a comparison of gas fermentation to thermochemical conversion routes. Environ Prog Sustain Energy. 2012;31:219–24.
4. Munasinghe PC, Khanal SK. Syngas fermentation to biofuel: evaluation of carbon monoxide mass transfer and analytical modeling using a composite hollow fiber (CHF) membrane bioreactor. Bioresour Technol. 2012;122:130–6.
5. Munasinghe PC, Khanal SK. Biomass-derived syngas fermentation into biofuels: opportunities and challenges. Bioresour Technol. 2010;101:5013–22.
6. Abubackar HN, Veiga MC, Kennes C, Coruña L. Biological conversion of carbon monoxide: rich syngas or waste gases to bioethanol. Biofuels Bioprod Biorefining. 2011;5:93–114.
7. Boateng A, Banowetz G, Steiner J, Barton T, Taylor D, Hicks K, et al. Gasification of Kentucky bluegrass (*Poa pratensis* l.) straw in a farm-scale reactor. Biomass Bioenergy. 2007;31:153–61.
8. Datar RP, Shenkman RM, Cateni BG, Huhnke RL, Lewis RS. Fermentation of biomass-generated producer gas to ethanol. Biotechnol Bioeng. 2004;86:587–94.
9. Maitlis PM, de Klerk A. Greener Fischer-Tropsch processes for fuels and Feedstocks. Greener Fischer-Tropsch process. Fuels feed. New York: Wiley; 2013.
10. de Klerk A, Li YW, Zennaro R. Fischer-Tropsch technology. Greener Fischer-Tropsch process. Fuels feed. New York: Wiley; 2013. p. 53–79.
11. Jaramillo P, Griffin WM, Matthews HS. Comparative analysis of the production costs and life-cycle GHG emissions of FT liquid fuels from coal and natural gas. Environ Sci Technol. 2008;42:7559–65.
12. Thackeray F, Leckie G. Stranded gas: a vital resource. Pet Econ. 2002;69(5):10–2.
13. Glomsrød S, Wei T, Aamaas B, Lund MT, Samset BH. A warmer policy for a colder climate: can China both reduce poverty and cap carbon emissions? Sci Total Environ. 2016;568:236–44.
14. Handler RM, Shonnard D, Palou-Rivera I, Lai A, Hallen RT, Zhu Y, et al. Life cycle assessments of jet fuel and co-products made from lanzatech biomass-based ethanol. AIChE Natl Meet. 2014.
15. Handler RM, Shonnard DR, Griffing EM, Lai A, Palou-Rivera I. Life cycle assessments of ethanol production via gas fermentation: anticipated greenhouse gas emissions for cellulosic and waste gas feedstocks. Ind Eng Chem Res. 2016;55:3253–61.
16. Lee SH, Kim HJ, Shin YA, Kim KH, Lee SJ. Single crossover-mediated markerless genome engineering in *Clostridium acetobutylicum*. J Microbiol Biotechnol. 2016;26(4):725–9.
17. Molitor B, Richter H, Martin ME, Jensen RO, Juminaga A, Mihalcea C, et al. TEMPORARY REMOVAL: Carbon recovery by fermentation of CO-rich off gases–turning steel mills into biorefineries. Bioresour Technol. 2016. **(In press)**.
18. Liew F, Martin E, Tappel R, Heijstra B, Mihalcea C, Köpke M. Gas fermentation–a flexible platform for commercial scale production of low carbon fuels and chemicals from waste and renewable feedstocks. Front Microbiol. 2016;7:694.
19. Bengelsdorf FR, Poehlein A, Linder S, Erz C, Hummel T, Hoffmeister S, et al. Industrial acetogenic biocatalysts: a comparative metabolic and genomic analysis. Front Microbiol. 2016;7:1–15.
20. Tracy BP, Jones SW, Fast AG, Indurthi DC, Papoutsakis ET. Clostridia: the importance of their exceptional substrate and metabolite diversity for biofuel and biorefinery applications. Curr Opin Biotechnol. 2011;23:1–18.
21. Fast AG, Papoutsakis ET. Stoichiometric and energetic analyses of non-photosynthetic CO_2-fixation pathways to support synthetic biology strategies for production of fuels and chemicals. Curr Opin Chem Eng. 2012;7:1–16.
22. Köpke M, Held C, Hujer S, Liesegang H, Wiezer A, Wollherr A, et al. *Clostridium ljungdahlii* represents a microbial production platform based on syngas. Proc Natl Acad Sci USA. 2010;107:13087–92.
23. Drake HL, Küsel K, Matthies C, Wood HG, Ljungdahl LG. Acetogenic Prokaryotes. In: Dworkin M, Falkow S, Rosenberg E, Schleifer K-H, Stackebrandt E, editors. The Prokaryotes. 3rd ed. New York: Springer; 2006. p. 354–420.
24. Köpke M, Mihalcea C, Liew F, Tizard JH, Ali MS, Conolly JJ, et al. 2,3-butanediol production by acetogenic bacteria, an alternative route to chemical synthesis, using industrial waste gas. Appl Environ Microbiol. 2011;77:5467–75.
25. Banerjee A, Leang C, Ueki T, Nevin KP, Lovley DR. A lactose-inducible system for metabolic engineering of *Clostridium ljungdahlii*. Appl Environ Microbiol. 2014;80:2410–6.
26. Mock J, Zheng Y, Mueller AP, Ly S, Tran L, Segovia S, et al. Energy conservation associated with ethanol formation from H_2 and CO_2 in *Clostridium autoethanogenum* involving electron bifurcation. J Bacteriol. 2015;197:2965–80.
27. Sakimoto KK, Wong AB, Yang P. Self-photosensitization of nonphotosynthetic bacteria for solar-to-chemical production. Science (80–). 2016;351:74–7.
28. Zhao Z, Zhang Y, Li Y, Zhao H, Quan X. Electrochemical reduction of carbon dioxide to formate with Fe-C electrodes in anaerobic sludge digestion process. Water Res. 2016;106:339–43.
29. Bajracharya S, Vanbroekhoven K, Buisman CJN, Pant D, Strik DP. Application of gas diffusion biocathode in microbial electrosynthesis from carbon dioxide. Environ Sci Pollut Res. 2016;23:22292–308.
30. Kattel S, Yan B, Yang Y, Chen JG, Liu P. Optimizing binding energies of key intermediates for CO_2 hydrogenation to methanol over oxide-supported copper. J Am Chem Soc. 2016;138:12440–50.
31. Buelens LC, Galvita VV, Poelman H, Detavernier C, Marin GB. Super-dry reforming of methane intensifies CO_2 utilization via Le Chatelier's principle. Science. 2016;354:449–52.
32. Bertsch J, Muller V. CO metabolism in the acetogen *Acetobacterium woodii*. Appl Environ Microbiol. 2015;81:5949–56.
33. Bredwell MD, Srivastava P, Worden RM. Reactor design issues for synthesis-gas fermentations. Biotechnol Prog. 1999;15:834–44.
34. Kimmel DE, Klasson KT, Clausen EC, Gaddy JL. Performance of trickle-bed bioreactors for converting synthesis gas to methane. Appl Biochem Biotechnol. 1991;28–29:457–69.
35. Orgill JJ, Atiyeh HK, Devarapalli M, Phillips JR, Lewis RS, Huhnke RL. A comparison of mass transfer coefficients between trickle-bed, hollow fiber membrane and stirred tank reactors. Bioresour Technol. 2013;133:340–6.
36. Ungerman AJ, Heindel TJ. Carbon monoxide mass transfer for syngas fermentation in a stirred tank reactor with dual impeller configurations. Biotechnol Prog. 2007;23:613–20.
37. Sikkema J, de Bont JA, Poolman B. Mechanisms of membrane toxicity of hydrocarbons. Microbiol Rev. 1995;59:201–22.
38. Zahn J. Scale-up of renewable chemical Manufacturing processes. Recent Adv Ferment Technol. 2015. https://sim.confex.com/sim/raft11/webprogram/Paper31231.html.
39. Shima S, Ataka K. Isocyanides inhibit [Fe]-hydrogenase with very high affinity. FEBS Lett. 2011;585:353–6.
40. Lane J. On the mend: Why INEOS Bio isn't producing ethanol in Florida. Biofuels Dig. 2014. http://www.biofuelsdigest.com/bdigest/2014/09/05/on-the-mend-why-ineos-bio-isnt-reporting-much-ethanol-production/.
41. Köpke M, Noack S, Dürre P. The past, present, and future of biofuels–biobutanol as promising alternative. Biofuel Prod Dev Prospect. 2011;451–86. http://www.intechopen.com/articles/show/title/the-past-present-and-future-of-biofuels-biobutanol-as-promising-alternative.

42. Mohammadi M, Najafpour GD, Younesi H, Lahijani P, Uzir MH, Mohamed AR. Bioconversion of synthesis gas to second generation biofuels: a review. Renew Sustain Energy Rev. 2011;15(9):4255–73.
43. Wang VC, Can M, Pierce E, Ragsdale SW, Armstrong FA. A unified electrocatalytic description of the action of inhibitors of nickel carbon monoxide dehydrogenase. J Am Chem Soc. 2013;135:2198–206.
44. Paterson N, Zhuo Y, Dugwell D, Kandiyoti R. Formation of hydrogen cyanide and ammonia during the gasification of sewage sludge and bituminous coal. Energy Fuels. 2005;19:1016–22.
45. Lin J-Y, Zhang S, Zhang L, Min Z, Tay H, Li C-Z. HCN and NH3 formation during coal/char gasification in the presence of NO. Environ Sci Technol. 2010;44:3719–23.
46. Ha SW, Korbas M, Klepsch M, Meyer-Klaucke W, Meyer O, Svetlitchnyi V. Interaction of potassium cyanide with the [Ni-4Fe-5S] active site cluster of CO dehydrogenase from *Carboxydothermus hydrogenoformans*. J Biol Chem. 2007;282:10639–46.
47. Ragsdale SW, Ljungdahl LG, DerVartanian DV. Isolation of carbon monoxide dehydrogenase from *Acetobacterium woodii* and comparison of its properties with those of the *Clostridium thermoaceticum* enzyme. J Bacteriol. 1983;155:1224–37.
48. Stern AM, Zhu J. An introduction to nitric oxide sensing and response in bacteria. Adv Appl Microbiol. 2014. doi:10.1016/B978-0-12-800261-2.00005-0.
49. Ahmed A, Lewis RS. Fermentation of biomass-generated synthesis gas: effects of nitric oxide. Biotechnol Bioeng. 2007;97:1080–6.
50. Hyman MR, Daniel A. Acetylene inhibition of metalloenzymes. Anal Biochem. 1988;173:207–20.
51. He SH, Woo SB, DerVartanian DV, Le Gall J, Peck HD. Effects of acetylene on hydrogenases from the sulfate reducing and methanogenic bacteria. Biochem Biophys Res Commun. 1989;161:127–33.
52. Maness PC, Weaver PF. Evidence for three distinct hydrogenase activities in *Rhodospirillum rubrum*. Appl Microbiol Biotechnol. 2001;57:751–6.
53. Oremland RS, Taylor BF. Inhibition of methanogenesis in marine sediments by acetylene and ethylene: validity of the acetylene reduction assay for anaerobic microcosms. Appl Microbiol. 1975;30:707–9.
54. Karnholz A, Kusel K, Goner A, Schramm A, Drake HL, Küsel K, et al. Tolerance and metabolic response of acetogenic bacteria toward oxygen. Appl Environ Microbiol. 2002;68:1005–9.
55. Kawasaki S, Sakai Y, Takahashi T, Suzuki I, Niimura Y. O_2 and reactive oxygen species detoxification complex, composed of O_2-responsive NADH:rubredoxin oxidoreductase-flavoprotein A2-desulfoferrodoxin operon enzymes, rubperoxin, and rubredoxin, in *Clostridium acetobutylicum*. Appl Environ Microbiol. 2009;75:1021–9.
56. Whitham JM, Tirado-Acevedo O, Chinn MS, Pawlak JJ, Grunden AM. Metabolic response of *Clostridium ljungdahlii* to oxygen exposure. Appl Environ Microbiol. 2015;81:AEM.02491.
57. Yim H, Haselbeck R, Niu W, Pujol-Baxley C, Burgard A, Boldt J, et al. Metabolic engineering of *Escherichia coli* for direct production of 1,4-butanediol. Nat Chem Biol. 2011;7:445–52.
58. Haselbeck R, Trawick JD, Niu W, Burgard AP. Microorganisms for the production of 1,4-butanediol, 4-hydroxybutanal, 4-hydroxybutyryl-coa, putrescine and related compounds, and methods related thereto. US 20110229946 A1. 2011.
59. Paddon CJ, Keasling JD. Semi-synthetic artemisinin: a model for the use of synthetic biology in pharmaceutical development. Nat Rev Microbiol. 2014;12:355–67.
60. Lane J. Amyris inks 5-year $ 100 M + biofene supply pact for nutraceutical market. Biofuels Dig. 2016. http://www.biofuelsdigest.com/bdigest/2016/04/28/amyris-inks-5-year-100m-biofene-supply-pact-for-nutraceutical-market/.
61. Marcellin E, Behrendorff JB, Nagaraju S, DeTissera S, Segovia S, Palfreyman R, et al. Low carbon fuels and commodity chemicals from waste gases–systematic approach to understand energy metabolism in a model acetogen. Green Chem. 2016;18:3020–8.
62. Leang C, Ueki T, Nevin KP, Lovley DR. A genetic system for *Clostridium ljungdahlii*: a chassis for autotrophic production of biocommodities and a model homoacetogen. Appl Environ Microbiol. 2013;79:1102–9.
63. Nagarajan H, Sahin M, Nogales J, Latif H, Lovley DR, Ebrahim A, et al. Characterizing acetogenic metabolism using a genome-scale metabolic reconstruction of *Clostridium ljungdahlii*. Microb Cell Fact. 2013;12:118.
64. Utturkar SM, Klingeman DM, Bruno-Barcena JM, Chinn MS, Grunden AM, Köpke M, et al. Sequence data for *Clostridium autoethanogenum* using three generations of sequencing technologies. Sci Data. 2015;2:150014.
65. Poehlein A, Cebulla M, Ilg MM, Bengelsdorf FR, Schiel-Bengelsdorf B, Whited G, et al. The complete genome sequence of *Clostridium aceticum*: a missing link between rnf- and cytochrome-containing autotrophic acetogens. MBio. 2015;6:e01168.
66. Poehlein A, Schmidt S, Kaster A-K, Goenrich M, Vollmers J, Thürmer A, et al. An ancient pathway combining carbon dioxide fixation with the generation and utilization of a sodium ion gradient for ATP synthesis. PLoS ONE. 2012;7:e33439.
67. Pierce E, Xie G, Barabote RD, Saunders E, Han CS, Detter JC, et al. The complete genome sequence of *Moorella thermoacetica (f. Clostridium thermoaceticum)*. Environ Microbiol. 2008;10:2550–73.
68. Roh H, Ko HJ, Kim D, Choi DG, Park S, Kim S, et al. Complete genome sequence of a carbon monoxide-utilizing acetogen, *Eubacterium limosum* KIST612. J Bacteriol. 2011;193:307–8.
69. Tan Y, Liu J, Chen X, Zheng H, Li F. RNA-seq-based comparative transcriptome analysis of the syngas-utilizing bacterium *Clostridium ljungdahlii* DSM 13528 grown autotrophically and heterotrophically. Mol BioSyst. 2013;9:2775–84.
70. Islam MA, Zengler K, Edwards EA, Mahadevan R, Stephanopoulos G. Investigating *Moorella thermoacetica* metabolism with a genome-scale constraint-based metabolic model. Integr Biol (Camb). 2015;7:869–82.
71. Brown SD, Nagaraju S, Utturkar S, De Tissera S, Segovia S, Mitchell W, et al. Comparison of single-molecule sequencing and hybrid approaches for finishing the genome of *Clostridium autoethanogenum* and analysis of CRISPR systems in industrial relevant Clostridia. Biotechnol Biofuels. 2014;7:40.
72. Strätz M, Sauer U, Kuhn A, Dürre P. Plasmid transfer into the homoacetogen *Acetobacterium woodii* by electroporation and conjugation. Appl Environ Microbiol. 1994;60:1033–7.
73. Köpke M, Liew F. Recombinant microorganisms and methods for production thereof. US 2011/0236941A1. 2011.
74. Straub M, Demler M, Weuster-Botz D, Dürre P. Selective enhancement of autotrophic acetate production with genetically modified *Acetobacterium woodii*. J Biotechnol. 2014;178:67–72.
75. Schiel-Bengelsdorf B, Dürre P. Pathway engineering and synthetic biology using acetogens. FEBS Lett. 2012;586:2191–8.
76. Kita A, Iwasaki Y, Sakai S, Okuto S, Takaoka K, Suzuki T, et al. Development of genetic transformation and heterologous expression system in carboxydotrophic thermophilic acetogen *Moorella thermoacetica*. J Biosci Bioeng. 2013;115:347–52.
77. Ueki T, Nevin KP, Woodard TL, Lovley DR. Converting carbon dioxide to butyrate with an engineered strain of Clostridium ljungdahlii. MBio. 2014;5.
78. Mermelstein LD, Papoutsakis ET. In vivo methylation in *Escherichia coli* by the *Bacillus subtilis* phage ?3T I methyltransferase to protect plasmids from restriction upon transformation of *Clostridium acetobutylicum* ATCC 824. Appl Environ Microbiol. 1993;59:1077–81.
79. Pyne ME, Moo-Young M, Chung DA, Chou CP. Development of an electrotransformation protocol for genetic manipulation of *Clostridium pasteurianum*. Biotechnol Biofuels. 2013;6:50.
80. Cui GZ, Hong W, Zhang J, Li WL, Feng Y, Liu YJ, et al. Targeted gene engineering in *Clostridium cellulolyticum* H10 without methylation. J Microbiol Methods. 2012;89:201–8.
81. Clark TA, Murray IA, Morgan RD, Kislyuk AO, Spittle KE, Boitano M, et al. Characterization of DNA methyltransferase specificities using single-molecule, real-time DNA sequencing. Nucleic Acids Res. 2012;40:e29.
82. Murray IA, Clark TA, Morgan RD, Boitano M, Anton BP, Luong K, et al. The methylomes of six bacteria. Nucleic Acids Res. 2012;40:11450–62.
83. Stratz M, Sauer U, Kuhn A, Durre P. Plasmid transfer into the homoacetogen *Acetobacterium woodii* by electroporation and conjugation. Appl Environ Microbiol. 1994;60:1033–7.
84. Al-Hinai MA, Fast AG, Papoutsakis ET. Novel system for efficient isolation of Clostridium double-crossover allelic exchange mutants enabling markerless chromosomal gene deletions and DNA integration. Appl Environ Microbiol. 2012;78:8112–21.

85. Ng YK, Ehsaan M, Philip S, Collery MM, Janoir C, Collignon A, et al. Expanding the repertoire of gene tools for precise manipulation of the clostridium difficile genome: allelic exchange using pyrE alleles. PLoS ONE. 2013;8:e56051.
86. Heap JT, Ehsaan M, Cooksley CM, Ng Y-K, Cartman ST, Winzer K, et al. Integration of DNA into bacterial chromosomes from plasmids without a counter-selection marker. Nucleic Acids Res. 2012;40:e59.
87. Walker DJF, Koepke M. Method of producing a recombinant microorganism. US9315830B2. 2016.
88. Liew FM, Martin ME, Tappel RC, Heijstra BD, Mihalcea C, Köpke M. Gas fermentation-a flexible platform for commercial scale production of low-carbon-fuels and chemicals from waste and renewable feedstocks. Front Microbiol. 2016;7:694.
89. Lambowitz AM, Zimmerly S. Mobile group II introns. Annu Rev Genet. 2004;38:1–35.
90. Mei Y, Wang Y, Chen H, Sun ZS, Da JuX. Recent progress in CRISPR/Cas9 technology. J Genet Genom. 2016;43:63–75.
91. Heap JT, Pennington OJ, Cartman ST, Carter GP, Minton NP. The ClosTron: a universal gene knock-out system for the genus Clostridium. J Microbiol Methods. 2007;70:452–64.
92. Jiang Y, Chen B, Duan C, Sun B, Yang J, Yang S. Multigene editing in the *Escherichia coli* genome via the CRISPR-Cas9 system. Appl Environ Microbiol. 2015;81:2506–14.
93. Jiang Wenyan, Bikard David, Cox David. Feng Zhang and LAM. CRISPR-assisted editing of bacterial genomes. Nat Biotechnol. 2013;31:233–9.
94. Dicarlo JE, Norville JE, Mali P, Rios X, Aach J, Church GM. Genome engineering in *Saccharomyces cerevisiae* using CRISPR-Cas systems. Nucleic Acids Res. 2013;41:4336–43.
95. Huang H, Chai C, Li N, Rowe P, Minton NP, Yang S, et al. CRISPR/Cas9-based efficient genome editing in *Clostridium ljungdahlii*, an autotrophic gas-fermenting bacterium. ACS Synth Biol. 2016. doi:10.1021/acssynbio.6b00044.
96. Nagaraju S, Davies NK, Walker DJF, Köpke M, Simpson SD. Genome editing of *Clostridium autoethanogenum* using CRISPR/Cas9. Biotechnol Biofuels. 2016;9:219.
97. Wang Y, Zhang ZT, Seo SO, Lynn P, Lu T, Jin YS, et al. Bacterial genome editing with CRISPR-Cas9: deletion, Integration, single nucleotide modification, and desirable "clean" mutant selection in *Clostridium beijerinckii* as an example. ACS Synth Biol. 2016;5:721–32.
98. Salis HM, Mirsky EA, Voigt CA. Automated design of synthetic ribosome binding sites to control protein expression. Nat Biotechnol. 2009;27:946–50.
99. Salis HM. The ribosome binding site calculator. Methods Enzymol. 2011;498:19–42.
100. Mukherjee A, Schroeder CM. Flavin-based fluorescent proteins: emerging paradigms in biological imaging. Curr Opin Biotechnol. 2015;31:16–23.
101. Teng L, Wang K, Xu J, Xu C. Flavin mononucleotide (FMN)-based fluorescent protein (FbFP) as reporter for promoter screening in *Clostridium cellulolyticum*. J Microbiol Methods. 2015;119:37–43.
102. Köpke M, Nagaraju S, Chen W. Recombinant microorganisms and methods of use thereof. WO 2013/115659 A2. 2013.
103. Heveling J, Nicolaides CP, Scurrell MS. Catalysts and conditions for the highly efficient, selective and stable heterogeneous oligomerisation of ethylene. Appl Catal A Gen. 1998;173:1–9.
104. Jones SW, Fast AG, Carlson ED, Wiedel CA, Au J, Antoniewicz MR, et al. CO_2 fixation by anaerobic non-photosynthetic mixotrophy for improved carbon conversion. Nat Commun. 2016;7:12800.
105. Köpke M, Gerth ML, Maddock DJ, Mueller AP, Liew F, Simpson SD, et al. Reconstruction of an acetogenic 2,3-butanediol pathway involving a novel NADPH-dependent primary-secondary alcohol dehydrogenase. Appl Environ Microbiol. 2014;80:3394–403.
106. Köpke M, Chen WY. Recombinant microorganisms and uses therefor. US20130323806 A1. 2013.
107. Mueller A, Koepke M, Nagaraju S. Recombinant microorganisms and uses therefor. US20130330809 A1. 2013.
108. Liew FM, Köpke M, Simpson SD. Gas fermentation for commercial biofuels production. In: Fang Z, editor. Biofuel Prod Dev Prospect. Rijeka: InTech; 2013. p. 125–74.
109. Boock JT, Gupta A, Prather KLJ. Screening and modular design for metabolic pathway optimization. Curr Opin Biotechnol. 2015;36:189–98.
110. Keasling JD. Manufacturing molecules through metabolic engineering. Science. 2010; 330:1355–8.
111. Stephanopoulos G. Metabolic fluxes and metabolic engineering. Metab Eng. 1999;1:1–11.
112. Lechner A, Brunk E, Keasling JD. The need for integrated approaches in metabolic engineering. Cold Spring Harb Perspect Biol. 2016;8:a023903.
113. Barton NR, Burgard AP, Burk MJ, Crater JS, Osterhout RE, Pharkya P, et al. An integrated biotechnology platform for developing sustainable chemical processes. J Ind Microbiol Biotechnol. 2015;42:349–60.
114. Biggs BW, De Paepe B, Santos CNS, De Mey M, Kumaran Ajikumar P. Multivariate modular metabolic engineering for pathway and strain optimization. Curr Opin Biotechnol. 2014;29:156–62.
115. Liu W, Jiang R. Combinatorial and high-throughput screening approaches for strain engineering. Appl Microbiol Biotechnol. 2015;99:2093–104.
116. Landels A, Evans C, Noirel J, Wright PC. Advances in proteomics for production strain analysis. Curr Opin Biotechnol. 2015;35:111–7.
117. Brunk E, George KW, Alonso-Gutierrez J, Keasling JD, Palsson BO, Lee TS, et al. Characterizing strain variation in engineered *E. coli* using a multi-omics-based workflow. Cell Syst. 2016;2:335–46.
118. Baidoo EE, Benke PI, Keasling JD. Mass spectrometry-based microbial metabolomics. Microb Syst Biol Methods Prot. 2012. doi:10.1007/978-1-61779-827-6_9.
119. Batth TS, Singh P, Ramakrishnan VR, Sousa MML, Chan LJG, Tran HM, et al. A targeted proteomics toolkit for high-throughput absolute quantification of *Escherichia coli* proteins. Metab Eng. 2014;26:48–56.
120. Oberortner E, Densmore D. Web-based software tool for constraint-based design specification of synthetic biological systems. ACS Synth. Biol. 2015;4:757–60.
121. Hillson NJ. j5 DNA assembly design automation. Methods Mol Biol. 2014; 1116:245-69. doi:10.1007/978-1-62703-764-8_17.
122. Quinn JY, Cox RS, Adler A, Beal J, Bhatia S, Cai Y, et al. SBOL Visual: a Graphical Language for Genetic Designs. PLOS Biol. 2015;13:e1002310.
123. Luo Y, Enghiad B, Zhao H. New tools for reconstruction and heterologous expression of natural product biosynthetic gene clusters. Nat Prod Rep. 2016;33(2):174–82.
124. Baek CH, Liss M, Clancy K, Chesnut J, Katzen F. DNA assembly tools and strategies for the generation of plasmids. Microbiol Spectr. 2014. doi:10.1128/microbiolspec.PLAS-0014-2013.
125. Kosuri S, Church GM. Large-scale de novo DNA synthesis: technologies and applications. Nat Methods. 2014;11:499–507.
126. Freestone TS, Zhao H. Combinatorial pathway engineering for optimized production of the anti-malarial FR900098. Biotechnol Bioeng. 2016;113:384–92.
127. Jeschek M, Gerngross D, Panke S. Rationally reduced libraries for combinatorial pathway optimization minimizing experimental effort. Nat Commun. 2016;7:11163.
128. Feng X, Zhuang W-Q, Colletti P, Tang YJ. Metabolic pathway determination and flux analysis in nonmodel microorganisms through ^{13}C-isotope labeling. In: Navid A, editor. Microbial systems biology: methods and protocols. Totowa: Humana Press; 2012. p. 309–30. doi:10.1007/978-1-61779-827-6_11.
129. Du Y, Jiang W, Yu M, Tang IC, Yang ST. Metabolic process engineering of Clostridium tyrobutyricum ?ack-adhE2 for enhanced n-butanol production from glucose: effects of methyl viologen on NADH availability, flux distribution, and fermentation kinetics. Biotechnol Bioeng. 2015;112:705–15.
130. Du J, McGraw A, Hestekin JA. Modeling of *Clostridium tyrobutyricum* for butyric acid selectivity in continuous fermentation. Energies. 2014;7:2421–35.
131. Chen J, Gomez JA, Höffner K, Phalak P, Barton PI, Henson MA. Spatiotemporal modeling of microbial metabolism. BMC Syst Biol. 2016;10:21.
132. Chen J, Henson MA. In silico metabolic engineering of *Clostridium ljungdahlii* for synthesis gas fermentation. Metab Eng. 2016;38:389–400.
133. Lane J. Coskata's technology re-emerges as Synata Bio : biofuels digest. 2016. http://www.biofuelsdigest.com/bdigest/2016/01/24/coskatas-technology-re-emerges-as-synata-bio/.
134. Sapp M. INEOS Bio selling 8 MGY demo plant in Florida : biofuels digest. 2016. http://www.biofuelsdigest.com/bdigest/2016/09/06/ineos-bio-selling-8-mgy-demo-plant-in-florida/.

135. Schill SR. Ethanol producer magazine–the latest news and data about ethanol production. 2013. http://www.ethanolproducer.com/articles/10096/ineos-declares-commercial-cellulosic-ethanol-online-in-florida.
136. Global SCS. Beijing Shougang LanzaTech New Energy Science & Technology Company Earns Roundtable on Sustainable Biomaterials (RSB) Certification. Newsroom. 2013. https://www.scsglobalservices.com/beijing-shougang-lanzatech-new-energy-science-technology-company-earns-roundtable-on-sustainable.
137. Lane J. Virgin ♥ LanzaJet fuel: "A real game changer for aviation," says Branson. Biofuels Dig. 2016. www.biofuelsdigest.com/bdigest/2016/09/15/virgin-%E2%99%A5-lanzajet-fuel-a-real-game-changer-for-aviation-says-branson/.
138. Lane J. China steel green-lights commercial-scale LanzaTech advanced biofuels project. Biofuels Dig. 2015; www.biofuelsdigest.com/bdigest/2015/04/22/china-steel-green-lights-46m-for-commercial-scale-lanzatech-advanced-biofuels-project/.
139. Lane J. Steel's Big Dog jumps into low carbon fuels: ArcelorMittal, LanzaTech, Primetals Technologies to construct $96 M biofuel production facility. Biofuels Dig. 2015. http://www.biofuelsdigest.com/bdigest/2015/07/13/steels-big-dog-jumps-into-low-carbon-fuels-arcelormittal-lanzatech-primetals-technologies-to-construct-96m-biofuel-production-facility/.

3

Transcriptional activator Cat8 is involved in regulation of xylose alcoholic fermentation in the thermotolerant yeast *Ogataea (Hansenula) polymorpha*

Justyna Ruchala[1], Olena O. Kurylenko[2], Nitnipa Soontorngun[3], Kostyantyn V. Dmytruk[2] and Andriy A. Sibirny[1,2*]

Abstract

Background: Efficient xylose alcoholic fermentation is one of the key to a successful lignocellulosic ethanol production. However, regulation of this process in the native xylose-fermenting yeasts is poorly understood. In this work, we paid attention to the transcriptional factor Cat8 and its possible role in xylose alcoholic fermentation in *Ogataea (Hansenula) polymorpha*. In *Saccharomyces cerevisiae*, organism, which does not metabolize xylose, gene *CAT8* encodes a Zn-cluster transcriptional activator necessary for expression of genes involved in gluconeogenesis, respiration, glyoxylic cycle and ethanol utilization. Xylose is a carbon source that could be fermented to ethanol and simultaneously could be used in gluconeogenesis for hexose synthesis. This potentially suggests involvement of *CAT8* in xylose metabolism.

Results: Here, the role of *CAT8* homolog in the natural xylose-fermenting thermotolerant yeast *O. polymorpha* was characterized. The *CAT8* ortholog was identified in *O. polymorpha* genome and deleted both in the wild-type strain and in advanced ethanol producer from xylose. Constructed *cat8Δ* strain isolated from wild strain showed diminished growth on glycerol, ethanol and xylose as well as diminished respiration on the last substrate. At the same time, *cat8Δ* mutant isolated from the best available *O. polymorpha* ethanol producer showed only visible defect in growth on ethanol. *CAT8* deletant was characterized by activated transcription of genes *XYL3, DAS1* and *RPE1* and slight increase in the activity of several enzymes involved in xylose metabolism and alcoholic fermentation. Ethanol production from xylose in *cat8Δ* mutants in the background of wild-type strain and the best available ethanol producer from xylose increased for 50 and 30%, respectively. The maximal titer of ethanol during xylose fermentation was 12.5 g ethanol/L at 45 °C. Deletion of *CAT8* did not change ethanol production from glucose. Gene *CAT8* was also overexpressed under control of the strong constitutive promoter *GAP* of glyceraldehyde-3-phosphate dehydrogenase. Corresponding strains showed drop in ethanol production in xylose medium whereas glucose alcoholic fermentation remained unchanged. Available data suggest on specific role of Cat8 in xylose alcoholic fermentation.

Conclusions: The *CAT8* gene is one of the first identified genes specifically involved in regulation of xylose alcoholic fermentation in the natural xylose-fermenting yeast *O. polymorpha*.

Keywords: Transcriptional activator, Xylose, High-temperature alcoholic fermentation, Yeast, *Ogataea (Hansenula) polymorpha*

*Correspondence: sibirny@cellbiol.lviv.ua
[2] Department of Molecular Genetics and Biotechnology, Institute of Cell Biology, Drahomanov Str., 14/16, Lviv 79005, Ukraine
Full list of author information is available at the end of the article

Background

Fermentation is the largest field of industrial biotechnology. In 2014, near 95 billion liters of ethanol were produced [1]. Currently, most of industrial ethanol is produced from starch and sucrose (1st generation ethanol), however, due to limited feedstock abundance, further increase in fuel ethanol production will depend on development of feasible technology of alcoholic fermentation from lignocellulose (2nd generation ethanol). One of the most important goals in the development of such technology is construction of strains capable of efficient fermentation of lignocellulosic pentoses, especially xylose, which constitutes about 30% of all sugars in lignocellulosic hydrolyzates [2, 3]. It would also be useful to carry out fermentation of xylose and other lignocellulosic sugars under elevated temperatures (around 50 °C), which would allow optimal activities of cellulases and hemicellulases necessary for the process known as Simultaneous Saccharification and Fermentation (SSF) [4]. In such a process, free sugars liberated by enzymatic hydrolysis do not exert product inhibition on hydrolyzing enzymes, since they are simultaneously converted to ethanol by thermotolerant microorganisms in the same vessel. Very few yeast organisms are capable of high-temperature alcoholic fermentation, namely *Kluyveromyces marxianus* [5] and *Ogataea (Hansenula) polymorpha* [6, 7]. Current work focuses on *O. polymorpha* which is the most thermotolerant yeast species known to date, with maximal growth and fermentation temperatures of 50 °C or even higher [8, 9]. It has been reported that *O. polymorpha* produces ethanol from glucose, cellobiose, glycerol and xylose at elevated temperatures [7, 10], however, ethanol yield and productivity from xylose by the wild-type strains is very low [11]. *O. polymorpha* can also produce ethanol directly from starch and xylan after expression of heterologous genes encoding corresponding hydrolytic enzymes [12]. Several methods of metabolic engineering, both original and those developed for other yeast species, were successfully used for improvements of ethanol synthesis from xylose in *O. polymorpha*. They include heterologous expression of bacterial xylose isomerases and overexpression of native xylulokinase [13] and, alternatively, overexpression of engineered xylose reductase with decreased affinity to NADPH as well as native xylitol dehydrogenase and xylulokinase [14] and overexpression of pyruvate decarboxylase in the strain unable to utilize ethanol as sole carbon source [15]. Combination of metabolic engineering (overexpression of engineered xylose reductase and native xylitol dehydrogenase and xylulokinase) with classical selection approaches (selection for strains unable to utilize ethanol as sole carbon source and resistant to glycolysis inhibitor 3-bromopyruvate), allowed isolation of strains that accumulate 15–20 times more ethanol from xylose relative to the wild-type strain, i.e. around 10 g ethanol/L at 45 °C [16]. While mutation(s) causing resistance to 3-bromopyruvate in the ethanol overproducing strain remain to be identified, we have recently mapped a corresponding mutation in the strain with the wild-type background and showed that it disrupted an autophagy-related gene *ATG13*. This mutation led to a 50% increase in ethanol production from xylose [17]; Dmytruk, Sibirny, in preparation]. Still, the achieved yield and productivity of ethanol synthesis from xylose are lower than that described for engineered *Saccharomyces cerevisiae* and several native xylose-fermenting yeasts (which however are mesophilic and therefore could not be useful for the SSF process). Further possible increase in ethanol synthesis by *O. polymorpha* from xylose is hampered due to absence of the knowledge on regulation of xylose metabolism and fermentation. Therefore, it is important to identify the corresponding genes and, depending on their functions, activate or repress them. Described functions of a transcription factor Cat8 (encoded by *CAT8* gene) in activating multiple metabolic processes in *S. cerevisiae*, mostly gluconeogenesis and ethanol utilization [18, 19], led us to hypothesize that it might also be involved in regulation of xylose metabolism in *O. polymorpha*. One of the reasons that just *CAT8* was selected among multiple genes coding for transcription factors involved in carbon metabolism [20] was that knock out of *CAT8* activated glucose alcoholic fermentation in *S. cerevisiae* [21] and non-conventional yeast *Pichia guilliermondii* [22]. Xylose is a unique carbon source as it could be fermented to ethanol, similarly to glucose, and simultaneously it has to be converted to glucose and other hexoses, mostly in pentose phosphate pathway though partial contribution of gluconeogenesis in hexose synthesis from xylose cannot be neglected. We hypothesized that for these reasons the mutants of *O. polymorpha* with knock out of the ortholog of *CAT8* gene will have impairments in xylose respiration and gluconeogenesis, so the flux of this sugar will be activated instead into fermentation direction.

Roles of *CAT8* gene in regulation of cell metabolism are quite well understood in *S. cerevisiae*. It encodes a Zn-cluster transcriptional activator necessary for expression of genes involved in gluconeogenesis, ethanol utilization and diauxic shift from fermentation to respiration [18, 19]. Strains with deletion of *CAT8* show defects in growth on ethanol, glycerol and other gluconeogenic substrates whereas disaccharides are utilized normally. Mechanistically, Cat8 exerts transcriptional activation of its target genes by binding to carbon source-responsive elements in their regulatory promoters [20, 23]. However, the limited data available on the functions of *CAT8* in non-*Saccharomyces* yeasts show differences in functions of the corresponding orthologs. Thus, *Kluyveromyces lactis* mutant defective in *CAT8* showed defects in ethanol utilization, whereas growth on glycerol was normal [24]. *cat8Δ* mutant of *Candida albicans* normally utilized all carbon substrates tested [25], whereas growth patterns of the mutant with knock out of *CAT8* in *Pichia guilliermondii* were not assayed at all [22]. Role of *CAT8* in regulation of xylose metabolism is poorly understood. Transcriptome analysis of the natural xylose-metabolizing yeast *O. polymorpha* did not find changes in *CAT8* expression between xylose- and glucose-containing media [26]. In recombinant *S. cerevisiae* capable of xylose metabolism, xylose caused only weak repression of *CAT8* relative to glucose suggesting xylose growing cells are in between totally repressed and derepressed state regarding catabolite repression [27].

To test our hypothesis on the role of Cat8 transcription factor in xylose fermentation, we have isolated *CAT8* knock-out mutants in *O. polymorpha* on either wild-type or ethanol overproducing (from xylose) strain [16]. We also overexpressed *CAT8*. In favor of our hypothesis, we found that strains with deletions of *CAT8Δ* accumulate more ethanol during xylose fermentation, while ethanol production from glucose was not changed. Mutant *O. polymorpha cat8* isolated from the advanced ethanol producer accumulated up to 12.5 g ethanol/L at 45 °C, which is the highest ethanol titer for high-temperature xylose fermentation. Inversely, strain of *O. polymorpha* with overexpression of *CAT8* accumulated less ethanol relative to the parental wild-type strain.

Results

Isolation and growth characteristics of *cat8Δ* mutants

We decided to delete the *O. polymorpha CAT8* ortholog in both wild-type strain and the best ethanol producer (BEP) from xylose [16] and to study the properties of the resulted deletants. In particular, we focused on growth patterns, respiration, activity of some enzymes, expression of selected genes and ethanol production in xylose and glucose media. Genome of *O. polymorpha* strain NCYC495 is sequenced and is publicly available [28]. It contains single ortholog of *S. cerevisiae* gene *CAT8*, which shows 31% identity and 53% similarity to *CAT8* gene of *S. cerevisiae*. To knock it out in *O. polymorpha*, a deletion cassette was constructed, which contained *natNT2* gene conferring resistance to nourseothricin as a selection marker, flanked with non-coding regions of the *CAT8* gene ortholog (see "Methods" section; Additional file 1A). Homologous recombination resulted in isolation of the *cat8Δ* strain. In total, near 1000 nourseothricin-resistant transformants were analyzed and 9 of them appeared to be *cat8Δ* mutants. Our attempts to isolate *cat8Δ* on the background of strain BEP were unsuccessful. In total, we analyzed near 2000 transformants and invariably without success. It is known that the selection marker has strong impact on the efficiency of homologous recombination [29]. Therefore, we decided to construct a deletion cassette using a selection marker gene *hphNT1*, conferring resistance to hygromycin (see "Methods" section; Additional file 1B). In this case, 10 *CAT8* knock out mutants were identified among 400 analyzed hygromycin-resistant transformants.

The isolated *cat8Δ* mutants on the background of the wild-type and BEP strains were assayed for growth, biochemical and physiological characteristics. Growth of these mutants was analyzed in YNB solid and liquid media supplemented with different carbon sources and compared with that of the corresponding parental strain. It was found that isolated mutants normally grew in media with glucose, whereas growth of *cat8Δ* mutant isolated from wild-type strain on glycerol and ethanol was retarded but not totally abolished. Growth of BEP *cat8Δ* was very similar to that of BEP in glycerol containing medium, while BEP *cat8Δ* was unable to growth in ethanol, unlike to BEP (Additional file 2). It is remarkable that growth of *cat8Δ* strain isolated from the wild-type strain on xylose was also partially retarded, whereas no significant difference in growth on xylose was observed between BEP and BEP *cat8Δ* strains. However, the BEP strain much better grows on xylose relative to the wild-type strain apparently due to overexpression of genes *XYL1*, *XYL2* and *XYL3* involved in primary xylose metabolism [16] (Fig. 1; Additional file 2). It has to be pointed out that ethanol overproducing strain BEP poorly grows on ethanol [16], whereas its derivative BEP *cat8Δ* mutant did not grow on this substrate at all. We suggest that function of *CAT8* in *O. polymorpha* is similar to that in *S. cerevisiae* as corresponding deletants grow poorly on ethanol and glycerol.

Isolation and growth characteristics of the strains with overexpression of *CAT8*

Transformants of *O. polymorpha* wild-type strain, which express *CAT8* under control of the strong constitutive *GAP* promoter of glyceraldehyde-3-phosphate dehydrogenase gene, were isolated (WT *CAT8*). Overexpression of *CAT8* was proved by qRT-PCR. It was found that indeed, the analyzed strain with *CAT8* gene under *GAP* promoter showed increase in *CAT8* expression for 2.65 times (Additional file 3). It was found that WT *CAT8* did not differ from the wild-type strain regarding growth on the tested substrates: glucose, xylose, glycerol and ethanol (Fig. 1).

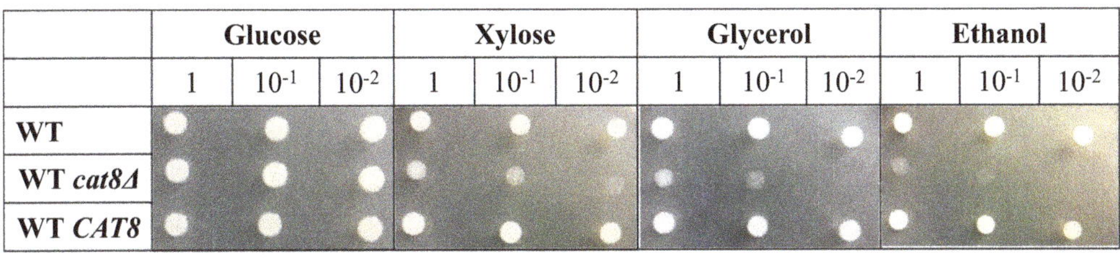

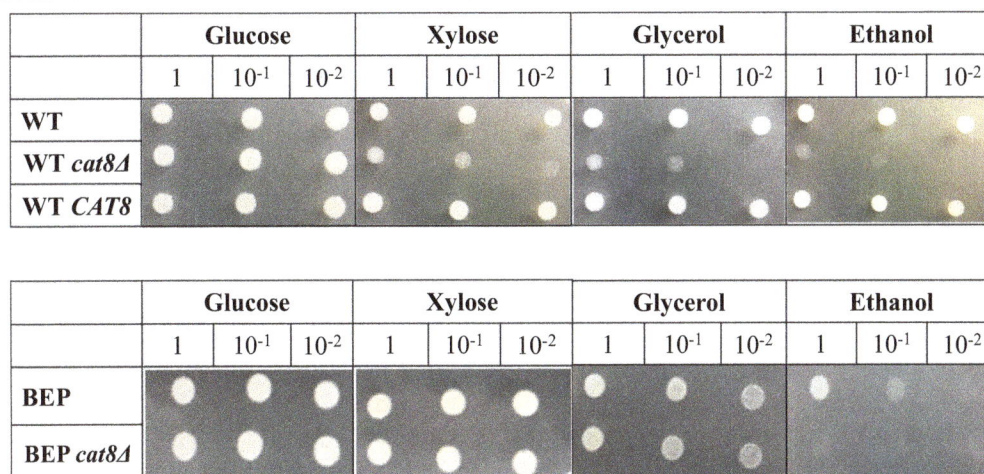

	Glucose			Xylose			Glycerol			Ethanol		
	1	10^{-1}	10^{-2}	1	10^{-1}	10^{-2}	1	10^{-1}	10^{-2}	1	10^{-1}	10^{-2}
BEP												
BEP *cat8Δ*												

Fig. 1 Growth of the strains with deletion (*cat8Δ*) or overexpression of *CAT8* gene (*CAT8**) on different carbon sources (glucose, xylose, glycerol, ethanol) as compared to the parental strains

Respiration, enzymatic profiles and transcription of selected genes in the isolated mutants

More detailed physiological, biochemical and genetic analyses were carried out on constructed deletion mutants *cat8Δ* and BEP *cat8Δ*. To reveal the role of *CAT8* gene in the metabolism of *O. polymorpha*, cell respiration of *cat8Δ* cells in glucose- and xylose-containing media was studied. It was found that cells of both *cat8Δ* and BEP *cat8Δ* strains showed up to 40% decrease in respiration with xylose as a substrate. Respiration of *cat8Δ* but not that of BEP *cat8Δ* cells also was decreased using glucose as a substrate (Table 1). These data confirm our suggestion on the similar role of *CAT8* in *O. polymorpha* and *S. cerevisiae*. The observed small increase in glucose respiration of BEP *cat8Δ* cells apparently depends on unidentified mutations introduced in BEP strain during its selection [16]. In the following experiments, we analyzed specific activities of several enzymes involved in xylose metabolism and ethanol synthesis in cells cultivated in xylose medium. It was found that deletion of *CAT8* led to moderate increase in specific activities of most of the analyzed enzymes involved in xylose metabolism and alcoholic fermentation: xylose reductase, xylulokinase, transketolase, pyruvate decarboxylase and alcohol dehydrogenase. Activity of fructose-1,6-bisphosphatase in *cat8Δ* mutants was slightly increased whereas xylitol dehydrogenase activity was, inversely, decreased as compared to that of the parental strains (Table 2). Cat8 protein is apparently involved in the regulation of the corresponding gene expression. To test this hypothesis, transcription profiles of several potentially involved genes were studied using quantitative reverse-transcription PCR (qRT-PCR). It was found that *cat8Δ* mutant isolated from the wild-type strain cultivated in xylose medium showed higher level of *XYL3, DAS1* and *RPE1* mRNAs whereas expression of the other analyzed genes (*XYL1, XYL2, PDC1, TKL1, TAL1, TAL2, FBP1, PCK1*) was quite similar as compared to that of the wild-type strain (Table 3). Strain BEP *cat8Δ* revealed increased expression of *RPE1*, decreased expression of *XYL1, XYL2* and *DAS1* while the expression of other tested genes possessed minor fluctuations relative to that of BEP strain on xylose containing medium (Table 3). Expression of *RPE1* was increased for both deletion mutants to infer this gene as a promising target for overexpression, aiming to increase performance of xylose alcoholic fermentation. We also assayed the relative expression of the studied genes between *O. polymorpha* wild-type strain NCYC495 and the BEP strain as it was not done previously [16]. It showed a substantial enhancement of the expression of genes involved in xylose metabolism in ethanol

Table 1 Respiration activity of analyzed *O. polymorpha* strains

Strain	Respiration (nanomoles of O_2 consumed per minute per mg of cells at 30 °C)	
	Glucose as substrate	Xylose as substrate
WT	11.81 ± 0.52	11.35 ± 0.56
cat8Δ	8.49 ± 0.07	7.08 ± 0.14
BEP	11.87 ± 0.59	17.17 ± 0.86
BEP *cat8Δ*	13.98 ± 0.70	10.53 ± 0.02

Determinations were performed in distilled air-saturated water with the concentration of cells 0.5 g/L of dry weight and started by addition of 1% carbon substrate (glucose or xylose). The respiratory rate was expressed as nanomoles of O_2 consumed per minute per mg of cells (dry weight)

Table 2 Specific activities of XR (xylose reductase), XDH (xylitol dehydrogenase), XK (xylulokinase), ADH (alcohol dehydrogenase), PDC (pyruvate decarboxylase), FBP (fructose-1,6-bisphosphatase), and TKL (transketolase) in the cells of analyzed *O. polymorpha* strains from third day of xylose alcoholic fermentation at 45 °C

Strain	Activity U/mg of protein						
	XR	XDH	XK	PDC	ADH	TKL	FBP
WT	0.012 ± 0.001	0.011 ± 0.001	–	0.165 ± 0.008	0.103 ± 0.005	0.005 ± 0.002	0.012 ± 0.001
cat8Δ	0.014 ± 0.002	0.006 ± 0.001	–	0.183 ± 0.012	0.119 ± 0.001	0.008 ± 0.001	0.014 ± 0.003
BEP	0.023 ± 0.003	0.335 ± 0.004	0.494 ± 0.031	0.323 ± 0.018	0.119 ± 0.020	0.012 ± 0.004	0.011 ± 0.001
BEP *cat8Δ*	0.028 ± 0.002	0.255 ± 0.018	0.629 ± 0.038	0.346 ± 0.006	0.189 ± 0.015	0.019 ± 0.003	0.015 ± 0.002

– Not determined

overproducing strain with especially high increase in expression of *RPE1* gene (Table 3). We speculate that this was achieved by metabolic engineering of the first steps of xylose metabolism but also possibly as a result of classical selection [16].

Ethanol production by mutants with deletion and overexpression of *CAT8* gene in xylose and glucose media

Xylose and glucose fermentation of the isolated *cat8Δ* and BEP *cat8Δ* strains was studied under semi-anaerobic conditions (see "Methods" section). It was found that defects of *CAT8* gene leads to 1.5-fold increase in ethanol accumulation on the background of the wild-type strain though concentration of the accumulated ethanol was quite low (Table 4; Fig. 2). At the same time, overexpression of *CAT8* led to decrease in ethanol production from xylose (Fig. 3). Effect of *CAT8* overexpression on glucose fermentation was insignificant (Additional file 4).

It is important to note that deletion of *CAT8* in BEP strain also had a positive effect on ethanol accumulation, which increased by 30% and reached 12.5 g ethanol/L. Increased ethanol production from xylose was accompanied by activated xylose consumption from the medium (Fig. 2). Data of Table 4 show that the strain BEP *cat8Δ* possessed increase in ethanol yield and productivity in xylose medium relative to the parental overproducing strain BEP for 13 and 21%, respectively. Strain BEP *cat8Δ* did not accumulate xylitol (data not shown) similar to that of the parental strain BEP [16]. Thus, we conclude that Cat8 transcription factor is involved in the control of xylose alcoholic fermentation and the deficiency of this protein activates ethanol production from xylose. In contrast, *CAT8* deletion did not have pronounced effect on ethanol production during glucose fermentation both in the wild-type and the BEP strains (Additional file 4). Deletion of *CAT8* gene on both wild-type and BEP backgrounds also did not have effects on alcoholic fermentation of sucrose (data not shown).

Discussion

The natural xylose-utilizing thermotolerant yeast *O. polymorpha* ferments xylose and glucose at highest temperatures known for yeasts, i.e. at 50 °C [6, 8]. The current work introduces *CAT8* as a gene involved in the regulation of xylose metabolism and alcoholic fermentation in this organism. Prior to this study, the role of *CAT8* in xylose alcoholic fermentation had not been addressed. It was shown that the deletion of this gene in *S. cerevisiae* slightly activated glucose alcoholic fermentation [30]. In contrast, strong activation was observed in *P. guilliermondii* [22] though maximally achieved level of ethanol in the latter species was still very low. In *O. polymorpha CAT8* deletion did not lead to any significant changes in ethanol production from glucose, while a considerable increase in xylose alcoholic fermentation was observed. The reasons for this difference remain to be elucidated; quite possibly the enzymes involved in ethanol production are not activated in *cat8Δ* mutants during glucose fermentation. It has also to be pointed out that cell respiration of *cat8Δ* mutants on xylose was impaired in much higher extent relative to that on glucose as a substrate where BEP *cat8Δ* showed some increase in glucose respiration (Table 1), assuming xylose redirection from the Krebs cycle and oxidative phosphorylation towards ethanol production. The reason of the increase of ethanol production from xylose by *cat8Δ* strains could be explained by activation of xylulokinase, alcohol dehydrogenase and ribulosephosphate epimerase (Tables 2, 3) which could be the limiting factors during xylose alcoholic fermentation.

We observed impaired ethanol and glycerol utilization in *cat8Δ* mutants, suggesting the involvement of *CAT8* in regulation of gluconeogenesis in *O. polymorpha*, similar to that in *S. cerevisiae*. Remarkably, growth on xylose of *cat8Δ* mutant isolated from the wild-type strain was also partially impaired which suggests that xylose can be considered, at least partially, as gluconeogenic substrate. i.e., if hexoses are to some extent synthesized from xylose

Table 3 The relative expression levels of the particular genes in the parental strains and *cat8Δ* mutants at the third day of xylose alcoholic fermentation at 45 °C

ΔΔCt	Genes										
	XYL1	*XYL2*	*XYL3*	*PDC1*	*TKL1*	*DAS1*	*TAL1*	*TAL2*	*RPE1*	*FBP1*	*PCK1*
cat8Δ/WT	1.13 ± 0.300	1.05 ± 0.577	2.82 ± 0.438	0.64 ± 0.400	1.10 ± 0.361	2.39 ± 0.342	1.23 ± 0.360	1.21 ± 0.193	2.60 ± 0.486	0.67 ± 0.165	1.10 ± 0.435
BEP *cat8Δ*/BEP	0.36 ± 0.300	0.62 ± 0.085	1.07 ± 0.086	0.88 ± 0.479	0.79 ± 0.175	0.57 ± 0.195	1.11 ± 0.091	1.12 ± 0.067	1.54 ± 0.052	0.74 ± 0.392	0.76 ± 0.140
BEP/WT	7.66 ± 0.971	18.03 ± 0.045	2.76 ± 0.158	1.96 ± 0.380	1.71 ± 0.178	1.70 ± 0.670	9.29 ± 0.138	3.59 ± 0.138	47.60 ± 0.301	1.02 ± 0.274	0.19 ± 0.414

The mRNA quantification was normalized to *ACT1* mRNA

Genes encode: *XYL1*, xylose reductase; *XYL2*, xylitol dehydrogenase; *XYL3*, xylulokinase; *PDC1*, pyruvate decarboxylase; *TKL1*, transketolase; *DAS1*, dihydroxyacetone phosphate synthase or peroxisomal transketolase; *TAL1*, transaldolase; *TAL2*, peroxisomal transaldolase; *RPE1*, ribulosephosphate epimerase; *FBP1*, fructose-1,6-bisphosphatase; *PCK1*, phosphoenolpyruvate carboxykinase

Table 4 Main parameters of xylose fermentation at 45 °C by the *O. polymorpha* strains tested

Strain	Ethanol (g/L)	Ethanol yield (g/g consumed xylose)	Rate of ethanol synthesis (g/g biomass/h)	Productivity of ethanol synthesis (g/L/h)
WT[a]	0.523 ± 0.054	0.029 ± 0.010	0.009 ± 0.001	0.022 ± 0.001
cat8Δ[b]	0.780 ± 0.083	0.034 ± 0.002	0.012 ± 0.001	0.026 ± 0.001
BEP[c]	9.620 ± 0.102	0.300 ± 0.011	0.082 ± 0.002	0.169 ± 0.007
BEP *cat8Δ*[c]	12.51 ± 0.134	0.340 ± 0.015	0.091 ± 0.003	0.205 ± 0.009

[a] Data of ethanol yield and ethanol (g/L) are represented on YNB medium supplemented with 9% of xylose on the first day (24 h) of fermentation

[b] 48 h of fermentation

[c] 72 h of fermentation

in gluconeogenesis (from glyceraldehyde-3-phosphate which is synthesized in pentose phosphate pathway), this, together with defects in respiration, especially strong on xylose, could cause the redirection of xylose flux of *cat8Δ* mutants to catabolism and thus the redirection of xylose metabolism to the fermentation mode. Enhanced ethanol production from xylose by *cat8Δ* mutants could also be explained by the observed increase in enzyme activities and transcriptions of genes involved in xylose utilization and alcoholic fermentation (Tables 2, 3). Contrary, the slight increase in specific activity of fructose-1,6-bisphosphatase in *cat8Δ* mutants was observed suggesting differences in Cat8 action between *S. cerevisiae* and *O. polymorpha*. We suggest that growth impairments of *O. polymorpha cat8Δ* mutants on glycerol and ethanol are determined by partial defects in respiration which is

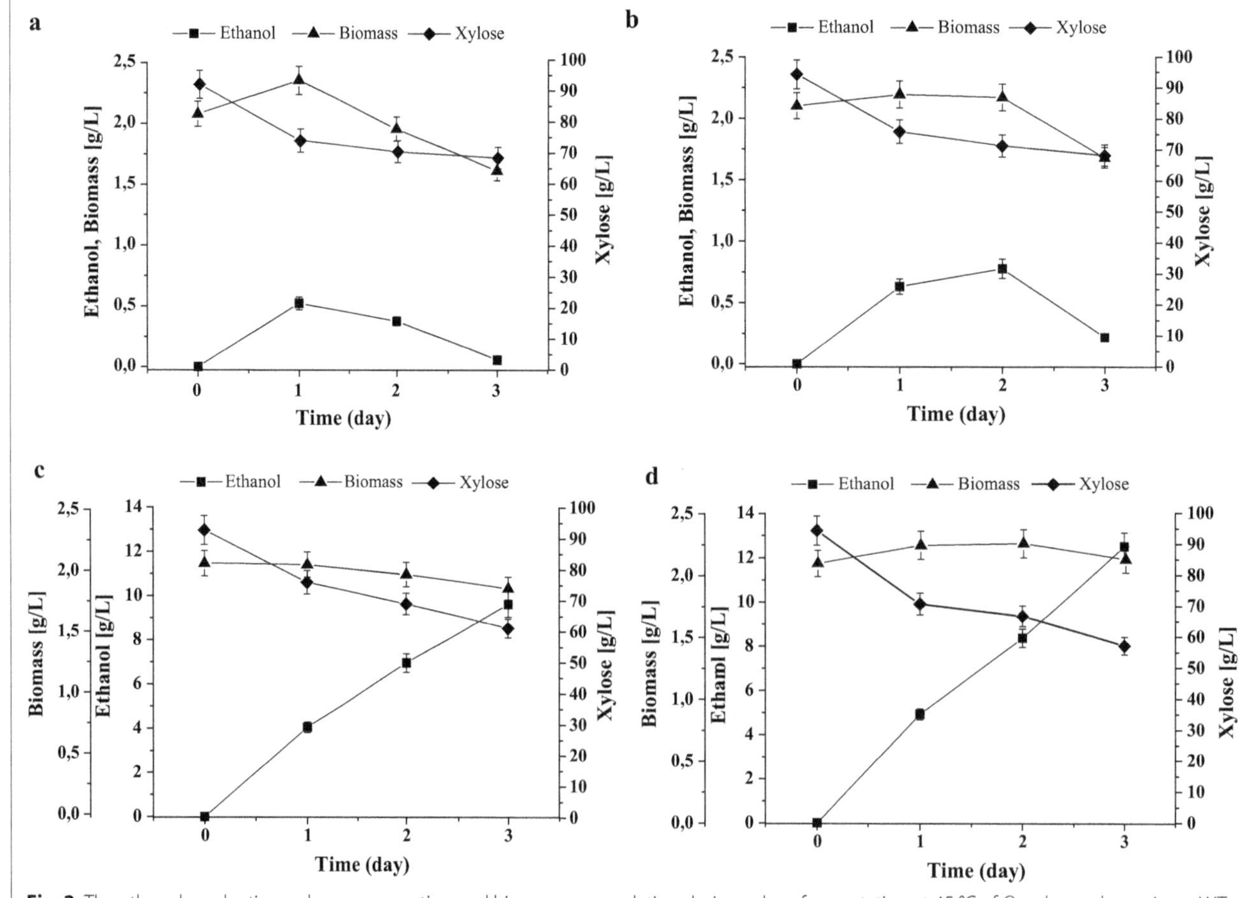

Fig. 2 The ethanol production, xylose consumption and biomass accumulation during xylose fermentation at 45 °C of *O. polymorpha* strains: **a** WT, **b** *cat8Δ*, **c** BEP, **d** BEP *cat8Δ*

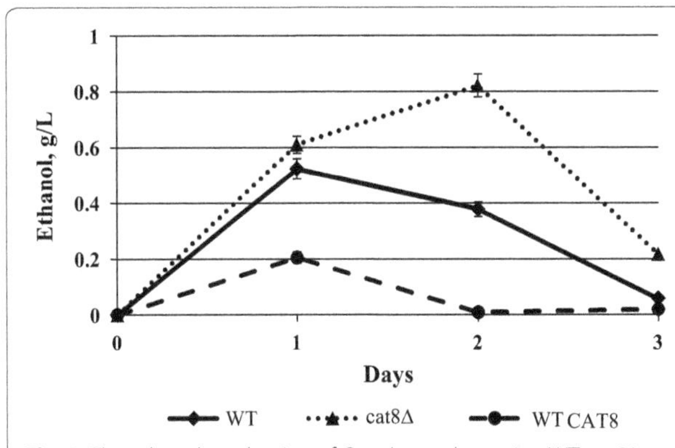

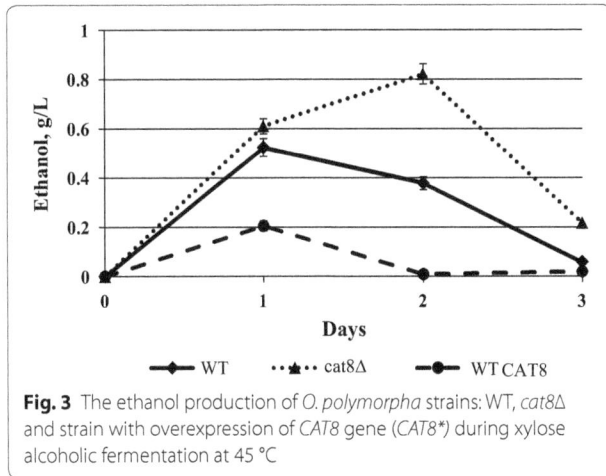

Fig. 3 The ethanol production of *O. polymorpha* strains: WT, *cat8Δ* and strain with overexpression of *CAT8* gene (*CAT8**) during xylose alcoholic fermentation at 45 °C

critical for growth on gluconeogenic substrates. Quite possible that this is also the reason of xylose growth retardation of *cat8Δ* mutant isolated from the wild-type strain. In spite activity of xylitol dehydrogenase is lowered in *cat8Δ* mutants, it is unlikely that Xyl2 is the limiting enzyme during growth on xylose as our earlier studies showed that deletion of the main paralog *XYL2* (assayed in current manuscript) did not impair growth on xylose at all and deletion of two paralogs of *XYL2* impaired growth still not completely [13].

It is interesting to note that overexpression of *CAT8* has opposite effect on xylose alcoholic fermentation as compared to that in *cat8Δ* mutants as transformants overexpressing *CAT8* gene were characterized by decrease in ethanol production from xylose (Fig. 3). Apparently high amounts of Cat8 activate xylose gluconeogenesis and respiration while inhibit fermentation of this pentose. Deletion or overexpression of *CAT8* had no effect on glucose fermentation suggesting specific involvement of Cat8 protein in regulation of xylose alcoholic fermentation.

Thus, the *CAT8* gene is one of the first identified genes specifically involved in regulation of xylose alcoholic fermentation in the natural xylose-fermenting yeasts. Inactivation of this gene (its knock out) increased ethanol production on backgrounds of the wild-type strain and of the advanced ethanol producer from xylose (BEP). The best ethanol producer from xylose described here, accumulated 30% more ethanol relative to the BEP strain from xylose reported previously and 20–25 times more compared to the wild-type strain [16]. The yield and productivity of ethanol synthesis in BEP *cat8Δ* strain, constructed in this work, for 13 and 21% exceeds those in the reported *O. polymorpha* ethanol overproducer from xylose. Ethanol yield in the BEP *cat8Δ* strain (0.34 g/g xylose) is close to that described for *S. stipitis* (0.35–0.44 g/g xylose) [31] and *S. passalidarum* (0.42 g/g xylose) [32], however, it was achieved for *O. polymorpha* at 45 °C whereas the compared organisms are mesophilic and thus cannot grow and ferment at so high temperature. Among thermotolerant ethanol producing strains the promising one is engineered *K. marxianus* strain with ethanol yield 0.38 g/g xylose at 42 °C, but lower yield at 45 °C (0.27 g/g xylose) [33]. In contrast to recombinant *K. marxianus* strain [33], BEP *cat8Δ* did not accumulate byproduct xylitol at all. Still, the level of increase in ethanol synthesis achieved in this work is not enough for feasible ethanol production from xylose. However, we suggest that the described approach could be useful, in combination with other ones, for future construction of the efficient thermotolerant ethanol producers from xylose.

One may assume that *cat8Δ* mutants of xylose-utilizing recombinant *S. cerevisiae* could be also characterized by an increase in ethanol production from this pentose. It would also be interesting to check the effects of *CAT8* deletion on xylose alcoholic fermentation in the species of natural xylose fermenting yeasts, such as *S. stipitis, S. passalidarum* and others. We hypothesize that the deletion of *CAT8* gene could become a standard approach for development of effective xylose fermenting strains. It would also be of interest to check the role of transcription factors Adr1 [30], and Znf1 [34], Rds2, Sip4 and others [20], in xylose alcoholic fermentation in *O. polymorpha* and other yeast species. Recently, we checked the effects of the knock-out of two *O. polymorpha* homologs of transcriptional regulator *HAP4*, *HAP4-A* and *HAP4-B*, on xylose growth and fermentation and found only a slight increase in ethanol production from xylose in *hap4-AΔ* mutant [35].

We envisage that there are new efficient strategies for additional increase in ethanol production from xylose in *O. polymorpha*. They include autophagy initiation gene *ATG13* [17; Dmytruk, Sibirny, in preparation] and several genes coding for peroxisomal proteins [17; Kurylenko, Ruchala, Vasylyshyn, Dmytruk, Sibirny, in preparation]. Change of expression of the mentioned genes leads to significant and specific increase in ethanol yield from xylose on the background of the wild-type strain. We hope that the manipulation with these gene expression could also be useful for further increase of ethanol production in the described here ethanol overproducer from xylose. Currently our attention is focused to the fermentation of lignocellulosic hydrolysates by constructed xylose fermenting strains. This could constitute an important step towards the establishment of *O. polymorpha* as a promising high-temperature ethanol producer from xylose and other lignocellulosic sugars.

Conclusions

The mutants of the methylotrophic yeast *Ogataea (Hansenula) polymorpha* with knock out and overexpression of the ortholog of *CAT8* gene coding for transcriptional activator, have been constructed. The *cat8Δ* mutants showed 30–50% increase in ethanol synthesis from xylose. No effect of *CAT8* knock out on ethanol production from glucose was observed. The best strain accumulated 12.5 g of ethanol/L from xylose at 45 °C. Inversely, overexpression of *CAT8* resulted in decrease of ethanol production from this pentose.

Methods

Strains, vectors, cultivation condition

The following strains of *O. polymorpha* were used: NCYC495 *leu1-1* (wild-type strain), 2EtOH/XYL1m/XYL2/XYL3/BrPA (designated as BEP from best ethanol producer) which is advanced ethanol producer from xylose isolated by combination of the methods of metabolic engineering and classical selection [16]. Yeast cells were grown on YPD (10 g/L yeast extract, 10 g/L peptone, 20 g/L glucose) or mineral medium (6.7 g/L YNB without amino acids, 20 g/L of carbon source—glucose, xylose, glycerol, ethanol) at 37 °C. For the NCYC495 *leu1-1* strain, leucine (40 mg/L) was added to the medium. For the selection of yeast transformants on YPD 0.1 g/L of nourseothricin or 0.35 g/L of hygromycin were added. Alcoholic fermentation of yeast strains was fulfilled by cultivation in liquid mineral medium at oxygen-limited conditions at 37 and 45 °C. The conditions were provided by agitation at 140 rpm. 9% xylose or 9% glucose was added into the medium used for the fermentation. The cells were pregrown in 100 mL of liquid YPX medium (1% yeast extract, 2% peptone and 4% xylose) in 300 mL Erlenmeyer flasks at 220 rpm till the mid-exponential growth phase. Than the cells were precipitated by centrifugation, washed by water and inoculated into 40 mL of the fermentation medium in 100 mL Erlenmeyer flasks covered with cotton plugs. The initial biomass concentration for fermentation experiments was 2 g (dry weight)/L. Fermentations were repeated at least in three independent experiments, each performed in triplicate to ensure the results are reproducible. The bars in the figures indicate the ranges of the standard deviation.

The *E. coli* DH5α strain (Φ80d*lacZ*ΔM15, *recA*1, *endA*1, *gyrA*96, *thi*-1, *hsdR*17(r_K^-, m_K^+), *supE*44, *relA*1, *deoR*, Δ(*lacZYA-argF*)U169) was used as a host for plasmid propagation. Strain DH5α was grown at 37 °C in LB medium as described previously [36]. Transformed *E. coli* cells were maintained on a medium containing 100 mg/L of ampicillin.

Molecular-biology techniques

Standard cloning techniques were carried out as described [36]. Genomic DNA of *O. polymorpha* was isolated using the Wizard® Genomic DNA Purification Kit (Promega, Madison, WI, USA). Restriction endonucleases and DNA ligase (Fermentas, Vilnius, Lithuania) were used according to the manufacturer specifications. Plasmid isolation from *E. coli* was performed with the Wizard® *Plus* SV Minipreps DNA Purification System (Promega, Madison, WI, USA). DNA fragments were separated on a 0.8% agarose (Fisher Scientific, Fair Lawn, NJ, USA) gel. Isolation of fragments from the gel was carried out with a DNA Gel Extraction Kit (Millipore, Bedford, MA, USA). PCR-amplification of the fragments of interest was done with Platinum® *Taq* DNA Polymerase High Fidelity (Invitrogen, Carlsbad, CA, USA) according to the manufacturer specification. PCRs were performed in GeneAmp® PCR System 9700 thermocycler (Applied Biosystems, Foster City, CA, USA). Transformation of the yeast *O. polymorpha* was carried out as described previously [37].

Construction and analysis of *cat8Δ O. polymorpha* deletion mutants

Genomic DNA of *O. polymorpha* NCYC495 *leu 1-1* strain was used as template for isolation of 5′ and 3′ uncoding regions of *CAT8* gene by PCR amplifications using primers 5′CAT8 FW/5′CAT8 RW and 3′CAT8 FW/3′CAT8 RW (Sequences of all primers represented in Additional file 5). The resulted 5′CAT8 (671 bp) and 3′CAT8 (697 bp) fragments were EcoRI/BglII or BglII/PstI digested and cloned into EcoRI/PstI linearized vector pUC57. The resulted recombinant was named pUC57-CAT8. Gene *natNT2* (1318 bp) conferring resistance to nourseothricin was amplified using vector pRS41N [38] as a template and primers OK19 and OK20. Obtained fragment was BglII-digested and subcloned into BglII-linearized plasmid pUC57-CAT8. As a result of further genetic manipulations recombinant plasmid pUC57-ΔCAT8-natNT2 was constructed (Additional file 1A). After that, plasmid pUC57-ΔCAT8-natNT2 was NdeI-linearized and transformed into *O. polymorpha* NCYC495 *leu1-1* recipient strain using electroporation method. Transformants were selected on the solid YPD medium supplemented with 0.1 g/L of nourseothricin after three days of incubation at 37 °C. Obtained transformants were examined by PCR using genomic DNA of recombinant strains as a template. Transformants with confirmed deletion of *CAT8* were stabilized by altering cultivation in nonselective and selective media and once again examined by PCR. Fragments with predicted size were amplified using pairs of primers homologous to the sequence of selective marker and regions outside from the fragments used for recombination (JR_CAT8_FW/OK20 and OK19/JR_CAT8_RW) (Additional file 1C).

Deletion cassette for isolation of *cat8Δ* mutant on the background of strain BEP was constructed as follows. Genomic DNA of *O. polymorpha* NCYC495 *leu1-1* strain was used as template for isolation of 5′ and 3′ uncoding regions of *CAT8* gene by PCR amplifications using primers 5′C8_FW/5′C8_RW and 3′C8_FW/3′C8_RW. The resulted 5′CAT8 (878 bp) and 3′CAT8 (780 bp) fragments were EcoRI/BglII and BglII/PstI double-digested and cloned into EcoRI/PstI linearized vector pUC57. The resulted recombinant was named pUC57-C8. Gene *hphNT1* (1777 bp) conferring resistance to hygromycin was amplified from plasmid pRS42H [38] as a template and primers Hyg_FW and Hyg_RW. Obtained fragment was BglII-digested and subcloned into BglII-linearized plasmid pUC57-C8. Resulted plasmid was designated as pUC57-ΔCAT8-hphNT1 (Additional file 1B). Plasmid pUC57-ΔCAT8-hphNT1 was XbaI-linearized and transformed into BEP strain by electroporation. Transformants were selected on the solid YPD medium supplemented with 0.35 g/L of hygromycin after four day of incubation at 37 °C. Homologous recombination of the deletion cassette with target site was verified by PCR applying the same approach as that described above using pairs of primers JR_CAT8_FW/Hyg RW and Hyg FW/JR_CAT8_RW (Additional file 1D).

Construction and analysis of *O. polymorpha* strains with overexpression of *CAT8* gene

Plasmid puc19-GAPp-GAPt-natNT2 [39] was used as the basic one for overexpression of *CAT8*. Promoter *GAP* of the gene coding for glyceraldehyde-3-phosphate dehydrogenase was used for *CAT8* overexpression. Genomic DNA of *O. polymorpha* NCYC495 *leu1-1* strain was used as template for isolation of *CAT8* gene by PCR amplifications using primers C8_F/C8E_R. After that, gene was XbaI/NotI double-degisted and cloned into XbaI/NotI linearized vector puc19-GAPp-NTC. The resulting plasmid was named p19-GAPp-CAT8-GAPt-natNT2 (Additional file 6). Plasmid p19-GAPp-CAT8-GAPt-natNT2 was ScaI-linearized and transformed into NCYC495 *leu1-1* strain by electroporation. Transformants were selected on the solid YPD medium supplemented with 0.1 g/L of nourseothricin after three days of incubation at 37 °C. The transformants were stabilized by cultivation in non-selective media with further shifting to the selective media with nourseothricin. The presence of recombinant *CAT8* gene driven by the *HpGAP* promoter in genomic DNA of stable transformants was confirmed by PCR using primers K_O644/C8E_R. Overexpression of *CAT8* in the resulted strain was confirmed by qRT-PCR (Additional file 3).

Respiration activity assay

Cells were grown to the late exponential phase in mineral medium with glucose or xylose, collected, washed in distilled water and starved in mineral medium without carbon source for 16–18 h. Viability of the starved cells was found to be around 70% of that of the non-starved cells by plate count of colony forming units (data not shown). The respiration rate was measured at 30 °C by Yellow Springs Instrument Co. Clark oxygen electrode (model YSI 5300) in a 5 mL reaction vessel. Determinations were performed in distilled air-saturated water with the concentration of cells 0.5 g/L of dry weight from 5 independent cultivations and started by addition of 1% carbon substrate (glucose, xylose). The respiratory rate was expressed as nanomoles of O_2 consumed per minute per mg of cells (dry weight).

Biochemical methods

Samples for enzyme activity measurements were taken from the cultures on the third day of xylose fermentation at 45 °C. The enzyme activity was measured directly after the preparation of cell-free extracts. Protein concentration was determined with Folin reagent [40]. The specific activities of XR, XDH and XK in cell extracts were determined spectrophotometrically as described before [14].

TKL activity was assayed spectrophotometrically at 278 nm as previously described with some modifications [41]. In brief, the reaction mixture contained: 50 mM Tris–HCl buffer (pH 7.5), 2.5 mM $MgCl_2$, 60 μM TPP, cell extract (0.4 mg of protein). The reaction was started by addition of 100 mM glycol aldehyde.

The PDC activity in cell extracts was determined spectrophotometrically according to the method described earlier [15]. The ADH activity was measured by following the reduction of NAD at 340 nm using 96% ethanol as a substrate as described previously [42]. Briefly, the assay mixture contained 100 mM Tris–HCl (pH 8.0), 2 mM NAD, 100 mM ethanol. The reaction was initiated with the addition of cell extract (0.1 mg of protein).

FBP activity was measured spectrophotometrically in cell extracts as described elsewhere with some modifications [43]. Briefly, the FBP assay was performed in a reaction mixture containing 100 mM Tris–HCl buffer (pH 8.5), 1 mM EDTA, 5 mM $MgCl_2$, 2 mM fructose-1,6- diphosphate, 0.4 mM NADP and 1 units of glucose-6-phosphate isomerase and glucose-6-phosphate dehydrogenase. The reaction was initiated with the addition of cell extract (0.4 mg of protein).

All assay experiments were repeated at least twice.

Quantitative real-time PCR (qRT-PCR)

Expression of the *XYL1, XYL2, XYL3, DAS1, TAL2, RPE1, TAL1, PDC1, FBP1* and *PCK1* genes was analyzed by real-time PCR. Total RNA was extracted using the GeneMATRIX Universal RNA Purification Kit with DNAse I (EURx Ltd., Gdansk, Poland). RNA was quantified using Picodrop Microliter UV/Vis Spectrophotometer and diluted in RNAse free water. The qRT-PCR was performed by 7500 Fast Real-Time PCR System (The Applied Biosystems, USA) with SG OneStep qRT-PCR kit (EURx Ltd., Gdansk, Poland) using gene-specific primer pairs, RNA as a template and ROX reference passive dye according to the manufacturer's instructions. The primers pairs used for qRT-PCR are listed in Additional file 5: Table S1. Sequences of tested genes were taken from *O. polymorpha* genome database [28]. In brief, normalized amount of RNA (100 ng) and 0.4 μM of each of the two primers were used in a total reaction volume of 20 μL. The amplification was performed with the following cycling profile: reverse transcription step at 50 °C for 30 min; initial denaturation at 95 °C for 3 min at preparation step; followed by 40 cycles of 15 s at 94 °C and 30 s at 60 °C. Melting curve analysis was performed to verify the specificity and identity of PCR products from 65 to 95 °C in the software of real-time cycler. The amplification for over 35–45 cycles gave abundance of PCR product indicating saturation phase. The fold change of each amplicon in each sample relative to the control sample was normalized to the internal control gene *ACT1* and calculated according to the comparative Ct (ΔΔCt) method. All data points were analyzed in triplicate.

Analyses

The biomass was determined turbidimetrically with a Helios Gamma spectrophotometer (OD, 590 nm; cuvette, 10 mm) with gravimetric calibration. Concentrations of xylose and ethanol from fermentation in medium broth were analyzed by HPLC (PerkinElmer, Series 2000, USA) with an Aminex HPX-87H ion-exchange column (Bio-Rad, Hercules, USA). A mobile phase of 4 mM H_2SO_4 was used at a flow rate 0.6 mL/min and the column temperature was 35 °C. Alternatively, concentrations of ethanol in the medium were determined using alcohol oxidase/peroxidase-based enzymatic kit "Alcotest" [44]. Experiments were performed at least twice.

Additional files

Additional file 1. A, B) Scheme of *CAT8* deletion cassettes (*natNT2*—gene conferring resistance to nourseothricin, *hphNT1*- gene conferring resistance to hygromycin); C) PCR verification of the correct cassette integration into genome of the wild-type strain using primers JR_CAT8 FW/OK20 or OK19/JR_CAT8 RW and genomic DNA of transformants as a template (*cat8Δ*—constructed deletion strains; WT—recipient strain NCYC495 *leu 1-1*; L—ladder); D) PCR verification of the correct cassette integration into genome of the BEP strain using primers JR_CAT8 FW/Hyg_RW or Hyg_FW/JR-CAT8 RW and genomic DNA of transformants as a template (BEP *cat8Δ*—constructed deletion strains; BEP—recipient strain, L—ladder).

Additional file 2. Growth of the mutants with deletion of *CAT8* gene on different carbon sources as compared to the parental strains.

Additional file 3. The relative expression levels of the *CAT8* gene in the parental strains and strain with overexpressed *CAT8* gene (*CAT8**) at the third day of xylose alcoholic fermentation at 45 °C. The mRNA quantification was normalized to *ACT1* mRNA.

Additional file 4. Ethanol production by parental and recombinant strains of *O. polymorpha*: (A) *cat8Δ* and strain with overexpression of *CAT8* gene (WT *CAT8*); (B) BEP *cat8Δ* during glucose fermentation at 45 °C.

Additional file 5. List of primers used in this study (restriction sites are underlined).

Additional file 6. Linear scheme of plasmid pUC19-GAPp-CAT8-GAPt-natNT2.

Abbreviations

BEP: best ethanol producer; SSF: Simultaneous Saccharification and Fermentation; qRT-PCR: quantitative reverse-transcription PCR; XR: xylose reductase; XDH: xylitol dehydrogenase; XK: xylulokinase; TKL: transketolase; ADH: alcohol dehydrogenase; PDC: pyruvate decarboxylase; FBP: fructose-1,6-bisphosphatase.

Authors' contributions

JR carried out strains construction, evaluation of enzymes activity, respiration activity and qRT-PCR, performed fermentation experiments and co-drafted the manuscript. OOK participated in design of cloning, evaluation of enzymes activity, respiration activity and co-drafted the manuscript. NS commented and approved the manuscript. KVD participated in design of cloning and strains construction, analyzed the date and co-drafted the manuscript. AAS provided guidance and suggestions for experimental design, wrote and edited the manuscript. All authors participated in finalizing the manuscript. All authors read and approved the final manuscript.

Author details

[1] Department of Biotechnology and Microbiology, University of Rzeszow, Zelwerowicza 4, 35-601 Rzeszow, Poland. [2] Department of Molecular Genetics and Biotechnology, Institute of Cell Biology, Drahomanov Str., 14/16, Lviv 79005, Ukraine. [3] King Mongkut Technical University, Thonbury, Thailand.

Acknowledgements

Not applicable.

Competing interests

The authors declare that they have no competing interests.

Funding

This work was supported in part by Polish Grant of National Scientific Center (NCN) DEC-2012/05/B/NZ1/01657 awarded to A. Sibirny, FEMS Research Grant FEMS-RG-2015-0096.R1 awarded to J. Ruchala; National Academy of Sciences of Ukraine (Grants 5-17, 6-17 and 35-17) and Science and Technology Center in Ukraine (STCU) (Grant 6188).

References

1. Alternative Fuels Data Center. http://www.afdc.energy.gov/data/10331. Accessed 29 Oct 2016.
2. Hahn-Hägerdal B, Galbe M, Gorwa-Grauslund MF, Lidén G, Zacchi G. Bio-ethanol–the fuel of tomorrow from the residues of today. Trends Biotechnol. 2006;24:549–56.
3. Jeffries TW, Jin YS. Metabolic engineering for improved fermentation of pentoses by yeasts. Appl Microbiol Biotechnol. 2004;63:495–509.
4. Olofsson K, Bertilsson M, Lidén G. A short review on SSF—an interesting process option for ethanol production from lignocellulosic feedstocks. Biotechnol Biofuels. 2008;1(1):7.
5. Nonklang S, Abdel-Banat BMA, Cha-aim K, Moonjai N, Hoshida H, Limtong S, Yamada M, Akada R. High-temperature ethanol fermentation and transformation with linear DNA in the thermotolerant yeast *Kluyveromyces marxianus* DMKU3-1042. Appl Environ Microb. 2008;74:7514–21.
6. Radecka D, Mukherjee V, Mateo RQ, Stojiljkovic M, Foulquié-Moreno MR, Thevelein JM. Looking beyond *Saccharomyces*: the potential of non-conventional yeast species for desirable traits in bioethanol fermentation. FEMS Yeast Res. 2015;15(6):053.
7. Ryabova OB, Chmil OM, Sibirny AA. Xylose and cellobiose fermentation to ethanol by the thermotolerant methylotrophic yeast *Hansenula polymorpha*. FEMS Yeast Res. 2003;4:157–64.
8. Ishchuk OP, Voronovsky AY, Abbas CA, Sibirny AA. Construction of *Hansenula polymorpha* strains with improved thermotolerance. Biotechnol Bioeng. 2009;104:911–9.
9. Peter G, Tornai-Lehoczki J, Shin K-S, Dlauchy D. *Ogataea thermophile* sp. nov., the teleomorph of *Candida thermophila*. FEMS Yeast Res. 2007;7:494–6.
10. Kata I, Semkiv MV, Ruchala J, Dmytruk KV, Sibirny AA. Overexpression of the genes *PDC1* and *ADH1* activates glycerol conversion to ethanol in the thermotolerant yeast *Ogataea (Hansenula) polymorpha*. Yeast. 2016;33(8):471–8.
11. Voronovsky AY, Ryabova OB, Verba OV, Ishchuk OP, Dmytruk KV, Sibirny AA. Expression of xylA genes encoding xylose isomerases from *Escherichia coli* and *Streptomyces coelicolor* in the methylotrophic yeast *Hansenula polymorpha*. FEMS Yeast Res. 2005;5:1055–62.
12. Voronovsky AY, Rohulya OV, Abbas CA, Sibirny AA. Development of strains of the thermotolerant yeast *Hansenula polymorpha* capable of alcoholic fermentation of starch and xylan. Metab Eng. 2009;11:234–42.
13. Dmytruk OV, Voronovsky AY, Abbas CA, Dmytruk KV, Ishchuk OP, Sibirny AA. Overexpression of bacterial xylose isomerase and yeast host xylulokinase improves xylose alcoholic fermentation in the thermotolerant yeast *Hansenula polymorpha*. FEMS Yeast Res. 2008;8:165–73.
14. Dmytruk OV, Dmytruk KV, Abbas CA, Voronovsky AY, Sibirny AA. Engineering of xylose reductase and overexpression of xylitol dehydrogenase and xylulokinase improves xylose alcoholic fermentation in the thermotolerant yeast *Hansenula polymorpha*. Microb Cell Fact. 2008;7:21.
15. Ishchuk OP, Voronovsky AY, Stasyk OV, Gayda GZ, Gonchar MV, Abbas CA, Sibirny AA. Overexpression of pyruvate decarboxylase in the yeast *Hansenula polymorpha* results in increased ethanol yield in high-temperature fermentation of xylose. FEMS Yeast Res. 2008;8:1164–74.
16. Kurylenko OO, Ruchala J, Hryniv OB, Abbas CA, Dmytruk KV, Sibirny AA. Metabolic engineering and classical selection of the methylotrophic thermotolerant yeast *Hansenula polymorpha* for improvement of high-temperature xylose alcoholic fermentation. Microb Cell Fact. 2014;13:122.
17. Kurylenko O, Semkiv M, Ruchala J, Hryniv O, Kshanovska B, Abbas C, Dmytruk K, Sibirny A. New approaches for improving the production of the 1st and 2nd generation ethanol by yeast. Acta Biochim Pol. 2016;63:31–8.
18. Haurie V, Perrot M, Mini T, Jenö P, Sagliocco F, Boucherie H. The transcriptional activator Cat8p provides a major contribution to the reprogramming of carbon metabolism during the diauxic shift in *Saccharomyces cerevisiae*. J Biol Chem. 2001;276:76–85.
19. Hedges D, Proft M, Entian KD. *CAT8*, a new zinc cluster-encoding gene necessary for derepression of gluconeogenic enzymes in the yeast *Saccharomyces cerevisiae*. Mol Cell Biol. 1995;15:1915–22.
20. Turcotte B, Liang XB, Robert F, Soontorngun N. Transcriptional regulation of nonfermentable carbon utilization in budding yeast. FEMS Yeast Res. 2010;10(1):2–13.
21. Watanabe T, Srichuwong S, Arakane M, Tamiya S, Yoshinaga M, Watanabe I, Yamamoto M, Ando A, Tokuyasu K, Nakamura T. election of stress-tolerant yeasts for Simultaneous Saccharification and Fermentation (SSF) of very high gravity (VHG) potato mash to ethanol. Bioresour Technol. 2010;101(24):9710–4.
22. Qi K, Zhong JJ, Xia XX. Triggering respirofermentative metabolism in the Crabtree-negative yeast *Pichia guilliermondii* by disrupting the *CAT8* gene. Appl Environ Microbiol. 2014;80:3879–87.
23. Randez-Gil F, Bojunga N, Proft M, Entian KD. Glucose derepression of gluconeogenic enzymes in *Saccharomyces cerevisiae* correlates with phosphorylation of the gene activator Cat8p. Mol Cell Biol. 1997;17:2502–10.
24. Georis I, Krijger JJ, Breunig KD, Vandenhaute J. Differences in regulation of yeast gluconeogenesis revealed by Cat8p-independent activation of *PCK1* and *FBP1* genes in *Kluyveromyces lactis*. Mol Gen Genet. 2000;264:193–203.
25. Ramírez MA, Lorenz MC. The transcription factor homolog *CTF1* regulates β-oxidation in *Candida albicans*. Eukaryot Cell. 2009;8:1604–14.
26. Kim OC, Suwannarangsee S, Oh DB, Kim S, Seo JW, Kim CH, Kang HA, Kim JY, Kwon O. Transcriptome analysis of xylose metabolism in the thermotolerant methylotrophic yeast *Hansenula polymorpha*. Bioprocess Biosyst Eng. 2013;36:1509–18.
27. Salusjärvi L, Kankainen M, Soliymani R, Pitkänen JP, Penttilä M, Ruohonen L. Regulation of xylose metabolism in recombinant *Saccharomyces cerevisiae*. Microb Cell Fact. 2008;7:18.
28. *Hansenula polymorpha* NCYC 495 leu1.1 v2.0—JGI Genome Portal. http://genome.jgi-psf.org/Hanpo2/Hanpo2.home.html. Accessed 27 Oct 2016.
29. Saraya R, Krikken AM, Kiel JA, Baerends RJ, Veenhuis M, van der Klei IJ. Novel genetic tools for *Hansenula polymorpha*. FEMS Yeast Res. 2012;12:271–8.
30. Watanabe D, Hashimoto N, Mizuno M, Zhou Y, Akao T, Shimoi H. Accelerated alcoholic fermentation caused by defective gene expression related to glucose derepression in *Saccharomyces cerevisiae*. Biosci Biotechnol Biochem. 2013;77:2255–62.
31. Jeffries TW, Grigoriev IV, Grimwood J, Laplaza JM, Aerts A, Salamov A, Schmutz J, Lindquist E, Dehal P, Shapiro H, Jin YS, Passoth V, Richardson PM. Genome sequence of the lignocellulose-bioconverting and xylose-fermenting yeast *Pichia stipitis*. Nat Biotechnol. 2007;25(3):319–26.
32. Long TM, Su YK, Headman J, Higbee A, Willis LB, Jeffries TW. Cofermentation of glucose, xylose and cellobiose by the beetle-associated yeast *Spathaspora passalidarum*. Appl Environ Microbiol. 2012;78(16):5492–500.
33. Zhang J, Zhang B, Wang D, Gao X, Sun L, Hong J. Rapid ethanol production at elevated temperatures by engineered thermotolerant *Kluyveromyces marxianus* via the NADP(H)-preferring xylose reductase-xylitol dehydrogenase pathway. Metab Eng. 2015;31:140–52.
34. Tangsombatvichit P, Semkiv MV, Sibirny AA, Jensen LT, Ratanakhanokchai K, Soontorngun N. Zinc cluster protein Znf1, a novel transcription factor of non-fermentative metabolism in *Saccharomyces cerevisiae*. FEMS Yeast Res. 2015;15(2):fou002.
35. Petryk N, Zhou YF, Sybirna K, Mucchielli MH, Guiard B, Bao WG, Stasyk OV, Stasyk OG, Krasovska OS, Budin K, Reymond N, Imbeaud S, Coudouel S, Delacroix H, Sibirny A, Bolotin-Fukuhara M. Functional study of the Hap4-like genes suggests that the key regulators of carbon metabolism *HAP4* and oxidative stress response *YAP1* in yeast diverged from a common ancestor. PLoS ONE. 2014;9(12):e112263.
36. Sambrook J, Fritsh EF, Maniatis T. Molecular cloning: a laboratory manual. Cold Spring Harbor: Cold Spring Harbor Press; 1989.
37. Faber KN, Haima P, Harder W, Veenhuis M, Ab G. Highly-efficient electrotransformation of the yeast *Hansenula polymorpha*. Curr Genet. 1994;25:305–10.
38. Taxis C, Knop M. System of centromeric, episomal, and integrative vectors based on drug resistance markers for *Saccharomyces cerevisiae*. Biotechnique. 2006;40:73–8.

39. Yurkiv M, Kurylenko O, Vasylyshyn R, Dmytruk K, Sibirny A. Construction of the efficient glutathione producers in the yeast *Hansenula polymorpha,* p 323–332. In: Sibirny A, Fedorovyvh D, Gonchar M, Grabek-Lejko D, editors. Living organisms and bioanalytical approaches for detoxification and monitoring of toxic compounds. Rzeszow: University of Rzeszow; 2015. p. 323–33.
40. Lowry OH, Rosebrough NJ, Farr AL, Randall RJ. Protein measurement with the Folin phenol reagent. J Biol Chem. 1951;193:265–75.
41. Sevostyanova IA, Solovjeva ON, Kochetov GA. Two methods for determination of transketolase activity. Biochemistry (Mosc). 2006;71(5):560–2.
42. Postma E, Verduyn C, Scheffers WA, Van Dijken JP. Enzymic analysis of the crabtree effect in glucose-limited chemostat cultures of *Saccharomyces cerevisiae.* Appl Environ Microbiol. 1989;55:468–77.
43. Gancedo JM, Gancedo C. Fructose-1,6-diphosphatase, phosphofructokinase and glucose-6-phosphate dehydrogenase from fermenting and non-fermenting yeasts. Arch Microbiol. 1971;76(2):132–8.
44. Gonchar MV, Maidan MM, Sibirny AA. A new oxidase-peroxidase kit "Alcotest" for ethanol assays in alcoholic beverages. Food Technol Biotechnol. 2001;39:37–42.

4

Implications of evolutionary engineering for growth and recombinant protein production in methanol-based growth media in the yeast *Pichia pastoris*

Josef W. Moser[1,2], Roland Prielhofer[2,3], Samuel M. Gerner[4], Alexandra B. Graf[2,4], Iain B. H. Wilson[1], Diethard Mattanovich[2,3] and Martin Dragosits[1*]

Abstract

Background: *Pichia pastoris* is a widely used eukaryotic expression host for recombinant protein production. Adaptive laboratory evolution (ALE) has been applied in a wide range of studies in order to improve strains for biotechnological purposes. In this context, the impact of long-term carbon source adaptation in *P. pastoris* has not been addressed so far. Thus, we performed a pilot experiment in order to analyze the applicability and potential benefits of ALE towards improved growth and recombinant protein production in *P. pastoris*.

Results: Adaptation towards growth on methanol was performed in replicate cultures in rich and minimal growth medium for 250 generations. Increased growth rates on these growth media were observed at the population and single clone level. Evolved populations showed various degrees of growth advantages and trade-offs in non-evolutionary growth conditions. Genome resequencing revealed a wide variety of potential genetic targets associated with improved growth performance on methanol-based growth media. Alcohol oxidase represented a mutational hotspot since four out of seven evolved *P. pastoris* clones harbored mutations in this gene, resulting in decreased Aox activity, despite increased growth rates. Selected clones displayed strain-dependent variations for AOX-promoter based recombinant protein expression yield. One particularly interesting clone showed increased product titers ranging from a 2.5-fold increase in shake flask batch culture to a 1.8-fold increase during fed batch cultivation.

Conclusions: Our data indicate a complex correlation of carbon source, growth context and recombinant protein production. While similar experiments have already shown their potential in other biotechnological areas where microbes were evolutionary engineered for improved stress resistance and growth, the current dataset encourages the analysis of the potential of ALE for improved protein production in *P. pastoris* on a broader scale.

Keywords: *Pichia pastoris*, Experimental evolution, Methanol, Recombinant protein

Background

In the field of microbial biotechnology, the methylotrophic yeast *Pichia pastoris* (*Komagataella phaffii* [1]) is a commonly used host organism for recombinant protein production, in small scale lab applications as well as larger scale protein production. Several factors, including its ability to grow on methanol, the availability of constitutive as well as inducible promoter systems, its minimal growth requirements, high biomass yields and eukaryotic-type post-translational modifications led to the establishment of this yeast as a heterologous host [2]. More recently, whole-cell glyco-engineering has achieved humanized N-glycosylation patterns in this yeast [3, 4]. Furthermore, in the last decade several studies on the systems-level led to an improved understanding of *P. pastoris* physiology

*Correspondence: martin.dragosits@boku.ac.at
[1] Department of Chemistry, University of Natural Resources and Life Sciences, Muthgasse 11, 1190 Vienna, Austria
Full list of author information is available at the end of the article

and the correlations of recombinant protein production and process-relevant environmental factors [5–7]. Thus, classical approaches, such as co-chaperone overexpression in order to increase production efficiency in bacteria and yeasts [8–10], can be complemented by systems-wide analysis in order to identify novel targets for strain and process engineering.

Although glucose- and glycerol based expression systems are also widely applied for *P. pastoris* [11], methanol-induced expression of recombinant proteins mediated by strongly inducible promoters such as the alcohol oxidase 1 (Aox1) gene promoter [12] can be considered as one of the core features of *P. pastoris*. Despite the toxicity and flammability of methanol, leading to the development of novel methanol-free expression systems in *P. pastoris* [13], recent approaches led to the identification of new potential methanol-inducible promoters and expression strategies for this expression system [14]. Consequently, methanol utilization and the Aox expression system have been studied in detail in recent years. The *P. pastoris* methanol utilization (Mut) phenotypes are commonly known as methanol utilization Mut^+ (fast growth), Mut^s (slow growth) and Mut^- (no growth) and depend on the presence of one or two functional copies of the alcohol oxidase gene (*AOX1* and *AOX2*, respectively). Furthermore, the implications of the Mut^+ and Mut^s phenotypes in recombinant production have been investigated but led to indistinct results in terms of recombinant protein productivity [15, 16]. The effect of co-overexpression of methanol-pathway genes was analyzed [15] and an *AOX* promoter mutant library was tested for protein production [17]. Recent studies also led to a better understanding of the methanol utilization process in general; e.g. methanol utilization is predominantly regulated on the transcriptional rather than on the translational level in *P. pastoris* [18], whereas other data show that methanol metabolic processes are confined to peroxisomes [19]. Additionally, transcriptional regulators involved in the expression of Mut proteins have been identified, clarifying how they work in concert to promote important steps such as Mut protein expression and peroxisome proliferation [20–22].

The investigation of evolutionary processes on a molecular level has become an intriguing topic in recent years. Thus, adaptive laboratory evolution (ALE) experiments with microbial cells led to important insights into evolutionary processes including the rate and progression of adaptation [23], pleiotropic effects [24], growth trade-offs and benefits in non-evolutionary growth conditions [25] among others. Harnessing the power of systems-level analysis and affordable genome sequencing technologies, it became also clear that such studies have significant potential for biotechnology, e.g. in the area of metabolic engineering and white biotechnology. Thus, the implications of ALE for biotechnology have been subject of recent reviews [26, 27].

To date it is not clear to which extent ALE experiments with *P. pastoris* can lead to improved growth on methanol and provide data that lead to a better understanding of host cell physiology and new strategies for host cell engineering. Furthermore, the impact of potentially increased growth rates due to long-term environmental adaptation on recombinant protein production is not understood. To address this question, we performed a serial passaging ALE experiment with consecutive transfers in batch cultures to maintain replicate *P. pastoris* populations in rich- and minimal medium with methanol as carbon source for 250 generations. Whereas there are several different methods for experimental evolution, serial passaging and chemostat selection are two highly preferred methods [27, 28]. Each of these methods has its benefits and trade-offs. Although microbial populations experience fluctuating conditions in serial batch cultures, the gradual reduction of lag phase and long exponential growth phase can easily lead to the selection for improved growth rates [28]. On the other hand, chemostat cultivation allows growth under strictly nutrient-limiting conditions and tight control of other environmental parameters, such as pH [28]. Nevertheless, this nutrient-limiting growth does not necessarily lead to improved growth rates in non-limiting conditions and due to the activation of nutrient scavenging mechanisms it has been reported to result in overall decreased stress resistance of microbial cells [29, 30]. Thus, in order to select for improved growth rates during growth on methanol in two different environmental settings, serial dilution was the method of choice. After ALE we performed growth tests with evolved populations on a broader scale and selected individual clones from both growth environments for whole genome sequencing in order to identify mutations associated with the observed growth phenotypes. Finally, selected clones were tested for recombinant protein production in small-scale and larger scale fed batch cultures.

Methods

P. pastoris and *Escherichia coli* strains

Pichia pastoris X-33 was used as model strain in the current study. Cloning steps were performed using *E. coli* JM109 cells.

Long-term cultivation on methanol as carbon source

Pichia pastoris cells were adapted to growth on either YPM medium (YP medium pH 7.4, 1% methanol) or BMM (buffered minimal medium [12], 1% methanol). For each condition, four populations were cultivated in 24-deep well plates (10 mL total reservoir volume, 2 mL culture volume) at 28 °C, 200 rpm on an orbital shaker (Thermo MaxQ 4000, orbit diameter 1.9 cm). For both growth environments, populations were transferred to a fresh deep-well plate in 24-h intervals. Daily dilutions of 1:20 and 1:10 were used for YPM and BMM medium, respectively. Due to a lower growth rate and higher fluctuations of cell densities, the lower dilution rate of 1:10 was necessary for BMM in order to maintain a stable serial transfer. The OD_{600} of each culture was determined on a daily basis (Tecan M200 plate reader) in order to calculate daily generations and the cumulative number of cell divisions (CCD). Long-term cultivation was performed for a total of approximately 250 generations. Cultures were checked for contamination by microscopy and plating samples on YPD-agar plates (plates were incubated at 28 °C for 2 days) every 50 generations. The calculation of CCD was essentially performed as described previously by Lee and co-workers [31], assuming 5×10^7 cells mL^{-1} per OD_{600} unit [12].

Growth tests in deep-well plates

For growth profiling in various growth media, yeast populations were grown in glass tubes (2 mL culture volume) in the respective growth medium at 28 °C, 200 rpm over night. On the next day, 2 mL cultures in deep-well plates were inoculated at a starting OD_{600} between 0.05 and 0.1 and growth was monitored for 8–12 h in order to calculate growth rates (μ_{max}) and after 24 h (48 h for BMM cultures) for final OD_{600} (Tecan M200 plate reader) values.

Genome re-sequencing and analysis

Genomic DNA of selected clones was isolated from o/n cultures grown in YPD (28 °C, 200 rpm) using the Masterpure Yeast DNA isolation kit (Epicentre). Illumina MiSeq paired-end sequencing with 300 bp read length was performed according to standard procedures using chemistry v3 at Eurofins Genomics (Eurofins Genomics NGS laboratory, Ebersberg, Germany). Cutadapt [32] was used for adapter-removal and quality filtering of the reads. For the genome assembly of the ancestral strain, Meraculous was applied [33]. The final assembly of the ancestral strain consisted of 48 scaffolds with a total of 9,310,711 bp and a N50 score of 910,678. Out of 4,048,141 quality trimmed input reads with an average read length of 260,48 bp, 3,581,438 (88.47%) were used in the assembly. The mean coverage amounted to 97.6.

The assembly scaffolds were further mapped to the CBS 7435 reference strain [34] in order to obtain an ordered assignment to chromosomes. CONTIGuator [35] was used for this purpose. This resulted in four large sequences containing 44 joined scaffolds and 3 scaffolds with 2506, 1439, and 1035 bp length, which could not be assigned to the reference.

Augustus [36] was used for gene prediction. Annotation was performed by matching predicted genes against the CBS 7435 reference strain using blast. Variant-calling was performed using kSNP3 [37] and GATK [38]; Magnolya [39] and cn.mops [40] were applied in order to test for potential CNVs in the evolved strains. Co-assembly for Magnolya was performed with velvet [41]. Alignment files were manually reviewed using the Integrative Genomics Viewer [42]. Potential mutations identified in the evolved clones were verified by subsequent Sanger-sequencing (primer list in Additional file 1: Table S1).

Alcohol oxidase activity

2 mL of yeast cultures were grown in deep-well plates in YP or buffered minimal medium containing 1% glucose at 28 °C, 200 rpm over night. Cultures on YPM and BMM were started at an OD_{600} of 0.2 and samples for AOX activity were taken after 6 h during exponential growth. The collected cell pellets were washed with 1× PBS and stored at −20 °C until further use. For AOX activity assays, pellets were resuspended in ice-cold 1× PBS and treated by two freeze–thaw cycles. The suspension was used directly for AOX activity assays. Alcohol oxidase assays were performed as follows: 40 µL of cell suspension with an OD_{600} of 2.5 were combined with 10 µL HRP solution (2 mg mL^{-1} in HQ-H_2O), 50 µL 1% methanol and 200 µL ABTS [2,2′-azino-bis-(3-ethylbenzothiazoline-6-sulfonic acid), Sigma Aldrich, dissolved in 100 mM KH_2PO_4, pH 7.5]. Assays were incubated at 28 °C for 45–90 min and absorbance was read at 405 nm.

Cloning and overexpression of PAS_chr2-1_0445

For the overexpression of the potential GAL4-like protein a DNA Hifi Assembly kit was used (New England Biolabs). The open reading frame was amplified from *P. pastoris* X-33 genomic DNA using Q5 Polymerase (New England Biolabs) and primers listed in Additional file 1: Table S1. The PCR-product was cloned into a pGAPzB vector backbone. *E. coli* competent cells were transformed by heat shock transformation. The insert sequence of positive clones was verified by Sanger sequencing. *P. pastoris* X-33 competent cells were used for the transformation with linearized versions of pGAPzB and pGAPzB-PAS_chr2-1_0455. Growth tests of four random clones for each construct were performed as described above.

Recombinant gene expression

Selected clones were used for the expression of rHSA (recombinant human serum albumin, [43]) and recombinant *Drosophila* hexosaminidase (rFDL) [44]. Electro-competent *P. pastoris* cells from the various strains were prepared as described in literature [12] and used for transformation with a linearized pPICzαA_rFDL, harboring a codon-optimized FDL coding sequence or a linearized pPM2dZ30_pAOX_HSA plasmid for rHSA expression. Both genes were expressed using the AOX promoter.

High biomass batch cultures: Cultures were grown in YPD at 28 °C, 200 rpm for 24 h and used to inoculate 2 mL M2 (1% glucose [45]) minimal medium supplemented with slow-release glucose feed beads (Kuhner, Switzerland) at an OD_{600} of 2.0. These cultures were grown for further 22 h at 28 °C, 200 rpm. Afterwards cells were harvested and taken up in 1 mL M2 medium without carbon source. The suspension was used to inoculate M2 medium containing 0.5% methanol at an OD_{600} of approximately 4.0. Cultures were grown at 28 °C, 200 rpm for 48 h. Cultures were supplemented with additional 1% methanol at 6, 24 and 36 h.

Low biomass batch cultures: Cultures were grown in 2 mL YPD or BMD medium at 28 °C, 200 rpm over night. On the following day, 2 mL of YPM or BMM medium were inoculated at a starting OD_{600} of 0.2. Cultures were grown at 28 °C, 200 rpm for a total of 48 h. Feeding with 1% methanol occurred in 12 h intervals.

Fed batch cultivations

rHSA-expressing *P. pastoris* clones were used for fed batch cultivations in parallel 1.0 L bioreactors (DASGIP, Germany). The clones used were selected based on similar rHSA protein levels achieved in batch cultivation (X-33 wt, Y250 3a, M250 1a and M250 3b, respectively). 100 mL YP-medium, 2% glycerol, starter cultures were grown in shake flasks at 28 °C, 200 rpm for 48 h. Cells were harvested by centrifugation, resuspended in sterile 1× PBS and used to inoculate 400 mL bioreactor batch cultures at an $OD_{600} = 1$. After the initial batch phase with a target biomass yield of 20 g YDM L^{-1}, a glycerol fed-batch with culture-dependent feed rates of 2.2–5.0 g h^{-1} was performed for a target biomass yield of 40 g YDM L^{-1}. Cultures were pulsed with methanol fed batch medium with a total of 0.5% methanol before methanol feeds were started. For the methanol feed, feed rates of 0.5, 1.0, 1.5 and 2.0 g h^{-1} were applied. The feed was stop between constant feed phases and cultures were pulsed several times with methanol fed batch medium (MeOH concentrations ranging from 0.75 to 1.5%, Additional file 1: Figure S1). For all cultivations, temperature was set to 28 °C during batch and glycerol fed batch. Starting with the methanol feed phase the temperature was set to 25 °C. DO was controlled at 20%, pH was set to 5.85 and controlled by the addition of 25% ammonia. 5% (w/w) antifoam solution (Glanapon 2000, Bussetti, Austria) was added on demand to prevent excessive foaming. Samples were taken at regular intervals. Biomass was determined in triplicate, by drying culture aliquots in pre-weighed tubes to constant weight.

Batch medium composition was essentially as described previously [46, 47] and contained (L^{-1}): 28.1 g H_3PO_4, 0.6 g $CaSO_4{\cdot}2H_2O$, 9.5 g K_2SO_4, 7.8 g $MgSO_4{\cdot}7H_2O$, 2.6 g KOH, 40.0 g glycerol, 4.6 g PTM_0 trace salts stock solution and 2.0 g biotin solution (0.2 g L^{-1}). Glycerol fed batch medium consisted of (L^{-1}): 724 g glycerol (86%), 10 g PTM_0 and 1.7 g biotin solution (0.2 g L^{-1}). Methanol fed batch medium consisted of (L^{-1}): 988 g methanol, 12 g PTM_0 and 2 g biotin solution (0.2 g L^{-1}). The PTM_0 trace salts stock solution contained (L^{-1}) 6.0 g $CuSO_4{\cdot}5H_2O$, 0.08 g NaI, 3.0 g $MnSO_4{\cdot}H_2O$, 0.2 g $Na_2MoO_4{\cdot}2H_2O$, 0.02 g H_3BO_3, 0.5 g $CoCl_2$, 20.0 g $ZnCl_2$, 65.0 g $FeSO_4{\cdot}7H_2O$ and 5.0 mL H_2SO_4 (95–98%).

rHSA and rFDL quantification

HSA was quantified using a HSA quantitation set (Bethyl laboratories). ELISA plates were coated with coating antibody (1:100) at 4 °C over night. After washing plates were blocked with 1% BSA at room temperature on a rotary shaker for 30 min. After washing, HSA standards and samples were applied to the plates and incubated at room temperature for 1 h. Plates were washed and the detection antibody (HRP-conjugate, 1:30,000) was added. After 1 h, ELISA plates were washed and TMB substrate was added. The detection reaction was stopped by the addition of 2 M H_2SO_4 and absorbance was measured at 450 nm. For fed batch cultures, rHSA was quantified using a Caliper Labchip-DS microfluidic instrument (Perkin Elmer) with BSA as standard protein.

rFDL was quantified by determining enzymatic activity in culture supernatants as described previously [44]. In short, 2 µL of appropriately diluted culture supernatant were combined with 38 µL 5 mM *p*-Nitrophenyl-β-D-*N*-Acetylglucosaminide (Sigma Aldrich) in McIlvaine buffer pH 4.0 and incubated at 30 °C for 1 h. 200 µL stop solution (0.4 M glycine pH 10.4) were added and absorbance was read at 405 nm.

Results

Long-term adaptation towards growth on methanol

Long-term adaptation was performed in YPM (rich) and BMM (minimal growth conditions) medium in order to analyze the implications of environmental specialization to different growth contexts on growth and recombinant protein production. *P. pastoris* populations were cultivated on YPM and BMM medium with a transfer to fresh growth medium in 24-h intervals for 58 and 75 days respectively, yielding a total of approximately 250 generations (Additional file 1: Figure S2). Previous studies showed that improved phenotypes could be successfully selected after 100 generations [27]. On average, the four parallel populations in complex (YPM) medium achieved 4.3 generations per day, whereas the replicate populations on BMM achieved 3.4 generations per day. The final cumulative number of cell divisions (CCD) ranged from $10^{9.96}$ to $10^{10.02}$ for YPM cultures, whereas the populations adapted to BMM minimal medium underwent a total of $10^{9.63}$–$10^{9.69}$ cumulative cell divisions (Additional file 1: Table S2). Henceforth, populations evolved on YPM medium will be denoted Y250 1-4 and populations adapted to BMM medium M250 1-4, respectively.

In order to analyze the impact of the ALE experiment, all populations were tested in terms of growth rate (Table 1; Additional file 1: Table S3) and biomass yield (OD_{600}, Table 2) in comparison with the ancestral strain. Three out of four Y250 populations showed significant growth rate improvements on YPM medium and all M250 populations showed significant improvements on BMM. Surprisingly, all YPM-adapted populations also showed significantly increased growth rates on BMM, but only the second M250 population (M250 2) showed a significant increase of growth rate on YPM. Similarly, we observe significantly increased or decreased growth rates in non-evolutionary growth conditions, including different growth media with glucose or glycerol as carbon source and additional NaCl-induced salt stress (Table 1). Interestingly, most populations also showed increased growth rates on YPD and, with the exception of populations Y250 1 and 4, a trend towards decreased growth rates on BMD. The latter was more pronounced in the M250 populations. Regarding the biomass yield, as determined by the measurement of OD_{600} values of the cultures after 24 h of growth, mostly minor effects were observed. The largest effect was observed for M250 populations which, in addition to the increased growth rate on BMM, also showed significantly increased optical density of, on average, 177% as compared with the ancestral strain. Furthermore, Y250 populations showed

Table 1 Growth rates μ [h^{-1}] of *P. pastoris* populations in different growth conditions

	YPM	BMM	YPD	YPDN	BMD	BMDN	YPG	BMG
X-33	0.209 ± 0.001	0.089 ± 0.009	0.308 ± 0.002	0.281 ± 0.007	0.312 ± 0.005	0.264 ± 0.007	0.321 ± 0.008	0.269 ± 0.004
X-33 Y250 1	*0.227 ± 0.002*	*0.108 ± 0.001*	*0.340 ± 0.004*	0.278 ± 0.005	0.316 ± 0.007	0.269 ± 0.015	*0.353 ± 0.006*	*0.250 ± 0.003*
X-33 Y250 2	0.212 ± 0.006	*0.123 ± 0.004*	*0.372 ± 0.002*	0.282 ± 0.005	0.297 ± 0.008	*0.205 ± 0.005*	*0.355 ± 0.007*	0.265 ± 0.004
X-33 Y250 3	*0.217 ± 0.005*	*0.142 ± 0.001*	*0.359 ± 0.005*	0.279 ± 0.02	0.302 ± 0.005	*0.316 ± 0.014*	0.321 ± 0.004	*0.285 ± 0.001*
X-33 Y250 4	*0.216 ± 0.003*	*0.136 ± 0.005*	*0.364 ± 0.004*	0.291 ± 0.005	*0.348 ± 0.011*	*0.356 ± 0.005*	*0.354 ± 0.004*	0.260 ± 0.002
X-33 M250 1	*0.192 ± 0.007*	*0.129 ± 0.002*	*0.334 ± 0.002*	*0.183 ± 0.003*	0.286 ± 0.015	0.246 ± 0.017	0.327 ± 0.002	0.257 ± 0.007
X-33 M250 2	*0.219 ± 0.003*	*0.127 ± 0.001*	*0.328 ± 0.002*	0.260 ± 0.032	*0.258 ± 0.009*	0.240 ± 0.019	*0.363 ± 0.002*	*0.310 ± 0.003*
X-33 M250 3	0.199 ± 0.005	*0.125 ± 0.007*	*0.368 ± 0.008*	0.264 ± 0.006	*0.259 ± 0.010*	*0.200 ± 0.013*	*0.350 ± 0.004*	*0.302 ± 0.011*
X-33 M250 4	0.214 ± 0.001	*0.126 ± 0.005*	0.314 ± 0.03	0.270 ± 0.009	*0.256 ± 0.004*	*0.215 ± 0.005*	*0.278 ± 0.013*	*0.289 ± 0.008*

Growth tests were performed in 24-deep well plates as described in the "Methods" section. *X-33* ancestral strain, *X-33 Y250a-d* populations evolved on YPM medium, *X-33 M250a-d* populations evolved on BMM medium, *YPM* YP medium 1% MeOH, *BMM* buffered minimal medium 1% MeOH, *YPD* YP medium 2% glucose, *YPDN* YPD 500 mM NaCl, *BMD* buffered minimal medium 2% glucose, *BMDN* BMD 250 mM NaCl, *YPG* YP medium 2% glycerol, *BMG* buffered minimal medium 2% glycerol; values represent averages ± standard error ($n = 4$). Growth rates in italics differ significantly from X-33 ancestral growth rates ($p < 0.05$, Additional file 1: Table S3)

Table 2 The final OD_{600} of ancestral and evolved populations

	YPM	BMM	YPD	YPDN	BMD	BMDN	YPG	BMG
X-33 wt	2.65 ± 0.09	0.83 ± 0.01	6.46 ± 0.13	4.56 ± 0.22	4.07 ± 0.15	3.09 ± 0.26	6.05 ± 0.16	7.98 ± 0.32
X-33 Y250 1	2.67 ± 0.07	1.26 ± 0.02	7.24 ± 0.13	4.34 ± 0.15	4.20 ± 0.02	2.11 ± 0.09	4.23 ± 0.10	8.77 ± 0.22
X-33 Y250 2	2.94 ± 0.07	1.05 ± 0.02	7.27 ± 0.10	4.80 ± 0.07	4.78 ± 0.10	2.55 ± 0.09	4.01 ± 0.20	8.02 ± 0.17
X-33 Y250 3	2.96 ± 0.02	0.81 ± 0.01	7.69 ± 0.04	4.35 ± 0.04	4.33 ± 0.10	2.35 ± 0.11	4.87 ± 0.13	6.84 ± 0.09
X-33 Y250 4	3.13 ± 0.04	1.01 ± 0.02	7.60 ± 0.15	4.44 ± 0.05	4.31 ± 0.10	2.18 ± 0.08	4.74 ± 0.05	7 47 ± 0.15
X-33 M250 1	2.63 ± 0.0.3	1.42 ± 0.01	7.08 ± 0.20	4.57 ± 0.08	3.70 ± 0.22	4.04 ± 0.02	5.52 ± 0.16	7 26 ± 0.26
X-33 M250 2	2.59 ± 0.08	1.43 ± 0.02	6.86 ± 0.12	4 52 ± 0.04	3.53 ± 0.09	3 18 ± 0.15	5.96 ± 0.13	7.14 ± 0.22
X-33 M250 3	3.33 ± 0.05	1.62 ± 0.01	6.79 ± 0.14	4 59 ± 0.06	4.28 ± 0.12	3.52 ± 0.15	6.10 ± 0.17	6.08 ± 0.07
X-33 M250 4	2.61 ± 0.05	1.43 ± 0.04	7.25 ± 0.05	4 27 ± 0.05	4.28 ± 0.12	3.36 ± 0.17	5.19 ± 0.17	6.23 ± 0.20

Growth tests performed in 24-deep well plates as described in the "Methods" section. *X-33* ancestral strain, *X-33 Y250a-d* populations evolved on YPM medium, *X-33 M250a-d* populations evolved on BMM medium, *YPM* YP medium 1% MeOH, *BMM* buffered minimal medium 1% MeOH, *YPD* YP medium 2% glucose, *YPDN* YPD 500 mM NaCl, *BMD* buffered minimal medium 2% glucose, *BMDN* BMD 250 mM NaCl, *YPG* YP medium 2% glycerol, *BMG* buffered minimal medium 2% glycerol; Measurements were performed after 24 h of cultivation (BMM after 48 h). Values represent averages ± standard error ($n = 4$)

significantly reduced biomass yields on the non-evolutionary BMDN growth medium, yielding only 74% of the ancestral OD_{600}. For the remaining conditions tested, only minor population-specific differences were observed (Table 2).

Single clone growth characteristics

From three of the evolved populations three single clones were randomly selected and tested for growth rate on various growth media. As can be seen in Fig. 1a, b, most of these clones showed increased growth rate under growth conditions that were used for adaptation. Seven out of nine clones evolved on YPM medium showed increased growth rates on this growth medium. In contrast to the results for the population samples, five out of nine single clones showed decreased growth rates as compared with the ancestral strain on BMM medium (Fig. 1b). Unlike YPM-evolved clones, BMM-evolved clones showed higher growth rates on both media (140 and 135% as compared to the ancestral strain in contrast to 110 and 81% for the YPM-evolved clones) and significantly lower variance on YPM medium (Fig. 1a, b); thus, the growth rate of the single clones does not reflect the effects observed on the population level. Furthermore, growth of single clones was also tested on YPD growth medium with glucose as carbon source. Independent of growth medium type used for adaptation, evolved clones showed a decreased growth rate on YPD with, on average, 80% ancestral growth rate (Additional file 1: Figure S3).

Genomic mutations in evolved *P. pastoris* clones

To identify potential mutations underlying the observed growth phenotypes after long-term adaptation, Illumina MiSeq whole genome sequencing was applied. Based on the single clone characterization results described in the previous section, the single clone with the highest growth rate in adaptive conditions from populations Y250 1-3 and M250 1-3 was selected (Y250 1c, 2c, 3a and M250 1a, 2c, 3b, respectively). Additionally, the ancestral strain and clone Y250 3c (Fig. 1b) were selected for resequencing. Clone Y250 3c was included since this clone showed the highest growth trade-off (almost no detectable growth) on BMM medium.

Illumina MiSeq paired-end 300 bp reads were mapped to the de-novo assembled ancestral strain which is covering approximately 91.1% of the CBS 7435 reference genome. An average 94-fold coverage was obtained for each sequenced clone (Additional file 2: Table S4). Mutations were found in each of the adapted clones. In total 17 mutations were identified, with the number for each individual clone ranging from two to four mutations (Table 3). 16 mutations represented point mutations with three mutations occurring in intergenic regions between open reading frames. The remainder of the mutations were identified within predicted genes and led to one premature stop codon and twelve amino acid conversions. In line with previous observations regarding mutation frequencies [26], G to A (31%) and C to T (12%) mutations were most frequent. We also found one insertion/deletion (Indel) mutation in one of the clones (Y250 3c). Although we applied two different methods for copy number variation (CNV) detection (see "Methods" sections for details), no potential CNVs were detected.

Convergent mutational targets in independently evolved *P. pastoris* clones

Adaptive evolution experiments often lead to a high degree of convergence in the selection of mutational targets. This convergence may be observed for distinct

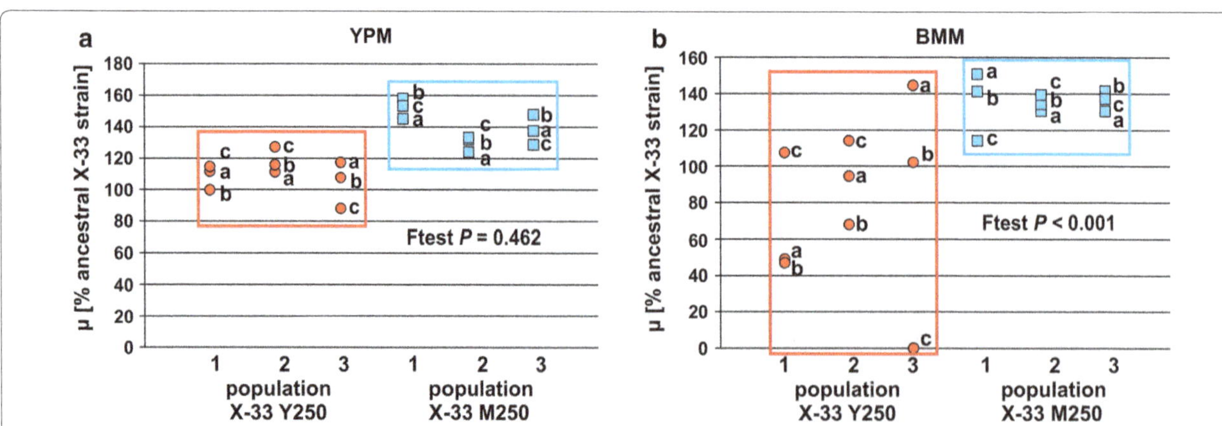

Fig. 1 Single clone growth rates in deep-well cultures. Three single clones from YPM (*red circles*) and BMM-evolved (*blue squares*) population 1–3 were randomly selected and growth rates were compared to the ancestral *P. pastoris* strain on YPM (**a**) and BMM (**b**). % growth rate relative to the ancestral strain is shown. Number of replicates per single clone, $n = 2$

Table 3 Mutations in methanol-adapted *P. pastoris* clones

Strain	chr	Position	Type	Ref	Alt	Gene/locus	Effect
Y250 1c	chr. 1	1,477,448	SNP	G	A	Upstream of PAS_chr1-4_0035 (*SPC110*)	
	chr. 2	1,572,979	SNP	C	G	PAS_chr2-1_0445 (Zn_cluster)	G142 to R142
	chr. 4	237,622	SNP	G	A	PAS_chr4_0821 (*AOX1*)	W190 to stop
	chr. 4	1,574,048	SNP	G	A	PAS_chr4_0108 (*YCT1*) weak homology	A64 to T64
Y250 2c	chr. 2	1,573,676	SNP	G	T	PAS_chr2-1_0445 (Zn_cluster)	Upstream of gene
	chr. 3	627,999	SNP	C	A	PAS_chr3_0836 (*ECM22*)	W95 to C95
Y250 3a	chr. 3	313,093	SNP	G	T	PAS_chr3_1001 (*TUP1*)	C285 to F285
	chr. 3	628,000	SNP	C	A	PAS_chr3_0836 (*ECM22*)	W95 to L95
Y250 3c	chr. 2	2,100,561	SNP	C	T	PAS_chr2-1_0162 (*SLN1*)	R336 to K336
	chr. 4	238,206	indel	AAGACAAGCC	A	PAS_chr4_0821 (*AOX1*)	3 amino acid deletion after E385
M250 1a	chr. 2	1,060,279	SNP	G	A	PAS_chr2-1_0701 (*PKH3*)	G354 to D354
	chr. 3	384,863	SNP	C	T	PAS_chr3_0956 (*RRP45*)	G206 to D206
	chr. 3	575,412	SNP	C	T	Downstream of PAS_chr3_1229 (*SEC5*) and downstream of PAS_chr3_0322 (tRNA-Thr7)	–
M250 2c	chr. 1	1,737,822	SNP	A	C	PAS_chr1-4_0181 (*NMA1*)	Q118 to H118
	chr. 4	238,443	SNP	G	A	PAS_chr4_0821 (*AOX1*)	R464 to K464
M250 3b	chr. 3	1,260,818	SNP	G	A	PAS_chr3_0512 (*PAH1*)	C304 to Y304
	chr. 4	238,309	SNP	C	G	PAS_chr4_0821 (*AOX1*)	F419 to L419

Mutations were identified by WGS (Illumina mi-Seq). The type of mutation (Single nucleotide polymorphism—SNP or insertion/deletion—indel) as well as the DNA sequence of the ancestral strain (ref) and the sequence of the evolved clone (alt) is shown. Chromosomal position (chr) and nucleotide position on contigs is shown with respect to the *P. pastoris* CBS7435 reference sequence

genetic loci or in terms of functional metabolic complexes [48]. In the current study, three genetic loci were affected in multiple independently evolved clones:

We found mutations in the alcohol oxidase 1 (*AOX1*) gene in four out of the seven evolved clones. Aox1 is the first enzyme in the methanol utilization pathway and catalyzes the conversion of methanol and oxygen into formaldehyde and hydrogen peroxide. Thus, it is not surprising that this key enzyme represents a potential selective target during the adaptation towards growth on methanol. Aox1 mutations were not limited to either rich or minimal growth conditions, but were found after long-term adaptation to both conditions (clones Y250 1c and 3c and M250 2c and 3b, respectively). Recently, the crystal structure of the *P. pastoris* Aox1 protein has been solved [49]. The enzyme monomer of the otherwise octameric protein structure has two main binding sites: the FAD cofactor binding site and the substrate binding site. The premature stop codon mutation in clone Y250 1c led to a heavily truncated protein, missing most of both sites. In clone Y250 3c, the deletion of amino acids 386–388 occurred in a loop region in proximity to the methanol binding site. The two amino acid conversions in the minimal medium evolved clones occurred in the methanol binding site (Fig. 2a). Since the truncated *AOX1* gene in clone Y250 1c most likely leads to a non-functional Aox1 enzyme, the alcohol oxidase activity was determined. Our data show that all four evolved clones with mutations in the *AOX1* gene exhibited lower alcohol oxidase activity as compared with the ancestral strain during exponential growth. This lower activity was irrespective of the growth medium, as all Y250 and M250 clones showed this reduced activity on both, YPM and BMM growth media (Fig. 2b). Aox activity was not abolished completely in mutated clones. As mentioned in the introduction, the *P. pastoris* genome encodes two *AOX* genes, *AOX1* and *AOX2*, respectively. Thus, the residual activity observed in the clones with mutated *AOX1* can be accounted for by *AOX2* expression (Fig. 2b).

A second convergent target was found in the open reading frame PAS_ch2-1_0445. Two evolved clones, both adapted to YPM growth medium, namely Y250 1c and 2c harbored a mutation linked to this predicted ORF. It encodes a putative Zn-cluster, GAL4-like transcription factor of unknown function. In order to analyze whether the encoded protein has any effect on the growth of *P. pastoris*, we overexpressed the gene in the X-33 wildtype background. Growth was assessed on glucose and methanol as carbon source in the context of both, YP and buffered minimal (BM) medium. Generally, the overexpression clones showed a tendency towards slightly decreased growth rates on both carbon sources on YP medium and a tendency towards increased growth rates on BMM medium (Additional file 1: Figure S4a, b). In terms of biomass yields, no differences were observed except from a significantly reduced biomass yield of the

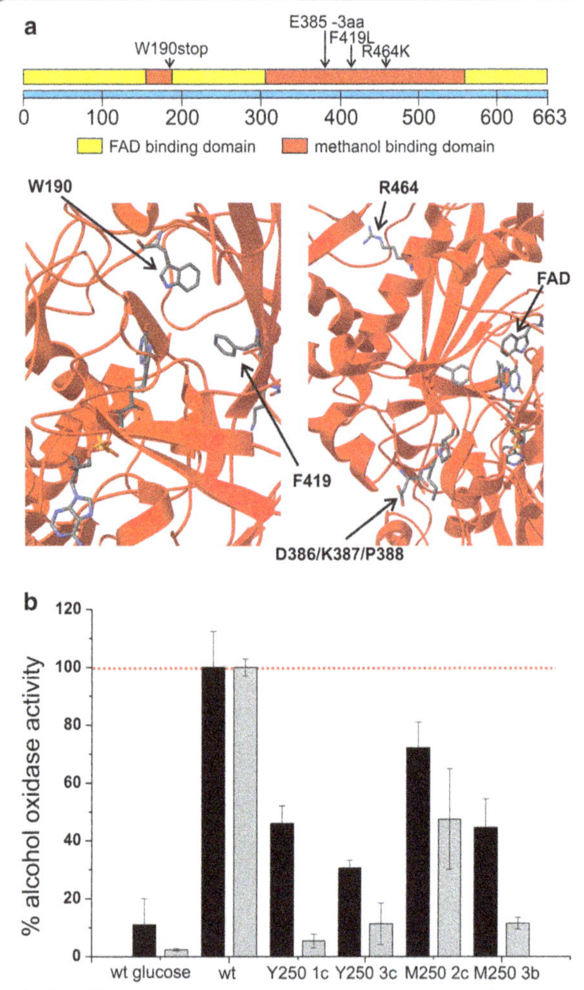

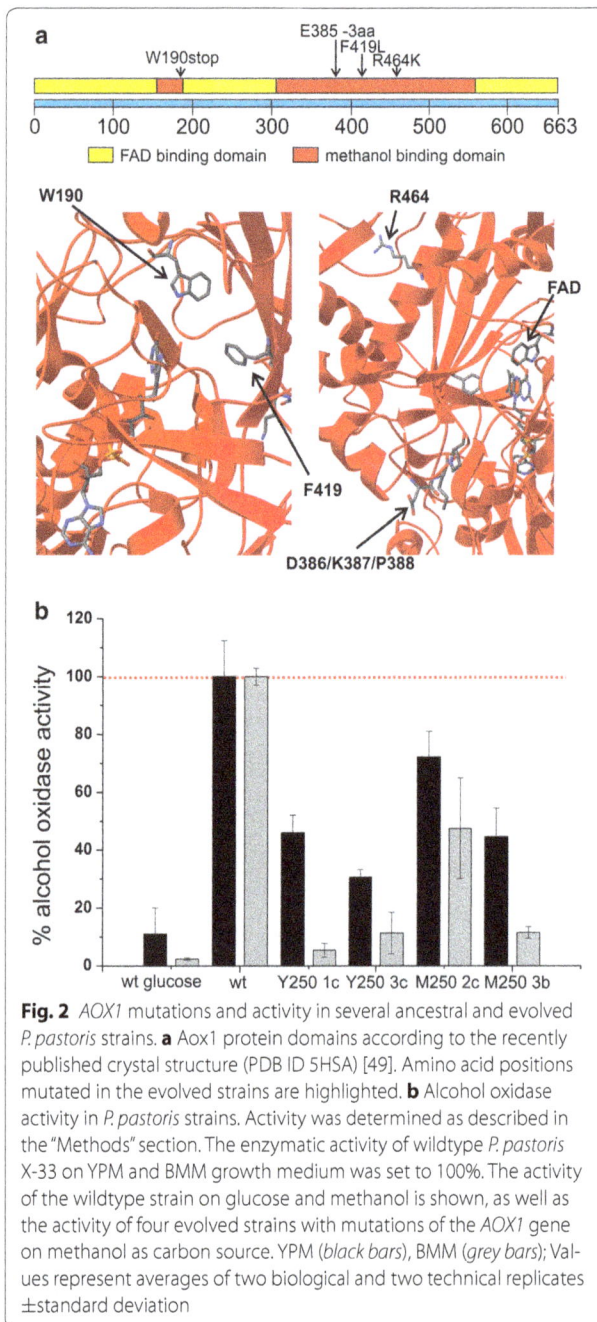

Fig. 2 *AOX1* mutations and activity in several ancestral and evolved *P. pastoris* strains. **a** Aox1 protein domains according to the recently published crystal structure (PDB ID 5HSA) [49]. Amino acid positions mutated in the evolved strains are highlighted. **b** Alcohol oxidase activity in *P. pastoris* strains. Activity was determined as described in the "Methods" section. The enzymatic activity of wildtype *P. pastoris* X-33 on YPM and BMM growth medium was set to 100%. The activity of the wildtype strain on glucose and methanol is shown, as well as the activity of four evolved strains with mutations of the *AOX1* gene on methanol as carbon source. YPM (*black bars*), BMM (*grey bars*); Values represent averages of two biological and two technical replicates ±standard deviation

PAS_ch2-1_0445 overexpression cells on BMM medium (Additional file 1: Figure S4c, d).

Another GAL4-like yeast specific transcription factor with similarity to *S. cerevisiae ECM22*, involved in the regulation of sterol biosynthesis, was mutated in two independent clones. Similar to the mutations related to the open reading frame PAS_chr2-1_0445, these mutations occurred after adaptation towards YPM medium in the clones Y250 2c and 3a.

Singular mutational targets

The remaining mutations were specific for the individual clones. The single mutations for YPM-evolved clones (Y250 1c, 3a and 3c) affected the genes *SPC110, PAS-chr4_0108, TUP1* and *SLN1*. SPC110 is a spindle pole body (SPB) component and PAS_chr4_0108, whose gene product belongs to the MFS general transporter family, shares weak homology with the *S. cerevisiae* gene for the Yct1 ER cysteine transporter. *TUP1* is an important transcriptional regulator involved in many physiological processes in yeast, including but not limited to stress response, mating as well as glucose sensing- and regulation [50]. Finally, clone Y250 3c had a SNP mutation in the transmembrane kinase *SLN1*, a transmembrane osmosensor critical for HOG pathway signaling [51].

Clone M250 1a harbored a SNP in PAS_chr2-1_0701, an ORF with similarity to the *S. cerevisiae PKH1/2/3* genes, whose gene products are regulators of protein serine/threonine kinase Pkc1. Furthermore this clone showed a SNP mutation in the *RRP45* gene, involved in cytoplasmic and nuclear RNA processing. The last mutation in this clone was intergenic between the genes for PAS-chr3_1229 (*SEC5*) and PAS-chr3_0322 (tRNA-Thr7) and also remains to be further analyzed in future studies. In addition to an *AOX1* mutation, the clone M250 2c differed from the ancestral strain in the *NMA1* gene, encoding a nicotinic acid mononucleotide adenylyltransferase involved in NAD synthesis. The evolved clone M250 3b had, in addition to an *AOX1* mutation, a nucleotide change in the *PAH1* locus, a phosphatidate phosphatase involved in the regulation of phospholipid biosynthesis.

The effect of methanol adaptation on recombinant protein production

In order to analyze the impact of MeOH-adaptation on recombinant protein expression, we used the model proteins rHSA [43] and rFDL [44] for expression in the ancestral X-33 and selected evolved clones. Whereas rHSA is a protein where relatively high product titers can be achieved, rFDL is a difficult-to-produce protein in *P. pastoris* [44]. In a first experiment, rHSA was expressed using a standardized screening protocol in minimal growth medium, mimicking a high cell density limited fed batch culture. For the expression procedure, cultures were grown in a glucose batch culture followed by an 18 h glucose fed batch phase using glucose feed beads. At a relatively high biomass concentration ($OD_{600} = 4$), cultures were induced with methanol for recombinant protein expression and grown for additional 48 h with methanol feeding in regular intervals. As summarized in Table 4, clones evolved on YPM medium showed significantly lower rHSA titers as well as lower biomass yield than the ancestral wildtype host strain. Contrary, clones evolved on minimal medium, namely M250 1a and 3b, showed on average a 28 and 15% increase in rHSA titers as well a trend towards higher biomass yields, whereas the rHSA yield per biomass (OD_{600}) was only higher for the clone M250 1a.

This screening protocol allows for rather low growth rates due the high initial biomass concentration prior to switching to methanol as carbon source. Thus, in order to further analyze potential growth rate effects during recombinant protein production, we also screened the clones in YPM and BMM medium with lower starting biomass concentrations at $OD_{600} = 0.2$, in order to maintain cells in exponential growth for a longer time during cultivation. On YPM, the wildtype showed similar rHSA titers to the first growth protocol, despite lower final OD_{600} values. The YPM-evolved clones showed largely reduced rHSA titers and biomass yields as compared with the ancestral strain. BMM-evolved clones showed a 20 and 78% increase in rHSA levels, which correlated with increased biomass yields. A similar trend, but with overall lower rHSA yield, was observed on BMM medium (Table 4). It has to be mentioned that both YPM-evolved clones did not show any significant growth in the BMM environment, although recombinant product could be detected, resulting in a high rHSA per OD_{600} yield for clone Y250 2c.

As a second model protein, a recombinant hexosaminidase (rFDL) was expressed in order to check whether the observed trends were protein specific. Overall, the results for rFDL were very similar to rHSA expression results, with BMM-evolved clones showing significantly increased product yields on both YPM and BMM medium, although e.g. in contrast to rHSA expression, the biomass yield for the BMM-evolved clones was not higher than for the wildtype clones on YPM (Table 4).

As mentioned above, recombinant FDL expression levels are low in general and are especially low in defined minimal medium fed batch cultures (own unpublished data). Consequently, rHSA-expressing clones were used to analyze growth in bioreactor fed batch cultures. The ancestral clone and evolved clones Y250 3a, M250 1a and M250 3b clone were selected based on the results summarized in Table 4. Except

Table 4 Recombinant gene expression in ancestral and evolved *P. pastoris* X-33 strains

Strain	rHSA ($mg\ L^{-1}$)	Final OD_{600}	rHSA OD_{600}^{-1}	Strain	rFDL ($U\ L^{-1}$)	Final OD_{600}	rFDL OD_{600}^{-1}
High biomass deep-well cultures[a]							
X-33 wt	13.8 ± 3.0	7.2 ± 0.2	1.9 ± 0.4	X-33 wt	nd	nd	nd
Y250 2c	0.9 ± 0.3	5.3 ± 0.2	0.2 ± 0.1	Y250 2c	nd	nd	nd
Y250 3a	4.7 ± 0.7	5.7 ± 0.3	0.9 ± 0.1	Y250 3a	nd	nd	nd
M250 1a	17.8 ± 2.0	7.3 ± 0.3	2.5 ± 0.3	M250 1a	nd	nd	nd
M250 3b	15.9 ± 2.9	8.4 ± 0.2	1.9 ± 0.3	M250 3b	nd	nd	nd
Strain	**YPM rHSA[b]**			**Strain**	**BMM rHSA[b]**		
	rHSA ($mg\ L^{-1}$)	**Final OD_{600}**	**rHSA OD_{600}^{-1}**		**rHSA ($mg\ L^{-1}$)**	**Final OD_{600}**	**rHSA OD_{600}^{-1}**
Low biomass deep-well cultures							
X-33 wt	13.8 ± 0.1	4.1 ± 0.2	3.4 ± 0.2	X-33 wt	3.9 ± 0.4	2.8 ± 0.0	1.3 ± 0.1
Y250 2c	0.4 ± 0.1	1.9 ± 0.1	0.2 ± 0.1	Y250 2c	0.7 ± 0.7	0.2 ± 0.0	3.9 ± 3.9
Y250 3a	2.0 ± 0.6	2.8 ± 0.2	0.7 ± 0.2	Y250 3a	0.1 ± 0.0	0.2 ± 0.0	0.6 ± 0.1
M250 1a	16.6 ± 2.0	5.9 ± 1.2	2.8 ± 0.2	M250 1a	7.0 ± 1.5	4.2 ± 0.2	1.6 ± 0.3
M250 3b	24.7 ± 5.7	7.4 ± 0.2	3.4 ± 0.9	M250 3b	4.7 ± 0.9	4.6 ± 0.1	1.0 ± 0.2
Strain	**YPM rFDL[c]**			**Strain**	**BMM rFDL[c]**		
	rFDL ($U\ L^{-1}$)	**Final OD_{600}**	**rFDL OD_{600}^{-1}**		**rFDL ($U\ L^{-1}$)**	**Final OD_{600}**	**rFDL OD_{600}^{-1}**
Low biomass deep-well cultures							
X-33 wt	179.4 ± 27.3	3.9 ± 0.1	38.4 ± 12.5	X-33 wt	8.5 ± 2.0	2.2 ± 0.3	3.5 ± 0.5
Y250 2c	10.3 ± 3.0	1.9 ± 0.0	5.6 ± 1.6	Y250 2c	6.0 ± 1.4	0.2 ± 0.0	36.1 ± 9.8
Y250 3a	16.9 ± 8.4	2.6 ± 0.0	6.7 ± 3.4	Y250 3a	16.4 ± 6.5	0.1 ± 0.0	122.7 ± 49.1
M250 1a	419.5 ± 26.5	3.8 ± 0,1	108.7 ± 7.5	M250 1a	65.0 ± 5.5	2.5 ± 0.1	25.6 ± 1.5
M250 3b	448.0 ± 57.3	4.1 ± 0.0	109.5 ± 13.9	M250 3b	40.0 ± 5.6	2.5 ± 0.1	17.8 ± 2.2

rHSA and rFDL were used as model proteins for expression in deep well cultures. For rHSA two screening protocols with high and low starting biomass prior to induction and growth on methanol were applied. For rFDL the low starting biomass protocol was applied. [a] and [b] number of replicates $n = 12$; [c] $n = 6$. Values represent averages ± standard deviation for Y2503a clone, which showed low productivity, the clones from the replicates of the batch profiling were selected based on comparable rHSA yields during deep well screening. For all fed batch cultivations, an initial glycerol batch with subsequent glycerol fed batch was applied, followed by a methanol pulse for adaptation and incrementally increased constant methanol feed phases (see "Methods" section for details). Furthermore, between incremental feed rate increases, methanol pulses were applied in order to analyze substrate uptake and respiratory behavior during exponential growth. Similar to deep well batch cultures, we find varying results in terms of rHSA production (Table 5). The YPM-evolved clone Y250 3a showed lower biomass yield as well as largely reduced rHSA yields compared to the ancestral strain. The M250 1a clones showed a biomass yield similar to the ancestral strain and specific (q_P) and volumetric productivity (Q_P) of only about 77% of the ancestral clone. The evolved clone M250 3b showed improved recombinant protein levels in all deep well screenings. During fed batch cultivation, the biomass yield was only approximately 5% higher than the ancestral strain, but a significant increase of rHSA production with a 55 and 76% increase for q_P and Q_P, was observed. Cell viability issues as potential cause for the observed differences were eliminated, since flow cytometric analysis showed high viability (>98%) throughout all fed cultivations (Additional file 1: Table S5).

However, differences were observed for other parameters, e.g. the length of the batch phase. For the ancestral and M250 3b strains, the glycerol batch phase was 28.9 and 29.8 h, whereas the initial glycerol batch took 35.8 and 34.3 h for clones Y250 3a and M250 1a respectively. Furthermore, a higher RQ was observed for these cultures during the glycerol batch (Table 6). The RQ values in combination with a significantly extended glycerol-batch phase are indicative of growth trade-offs on defined glycerol medium. Furthermore, increased RQ values for clones of the Y250 3a and M250 1a background during methanol pulse phases and constant feed phases were observed. Clone Y250 3a also showed a lower methanol consumption rate during the methanol pulse phases (Additional file 1: Table S6) although no *AOX* mutation was found. Thus, metabolic deficits might be linked to transcriptional rewiring caused by the *TUP1* or *ECM22* mutations found in this clone (Table 3). On the other hand, the high-producing clone M250 3b showed less pronounced differences in terms of RQ values for both glycerol and methanol phases. This clone also showed the fastest methanol consumption rate during methanol pulses (Table 6 and Additional file 1: Table S6). Besides these differences in RQ, the OTR during methanol growth was lower for methanol-adapted fed batch cultures than for the wildtype cultivation. Whereas this difference was negligible during the strongly limited constant feed phases, all three evolved clones showed OTRs ranging from 60 to 80% as compared with the wildtype fed batch culture during the MeOH pulse phases, indicating a reduced oxygen demand during exponential growth; in the case of the M250 3b clone this was despite

Table 5 Results of fed batch cultivations of selected rHSA-expressing clones

Clone	YDM (g L^{-1})	rHSA (mg L^{-1})	$Y_{x/s}$ glycerol (g g^{-1})	$Y_{x/s}$ methanol (g g^{-1})	q_P (mg g^{-1} h^{-1})	Q_P (mg L^{-1} h^{-1})
X33 rHSA	82.2 ± 0.8	198.4 ± 0.4	0.666 ± 0.006	0.139 ± 0.004	0.009	1.39
Y250 3a rHSA	73.3 ± 0.8	34.2 ± 0.89	0.578 ± 0.011	0.093 ± 0.004	0.002	0.24
M250 1a rHSA	83.0 ± 0.6	151.7 ± 1.88	0.610 ± 0.003	0.141 ± 0.003	0.007	1.06
M250 3b rHSA	86.3 ± 0.1	350.0 ± 3.0	0.686 ± 0.011	0.149 ± 0.0	0.014	2.45

For cultivations a glycerol batch and fed batch were performed and MeOH pulses as well as constant feed phases were applied as described in the "Methods" section. YDM, rHSA yield and biomass yields ($Y_{x/s}$) are shown. Values represent averages ± standard deviation. Specific (q_P) and volumetric productivity (Q_P) were calculated for the entire methanol feed process as an average for all methanol fed batch and pulse phases. Biomass yield ($Y_{x/s}$) for glycerol was calculated for the glycerol batch phase. Biomass yield for methanol shows the average over all MeOH phases

Table 6 Respiratory quotient (RQ) of fed batch cultures during different cultivation phases

Clone	RQ glycerol batch	RQ MeOH pulse phase	RQ MeOH constant feed
X33 rHSA	0.61 ± 0.03	0.48 ± 0.02	0.52 ± 0.02
Y250 3a rHSA	0.73 ± 0.05	0.60 ± 0.03	0.62 ± 0.02
M250 1a rHSA	0.81 ± 0.08	0.58 ± 0.03	0.64 ± 0.04
M250 3b rHSA	0.59 ± 0.05	0.53 ± 0.02	0.56 ± 0.02

For MeOH phases, value of all pulse phases and constant feed phases were combined. RQ values represent average values ± standard deviation of a faster methanol consumption rate (Additional file 1: Tables S6, S7).

Discussion

Phenotypic diversity

In order to provide an advantage in alternating environments and during niche exploration, population heterogeneity is a widely observed phenomenon among microbial populations; the basic driving force behind this heterogeneity being mechanisms such as epigenetic regulation and stochastic effects during gene regulation [52]. Thus, it is not surprising that we find differences in terms of growth rates among the individual clones that were isolated from a single population. It is interesting to observe higher growth rates of evolved populations on non-evolutionary carbon sources such as YPD (Table 1) but lower or decreased growth rates as compared to the ancestral strain for the randomly picked single clones (Additional file 1: Figure S3). It can be concluded that the single clones isolated from each population do not capture the geno- and phenotypic diversity present in a single population after 250 generations or approximately 10^{10} cumulative cell divisions. This observation is also supported by the fact that the two re-sequenced genomes from clones isolated from the same population (Y250 3a and 3c, Table 3) do not share any mutations and thus represent two independent lineages in this population.

Furthermore, YPM-evolved clones showed a significantly higher variance during growth on BMM (Fig. 1), whereas BMM-evolved clones did not display such a high degree of variance during growth on YPM medium. Thus, we conclude that long-term adaptation to growth on BMM results in the selection of phenotypes that are generally more robust and compatible with different environmental conditions as compared with methanol adaptation in rich (YP) medium. Generally our data indicate partially dual adaptation towards carbon source on the one hand and environment (nutrient rich vs. poor) on the other hand.

Genomic mutations upon adaptive evolution

In general, DNA replication is a high fidelity process and mutations are relatively rare. However, considering the relatively large population size of microbial cultures, mutations are frequent and can be fixed within a 100–200 generation experiment (for a review see [27]). On average, we identified 2.4 mutations per clone. This number is comparable to the number observed in similar experiments in pro- and eukaryotic microorganisms. A recent study for *E. coli* showed that cultures underwent approximately $10^{11.2}$ total cumulative cell divisions in order to produce a new stable phenotype [31]. Furthermore, these phenotypes were based on two to eight mutations for the evolved *E. coli* populations. The results are in agreement with the high phenotypic diversity observed at a lower CCD for *P. pastoris* cultures in the current study. Our data suggest similar numbers of fixed mutations and CCDs in order to achieve a stable phenotype across different microbial species.

Regarding the large number of recent studies dealing with laboratory evolution, an intriguing question is to which extent mutations are selective, deleterious or neutral. Under neutral selection, the occurrence of synonymous mutations might be readily expected but neutral and deleterious mutations can also hitchhike under selective conditions [53]. As already observed by Lenski et al. in his large-scale experiment with *E. coli* and in *S. cerevisiae* studies, the lack or underrepresentation of synonymous mutations suggests that the majority of mutations during ALE may be selective [30, 54]. In this context, our current study did not identify any synonymous mutations in the evolved clones, suggesting that most mutations may have been selected for and correlate with the observed growth phenotypes.

Mutational targets and functional complexes

Several mutations were discovered by genome sequencing and, to a certain degree, an overlap with respect to the targets was observed (Table 3; Fig. 3). Apart from the *AOX1* locus, we did not identify convergent genes or functional complexes when comparing YPM and BMM evolved populations:

For growth on YPM, mutations in *AOX1*, *ECM22* and a novel putative transcription factor (PAS_chr2-1_0445) seem to be of particular importance as they appear in multiple independently evolved clones. In the case of *ECM22*, even the same amino acid was affected, in one case leading to a W95C, in the second case to a W95L conversion. Regarding the PAS_chr2-1_0445, we overexpressed the corresponding gene in order to analyze whether a higher gene dosage has an effect on growth. Although only minor differences were observed in comparison with the control, our data indicate that this putative transcription factor might be involved in a general environmental response and is not directly linked to methanol-adaptation. Tendencies towards decreased growth rates in YPM and YPD medium upon gene overexpression were observed (Additional file 1: Figure S4). Thus, mutations emerging in the PAS_ch2-1_0445 locus may be under pleiotropic stabilizing selection with the additional mutations identified in the respective clones. A recent random mutagenesis study identified mutations in a different *P. pastoris* GAL4-like transcriptional regulator (*ATT1*) to confer improved fitness under thermal stress and in glyco-engineered *P. pastoris* strains [55]. *S. cerevisiae* Gal4 is involved in the regulation of galactose

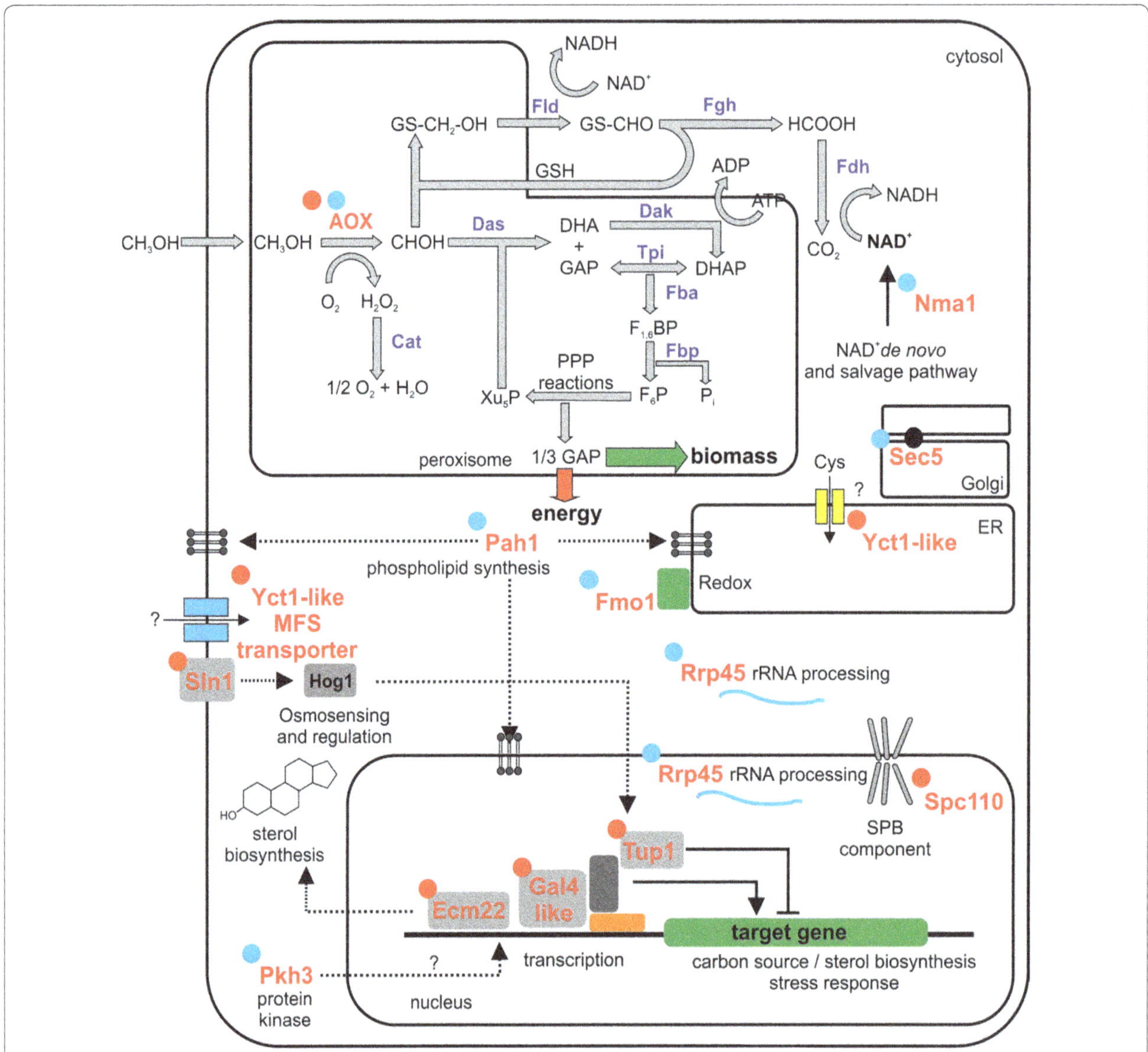

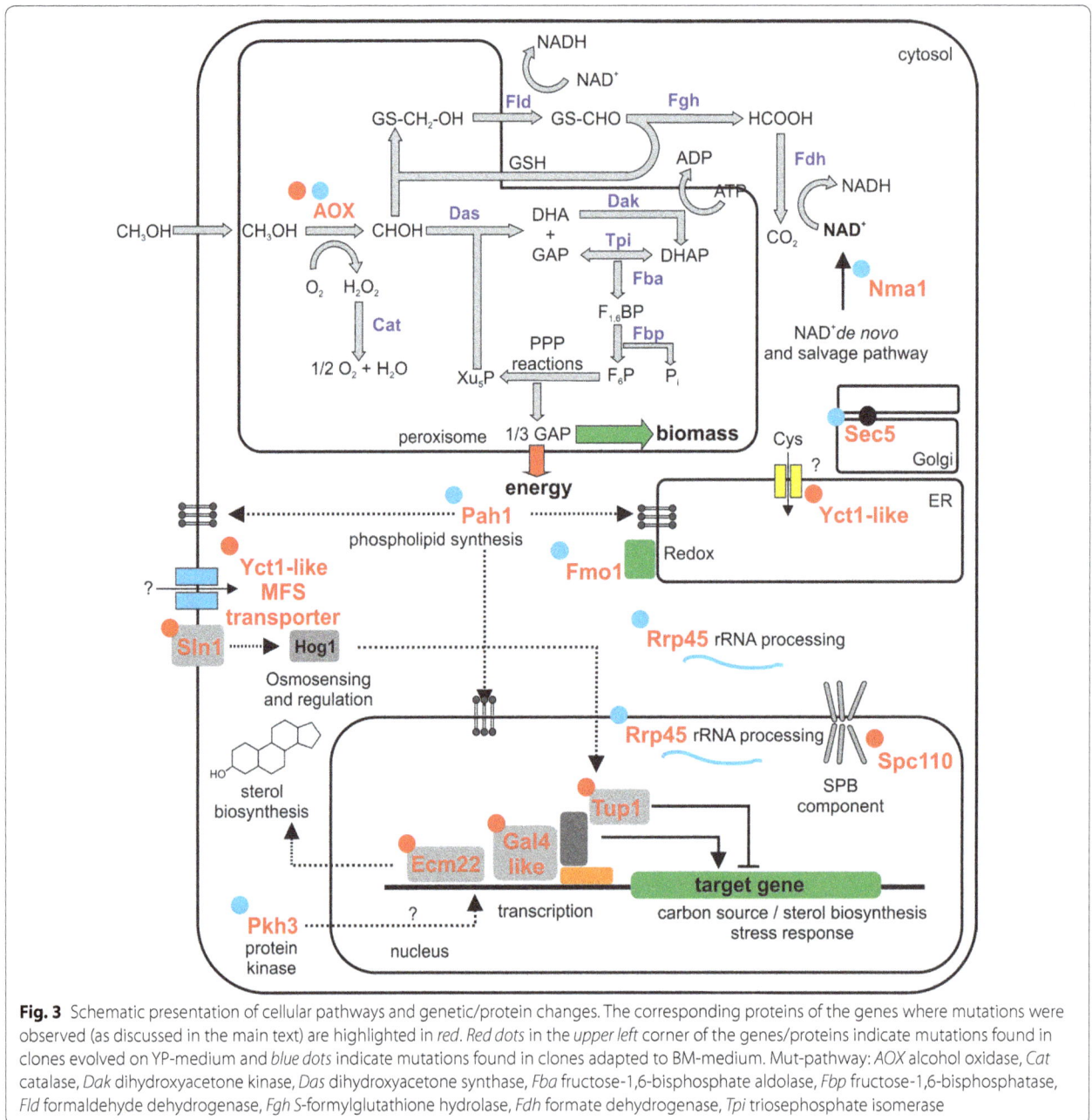

Fig. 3 Schematic presentation of cellular pathways and genetic/protein changes. The corresponding proteins of the genes where mutations were observed (as discussed in the main text) are highlighted in *red*. *Red dots* in the *upper left* corner of the genes/proteins indicate mutations found in clones evolved on YP-medium and *blue dots* indicate mutations found in clones adapted to BM-medium. Mut-pathway: *AOX* alcohol oxidase, *Cat* catalase, *Dak* dihydroxyacetone kinase, *Das* dihydroxyacetone synthase, *Fba* fructose-1,6-bisphosphate aldolase, *Fbp* fructose-1,6-bisphosphatase, *Fld* formaldehyde dehydrogenase, *Fgh* S-formylglutathione hydrolase, *Fdh* formate dehydrogenase, *Tpi* triosephosphate isomerase

metabolism [56], whereas the GAL4-like zinc cluster protein family is ubiquitous in fungi and serves several highly diverse functions including metabolic and stress response functions [57]. In the *P. pastoris* genome more than 20 GAL4-like proteins can be found. Considering the inability of this yeast to grow on galactose, these transcriptional regulators are likely to be involved in the regulation of other cellular pathways and may therefore represent critical selection targets in random mutagenesis and ALE approaches for improved growth.

Further mutations with probable consequences for cellular regulation were found for YPM in the *TUP1* transcriptional repressor and for BMM-evolved clones in *PKH3* and *RRP45*. Transcriptional modulation was previously exploited in *S. cerevisiae* for e.g. improved ethanol tolerance [58]; therefore, judging from multiple mutational targets related to regulatory processes, we conclude that transcriptional rewiring may represent a preferred route towards improved growth on methanol-based growth media for *P. pastoris*.

The *AOX1* gene as convergent target

Since the *AOX1* promoter is strongly induced during growth on methanol, it is one of the most prominent promoter systems for protein production in *P. pastoris*, although issues related to the toxicity and safety of methanol have been emphasized [13]. Therefore, it is not unusual that several attempts such as promoter engineering were previously evaluated [17]. *P. pastoris* harbors two genes for the alcohol oxidase gene, *AOX1* and *AOX2*. Whereas the wildtype strain with both of these genes grows fast on methanol (Mut$^+$ phenotype), impairing *AOX1* function leads to slower growth and the so-called Mut^s phenotype. As mentioned in the "Background" section, both cell types have been evaluated for recombinant protein production with inconclusive results. In this context, it is noteworthy that in the current study, long-term adaptation resulted in the selection of clones with mutations in the *AOX1* gene and subsequently reduced alcohol oxidase activity (Fig. 2). Since the loss of Aox1 activity generally results in a Mut^s growth phenotype it is surprising that three out of the four clones with *AOX1* mutations and reduced alcohol oxidase activity in this study showed increased growth rates as compared with the ancestor. We identified at least one additional mutation in each of these clones and it is very likely that these additional mutations are also adaptive and may be selective in conjunction with reduced Aox activity. For example, the evolved *P. pastoris* clone Y250 3c, with an additional mutation in the high osmolarity glycerol (HOG) pathway sensor *SLN1* [51], showed a decreased growth rate on YPM and BMM as compared with the ancestor but was still present in the population after 250 generations. This indicates that growth rate alone might not be an accurate denominator of fitness as the additional mutation might confer a selective advantage in other growth phases than exponential growth. It has been shown that the HOG pathway itself is involved in multiple cellular stress responses [59, 60]. Thus, whereas this mutation might provide a selective advantage in nutrient rich growth conditions, the modulation of HOG signaling may also be responsible for the poor growth performance in BMM.

Furthermore, previous studies already identified NADH recycling via methanol dissimilation as a major bottleneck during biotransformation in *P. pastoris* [61]. In this context, changes of both Aox activity and altered NAD synthesis rates (*NMA1* mutation) in clone M250 2c may confer an advantage during growth on minimal medium and methanol (Table 3; Fig. 3).

While Aox activity is certainly a prerequisite for growth on methanol as a carbon source, reduced Aox activity does not necessarily lead to reduced growth rates or fitness. Our data rather indicate higher-order pleiotropic effects with cellular processes such as signaling, transcriptional regulation and co-factor synthesis. Altogether, reduced Aox activity may lead to a reduction of excess intracellular formaldehyde, decreased formaldehyde-induced toxicity and an improved balance between assimilating and dissimilating methanol utilization processes.

Putative implications of ergosterol synthesis

In the current study, we also find mutations linked to the sterol regulatory element binding protein Ecm22, which is involved in the regulation of sterol biosynthesis. Research in the model yeast *S. cerevisiae* showed that the fungal sterol, ergosterol, plays an essential role for general membrane function, affecting membrane permeability and fluidity among others [62]. Many genes in this synthesis pathway are essential, whereas cells defective in the late steps of the biosynthesis are viable but show altered membrane structures and drug susceptibility [63]. In *P. pastoris*, a recent study showed that ergosterol synthesis was heavily up-regulated on a transcriptional level under hypoxic growth conditions [64]. Hypoxic conditions conferred a positive effect on recombinant protein production using the GAP-promoter system [65]. Furthermore, it was shown that mimicking hypoxia by fluconazole treatment, thereby inhibiting ergosterol synthesis, had a positive effect on recombinant protein production [66]. Furthermore, the growth of *P. pastoris* on methanol itself relies on peroxisomal proliferation and regulatory adaptation of cellular membrane synthesis, which includes ergosterol synthesis. *ECM22* mutations only occurred in YPM-evolved clones and selection experiments were performed in potentially oxygen-limited deep-well plates. Therefore, a closer investigation of ergosterol synthesis in various methanol growth conditions along with potential implications of hypoxic growth will be have to be done in future experiments.

Implications for recombinant protein production

Positive as well as negative effects were observed in terms of recombinant protein production in evolved *P. pastoris* clones. As shown in Tables 4 and 5, clones adapted to YPM (rich medium) showed decreased protein production performance in shake flask as well as fed batch cultivation. On the other hand, the adaptation to minimal methanol medium (BMM) generally resulted in improved recombinant protein production in comparison to the ancestral X-33 strain (Table 4) in shake flask cultures. Under methanol-limited fed batch conditions only one of the two tested clones (M250 3b) showed improved protein production, with a protein titer of approximately 180% compared to the wildtype. Moreover, our initial screening indicated that the YPM evolved clones subsequently selected for recombinant protein production retained the ability to grow on BMM medium in non-expressing conditions (Fig. 1). In contrast to these initial results, during cultivation and recombinant protein production, these clones showed almost no growth in BMM shake flask cultures. Additionally, clone Y250 2c also had an extended glycerol batch phase and very poor rHSA productivity in fed batch cultures. It has previously been noticed that recombinant protein production influences metabolic fluxes of the *P. pastoris* host cells and can lead to increased maintenance requirements [67]. Thus, an interesting question for further experiments will be, to which extend the metabolic rewiring in YPM-evolved clones in combination with the metabolic burden imposed by recombinant protein production lead to this diverging growth phenotypes of recombinant and non-recombinant evolved strains in rich and minimal growth environments. It has also been shown that the variation of environmental parameters in combination with adjusted feed strategies can lead to improved results [65]. In this study, an incrementally increased constant methanol feed protocol was applied, resulting in low growth rates (average μ during all methanol feeds ≤ 0.01). It will have to be addressed to which extent faster growth rates, in combination with the altered oxygen demand of methanol adapted clones, can be exploited for optimized and sustainable bioprocesses.

As discussed in the "Background" section there are several methods that can be implemented for an ALE study. In this context, it was interesting to observe that a clone selected for fast growth by serial dilution showed increased recombinant protein titers in batch as well as in methanol-limited fed batch cultivations. Future attempts may involve more elaborate selection methods such as nutrient-limited chemostats in order to improve our understanding of *P. pastoris* methanol metabolism but our data clearly indicate the suitability of relatively simple and efficient serial dilutions for the selection of clones with improved phenotypes across different cultivation methods.

Conclusion

We provide an initial dataset on the implications of evolutionary methanol adaptation and recombinant protein production for *P. pastoris*. Our data show that selected evolved *P. pastoris* cells, especially cells adapted to a minimal growth environment, can be directly applied as recombinant host strains as we observed increased protein production in shake flask and high cell density fed batch cultures. Our dataset also indicates a complex environment—carbon source correlation where optimized phenotypes are achieved by tuning methanol metabolism, via mutations related to alcohol oxidase, in combination with the modification of transcriptional regulators. As such, we provide a basis for extended studies of *P. pastoris* cell physiology in methanol growth environments.

Abbreviations

ALE: adaptive laboratory evolution; CCD: cumulative cell division; DO: dissolved oxygen; GAP: glyceraldehyde-3-phosphate; ER: endoplasmic reticulum; HOG: high osmolarity glycerol pathway; HRP: horseradish peroxidase; Indel: insertion/deletion; MeOH: methanol; OTR: oxygen transfer rate; PBS: phosphate buffered saline; RQ: respiratory quotient; SNP: single nucleotide polymorphism; WGS: whole genome sequencing.

Authors' contributions

J.M, SG and ABG performed genome sequence analysis; RP participated in fed batch fermentations and HSA quantification; MD, DM and IBHW contributed to study design and manuscript preparation; MD performed ALE, growth characterization, cloning experiments and conceived of the study. All authors read and approved the final manuscript.

Author details

[1] Department of Chemistry, University of Natural Resources and Life Sciences, Muthgasse 11, 1190 Vienna, Austria. [2] Austrian Centre of Industrial Biotechnology (ACIB), 1190 Vienna, Austria. [3] Department of Biotechnology, University of Natural Resources and Life Sciences, Vienna, Austria. [4] University of Applied Sciences FH-Campus Wien, Bioengineering, Vienna, Austria.

Acknowledgements

The authors would like to thank Fabian Schneider and Jonas Burgard for their support with HSA-ELISA and bioreactor experiments. Vienna Business Agency and EQ BOKU VIBT GmbH are acknowledged for providing fermentation equipment.

Competing interests

The authors declare that they have no competing interests.

Funding

This work was funded by the Austrian Science Fund (FWF) Project No. P26210 awarded to MD.

References

1. Kurtzman CP. Biotechnological strains of *Komagataella* (Pichia) *pastoris* are *Komagataella phaffii* as determined from multigene sequence analysis. J Ind Microbiol Biotechnol. 2009;36:1435–8.
2. Cregg J, Cereghino J, Shi J, Higgins D. Recombinant protein expression in *Pichia pastoris*. Mol Biotechnol. 2000;16:23–52.
3. Jacobs P, Geysens S, Vervecken W, Contreras R, Callewaert N. Engineering complex-type N-glycosylation in *Pichia pastoris* using GlycoSwitch technology. Nat Protoc. 2009;4:58–70.
4. Hamilton S, Gerngross T. Glycosylation engineering in yeast: the advent of fully humanized yeast. Curr Opin Biotechnol. 2007;18:387–92.

5. Baumann K, Carnicer M, Dragosits M, Graf AB, Stadlmann J, Jouhten P, Maaheimo H, Gasser B, Albiol J, Mattanovich D, Ferrer P. A multi-level study of recombinant *Pichia pastoris* in different oxygen conditions. BMC Syst Biol. 2010;4:141.
6. Dragosits M, Stadlmann J, Graf A, Gasser B, Maurer M, Sauer M, Kreil DP, Altmann F, Mattanovich D. The response to unfolded protein is involved in the osmotolerance of *Pichia pastoris*. BMC Genom. 2010;11:207.
7. Dragosits M, Stadlmann J, Albiol J, Baumann K, Maurer M, Gasser B, Sauer M, Altmann F, Ferrer P, Mattanovich D. The effect of temperature on the proteome of recombinant *Pichia pastoris*. J Proteome Res. 2009;8:1380–92.
8. Gasser B, Maurer M, Gach J, Kunert R, Mattanovich D. Engineering of *Pichia pastoris* for improved production of antibody fragments. Biotechnol Bioeng. 2006;94:353–61.
9. Xu P, Raden D, Doyle FR, Robinson A. Analysis of unfolded protein response during single-chain antibody expression in *Saccharomyces cerevisiae* reveals different roles for BiP and PDI in folding. Metab Eng. 2005;7:269–79.
10. Gupta P, Ghosalkar A, Mishra S, Chaudhuri T. Enhancement of over expression and chaperone assisted yield of folded recombinant aconitase in *Escherichia coli* in bioreactor cultures. J Biosci Bioeng. 2009;107:102–7.
11. Waterham H, Digan M, Koutz P, Lair S, Cregg J. Isolation of the *Pichia pastoris* glyceraldehyde-3-phosphate dehydrogenase gene and regulation and use of its promoter. Gene. 1997;186:37–44.
12. Invitrogen: Pichia expression kit, User manual. 2010.
13. Shen W, Xue Y, Liu Y, Kong C, Wang X, Huang M, Cai M, Zhou X, Zhang Y, Zhou M. A novel methanol-free *Pichia pastoris* system for recombinant protein expression. Microb Cell Fact. 2016;15:178.
14. Gasser B, Steiger MG, Mattanovich D. Methanol regulated yeast promoters: production vehicles and toolbox for synthetic biology. Microb Cell Fact. 2015;14:196.
15. Krainer FW, Dietzsch C, Hajek T, Herwig C, Spadiut O, Glieder A. Recombinant protein expression in *Pichia pastoris* strains with an engineered methanol utilization pathway. Microb Cell Fact. 2012;11:22.
16. Kim SJ, Lee JA, Kim YH, Song BK. Optimization of the functional expression of *Coprinus cinereus* peroxidase in *Pichia pastoris* by varying the host and promoter. J Microbiol Biotechnol. 2009;19:966–71.
17. Hartner FS, Ruth C, Langenegger D, Johnson SN, Hyka P, Lin-Cereghino GP, Lin-Cereghino J, Kovar K, Cregg JM, Glieder A. Promoter library designed for fine-tuned gene expression in *Pichia pastoris*. Nucleic Acids Res. 2008;36:e76.
18. Prielhofer R, Cartwright SP, Graf AB, Valli M, Bill RM, Mattanovich D, Gasser B. *Pichia pastoris* regulates its gene-specific response to different carbon sources at the transcriptional, rather than the translational, level. BMC Genom. 2015;16:167.
19. Rußmayer H, Buchetics M, Gruber C, Valli M, Grillitsch K, Modarres G, Guerrasio R, Klavins K, Neubauer S, Drexler H, et al. Systems-level organization of yeast methylotrophic lifestyle. BMC Biol. 2015;13:80.
20. Wang X, Wang Q, Wang J, Bai P, Shi L, Shen W, Zhou M, Zhou X, Zhang Y, Cai M. Mit1 transcription factor mediates methanol signaling and regulates the alcohol oxidase 1 (AOX1) promoter in *Pichia pastoris*. J Biol Chem. 2016;291:6245–61.
21. Sahu U, Krishna Rao K, Rangarajan PN. Trm1p, a Zn(II)(2)Cys(6)-type transcription factor, is essential for the transcriptional activation of genes of methanol utilization pathway, in *Pichia pastoris*. Biochem Biophys Res Commun. 2014;451:158–64.
22. Lin-Cereghino GP, Godfrey L, de la Cruz BJ, Johnson S, Khuongsathiene S, Tolstorukov I, Yan M, Lin-Cereghino J, Veenhuis M, Subramani S, Cregg JM. Mxr1p, a key regulator of the methanol utilization pathway and peroxisomal genes in *Pichia pastoris*. Mol Cell Biol. 2006;26:883–97.
23. Tenaillon O, Barrick JE, Ribeck N, Deatherage DE, Blanchard JL, Dasgupta A, Wu GC, Wielgoss S, Cruveiller S, Médigue C, et al. Tempo and mode of genome evolution in a 50,000-generation experiment. Nature. 2016;536:165–70.
24. Cooper V, Lenski R. The population genetics of ecological specialization in evolving *Escherichia coli* populations. Nature. 2000;407:736–9.
25. Hong KK, Nielsen J. Adaptively evolved yeast mutants on galactose show trade-offs in carbon utilization on glucose. Metab Eng. 2013;16:78–86.
26. Conrad TM, Lewis NE, Palsson B. Microbial laboratory evolution in the era of genome-scale science. Mol Syst Biol. 2011;7:509.
27. Dragosits M, Mattanovich D. Adaptive laboratory evolution—principles and applications for biotechnology. Microb Cell Fact. 2013;12:64.
28. Dunham MJ. Experimental evolution in yeast: a practical guide. Methods Enzymol. 2010;470:487–507.
29. Notley-McRobb L, King T, Ferenci T. rpoS mutations and loss of general stress resistance in *Escherichia coli* populations as a consequence of conflict between competing stress responses. J Bacteriol. 2002;184:806–11.
30. Kvitek DJ, Sherlock G. Whole genome, whole population sequencing reveals that loss of signaling networks is the major adaptive strategy in a constant environment. PLoS Genet. 2013;9:e1003972.
31. Lee DH, Feist AM, Barrett CL, Palsson B. Cumulative number of cell divisions as a meaningful timescale for adaptive laboratory evolution of *Escherichia coli*. PLoS ONE. 2011;6:e26172.
32. Martin M. Cutadapt removes adapter sequences from high-throughput sequencing reads. Embnet J. 2011;17:1
33. Chapman JA, Ho I, Sunkara S, Luo S, Schroth GP, Rokhsar DS. Meraculous: de novo genome assembly with short paired-end reads. PLoS ONE. 2011;6:e23501.
34. Küberl A, Schneider J, Thallinger GG, Anderl I, Wibberg D, Hajek T, Jaenicke S, Brinkrolf K, Goesmann A, Szczepanowski R, et al. High-quality genome sequence of *Pichia pastoris* CBS7435. J Biotechnol. 2011;154:312–20.
35. Galardini M, Biondi EG, Bazzicalupo M, Mengoni A. CONTIGuator: a bacterial genomes finishing tool for structural insights on draft genomes. Source Code Biol Med. 2011;6:11.
36. Stanke M, Morgenstern B. AUGUSTUS: a web server for gene prediction in eukaryotes that allows user-defined constraints. Nucleic Acids Res. 2005;33:W465–7.
37. Gardner SN, Slezak T, Hall BG. kSNP3.0: SNP detection and phylogenetic analysis of genomes without genome alignment or reference genome. Bioinformatics. 2015;31:2877–8.
38. McKenna A, Hanna M, Banks E, Sivachenko A, Cibulskis K, Kernytsky A, Garimella K, Altshuler D, Gabriel S, Daly M, DePristo MA. The genome analysis toolkit: a mapreduce framework for analyzing next-generation DNA sequencing data. Genome Res. 2010;20:1297–303.
39. Nijkamp JF, van den Broek MA, Geertman JM, Reinders MJ, Daran JM, de Ridder D. De novo detection of copy number variation by co-assembly. Bioinformatics. 2012;28:3195–202.
40. Klambauer G, Schwarzbauer K, Mayr A, Clevert DA, Mitterecker A, Bodenhofer U, Hochreiter S. cn.MOPS: mixture of Poissons for discovering copy number variations in next-generation sequencing data with a low false discovery rate. Nucleic Acids Res. 2012;40:e69.
41. Zerbino DR, Birney E. Velvet: algorithms for de novo short read assembly using de Bruijn graphs. Genome Res. 2008;18:821–9.
42. Robinson JT, Thorvaldsdóttir H, Winckler W, Guttman M, Lander ES, Getz G, Mesirov JP. Integrative genomics viewer. Nat Biotechnol. 2011;29:24–6.
43. Marx H, Sauer M, Resina D, Vai M, Porro D, Valero F, Ferrer P, Mattanovich D. Cloning, disruption and protein secretory phenotype of the GAS1 homologue of *Pichia pastoris*. FEMS Microbiol Lett. 2006;264:40–7.
44. Dragosits M, Yan S, Razzazi-Fazeli E, Wilson IB, Rendic D. Enzymatic properties and subtle differences in the substrate specificity of phylogenetically distinct invertebrate *N*-glycan processing hexosaminidases. Glycobiology. 2015;25:448–64.
45. Delic M, Mattanovich D, Gasser B. Repressible promoters—a novel tool to generate conditional mutants in *Pichia pastoris*. Microb Cell Fact. 2013;12:6.
46. Kobayashi K, Kuwae S, Ohya T, Ohda T, Ohyama M, Ohi H, Tomomitsu K, Ohmura T. High-level expression of recombinant human serum albumin from the methylotrophic yeast *Pichia pastoris* with minimal protease production and activation. J Biosci Bioeng. 2000;89:55–61.
47. Marx H, Mecklenbräuker A, Gasser B, Sauer M, Mattanovich D. Directed gene copy number amplification in *Pichia pastoris* by vector integration into the ribosomal DNA locus. FEMS Yeast Res. 2009;9:1260–70.
48. Tenaillon O, Rodríguez-Verdugo A, Gaut RL, McDonald P, Bennett AF, Long AD, Gaut BS. The molecular diversity of adaptive convergence. Science. 2012;335:457–61.
49. Koch C, Neumann P, Valerius O, Feussner I, Ficner R. Crystal structure of alcohol oxidase from *Pichia pastoris*. PLoS ONE. 2016;11:e0149846.
50. Smith RL, Johnson AD. Turning genes off by Ssn6-Tup1: a conserved system of transcriptional repression in eukaryotes. Trends Biochem Sci. 2000;25:325–30.

51. Hohmann S. Osmotic stress signaling and osmoadaptation in yeasts. Microbiol Mol Biol Rev. 2002;66:300–72.
52. Avery SV. Microbial cell individuality and the underlying sources of heterogeneity. Nat Rev Microbiol. 2006;4:577–87.
53. Notley-McRobb L, Seeto S, Ferenci T. Enrichment and elimination of mutY mutators in *Escherichia coli* populations. Genetics. 2002;162:1055–62.
54. Barrick J, Yu D, Yoon S, Jeong H, Oh T, Schneider D, Lenski R, Kim J. Genome evolution and adaptation in a long-term experiment with *Escherichia coli*. Nature. 2009;461:1243–7.
55. Jiang B, Argyros R, Bukowski J, Nelson S, Sharkey N, Kim S, Copeland V, Davidson RC, Chen R, Zhuang J, et al. Inactivation of a GAL4-like transcription factor improves cell fitness and product yield in glycoengineered *Pichia pastoris* strains. Appl Environ Microbiol. 2015;81:260–71.
56. Traven A, Jelicic B, Sopta M. Yeast Gal4: a transcriptional paradigm revisited. EMBO Rep. 2006;7:496–9.
57. MacPherson S, Larochelle M, Turcotte B. A fungal family of transcriptional regulators: the zinc cluster proteins. Microbiol Mol Biol Rev. 2006;70:583–604.
58. Alper H, Moxley J, Nevoigt E, Fink G, Stephanopoulos G. Engineering yeast transcription machinery for improved ethanol tolerance and production. Science. 2006;314:1565–8.
59. Hayashi M, Maeda T. Activation of the HOG pathway upon cold stress in *Saccharomyces cerevisiae*. J Biochem. 2006;139:797–803.
60. Bilsland E, Molin C, Swaminathan S, Ramne A, Sunnerhagen P. Rck1 and Rck2 MAPKAP kinases and the HOG pathway are required for oxidative stress resistance. Mol Microbiol. 2004;53:1743–56.
61. Schroer K, Peter Luef K, Stefan Hartner F, Glieder A, Pscheidt B. Engineering the *Pichia pastoris* methanol oxidation pathway for improved NADH regeneration during whole-cell biotransformation. Metab Eng. 2010;12:8–17.
62. Parks LW, Casey WM. Physiological implications of sterol biosynthesis in yeast. Annu Rev Microbiol. 1995;49:95–116.
63. Abe F, Hiraki T. Mechanistic role of ergosterol in membrane rigidity and cycloheximide resistance in *Saccharomyces cerevisiae*. Biochim Biophys Acta. 2009;1788:743–52.
64. Baumann K, Dato L, Graf AB, Frascotti G, Dragosits M, Porro D, Mattanovich D, Ferrer P, Branduardi P. The impact of oxygen on the transcriptome of recombinant S. cerevisiae and *P. pastoris*—a comparative analysis. BMC Genom. 2011;12:218.
65. Baumann K, Maurer M, Dragosits M, Cos O, Ferrer P, Mattanovich D. Hypoxic fed-batch cultivation of *Pichia pastoris* increases specific and volumetric productivity of recombinant proteins. Biotechnol Bioeng. 2008;100:177–83.
66. Baumann K, Adelantado N, Lang C, Mattanovich D, Ferrer P. Protein trafficking, ergosterol biosynthesis and membrane physics impact recombinant protein secretion in *Pichia pastoris*. Microb Cell Fact. 2011;10:93.
67. Jordà J, Jouhten P, Cámara E, Maaheimo H, Albiol J, Ferrer P. Metabolic flux profiling of recombinant protein secreting *Pichia pastoris* growing on glucose:methanol mixtures. Microb Cell Fact. 2012;11:57.

5

Heterologous biosynthesis and manipulation of crocetin in *Saccharomyces cerevisiae*

Fenghua Chai[1,2†], Ying Wang[1,2†], Xueang Mei[1,2†], Mingdong Yao[1,2], Yan Chen[1,2], Hong Liu[1,2], Wenhai Xiao[1,2*] and Yingjin Yuan[1,2]

Abstract

Background: Due to excellent performance in antitumor, antioxidation, antihypertension, antiatherosclerotic and antidepressant activities, crocetin, naturally exists in *Crocus sativus* L., has great potential applications in medical and food fields. Microbial production of crocetin has received increasing concern in recent years. However, only a patent from EVOVA Inc. and a report from Lou et al. have illustrated the feasibility of microbial biosynthesis of crocetin, but there was no specific titer data reported so far. *Saccharomyces cerevisiae* is generally regarded as food safety and productive host, and manipulation of key enzymes is critical to balance metabolic flux, consequently improve output. Therefore, to promote crocetin production in *S. cerevisiae*, all the key enzymes, such as CrtZ, CCD and ALD should be engineered combinatorially.

Results: By introduction of heterologous CrtZ and CCD in existing β-carotene producing strain, crocetin biosynthesis was achieved successfully in *S. cerevisiae*. Compared to culturing at 30 °C, the crocetin production was improved to 223 μg/L at 20 °C. Moreover, an optimal CrtZ/CCD combination and a titer of 351 μg/L crocetin were obtained by combinatorial screening of CrtZs from nine species and four CCDs from *Crocus*. Then through screening of heterologous ALDs from *Bixa orellana* (Bix_ALD) and *Synechocystis* sp. PCC6803 (Syn_ALD) as well as endogenous ALD6, the crocetin titer was further enhanced by 1.8-folds after incorporating Syn_ALD. Finally a highest reported titer of 1219 μg/L at shake flask level was achieved by overexpression of CCD2 and Syn_ALD. Eventually, through fed-batch fermentation, the production of crocetin in 5-L bioreactor reached to 6278 μg/L, which is the highest crocetin titer reported in eukaryotic cell.

Conclusions: *Saccharomyces cerevisiae* was engineered to achieve crocetin production in this study. Through combinatorial manipulation of three key enzymes CrtZ, CCD and ALD in terms of screening enzymes sources and regulating protein expression level (reaction temperature and copy number), crocetin titer was stepwise improved by 129.4-fold (from 9.42 to 1219 μg/L) as compared to the starting strain. The highest crocetin titer (6278 μg/L) reported in microbes was achieved in 5-L bioreactors. This study provides a good insight into key enzyme manipulation involved in serial reactions for microbial overproduction of desired compounds with complex structure.

Keywords: Metabolic engineering, Crocetin, *Saccharomyces cerevisiae*, Synthetic biology, Enzyme sources

*Correspondence: wenhai.xiao@tju.edu.cn
†Fenghua Chai, Ying Wang and Xueang Mei contributed equally to this work
[1] Key Laboratory of Systems Bioengineering (Ministry of Education), Tianjin University, 92, Weijin Road, Nankai District, Tianjin 300072, People's Republic of China
Full list of author information is available at the end of the article

Background

Crocetin, a kind of carotenoid existing in *Crocus sativus* L. [1], has great potential medical applications due to various pharmacological activities, such as antitumor [2, 3], antioxidation [4], antihypertension [5], antiatherosclerotic [6] and antidepressant [7]. Additionally, crocetin can be also used as edible pigment. Currently, since crocetin manufacture mainly relied on extraction and purification from *Crocus* stigmas, deficient resource and low extraction rate restricted the large-scale application for commercialization. De novo synthesis of crocetin from simple carbon (glucose etc.) in engineered heterologous hosts would be an important complement to traditional sources. For crocetin biosynthesis, the conversion of β-carotene to crocetin required three steps catalyzed by β-carotene hydroxylase (CrtZ), carotenoid cleavage dioxygenase (CCD) and aldehyde dehydrogenase (ALD), respectively (Fig. 1a) [8]. It is speculated that balancing metabolic flux mediated by the above three enzymes is a big challenge for high output. To date, only a patent from EVOVA Inc. [9] and a report from Lou et al. [10] have just illustrated the feasibility of heterologous biosynthesis of crocetin, and there was no specific titer data reported yet. For promoting crocetin production, combinatorial manipulation of the CrtZ, CCD and ALD would be a promising solution to overcome this challenge.

Screening enzymes sources and regulating protein expression level have been proved to be efficient strategies for manipulating the key enzymes for balancing metabolic flux, consequently improving production [11–13]. Cao et al. [14] once improved odd-chain fatty alcohols production in *Escherichia coli* through balancing the expression level of TesA, αDOX, AHRs and the genes involved in fatty acids metabolism pathway. Meanwhile, through combinatorially screening the carotenogenic enzymes (CrtE, CrtB and CrtI) from diverse organisms and fine-tuning the expression level of CrtI, an optimal enzymes combination with the highest lycopene yield was obtained in *Saccharomyces cerevisiae* [15]. In crocetin biosynthesis fields, CrtZ, CCD and ALD have been characterized separately in the last decades. Li et al. [16] once achieved zeaxanthin titer as 43.46 mg/L in a recombinant *E. coli* strain by integrating *Pantoea ananatis* CrtZ into a β-carotene producing strain. Meanwhile, *Crocus* ZCD was firstly annotated as 7, 8 (7′, 8′)-zeaxanthin cleavage dioxygenase in 2003 [17]. However, Frusciante et al. [18] demonstrated this enzyme could not achieve crocetin synthesis in *E. coli* and corn. Another two *Crocus* CCDs,

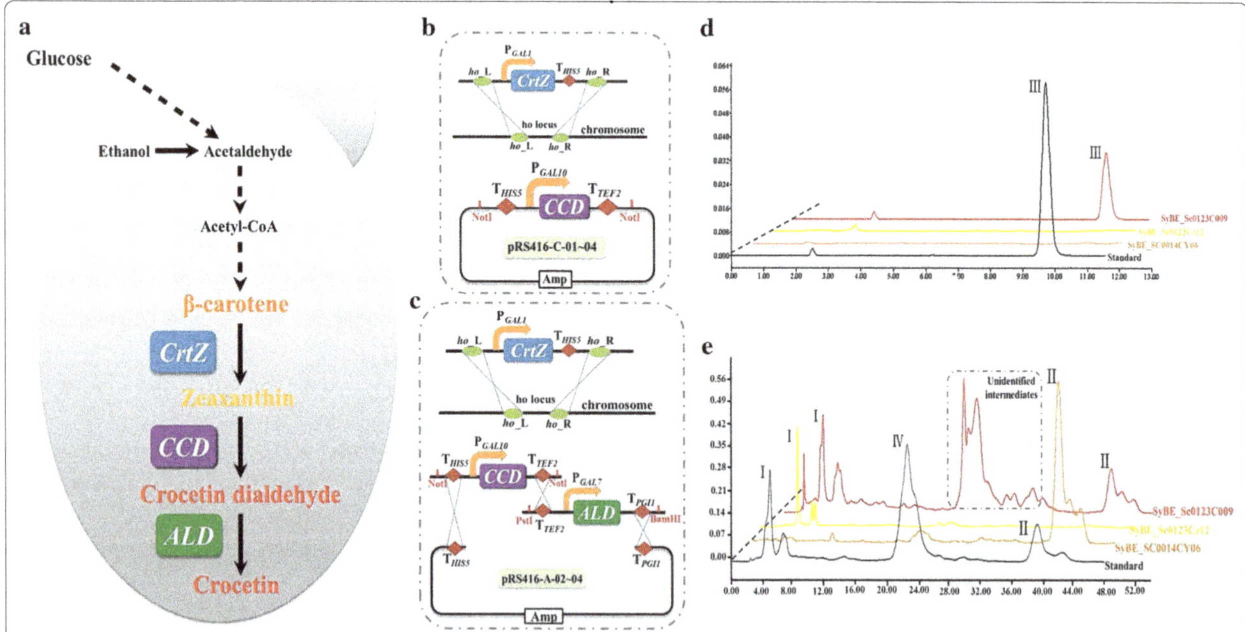

Fig. 1 Crocetin biosynthesis pathway construction in *S. cerevisiae*. **a** The paradigm of crocetin biosynthetic pathway in *S. cerevisiae*. The synthetic pathway to crocetin from β-carotene consists of three enzymes: CrtZ, β-carotene hydroxylase; CCD, carotenoid cleavage dioxygenase and ALD, aldehyde dehydrogenase. **b**, **c** Schematic representation of the engineering strategies for CrtZ, CCD and ALD expression cassette. CrtZ expression cassette was integrated into the *ho* locus of the chromosome, while CCD or CCD/ALD was carried by centromeric plasmid pRS416. *ho*_L, *ho* locus left homologous arm; *ho*_R, *ho* locus right homologous arm. **d**, **e** The HPLC profile of the parent strain SyBE_Sc0014CY06 (*orange*), zeaxanthin producing strain SyBE_Sc0123Cz12 (*yellow*), crocetin producing strain SyBE_Sc0123C009 (*red*), and standard (*black*). The signals for zeaxanthin (I), β-carotene (II) and lycopene (IV) were detected at 450 nm, while crocetin (III) was at 430 nm. The retention time of the unidentified intermediates which were boxed was close to that of lycopene

CCD2 [18] and ZCD1 [10], have been proved to cleavage of zeaxanthin at the 7, 8- and 7′, 8′-positions for forming crocetin dialdehyde in *E. coli* and *Chlorella vulgaris*, respectively. Moreover, even though EVOVA Inc. [9] and Lou et al. [10] realized crocetin synthesis by using endogenous ALD in yeast and algae, respectively, there was no titer data uncovered yet. It could guess that the complexity of fine-tuning *CrtZ, CCD and ALD* was the main obstacle. Therefore, it is urgent to explore *CrtZ, CCD* and *ALD* systematically for crocetin higher production.

Saccharomyces cerevisiae has been reported as a safe (Generally Recognized as Safe, GRAS) and robust host cell to produce heterologous carotenoids, including lycopene [19], β-carotene [20] and astaxanthin [21]. Thus, in our study, crocetin was successfully synthesized in *S. cerevisiae* through incorporating heterologous CrtZ and CCD in an existing β-carotene producing strain SyBE_Sc0014CY06 (with β-carotene titer of 220 mg/L) (Table 1). A higher crocetin titer was achieved by adjusting the culture temperature from 30 to 20 °C. The production of crocetin was further enhanced by 2.8-fold via screening of CrtZ/CCD combination and ALD sources. Moreover, the crocetin titer was reached to 1219 μg/L by increasing the copy numbers of *ccd* and *ald*. Finally, the highest reported crocetin titer as 6278 μg/L was archived in 5-L bioreactors. This study sets a good example of fine-tuning multiple enzymes systematically for heterologous biosynthesis of desired pharmaceuticals and chemicals.

Methods

Construction of plasmids and strains

Primers and plasmids used in this study were listed in Additional file 1: Table S1; Table 1, respectively. All the heterologous genes including *crtZ, ccd*, and *ald* were codon optimized (Additional file 1: Table S2) and synthesized by GENEWIZ (Suzhou, China). All these genes were delivered as pUC57-simple serious plasmids (Table 1). Promoters (P_{GAL1}, P_{GAL7} and P_{GAL10}), terminators (T_{HIS5}, T_{TEF2}, and T_{PGI1}) and integration homologous arms (*ho*_L and *ho*_R) were amplified from the genomic DNA of *S. cerevisiae* CEN.PK2-1C, as well as the auxotroph marker *URA3* was amplified from the plasmid pRS416. Cassette *ho*_L-P_{GAL1}-T_{HIS5}-*URA3-ho*_R was assembled by overlap extension PCR (OE-PCR) and cloned into pJET1.2, obtaining the plasmid pJET1.2-Z-01 (Table 1; Additional file 1: Figure S1). Genes *crtZ* were recovered by *Bsa*I digestion from pUC57-Simple-01–09 and inserted into the same site of pJET1.2-Z-01, generating pJET1.2-Z series plasmids (CrtZ expression cassette plasmids pJET1.2-Z-02–10, Table 1; Additional file 1: Figure S1). Then the CrtZ expression cassette *ho*_L-P_{GAL1}-CrtZ-T_{HIS5}-*URA3-ho*_R were cut from pJET1.2-Z series plasmids by *Pme*I and transformed into *S. cerevisiae* SyBE_SC0014CY06 for genomic integration (Fig. 1b) via the lithium acetate method [22]. Marker *URA3* was deleted according to Boeke et al. [23], obtaining zeaxanthin producing strains SyBE_Sc0123Cz10-18 (Table 1) as the host cell in our study.

For constructing the initial crocetin producing strain and screening CrtZ/CCD combination, only heterologous CCDs were carried by single copy plasmid pRS416 and introduced into zeaxanthin producing strains (Fig. 1b). Genes *ccd* were amplified from the plasmid pUC57-Simple-10–13 and assembled together with promoter P_{GAL10}, terminators T_{HIS5} and T_{TEF2} into CCD expression cassette T_{HIS5}-P_{GAL10}-CCD-T_{TEF2} by OE-PCR. The products were inserted into the *Not*I site of plasmid pRS416, obtaining pRS416-C serious plasmids (CCD expression plasmids pRS416-C-01-04, Table 1; Additional file 1: Figure S2). These plasmids were transferred into zeaxanthin producing strains according to Table 1, producing crocetin producing strains (Table 1).

For screening ALD sources, heterologous CCD and ALD were carried by centromeric plasmid pRS416 and introduced into zeaxanthin producing strain (Fig. 1c). Cassette T_{TEF2}-P_{GAL7}-T_{PGI1} was also assembled by OE-PCR and cloned into pRS425 K, obtaining the plasmid pRS425 K-A-01 at first (Table 1; Additional file 1: Figure S3). Genes *ald* were recovered by *Bsa*I digestion from pUC57-Simple-14–16 and inserted into the same site of pRS425 K-A-01, generating pRS425 K-A series plasmids (pRS425 K-A-02–04, Table 1; Additional file 1: Figure S3). Meanwhile, cassette T_{TEF2}-T_{PGI1} was assembled by OE-PCR. The product was incubated with *Xho*I/*Sac*I and inserted into the same sites of pRS416, producing pRS416-A-01. Then cassettes T_{HIS5}-P_{GAL10}-*ccd2*-T_{TEF2} (digested from pRS416-C-01 by *Not*I), T_{TEF2}-P_{GAL7}-*ald*-T_{PGI1} (digested from pRS425 K-A-02–04 by *Pst*I/*Bam*HI) and linearized vector pRS416-A-01 (digested by *Bam*HI) were assembled based on RADOM method in the particular zeaxanthin producing strain (producing strains SyBE_Sc0123C048–50 harboring plasmids pRS416-A-02–04 respectively, Table 1; Additional file 1: Figure S3) [24]. For adjusting the expression level of CCD and ALD, multiple plasmid pRS426, instead of pRS416, was employed to carry CCD and ALD expression cassettes. Similar procedures were taken as motioned above, which were presented in Additional file 1: Figure S3.

Strains and culture conditions

Escherichia coli DH5α or TransT1 was used for plasmid construction, which was cultured at 37 °C in Luria–Bertani medium [15] supplemented with 50 μg/mL kanamycin or 100 μg/mL ampicillin for selection. Meanwhile, all the engineered yeast strains summarized in Table 1 were based on an existing β-carotene producing strain, *S.*

Table 1 ***S. cerevisiae*** **strains and plasmids used in this study**

	Description	Source
Strain		
CEN.PK2-1C	*MATa, ura3-52, trp1-289, leu2-3,112, his3Δ1, MAL2-8C, SUC2*	EUROSCARF
SyBE_Sc0014CY06	CEN.PK2-1C, *Δgal1 Δgal7 Δgal10::HIS3, Δypl062w::KanMX, trp1::TRP1*_T_{CYC1}-BtCrtI-P_{GAL10}-P_{GAL1}-PaCrtB-T_{PGK1}, *leu2::LEU2*_T_{CYC1}-BtCrtI-P_{GAL7}-T_{ACT1}-*tHMG1*-P_{GAL10}-P_{GAL1}-TmCrtE-T_{GPM1}, *Δymrwdelta15::*$P_{UAS\text{-}GAL1}$-PaCrtY-T_{ADH1}, *Δynrcdelta9::* $P_{UAS\text{-}GAL1}$-PaCrtY-T_{ADH1}	This lab
SyBE_Sc0123Z001	SyBE_SC0014CY06, *Δho::*P_{GAL1}-Aa_*CrtZ*-T_{HIS5}-*URA3*	This study
SyBE_Sc0123Z002	SyBE_SC0014CY06, *Δho::*P_{GAL1}-As_*CrtZ*-T_{HIS5}-*URA3*	This study
SyBE_Sc0123Z003	SyBE_SC0014CY06, *Δho::*P_{GAL1}-Eu_*CrtZ*-T_{HIS5}-*URA3*	This study
SyBE_Sc0123Z004	SyBE_SC0014CY06, *Δho::*P_{GAL1}-Pa_*CrtZ*-T_{HIS5}-*URA3*	This study
SyBE_Sc0123Z005	SyBE_SC0014CY06, *Δho::*P_{GAL1}-Ps_*CrtZ*-T_{HIS5} -*URA3*	This study
SyBE_Sc0123Z006	SyBE_SC0014CY06, *Δho::*P_{GAL1}-Ss_*CrtZ*-T_{HIS5}-*URA3*	This study
SyBE_Sc0123Z007	SyBE_SC0014CY06, *Δho::*P_{GAL1}-B.SD_*CrtZ*-T_{HIS5}-*URA3*	This study
SyBE_Sc0123Z008	SyBE_SC0014CY06, *Δho::*P_{GAL1}-B.DC_*CrtZ*-T_{HIS5}-*URA3*	This study
SyBE_Sc0123Z009	SyBE_SC0014CY06, *Δho::*P_{GAL1}-Hp_*CrtZ*-T_{HIS5}-*URA3*	This study
SyBE_Sc0123Cz10	SyBE_SC0014CY06, *Δho::*P_{GAL1}-Aa_*CrtZ*-T_{HIS5}	This study
SyBE_Sc0123Cz11	SyBE_SC0014CY06, *Δho::*P_{GAL1}-As_*CrtZ*-T_{HIS5}	This study
SyBE_Sc0123Cz12	SyBE_SC0014CY06, *Δho::*P_{GAL1}-Eu_*CrtZ*-T_{HIS5}	This study
SyBE_Sc0123Cz13	SyBE_SC0014CY06, *Δho::*P_{GAL1}-Pa_*CrtZ*-T_{HIS5}	This study
SyBE_Sc0123Cz14	SyBE_SC0014CY06, *Δho::*P_{GAL1}-Ps_*CrtZ*-T_{HIS5}	This study
SyBE_Sc0123Cz15	SyBE_SC0014CY06, *Δho::*P_{GAL1}-Ss_*CrtZ*-T_{HIS5}	This study
SyBE_Sc0123Cz16	SyBE_SC0014CY06, *Δho::*P_{GAL1}-B.SD_*CrtZ*-T_{HIS5}	This study
SyBE_Sc0123Cz17	SyBE_SC0014CY06, *Δho::*P_{GAL1}-B.DC_*CrtZ*-T_{HIS5}	This study
SyBE_Sc0123Cz18	SyBE_SC0014CY06, *Δho::*P_{GAL1}-Hp_*CrtZ*-T_{HIS5}	This study
SyBE_Sc0123C001	SyBE_Sc0123Cz10 with pRS416-C-01 (pRS416-T_{HIS5}-P_{GAL10}-*CCD2*-T_{TEF2})	This study
SyBE_Sc0123C002	SyBE_Sc0123Cz10 with pRS416-C-02 (pRS416-T_{HIS5}-P_{GAL10}-*CCD3*-T_{TEF2})	This study
SyBE_Sc0123C003	SyBE_Sc0123Cz10 with pRS416-C-03 (pRS416-T_{HIS5}-P_{GAL10}-*ZCD*-T_{TEF2})	This study
SyBE_Sc0123C004	SyBE_Sc0123Cz10 with pRS416-C-04 (pRS416-T_{HIS5}-P_{GAL10}-*ZCD1*-T_{TEF2})	This study
SyBE_Sc0123C005	SyBE_Sc0123Cz11 with pRS416-C-01 (pRS416-T_{HIS5}-P_{GAL10}-*CCD2*-T_{TEF2})	This study
SyBE_Sc0123C006	SyBE_Sc0123Cz11 with pRS416-C-02 (pRS416-T_{HIS5}-P_{GAL10}-*CCD3*-T_{TEF2})	This study
SyBE_Sc0123C007	SyBE_Sc0123Cz11 with pRS416-C-03 (pRS416-T_{HIS5}-P_{GAL10}-*ZCD*-T_{TEF2})	This study
SyBE_Sc0123C008	SyBE_Sc0123Cz11 with pRS416-C-04 (pRS416-T_{HIS5}-P_{GAL10}-*ZCD1*-T_{TEF2})	This study
SyBE_Sc0123C009	SyBE_Sc0123Cz12 with pRS416-C-01 (pRS416-T_{HIS5}-P_{GAL10}-*CCD2*-T_{TEF2})	This study
SyBE_Sc0123C010	SyBE_Sc0123Cz12 with pRS416-C-02 (pRS416-T_{HIS5}-P_{GAL10}-*CCD3*-T_{TEF2})	This study
SyBE_Sc0123C011	SyBE_Sc0123Cz12 with pRS416-C-03 (pRS416-T_{HIS5}-P_{GAL10}-*ZCD*-T_{TEF2})	This study
SyBE_Sc0123C012	SyBE_Sc0123Cz12 with pRS416-C-04 (pRS416-T_{HIS5}-P_{GAL10}-*ZCD1*-T_{TEF2})	This study
SyBE_Sc0123C013	SyBE_Sc0123Cz13 with pRS416-C-01 (pRS416-T_{HIS5}-P_{GAL10}-*CCD2*-T_{TEF2})	This study
SyBE_Sc0123C014	SyBE_Sc0123Cz13 with pRS416-C-02 (pRS416-T_{HIS5}-P_{GAL10}-*CCD3*-T_{TEF2})	This study
SyBE_Sc0123C015	SyBE_Sc0123Cz13 with pRS416-C-03 (pRS416-T_{HIS5}-P_{GAL10}-*ZCD*-T_{TEF2})	This study
SyBE_Sc0123C016	SyBE_Sc0123Cz13 with pRS416-C-04 (pRS416-T_{HIS5}-P_{GAL10}-*ZCD1*-T_{TEF2})	This study
SyBE_Sc0123C017	SyBE_Sc0123Cz14 with pRS416-C-01 (pRS416-T_{HIS5}-P_{GAL10}-*CCD2*-T_{TEF2})	This study
SyBE_Sc0123C018	SyBE_Sc0123Cz14 with pRS416-C-02 (pRS416-T_{HIS5}-P_{GAL10}-*CCD3*-T_{TEF2})	This study
SyBE_Sc0123C019	SyBE_Sc0123Cz14 with pRS416-C-03 (pRS416-T_{HIS5}-P_{GAL10}-*ZCD*-T_{TEF2})	This study
SyBE_Sc0123C020	SyBE_Sc0123Cz14 with pRS416-C-04(pRS416-T_{HIS5}-P_{GAL10}-*ZCD1*-T_{TEF2})	This study
SyBE_Sc0123C021	SyBE_Sc0123Cz15 with pRS416-C-01 (pRS416-T_{HIS5}-P_{GAL10}-*CCD2*-T_{TEF2})	This study
SyBE_Sc0123C022	SyBE_Sc0123Cz15 with pRS416-C-02 (pRS416-T_{HIS5}-P_{GAL10}-*CCD3*-T_{TEF2})	This study
SyBE_Sc0123C023	SyBE_Sc0123Cz15 with pRS416-C-03 (pRS416-T_{HIS5}-P_{GAL10}-*ZCD*-T_{TEF2})	This study
SyBE_Sc0123C024	SyBE_Sc0123Cz15 with pRS416-C-04 (pRS416-T_{HIS5}-P_{GAL10}-*ZCD1*-T_{TEF2})	This study
SyBE_Sc0123C025	SyBE_Sc0123Cz16 with pRS416-C-01 (pRS416-T_{HIS5}-P_{GAL10}-*CCD2*-T_{TEF2})	This study
SyBE_Sc0123C026	SyBE_Sc0123Cz16 with pRS416-C-02 (pRS416-T_{HIS5}-P_{GAL10}-*CCD3*-T_{TEF2})	This study

Table 1 continued

	Description	Source
SyBE_Sc0123C027	SyBE_Sc0123Cz16 with pRS416-C-03 (pRS416-T_{HIS5}-P_{GAL10}-*ZCD*-T_{TEF2})	This study
SyBE_Sc0123C028	SyBE_Sc0123Cz16 with pRS416-C-04 (pRS416-T_{HIS5}-P_{GAL10}-*ZCD1*-T_{TEF2})	This study
SyBE_Sc0123C029	SyBE_Sc0123Cz17 with pRS416-C-01 (pRS416-T_{HIS5}-P_{GAL10}-*CCD2*-T_{TEF2})	This study
SyBE_Sc0123C030	SyBE_Sc0123Cz17 with pRS416-C-02 (pRS416-T_{HIS5}-P_{GAL10}-*CCD3*-T_{TEF2})	This study
SyBE_Sc0123C031	SyBE_Sc0123Cz17 with pRS416-C-03 (pRS416-T_{HIS5}-P_{GAL10}-*ZCD*-T_{TEF2})	This study
SyBE_Sc0123C032	SyBE_Sc0123Cz17 with pRS416-C-04 (pRS416-T_{HIS5}-P_{GAL10}-*ZCD1*-T_{TEF2})	This study
SyBE_Sc0123C033	SyBE_Sc0123Cz18 with pRS416-C-01 (pRS416-T_{HIS5}-P_{GAL10}-*CCD2*-T_{TEF2})	This study
SyBE_Sc0123C034	SyBE_Sc0123Cz18 with pRS416-C-02 (pRS416-T_{HIS5}-P_{GAL10}-*CCD3*-T_{TEF2})	This study
SyBE_Sc0123C035	SyBE_Sc0123Cz18 with pRS416-C-03 (pRS416-T_{HIS5}-P_{GAL10}-*ZCD*-T_{TEF2})	This study
SyBE_Sc0123C036	SyBE_Sc0123Cz18 with pRS416-C-04 (pRS416-T_{HIS5}-P_{GAL10}-*ZCD1*-T_{TEF2})	This study
SyBE_Sc0123C048	SyBE_Sc0123Cz14 with pRS416-A-02 (pRS416-T_{HIS5}-P_{GAL10}-*CCD2*-T_{TEF2}-P_{GAL7}-*ALD6*-T_{PGI1})	This study
SyBE_Sc0123C049	SyBE_Sc0123Cz14 with pRS416-A-03 (pRS416-T_{HIS5}-P_{GAL10}-*CCD2*-T_{TEF2}-P_{GAL7}- Bix_*ALD*-T_{PGI1})	This study
SyBE_Sc0123C050	SyBE_Sc0123Cz14 with pRS416-A-04 (pRS416-T_{HIS5}-P_{GAL10}-*CCD2*-T_{TEF2}-P_{GAL7}- Syn_*ALD*-T_{PGI1})	This study
SyBE_Sc0123C053	SyBE_Sc0123Cz14 with pRS426-A-02 (pRS426-T_{HIS5}-P_{GAL10}-*CCD2*-T_{TEF2}-P_{GAL7}- Syn_*ALD*-T_{PGI1})	This study
Plasmid		
pJET1.2	Blunt Cloning vector, resistant to ampicillin	Thermo scientific
pUC57-Simple	Blunt Cloning vector, resistant to ampicillin	GenScript
pRS416	Single copy plasmid in *S.cerevisiae* with *URA3* and Amp^r marker	This Lab
pRS426	Multiple copy plasmid in *S.cerevisiae* with *URA3* and Amp^r marker	This Lab
pRS425 K	Multiple copy plasmid in *S.cerevisiae* with *LEU2* and KanMX marker	This Lab
pUC57-Simple-01	*CrtZ* from *Agrobacterium aurantiacum* (Aa_*CrtZ*) was codon optimized, synthesized and cloned into pUC57-Simple	This study
pUC57-Simple-02	*CrtZ* from *Alcaligenes* sp. PC-1 (As_*CrtZ*) was codon optimized, synthesized and cloned into pUC57-Simple	This study
pUC57-Simple-03	*CrtZ* from *Erwinia uredovora* (Eu_*CrtZ*) was codon optimized, synthesized and cloned into pUC57-Simple	This study
pUC57-Simple-04	*CrtZ* from *Pantoea agglomerans* (Pa_*CrtZ*) was codon optimized, synthesized and cloned into pUC57-Simple	This study
pUC57-Simple-05	*CrtZ* from *Pantoea stewartii* (Ps_*CrtZ*) was codon optimized, synthesized and cloned into pUC57-Simple	This study
pUC57-Simple-06	*CrtZ* from *Sulfolobus solfataricus* P2 (Ss_*CrtZ*) was codon optimized, synthesized and cloned into pUC57-Simple	This study
pUC57-Simple-07	*CrtZ* from *Brevundimonas* sp. SD212 (B.SD_*CrtZ*) was codon optimized, synthesized and cloned into pUC57-Simple	This study
pUC57-Simple-08	*CrtZ* from *Brevundimonas vesicularis* DC263 (B.DC_*CrtZ*) was codon optimized, synthesized and cloned into pUC57-Simple	This study
pUC57-Simple-09	*CrtZ* from *Haematococcus pluvialis* (Hp_*CrtZ*) was codon optimized, synthesized and cloned into pUC57-Simple	This study
pUC57-Simple-10	*CCD2* from *Crocus* was codon optimized, synthesized and cloned into pUC57-Simple	This study
pUC57-Simple-11	*CCD3* from *Crocus* was codon optimized, synthesized and cloned into pUC57-Simple	This study
pUC57-Simple-12	*ZCD* from *Crocus* was codon optimized, synthesized and cloned into pUC57-Simple	This study
pUC57-Simple-13	*ZCD1* from *Crocus* was codon optimized, synthesized and cloned into pUC57-Simple	This study
pUC57-Simple-14	*ALD6* from *S. cerevisiae* was cloned into pUC57-Simple	This study
pUC57-Simple-15	*ALD* from *Bixa orellana* (Bix_*ALD*) was codon optimized, synthesized and cloned into pUC57-Simple	This study
pUC57-Simple-16	*ALD* from *Synechocystis* sp. PCC6803 (Syn_*ALD*) was codon optimized, synthesized and cloned into pUC57-Simple	This study
pJET1.2-Z-01	The cassette *ho*_F-P_{GAL1}-T_{HIS5}-*URA3*-*ho*_R was cloned and inserted into the pJET1.2	This study
pJET1.2-Z-02	Aa_*CrtZ* was digested from pUC57-Simple-01 by *BsaI* and inserted into the same site of pJET1.2-Z-01	This study
pJET1.2-Z-03	As_*CrtZ* was digested from pUC57-Simple-02 by *BsaI* and inserted into the same site of pJET1.2-Z-01	This study
pJET1.2-Z-04	Eu_*CrtZ* was digested from pUC57-Simple-03 by *BsaI* and inserted into the same site of pJET1.2-Z-01	This study
pJET1.2-Z-05	Pa_*CrtZ* was digested from pUC57-Simple-04 by *BsaI* and inserted into the same site of pJET1.2-Z-01	This study
pJET1.2-Z-06	Ps_*CrtZ* was digested from pUC57-Simple-05 by *BsaI* and inserted into the same site of pJET1.2-Z-01	This study
pJET1.2-Z-07	Ss_*CrtZ* was digested from pUC57-Simple-06 by *BsaI* and inserted into the Same site of pJET1.2-Z-01	This study
pJET1.2-Z-08	B.SD_*CrtZ* was digested from pUC57-Simple-07 by *BsaI* and inserted into the Same site of pJET1.2-Z-01	This study

Table 1 continued

	Description	Source
pJET1.2-Z-09	B.DC_*CrtZ* was digested from pUC57-Simple-08 by *BsaI* and inserted into the Same site of pJET1.2-Z-01	This study
pJET1.2-Z-10	Hp_*CrtZ* was digested from pUC57-Simple-09 by *BsaI* and inserted into the Same site of pJET1.2-Z-01	This study
pRS416-C-01	The cassette T_{HIS5}-P_{GAL10}-*CCD2*-T_{TEF2} was cloned and inserted into the *NotI* site of pRS416	This study
pRS416-C-02	The cassette T_{HIS5}-P_{GAL10}-*CCD3*-T_{TEF2} was cloned and inserted into the *NotI* site of pRS416	This study
pRS416-C-03	The cassette T_{HIS5}-P_{GAL10}-*ZCD*-T_{TEF2} was cloned and inserted into the *NotI* site of pRS416	This study
pRS416-C-04	The cassette T_{HIS5}-P_{GAL10}-*ZCD1*-T_{TEF2} was cloned and inserted into the *NotI* site of pRS416	This study
pRS425 K-A-01	The cassette T_{TEF2}-P_{GAL7}-T_{PGI1} was cloned and inserted into the *PstI*/*Bam*HI site of pRS425 K	This study
pRS425 K-A-02	*ALD6* was digested from pUC57-Simple-14 by *BsaI* and inserted into the same site of pRS425 K-A-01	This study
pRS425 K-A-03	Bix_*ALD* was digested from pUC57-Simple-15 by *BsaI* and inserted into the same site of pRS425 K-A-01	This study
pRS425 K-A-04	Syn_*ALD* was digested from pUC57-Simple-16 by *BsaI* and inserted into the same site of pRS425 K-A-01	This study
pRS416-A-01	The cassette T_{HIS5}–T_{PGI1} was cloned and inserted into the *XhoI*/*SacI* site of pRS416	This study
pRS416-A-02	The cassette T_{HIS5}-P_{GAL10}-*CCD2*-T_{TEF2} (digested from pRS416-C-01 by *NotI*), the cassette T_{TEF2}-P_{GAL7}-*ALD6*-T_{PGI1} (digested from pRS425 K-A-02 by *PstI*/*Bam*HI) and plasmid pRS416-A-01 (digested by *Bam*HI) were assembled based on RADOM method	This study
pRS416-A-03	The cassette T_{HIS5}-P_{GAL10}-*CCD2*-T_{TEF2} (digested from pRS416-C-01 by *NotI*), the cassette T_{TEF2}-P_{GAL7}-Bix_*ALD*-T_{PGI1} (digested from pRS425 K-A-03 by *PstI*/*Bam*HI) and plasmid pRS416-A-01 (digested by *Bam*HI) were assembled based on RADOM method	This study
pRS416-A-04	The cassette T_{HIS5}-P_{GAL10}-*CCD2*-T_{TEF2} (digested from pRS416-C-01 by *NotI*), the cassette T_{TEF2}-P_{GAL7}-Syn_*ALD*-T_{PGI1} (digested from pRS425 K-A-04 by *PstI*/*Bam*HI) and plasmid pRS416-A-01 (digested by *Bam*HI) were assembled based on RADOM method	This study
pRS426-A-01	The cassette T_{HIS5}–T_{PGI1} was cloned and inserted into the *XhoI*/*SacI* site of pRS426	This study
pRS426-A-02	The cassette T_{HIS5}-P_{GAL10}-*CCD2*-T_{TEF2} (digested from pRS416-C-01 by *NotI*), the cassette T_{TEF2}-P_{GAL7}-Syn_*ALD*-T_{PGI1} (digested from pRS425 K-A-04 by *PstI*/*Bam*HI) and plasmid pRS426-A-01 (digested by *Bam*HI) were assembled based on RADOM method	This study

cerevisiae SyBE_SC0014CY06. Engineered yeast strains were cultured on YPD medium or synthetic complete (SC) medium lacking appropriate nutrient component for selection [25]. When needed, 1% (w/v) D-(+)-galactose were used as the inducer in fermentations and supplied into YPD medium (generating YPDG medium).

For shake-flask cultivation, colonies on solid plates were picked up and cultured in 3 mL SC medium for overnight growth at 30 °C. Then the preculture was transferred into 25 mL fresh SC medium and grew until reaching to mid-log phase. After that, the seed culture was inoculated into 50 mL YPD medium with an initial OD600 of 0.1 and cultivated at 30 °C for 72 h or 20 °C for 96 h. All the fermentation experiments were performed in triplicate.

Fed-batch fermentation

The strain SyBE_Sc0123C053 was used for fed-batch fermentation. 100 μL glycerol-stock was inoculated into 25 mL SC medium and cultured at 30 °C, 250 rpm for overnight growth. Then the preculture was transferred to 200 mL fresh SC medium and grew until entering mid-exponential phase. Seed cultures were transferred to 1.8 L YPD medium (20 g/L glucose) in a 5 L bioreactor (BLBIO-5GJG-2, Shanghai, China) at a 10% (v/v) inoculum. The pH was automatically controlled at 5.5 with ammonia hydroxide (6 M). And the dissolved oxygen was kept at 40% by agitation cascade from 400 to 600 rpm, while the air flow was set at 2.5 vvm.

As the crocetin production modules were controlled by employed galactose-inducible system, the fed-batch fermentation should be divided into two stages: cell growth stage and crocetin accumulation stage. During the period of the cell growth stage, fermentation was carried out at 30 °C. The glucose concentration was monitored every 2 h and the glucose consumption rate was obtained accordingly. Based on this data, the glucose concentration was maintained less than 1 g/L by adding an appropriate volume of concentrated glucose solution (500 g/L) continuously. And 5 g yeast extract was added into the bioreactor every 12 h by feeding 400 g/L yeast extract stock solution. When the cell growth fell into stable phase, fermentation entered the second stage: crocetin accumulation stage. Then after fermentation temperature reduced to 20 °C, 10 g/L of D-(+)-galactose was fed to induce crocetin biosynthesis. As glucose was exhausted, cells begun to use ethanol as carbon source. The ethanol concentration was controlled below 5 g/L through adjusting the feeding rate of ethanol until harvest. Duplicate samples were collected to determine the cell density, glucose concentration, ethanol concentration and crocetin production. To avoid the spontaneous degradation from light, bioreactor should be covered with foils.

Extraction and analysis of carotenoids

To determine carotenoids accumulation, standards of lycopene, β-carotene and zeaxanthin were purchased from Sigma (Sigma-Aldrich, MO, USA), and standard of crocetin was purchased from Yuanye Bio-Technology (Shanghai, China). The procedures for extracting and analyzing carotenoids were modified according to Xie et al. [20]. To be specific, after harvested cells were washed with distilled water, the cell pellet was re-suspended in 3 N HCl and boiled for 2 min, and then immediately cooled in ice for 3 min. Then cells debris were harvested and resuspended in acetone containing 1% (w/v) butylated hydroxytoluene. The above mixture was vortexed until colorless. After centrifugation, the acetone phase containing the extracted carotenoid was collected and evaporated by nitrogen blow. The products were analyzed by high-performance liquid chromatography system (HPLC, Waterse2695, Waters Corp, USA) equipped with a BDS HYPERSIL C18 column (150 mm × 4.6 mm, 5 μm, Thermo Scientific) and a UV/VIS detector (Waters 2489). To characterize lycopene, β-carotene and zeaxanthin, the product was dissolved in acetone and the signals were detected at 450 nm. The mobile phase consisting of acetonitrile-methanol (65:35 v/v) was chosen with a flow rate of 0.8 mL/min and the column temperature was set at 25 °C. In the meanwhile, for crocetin analysis, sample was dissolved in methanol-dimethylformamide (7:1 v/v) and crocetin was detected at 430 nm. 70% (v/v) methanol–water (containing 2% formic acid) was utilized as the mobile phase with a flow rate of 1 mL/min at 40 °C. Notably, considering that carotenoids are extremely unstable and susceptible to light, brown centrifugal tubes were used in the above procedures to avoid exposure to light.

Bioinformatics and structural analysis of CCD

The protein identified sequences of the target CCD from different taxa were queried from protein knowledgebase (UniProtKB) available at http://www.uniprot.org/, using the key term "carotenoid cleavage dioxygenase", and subjected to a brief bioinformatics analysis to guarantee suitable diversity. Initially the CCD protein sequences were aligned by means of clustal W with default settings [26]. Phylogenetic tree of CCD gene family was conducted in MEGA7 [27] and inferred by Neighbor-Joining method [28]. The bootstrap consensus tree deduced from 1000 replicates was taken to represent the evolutionary history of the taxa analyzed [29].

The structures of the CCD2 and CCD3 were both modeled based on the target-template (PDB ID: 2biw) alignment using SWISS-MODEL [30, 31]. And the Coordinates which are conserved between the targets and the template are copied from the template to the model. Insertions and deletions are remodeled using a fragment library. Side chains are then rebuilt. Finally, the geometry of the resulting model is regularized by using a force field. The modeled structures of target proteins were resolved with PyMol software [32].

Results and discussion

Construction of inducible crocetin biosynthesis pathway

To realize crocetin biosynthesis, heterologous *crtZ* and *ccd* were codon optimized and introduced into an existing β-carotene producer (*S. cerevisiae* SyBE_SC0014CY06), which processed endogenous ALDs to catalyze the final step in crocetin synthesis pathway (Fig. 1a) [9]. At first, *crtZ* was integrated into the *ho* locus of the chromosome, while *ccd* was carried by centromeric plasmid pRS416. The expression of CrtZ and CCD were under the control of galactose-regulated GAL promoters GAL1 and GAL10, respectively (Fig. 1b). Because a highest zeaxanthin production was once achieved in yeast strain harboring CrtZ from *Erwinia uredovora* (Eu_CrtZ) among nine selected CrtZ species [33], Eu_CrtZ were also selected and intergraded into the chromosome of strain SyBE_SC0014CY06, generating strain SyBE_Sc0123Cz12 as a host cell in our study. In the meanwhile, CCD2 from *Crocus* was also selected to convert zeaxanthin to crocetin dialdehyde, obtaining strain SyBE_Sc0123C009. Strains SyBE_Sc0123C009 and SyBE_Sc0123Cz12 together with the parent strain SyBE_SC0014CY06 were cultured in shake-flask with YPDG medium at 30 °C and their products were analyzed by HPLC after 72 h incubation. As shown in Fig. 1d, crocetin (peak III) was successfully detected with a titer as 9.42 μg/L in strain SyBE_Sc0123C009, indicating that a functional crocetin biosynthesis pathway succeeded here. To be notably, there was no distinct β-carotene accumulation in zeaxanthin producing strain SyBE_Sc0123Cz12, while an amount of β-carotene (peak II), zeaxanthin (peak I), as well as other unidentified byproducts or intermediates were observed in crocetin producing strain SyBE_Sc0123C009 (Fig. 1e), suggesting that the step catalyzed by CCD was rate-limiting here and the selected CrtZ/CCD combination did not match well, which needed to be optimized further.

Optimization of cultivation temperature

It is reported by Shi et al. [34] that low temperature was benefit for carotenoids accumulation in *Phaffia rhodozyma*. In our study, by cultivation of series zeaxanthin producing strains at 20 and 30 °C respectively, it was also found that the production of zeaxanthin was higher at 20 °C than that at 30 °C (Additional file 1: Figure S4), indicating lower temperature benefited much for zeaxanthin production, which would provide more sufficient precursor supplies for higher crocetin production. Moreover, concerning that root development and flower emergence occur at low temperature for *Crocus* plants, and the expression of CCD were induced by low temperature in *Crocus* [35–37], the effect of culture temperature was also investigated here. Thus, for higher crocetin titer, the culture temperature for strain SyBE_Sc0123C009 was decreased from 30 °C, via 25 to 20 °C. The cell density, zeaxanthin accumulation and crocetin production were measured during the time course. As a result, in case of cell growth, there was a longer lag phase under lower temperature, compared to cultivating at 30 °C (Fig. 2a). Meanwhile, a dramatical increase on crocetin production along with a decrease on zeaxanthin accumulation was achieved by reducing cultivation temperature (Fig. 2b, c), suggesting 20 °C was the optimal temperature for converting zeaxanthin to crocetin. Javiera López et al. [38] once reported that β-ionone producing yeast strain processing CCD1, the homologue of CCD2, worked much better at low temperature, which showed similar results as our study. Finally, the crocetin titer reached 223 μg/L at 20 °C after 96 h fermentation in shake-flask (Fig. 2b). And 20 °C was used as the culture temperature in further study.

Optimal CrtZ/CCD combination by screening enzymes from diverse sources

As mentioned above, combinatorially screening enzymes from diverse sources has been proved to be a promising method to obtain the best combination in terms of substrate selectivity, catalytic activity and host cell compatibility, which would lead to higher productivity of the target compound [39–42]. Through blastp searching through NCBI database (https://blast.ncbi.nlm.nih.gov/Blast.cgi?PROGRAM=blastp&PAGE_TYPE=Blast Search&LINK_LOC = blasthome based on the sequence of CCD2, CCD3 showed a 97% identity with CCD2 (Fig. 4a). Hence besides three crocetin synthesis related CCDs (ZCD, ZCD1 and CCD2) described before [10, 17, 18], CCD3 was selected as potential candidate in our study. Here, these four CCDs together with nine CrtZs from *E. uredovora* (Eu_CrtZ), *Pantoea agglomerans* (Pa_CrtZ), *Sulfolobus solfataricus* P2 (Ss_CrtZ), *Pantoea stewartii* (Ps_CrtZ), *Brevundimonas* sp. SD212 (B.SD_CrtZ), *Brevundimonas vesicularis* DC263 (B.DC_CrtZ), *Haematococcus pluvialis* (Hp_CrtZ), *Agrobacterium aurantiacum* (Aa_CrtZ), *Alcaligenes* sp. PC-1 (As_CrtZ) were expressed in strain SyBE_SC0014CY06, generating 36 strains with diverse CrtZ/CCD combinations (Fig. 3a; Table 1). Nine strains carrying different CrtZs without CCDs introduced were used as the blank control (Fig. 3b, c; Table 1). All the above strains were cultured in YPDG medium to analyze the accumulation of zeaxanthin and crocetin. As illustrated in Fig. 3b, only the strain harboring CCD2 instead of other three CCDs could achieve crocetin accumulation in yeast, furtherly demonstrating CCD was a rate-limiting enzyme in crocetin synthesis pathway. Rather than CrtZ, CCD seemed to be more crucial for crocetin production. Finally, the combination as Ps_CrtZ/CCD2 achieved the highest crocetin titer as 351 μg/L in strain SyBE_Sc0123C017. This optimal combination would be a promising candidate for further optimization.

In this study, ZCD, ZCD1 and CCD3 could not achieve crocetin production in yeast, which required sequential cleavage at C7–C8 and C7′–C8′ double bonds adjacent to the 3-OH-β-ionone ring [43]. Even though there was no crocetin detected in strains carrying these three enzymes separately, zeaxanthin accumulations were consumed at varying degrees in these strains, suggesting their cleave activities in yeast might at other position or only at one side of the molecules. Among the five subfamilies of plant

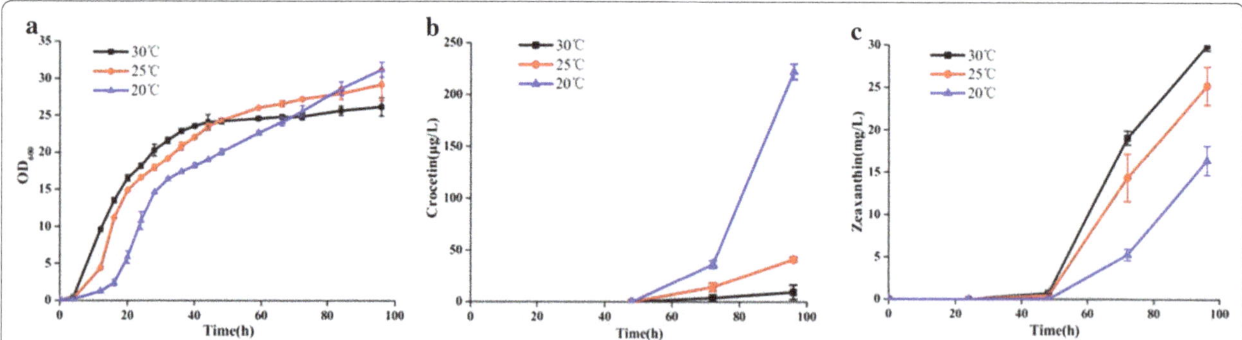

Fig. 2 The effects of culture temperature on cell growth (**a**), crocetin production (**b**), and zeaxanthin accumulation (**c**). *S. cerevisiae* strain SyBE_Sc0123C009 was cultivated in YPDG media under different cultivation temperature (30 °C shown in *squares*, 25 °C in *circles* and 20 °C in *triangles*), respectively, in shake-flasks for analysis by HPLC. The *error bars* represent standard deviation calculated from triplicate experiments

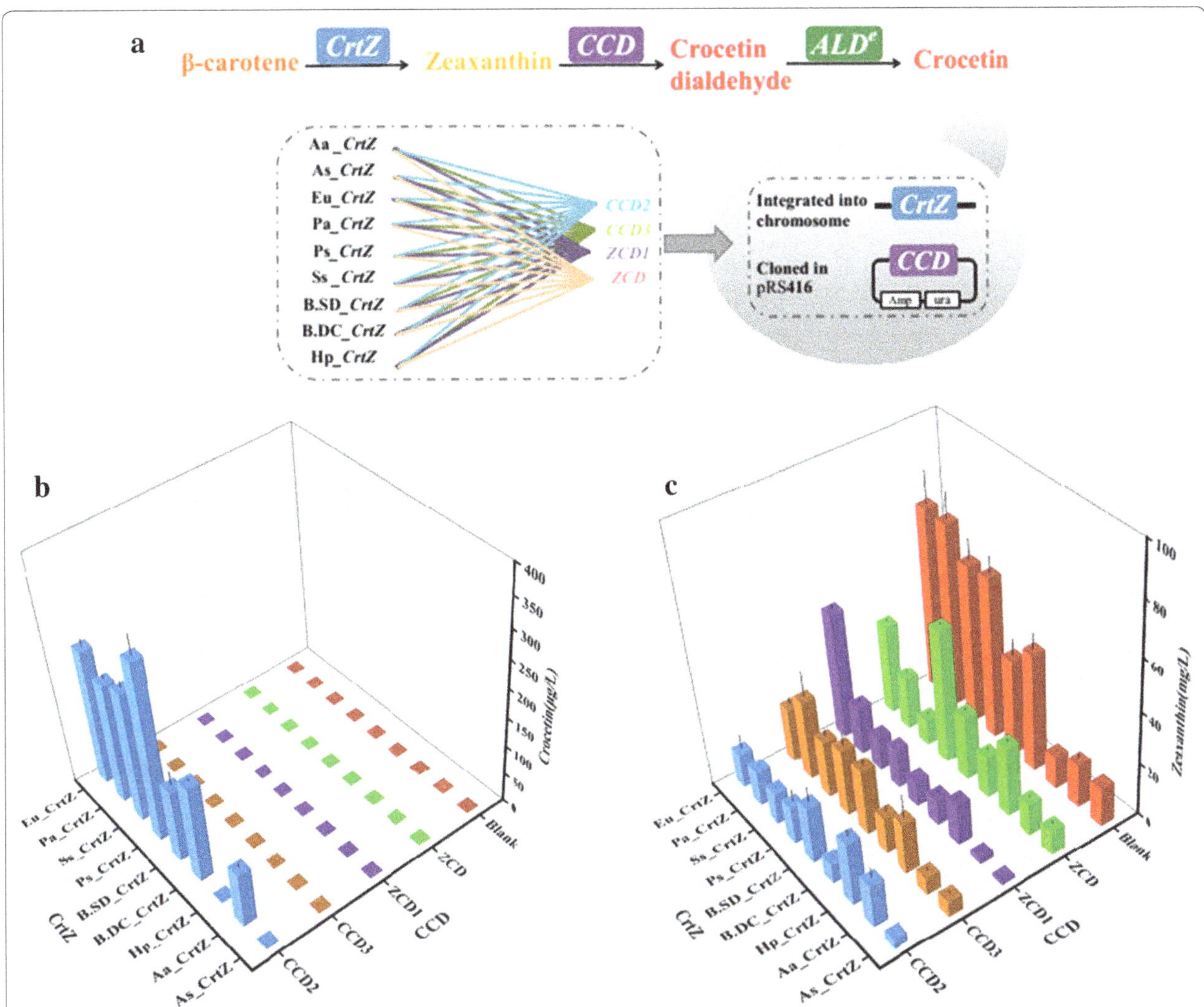

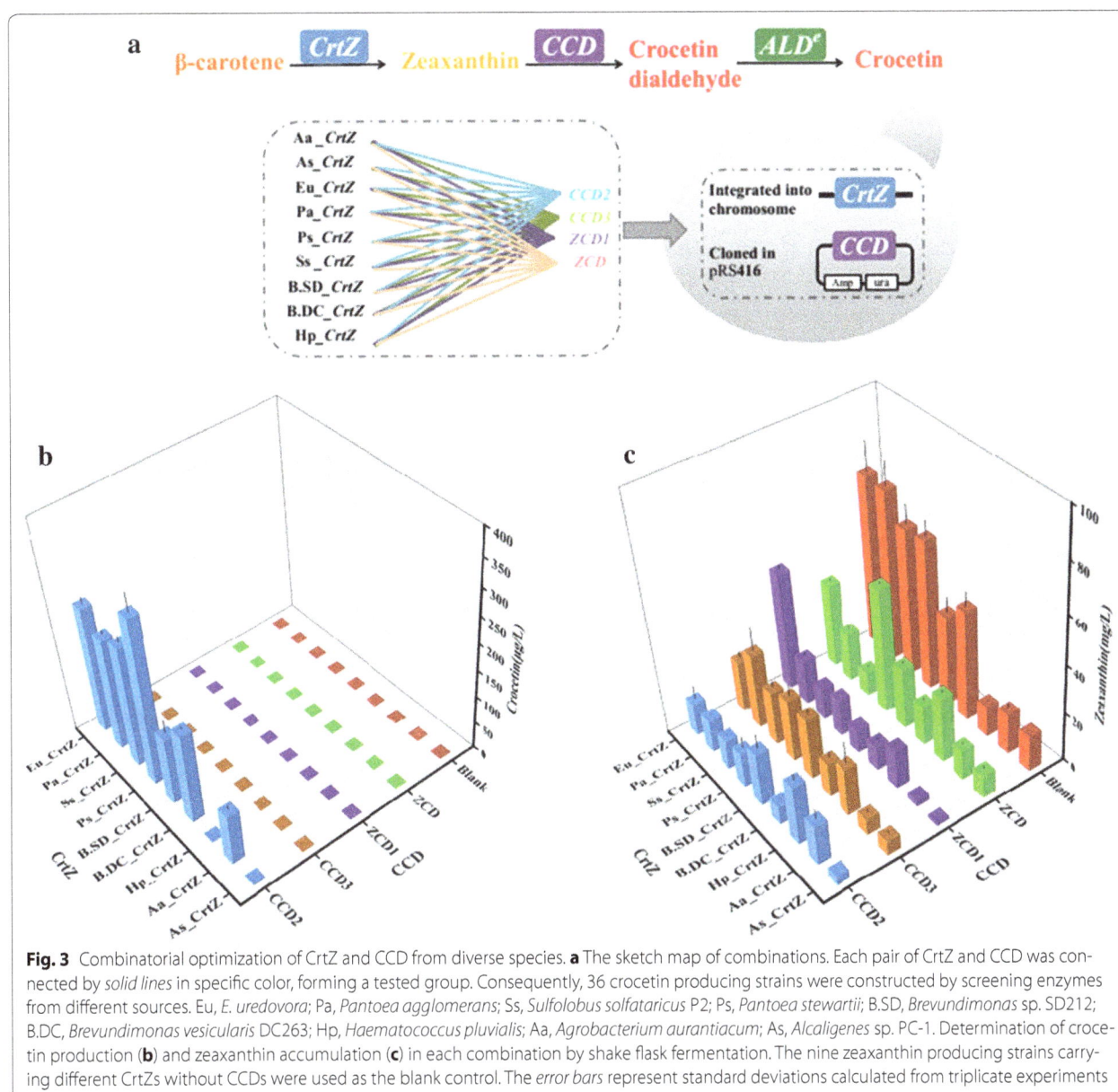

Fig. 3 Combinatorial optimization of CrtZ and CCD from diverse species. **a** The sketch map of combinations. Each pair of CrtZ and CCD was connected by *solid lines* in specific color, forming a tested group. Consequently, 36 crocetin producing strains were constructed by screening enzymes from different sources. Eu, *E. uredovora*; Pa, *Pantoea agglomerans*; Ss, *Sulfolobus solfataricus* P2; Ps, *Pantoea stewartii*; B.SD, *Brevundimonas* sp. SD212; B.DC, *Brevundimonas vesicularis* DC263; Hp, *Haematococcus pluvialis*; Aa, *Agrobacterium aurantiacum*; As, *Alcaligenes* sp. PC-1. Determination of crocetin production (**b**) and zeaxanthin accumulation (**c**) in each combination by shake flask fermentation. The nine zeaxanthin producing strains carrying different CrtZs without CCDs were used as the blank control. The *error bars* represent standard deviations calculated from triplicate experiments

CCDs, the CCD1 and CCD4 families were the only two involved in the cleave activities at 7, 8/7′, 8′ positions [44, 45]. A phylogenetic analysis of CCD sequences from diversity sources belonging to CCD1 and CCD4 families, illustrated that CCD2 and CCD3 belonged to CCD1 subfamily, while ZCD and ZCD1 were members of CCD4 subfamily (Additional file 1: Figure S5).

For CCD2 and CCD3, they shared 97% identities and exhibited dramatically diversity on enzyme activities. Through alignment of their protein sequences, seven dissimilar short fragments were detected (Fig. 4a). In order to further characterize these differences, the structural models of CCD2 and CCD3 were generated based on the crystal structure of their homological protein apocarotenoid cleavage oxygenase from *Synechocystis* (PDB accession ID: 2biw). As shown in Fig. 4b, CCD comprised seven bladed β-propellers, which is highly conserved among all CCDs and covered by a less rigid dome formed by a series of loops [46]. To be notably, there is a tunnel perpendicular to the propeller axis of CCD. As reported, the tunnel acted as a channel for the passage of their hydrophobic substrates to the active site, and was consisted of hydrophobic residues (mainly Phe, Val, Leu) interacting with their lipophilic substrates via hydrophobic forces to guarantee both the specificity and correct orientation of substrate for the cleavage reactions [47]. Thus, when

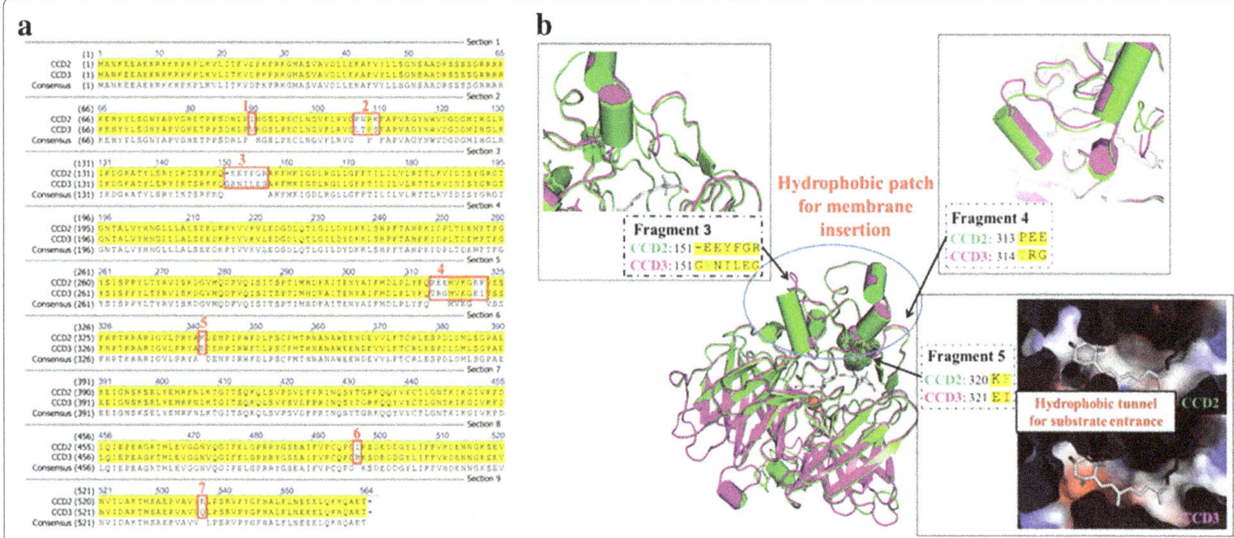

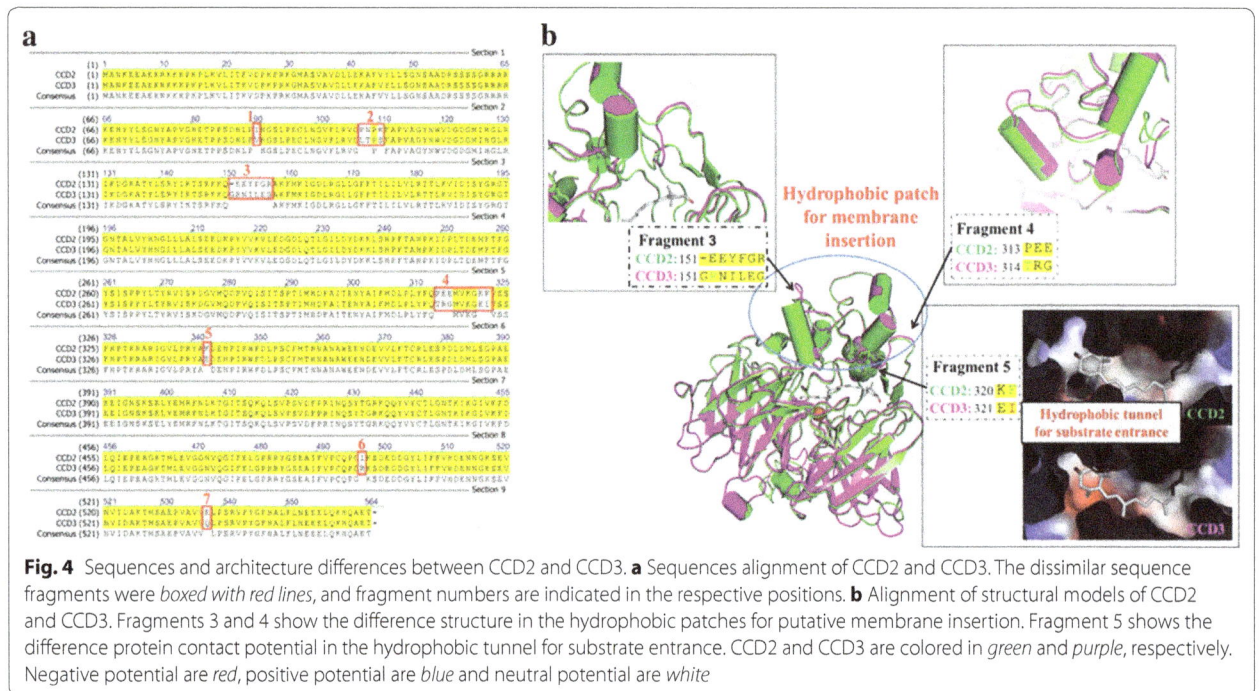

Fig. 4 Sequences and architecture differences between CCD2 and CCD3. **a** Sequences alignment of CCD2 and CCD3. The dissimilar sequence fragments were *boxed with red lines*, and fragment numbers are indicated in the respective positions. **b** Alignment of structural models of CCD2 and CCD3. Fragments 3 and 4 show the difference structure in the hydrophobic patches for putative membrane insertion. Fragment 5 shows the difference protein contact potential in the hydrophobic tunnel for substrate entrance. CCD2 and CCD3 are colored in *green* and *purple*, respectively. Negative potential are *red*, positive potential are *blue* and neutral potential are *white*

the high hydrophobicity of the tunnel was subsided by the alteration in Fragment 5 (which located at the tunnel) as the residues of K320-F321 from CCD2 and E321-I322 from CCD3 (Fig. 4b), the substrate entrance to CCD3 was impacted for change on substrate specificity consequently. Meanwhile, the entrance of the tunnel located in a large hydrophobic patch for membrane insertion, which provided an appropriate environment for lipophilic substrates accommodation and enzyme contraction. The function of this hydrophobic patch mainly depended on the stable α-helices region, which was involved in the Fragment 3 and Fragment 4. The structure of CCD3 in Fragment 3 and Fragment 4 showed the longer and more unstable loops than that of CCD2 (Fig. 4b). As illustrated in Fig. 3c, besides substrates selectivity, CCD3 exhibited lower cleavage activity on zeaxanthin than CCD2, no matter cooperated with what kind of CrtZ sources. These results could be explained by above descriptions. Moreover, there were still some variances between CCD3 and CCD2 which could not support above results by current protein model. Therefore, a more delicate phylogenetic analysis of CCD sequences only from CCD1 family members were performed and showed that those unexplained different residues were highly conserved among all the tested CCD1 subfamily members except CCD3 (Additional file 1: Figure S7), suggesting the alternation on these conserved regions which might be essential to CCD function would reduce enzyme activities.

For ZCD and ZCD1, they share 96% identities (Additional file 1: Figure S6), and both truncated at the N-terminal as lacking a blade of β-propeller and part of the dome in classic CCD4 subfamily members. The truncation was once proved to lead to loss on any cleavage activity for ZCD in *E. coli* [18]. ZCD1 was reported to once achieve crocetin production in *C. vulgaris* [10]. However, in our study, both these two enzymes could not sequentially cleave zeaxanthin on 7, 8/(7′, 8′) positions in yeast (Fig. 3b). These conflicting dates highlight the importance of host cell compatibility on the performance of heterologous enzymes, which were also corroborated by the reports from Greene et al. [48].

Screening ALD sources and fine-tuning of CCD/ALD

As so far, there is no ALD has been identified in *Crocus* for crocetin synthesis. Meanwhile, except endogenous ALDs in yeast (such as ALD6) and algae, none heterologous ALD has been reported yet to realize crocetin producing. Since the current crocetin titer, which was achieved by yeast endogenous ALDs, was still low, it is urgent to search and screen ALD isozymes from other organisms for higher crocetin production. Here, besides yeast endogenous ALD6 [49], two heterologous ALD originated from *Bixa orellana* (Bix_ALD) [50] and *Synechocystis* sp. PCC6803 (Syn_ALD) [51], whose substrates share the similar structure with crocetin dialdehyde, were selected and introduced together with CCD2 into the strain with Ps_CrtZ integrated in its chromosome (Fig. 5a). CCD2 and ALD were carried by single copy plasmid pRS416 and placed under the control of promoters GAL10 and GAL7, respectively (Fig. 1c). After growing in YPDG medium for 96 h, strain SyBE_Sc0123C050 harboring Syn_ALD achieved higher crocetin titer as 633 μg/L (Fig. 5b). Moreover, by increasing copy numbers of CCD2 and Syn_ALD via interchange of vector pRS416 into multicopy plasmid pRS426, the crocetin titer further improved to 1219 μg/L (Fig. 5b), obtaining strain SyBE_Sc0123C053 for bioreactor experiment.

Optimization of crocetin production in bioreactor

To evaluate the production performance of the engineered strain SyBE_Sc0123C053, fed-batch fermentation was performed at a 2 L scale using YPD as the medium (Fig. 6). During cell growth stage, based on carbon restriction strategy, glucose concentration was strictly restricted. Cell density reached 96 for 35 h cultivation at 30 °C. There was also no acetate observed in this stage (data was not shown). When the culture temperature reduced to 20 °C at 36 h, D-(+)-galactose was added to induce crocetin production. After the initial ethanol

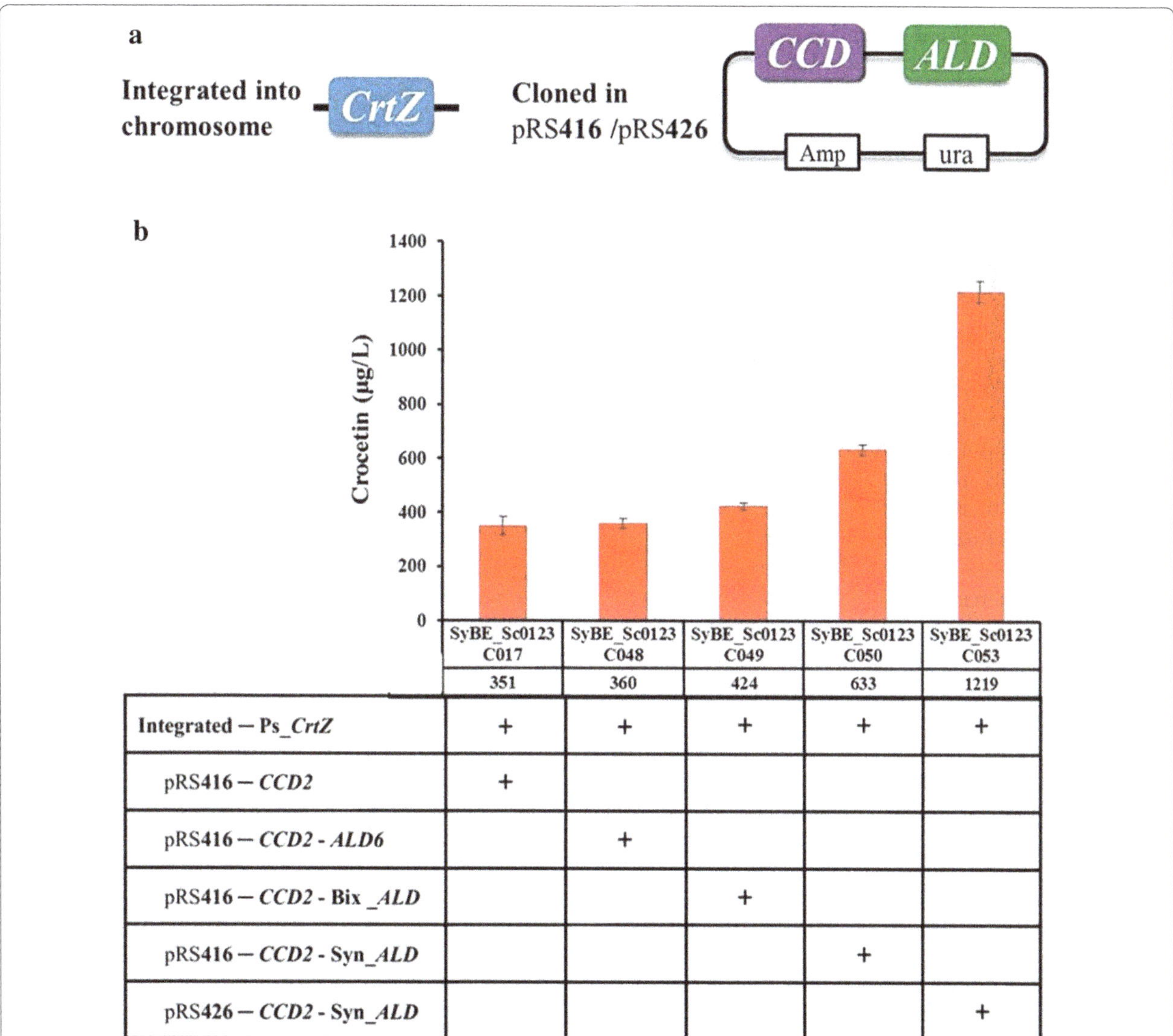

	SyBE_Sc0123 C017	SyBE_Sc0123 C048	SyBE_Sc0123 C049	SyBE_Sc0123 C050	SyBE_Sc0123 C053
	351	360	424	633	1219
Integrated — Ps_*CrtZ*	+	+	+	+	+
pRS416 — *CCD2*	+				
pRS416 — *CCD2* - *ALD6*		+			
pRS416 — *CCD2* - Bix _*ALD*			+		
pRS416 — *CCD2* - Syn_*ALD*				+	
pRS426 — *CCD2* - Syn_*ALD*					+

Fig. 5 Screening ALD sources and fine-tuning of CCD/ALD. **a** The diagrammatic sketch representation of the engineering strategies for CrtZ, CCD and ALD expression modular. CrtZ expression cassette was integrated into the *ho* locus of the chromosome, CCD2 and ALD were carried in plasmid pRS416/pRS426. **b** The effect of ALD sources and the expression level of CCD/ALD on crocetin production. Ps, *Pantoea stewartii*; Bix, *Bixa orellana*; Syn, *Synechocystis* sp. PCC6803. The *error bars* represent standard deviation calculated from triplicate experiments

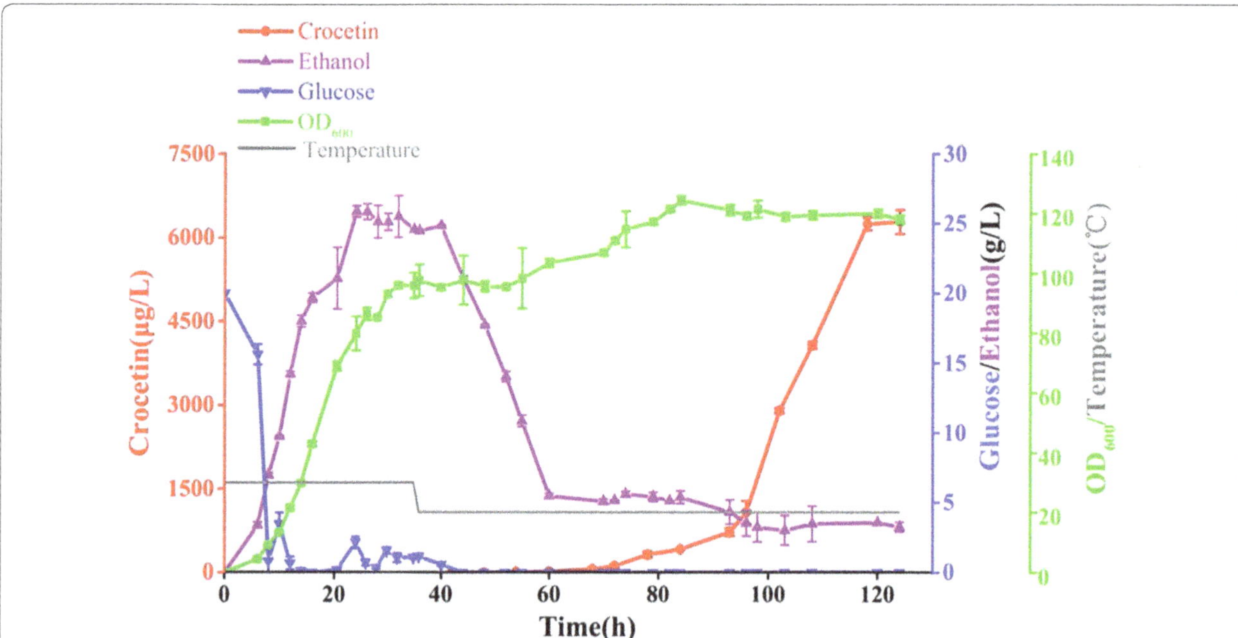

Fig. 6 Crocetin production in fed-batch fermentation. Profile of crocetin production (*red line*), glucose (*blue line*), ethanol (*purple line*), cell density (*green line*) and fermentation temperature (*gray line*) in strain SyBE_Sc0123C053 during fed-batch fermentation. The *error bars* represent standard deviation calculated from duplicate experiments

generated by glucose was consumed below 5 g/L, additional 100 mL ethanol was fed into the medium periodically to maintain ethanol concentration around at 5 g/L until harvest. Eventually, a crocetin titer of 6278 µg/L was obtained after 124 h cultivation (Fig. 6), which was the highest reported titer in eukaryotic cell to date. However, the absolute titer (6278 µg/L) and the production yield based on ethanol consumption ($Y_{P/S} = 0.012\%$) were far away from commercialization, strain engineering by metabolic engineering as well as synthetic biology and process innovation would be two basic but efficient aspects to promote crocetin output. In terms of strain engineering, increasing the catalysis activity of CCD and strain tolerance to product were the main challenges. Combinatorial engineering of *S. cerevisiae* and crocetin biosynthesis pathway in parallel would probably meet the demand [15]. As recent study in process optimization has demonstrated great potential in isoprene overproduction (up to 24 g/L) [52], we believe that crocetin production by our engineered strain would be further improved by continuous efforts in metabolic engineering, synthetic biology and fermentation optimization.

Conclusions

In our study, crocetin biosynthesis pathway was successfully established in *S. cerevisiae* through incorporating heterologous CrtZ and CCD in an existing β-carotene producing strain. Then the effects of culture temperature, combination of CrtZ/CCD, ALD from different species, as well as the expression level of CCD and ALD on crocetin were investigated respectively. Compared to culturing at 30 °C, the crocetin accumulation performed much better at 20 °C. The accumulation of crocetin was further promoted by 2.8-fold by screening of CrtZ/CCD combination and ALD sources. Moreover, the crocetin titer was reached to 1219 µg/L by overexpression of *ccd* and *ald*. Consequently, the highest reported crocetin titer of 6278 µg/L was obtained in 5-L bioreactors. This study promotes the opportunities for industrialization of crocetin and crocin. This study also sets a good reference for microbial production of pharmaceuticals and chemicals in complex structure by fine-tuning multiple enzymes systematically.

Additional file

Additional file 1: Table S1. Oligonucleotides used in this study. **Table S2.** The Codon-optimized sequences of *CrtZ*, *CCD* and *ALD* involved in this study. **Figure S1.** Schematic representation of the engineering strategies for CrtZ expression cassette. **Figure S2.** Schematic representation of the engineering strategies for CCD expression cassette. **Figure S3.** Schematic representation of the engineering strategies for ALD expression cassette. **Figure S4.** The effect of temperature on zeaxanthin production in zeaxanthin producing strains. **Figure S5.** Phylogenetic tree of CCD genes family was constructed and inferred by Neighbor-Joining method. **Figure S6.** Sequence alignment of ZCD and ZCD1 to identify the conserved region. **Figure S7.** Sequence alignment of CCD1 genes subfamily.

Abbreviations

CrtZ: β-carotene hydroxylase; CCD: carotenoid cleavage dioxygenase; ALD: aldehyde dehydrogenase; *Crocus*: *Crocus sativus* L.; Eu: *E. uredovora*; Pa: *Pantoea agglomerans*; Ss: *Sulfolobus solfataricus* P2; Ps: *Pantoea stewartii*; B.SD: *Brevundimonas* sp. SD212; B.DC: *Brevundimonas vesicularis* DC263; Hp: *Haematococcus pluvialis*; Aa: *Agrobacterium aurantiacum*; As: *Alcaligenes* sp. PC-1; Bix: *Bixa orellana*; Syn: *Synechocystis* sp. PCC6803.

Authors' contributions

FC, WX and YY conceived of the study and drafted the manuscript. FC and XM carried out the molecular genetic studies. FC and WX carried out the fed-batch fermentation experiments. YW participated in design and coordination of the study and helped to draft the manuscript. MY carried out the protein analysis. YC and HL participated in strain construction and HPLC analysis respectively. WX supervised the whole research and revised the manuscript. All authors read and approved the final manuscript.

Author details

[1] Key Laboratory of Systems Bioengineering (Ministry of Education), Tianjin University, 92, Weijin Road, Nankai District, Tianjin 300072, People's Republic of China. [2] SynBio Research Platform, Collaborative Innovation Center of Chemical Science and Engineering (Tianjin), School of Chemical Engineering and Technology, Tianjin University, Tianjin 300072, People's Republic of China.

Acknowledgements

The authors are grateful for the financial support from the International S&T Cooperation Program of China (2015DFA00960), the National Natural Science Foundation of China (31600052 and 21676192) and Innovative Talents and Platform Program of Tianjin (16PTSYJC00050).

Competing interests

The authors declare that they have no competing interests.

Funding

The International S&T Cooperation Program of China (2015DFA00960), the National Natural Science Foundation of China (31600052 and 21676192) and Innovative Talents and Platform Program of Tianjin (16PTSYJC00050) supported this work.

References

1. Winterhalter P. Carotenoid-derived aroma compounds: an introduction. 2001.
2. Dhar A, Mehta S, Dhar G, et al. Crocetin inhibits pancreatic cancer cell proliferation and tumor progression in a xenograft mouse model. Mol Cancer Ther. 2009;8(2):315–23.
3. Bathaie SZ, Hoshyar R, Miri H, et al. Anticancer effects of crocetin in both human adenocarcinoma gastric cancer cells and rat model of gastric cancer. Biochem Cell Biol. 2013;91(6):397–403.
4. Hsu JD, Chou FP, Lee MJ, et al. Suppression of the TPA-induced expression of nuclear-protooncogenes in mouse epidermis by crocetin via antioxidant activity. Anticancer Res. 1999;19(5B):4221–7.
5. Higashino S, Sasaki Y, Giddings JC, et al. Crocetin, a carotenoid from Gardenia jasminoides Ellis, protects against hypertension and cerebral thrombogenesis in stroke-prone spontaneously hypertensive rats. Phytother Res. 2014;28(9):1315–9.
6. Zheng S, Qian Z, Tang F, et al. Suppression of vascular cell adhesion molecule-1 expression by crocetin contributes to attenuation of atherosclerosis in hypercholesterolemic rabbits. Biochem Pharmacol. 2005;70(8):1192–9.
7. Amin B, Nakhsaz A, Hosseinzadeh H. Evaluation of the antidepressant-like effects of acute and sub-acute administration of crocin and crocetin in mice. Avicenna J Phytomedicine. 1900;5(5):458–68.
8. Pfander H, Schurtenberger H. Biosynthesis of C20-carotenoids in *Crocus sativus*. Phytochemistry. 1982;21(5):1039–42.
9. Raghavan S, Hansen J, Sonkar S, et al. Methods and materials for recombinant production of saffron compounds: WO, WO/2013/021261[P]. 2013.
10. Lou S, Wang L, He L, et al. Production of crocetin in transgenic *Chlorella vulgaris* expressing genes crtRB and ZCD1. J Appl Phycol. 2016;28(3):1657–65.
11. Ajikumar PK, Xiao WH, Tyo KE, et al. Isoprenoid pathway optimization for taxol precursor overproduction in *Escherichia coli*. Science. 2010;330(6000):70–4.
12. Song MC, Kim EJ, Kim E, et al. Microbial biosynthesis of medicinally important plant secondary metabolites. Nat Product Rep. 2014;31(11):1497–509.
13. Yao YF, Wang CS, Qiao J, et al. Metabolic engineering of *Escherichia coli* for production of salvianic acid A via an artificial biosynthetic pathway. Metab Eng. 2013;19(5):79–87.
14. Cao YX, Xiao WH, Liu D, et al. Biosynthesis of odd-chain fatty alcohols in *Escherichia coli*. Metab Eng. 2015;29:113–23.
15. Chen Y, Xiao W, Wang Y, et al. Lycopene overproduction in *Saccharomyces cerevisiae* through combining pathway engineering with host engineering. Microb Cell Fact. 2016;15(1):1–13.
16. Li XR, Tian GQ, Shen HJ, et al. Metabolic engineering of *Escherichia coli* to produce zeaxanthin. J Ind Microbiol Biotechnol. 2015;42(4):627–36.
17. Bouvier F, Suire C, Mutterer J, et al. Oxidative remodeling of chromoplast carotenoids identification of the carotenoid dioxygenase *CsCCD* and *CsZCD* genes involved in crocus secondary metabolite biogenesis. Plant Cell. 2003;15(1):47–62.
18. Frusciante S, Diretto G, Bruno M, et al. Novel carotenoid cleavage dioxygenase catalyzes the first dedicated step in saffron crocin biosynthesis. Proc Natl Acad Sci USA. 2014;111(33):12246–51.
19. Xie W, Lv X, Ye L, et al. Construction of lycopene-overproducing *Saccharomyces cerevisiae* by combining directed evolution and metabolic engineering. Metab Eng. 2015;30:69–78.
20. Xie W, Liu M, Lv X, et al. Construction of a controllable β-carotene biosynthetic pathway by decentralized assembly strategy in *Saccharomyces cerevisiae*. Biotechnol Bioeng. 2014;111(1):125–33.
21. Zhou P, Ye L, Xie W, et al. Highly efficient biosynthesis of astaxanthin in *Saccharomyces cerevisiae* by integration and tuning of algal *crtZ* and *bkt*. Appl Microbiol Biotechnol. 2015;99(20):8419–28.
22. Gietz RD, Schiestl RH. High-efficiency yeast transformation using the LiAc/SS carrier DNA/PEG method. Nat Protoc. 2007;2:31–4.
23. Boeke JD, Lacroute F, Fink GR. A positive selection for mutants lacking orotidine-5′-phosphate decarboxylase activity in yeast: 5-fluoro-orotic acid resistance. Mol Genet Genom. 1984;197(2):345–6.
24. Lin Q, Jia B, Mitchell LA, et al. RADOM, an efficient in vivo method for assembling designed DNA fragments up to 10 kb long in *Saccharomyces cerevisiae*. Acs Synth Biol. 2015;4(3):213–20.
25. Gietz RD, Woods RA. Yeast transformation by the LiAc/SS carrier DNA/PEG method. Methods Mol Biol. 2014;313:107–20.
26. Thompson JD, Higgins DG, Gibson TJ. CLUSTAL W: improving the sensitivity of progressive multiple sequence alignment through sequence weighting, positions-specific gap penalties and weight matrix choice. Nucleic Acids Res. 1994;22:4673–80.
27. Kumar S, Stecher G, Tamura K. MEGA7: molecular evolutionary genetics analysis version 7.0 for bigger datasets. Mol Biol Evol. 2016;33(7):1870–4.
28. Saitou N, Nei M. The neighbor-joining method: a new method for reconstructing phylogenetic trees. Mol Biol Evol. 1987;4:406–25.
29. Sanderson MJ, Wojciechowski MF. Improved bootstrap confidence limits in large-scale phylogenies, with an example from Neo-Astragalus (Leguminosae). Syst Biol. 2000;49(4):671–85.
30. Biasini M, Bienert S, Waterhouse A, et al. SWISS-MODEL: modelling protein tertiary and quaternary structure using evolutionary information. Nucleic Acids Res. 2014;42(w1):252–8.
31. Arnold K, Bordoli L, Kopp J, et al. The SWISS-MODEL workspace: a web-based environment for protein structure homology modelling. Bioinformatics. 2006;22(2):195–201.
32. Delano WL. The PyMOL molecular graphics system. My publications, 2010.

33. Mei XA, Chen Y, Wang R, et al. Construction of zeaxanthin pathway in *Saccharomyces cerevisiae*. China Biotechnol. 2016;36(8):64–72.
34. Shi F, Zhan W, Li Y, et al. Temperature influences β-carotene production in recombinant *Saccharomyces cerevisiae* expressing carotenogenic genes from *Phaffia rhodozyma*. World J Microbiol Biotechnol. 2014;30(1):125–33.
35. Ahrazem O, Rubiomoraga A, Nebauer SG, et al. Saffron: its phytochemistry, developmental processes, and biotechnological prospects. J Agric Food Chem. 2015;63(40):8751–64.
36. Molina RV, Valero M, Navarro Y, et al. Low temperature storage of corms extends the flowering season of saffron (*Crocus sativus* L.). J Hortic Sci Biotechnol. 2005;80(3):319–26.
37. Molina RV, Valero M, Navarro Y, et al. Temperature effects on flower formation in saffron (*Crocus sativus* L.). Sci Hortic. 2005;103(3):361–79.
38. López J, Essus K, Kim IK, et al. Production of β-ionone by combined expression of carotenogenic and plant *CCD1* genes in *Saccharomyces cerevisiae*. Microb Cell Fact. 2015;14(1):1–13.
39. Kim E, Moore BS, Yoon YJ. Reinvigorating natural product combinatorial biosynthesis with synthetic biology. Nat Chem Biol. 2015;11(9):649–59.
40. Ding MZ, Yan HF, Li LF, et al. Biosynthesis of taxadiene in *Saccharomyces cerevisiae*: selection of geranylgeranyl diphosphate synthase directed by a computer-aided docking strategy. PLoS ONE. 2014;9(10):e109348.
41. Chang JJ, Thia C, Lin HY, et al. Integrating an algal β-carotene hydroxylase gene into a designed carotenoid-biosynthesis pathway increases carotenoid production in yeast. Bioresour Technol. 2015;184:2–8.
42. Sarria S, Wong B, Martín HG, et al. Microbial synthesis of pinene. Acs Synth Biol. 2014;3(7):466–75.
43. Moraga AR, Nohales PF, Pérez JAF, et al. Glucosylation of the saffron apocarotenoid crocetin by a glucosyltransferase isolated from *Crocus sativus* stigmas. Planta. 2004;219(6):955–66.
44. Ahrazem O, Rubio-Moraga A, Berman J, et al. The carotenoid cleavage dioxygenase CCD2 catalysing the synthesis of crocetin in spring crocuses and saffron is a plastidial enzyme. New Phytol. 2016;209(2):650–63.
45. Priya R, Siva R. Phylogenetic analysis and evolutionary studies of plant carotenoid cleavage dioxygenase gene. Gene. 2014;548(2):223–33.
46. Oussama Ahrazem, Gómez-Gómez Lourdes, Rodrigo María J, et al. Carotenoid cleavage oxygenases from microbes and photosynthetic organisms: features and functions. Int J Mol Sci. 2016;17:1781.
47. Sui X, Kiser PD, Lintig JV, et al. Structural basis of carotenoid cleavage: from bacteria to mammals. Arch Biochem Biophys. 2011;539(2):203–13.
48. Greene JJ. Host cell compatibility in protein expression. Methods Mol Biol. 2004;267:3–14.
49. Saint-Prix F, Bönquist L, Dequin S. Functional analysis of the *ALD* gene family of *Saccharomyces cerevisiae* during anaerobic growth on glucose: the $NADP^+$-dependent Ald6p and Ald5p isoforms play a major role in acetate formation. Microbiology. 2004;150(7):2209–20.
50. Bouvier F, Dogbo O, Camara B. Biosynthesis of the food and cosmetic plant pigment bixin (annatto). Science. 2003;300(5628):2089–91.
51. Trautmann D, Beyer P, Al-Babili S. The ORF *slr0091* of *Synechocystis* sp. PCC6803 encodes a high-light induced aldehyde dehydrogenase converting apocarotenals and alkanals. FEBS J. 2013;280(15):3685–96.
52. Chen Y, Xiang G, Yu J, et al. Synergy between methylerythritol phosphate pathway and mevalonate pathway for isoprene production in *Escherichia coli*. Metab Eng. 2016;37:79–91.

6

Using a vector pool containing variable-strength promoters to optimize protein production in *Yarrowia lipolytica*

Rémi Dulermo*, François Brunel, Thierry Dulermo, Rodrigo Ledesma-Amaro, Jérémy Vion, Marion Trassaert, Stéphane Thomas, Jean-Marc Nicaud* and Christophe Leplat*

Abstract

Background: The yeast *Yarrowia lipolytica* is an increasingly common biofactory. To enhance protein expression, several promoters have been developed, including the constitutive *TEF* promoter, the inducible *POX2* promotor, and the hybrid hp4d promoter. Recently, new hp4d-inspired promoters have been created that couple various numbers of UAS1 tandem elements with the minimal *LEU2* promoter or the *TEF* promoter. Three different protein-secretion signaling sequences can be used: preLip2, preXpr2, and preSuc2.

Results: To our knowledge, our study is the first to use a set of vectors with promoters of variable strength to produce proteins of industrial interest. We used the more conventional *TEF* and hp4d promoters along with five new hybrid promoters: 2UAS1-p*TEF*, 3UAS1-p*TEF*, 4UAS1-p*TEF*, 8UAS1-p*TEF*, and hp8d. We compared the production of RedStar2, glucoamylase, and xylanase C when strains were grown on three media. As expected, levels of RedStar2 and glucoamylase were greatest in the strain with the 8UAS1-p*TEF* promoter, which was stronger. However, surprisingly, the 2UAS1-p*TEF* promoter was associated with the greatest xylanase C production and activity. This finding underscored that stronger promoters are not always better when it comes to protein production. We therefore developed a method for easily identifying the best promoter for a given protein of interest. In this gateway method, genes for YFP and α-amylase were transferred into a pool of vectors containing different promoters and gene expression was then analyzed. We observed that, in most cases, protein production and activity were correlated with promoter strength, although this pattern was protein dependent.

Conclusions: Protein expression depends on more than just promoter strength. Indeed, promoter suitability appears to be protein dependent; in some cases, optimal expression and activity was obtained using a weaker promoter. We showed that using a vector pool containing promoters of variable strength can be a powerful tool for rapidly identifying the best producer for a given protein of interest.

Keywords: *Yarrowia lipolytica*, Protein production, RedStar2, Glucoamylase, Xylanase, Hybrid promoters

Background

Increasing the efficiency of heterologous gene expression is a major goal for the agrifood, bioconversion, and pharmaceutical industries as they have a growing need for recombinant proteins. Expression systems using yeasts present several advantages: yeasts are easy to manipulate, they are unicellular organisms with rapid growth rates, and they are eukaryotes that can incorporate post-translational modifications. In addition to the more conventional *Saccharomyces cerevisiae* [1], alternative model species are also used as biofactories, including *Pichia pastoris, Hansanula polymorpha, Kluyveromyces lactis, Kluyveromyces marxianus* [2–5], and *Yarrowia lipolytica* [6, 7].

*Correspondence: remi.dulermo@ifremer.fr; jean-marc.nicaud@inra.fr; christophe.leplat.01@gmail.com
Micalis Institute, INRA-AgroParisTech, UMR1319, Team BIMLip: Integrative Metabolism of Microbial Lipids, Université Paris-Saclay, domaine de Vilvert, 78350 Jouy-en-Josas, France

Production systems exploiting *Y. lipolytica* have several advantages [7, 8]. First, *Y. lipolytica* is a non-pathogenic organism that can grow on a diversity of substrates. Second, the products of several *Y. lipolytica*-based processes have received the "generally recognized as safe" (GRAS) designation from the FDA. Third, *Y. lipolytica* has a naturally strong secretory ability [7, 8] and demonstrates weak protein glycosylation [9].

Several genetic tools are available to enhance protein expression in *Y. lipolytica*. Indeed, integrative expression cassettes containing different markers, such as *LEU2*, *URA3*, *ADE2*, and *LYS5*, have been constructed. They can be used to transform competent auxotrophic strains of *Y. lipolytica*. Moreover, several promoters are also available, including the constitutive *TEF* promoter, the constitutive and hybrid hp4d promoter, and the inducible *POX2* and *LIP2* promoters [10–14]. In addition, several transformation methods have been developed to optimize the transformation rate [15–17]. Currently, the lithium-acetate method is the most common, whether the goal is to inactivate endogenous genes or to transform expression cassettes [18]. All of these tools have been successfully used in *Y. lipolytica* to produce such proteins as xylanase, lipase, leucine aminopeptidase II, human interferon, α2b endoglucanase II, and cellobiohydrolase II [6, 9, 14, 19, 20]. Past studies have also identified at least three sequences that can be used to optimize protein secretion in *Y. lipolytica*: preLip2, preXpr2, and preSuc2 [6, 14, 21–24].

Several studies have suggested that *Y. lipolytica* is better than *P. pastoris* at producing heterologous proteins [20, 25]. Indeed, Nars and colleagues [25] found that, as opposed to *P. pastoris*, *E. coli*, or simple free cells, *Y. lipolytica* was the best candidate for generating extracellular Lip2 because it can form a stable isotope-labeled version of the protein. Boonvitthya and colleagues [19] compared endoglucanase II and cellobiohydrolase II production in *Y. lipolytica* and *P. pastoris*. In YT medium, *Y. lipolytica* produced up to 15 mg/L of endoglucanase and 50 mg/L of cellobiohydrolase. Furthermore, the enzymes produced by *Y. lipolytica* had higher levels of specific activity than did their counterparts in *P. pastoris*. Finally, it has been found that *Y. lipolytica* has weaker protein glycosylation than does *P. pastoris* [9].

One of the first strong constitutive promoters was developed by Novo, using the *TEF1* gene, which encodes the translation elongation factor-1α [10]. Later, Madzak and colleagues [26] identified the upstream activating sequence UAS1 in the *XPR2* gene (which encodes the secreted alkaline extracellular protease). This discovery led to the development of the hp4d promoter, which is based on the minimal *LEU2* promoter and contains four UAS1 tandem elements; with this promoter, expression increases as the number of UAS1 tandem elements increases. More recently, several research groups have used this basic model (i.e., multiple UAS tandem elements associated with a core promoter) to develop improved promoters [27–29]. It has been found that the core promoter and the upstream activating sequence (i.e., the UAS1 tandem elements) act independently and that, as previously noted, promoter strength increases with the number of UAS1 tandem elements. Shabbir Hussain and colleagues [29] showed that promoter strength can be fine-tuned by engineering the sequences of the TATA box, the core promoter, or the upstream activating region. To quantify promoter strength, they used fluorescent proteins and β-galactosidase assays.

However, to our knowledge, no study to date has used these UAS1-based promoters to produce proteins of industrial interest. Here, we used two conventional promoters, p*TEF* and hp4d, as well as five new hybrid promoters of our own construction. To create the latter, we added two, three, four, or eight UAS1 tandem elements to p*TEF*; we also added four tandem elements to hp4d. Promoter strength in transformed *Y. lipolytica* strains was quantified using RedStar2, a fluorescent protein, as a reporter; we also analyzed the production of secreted *Aspergillus niger* glucoamylase (GA) and xylanase C (XlnC). GA is a glucan 1,4-alpha-glucosidase that belongs to the glycosyl hydrolase family. It catalyzes the degradation of starch and other complex sugars, releasing D-glucose. GA is largely used to produce biolipids and bioethanol from starch or lignocellulosic materials [30, 31]. XlnC is a beta-1,4-beta-xylanase that breaks down hemicellulose, a component of plant cell walls, releasing xylose. The paper, textile, and pet-food industries are major consumers of xylanase.

Our results revealed that optimal protein expression, secretion, and activity are not always correlated with promoter strength. Consequently, we developed a simple method for improving protein expression that involves the use of a pool of vectors containing promoters of variable strength.

Methods

Yeast strains, growth media, and culture conditions

The *Y. lipolytica* wild-type strain W29 (ATCC20460) was used as the basis for all the *Y. lipolytica* strains built in this study (see Additional file 1: Table S1 for the full list). The auxotrophic strain Po1d (Leu^- Ura^-) has previously been described by Barth and Gaillardin [19]. *Escherichia coli* strain DH5α was used to construct the plasmids, except in the case of vectors containing *ccdB*, for which *E. coli* strain DB3.1 was used. *E. coli* growth media and culture conditions have been previously described by Sambrook and colleagues [32], and those for *Y. lipolytica* have been described by Barth and Gaillardin [15]. Rich medium (YPD) and minimal glucose medium (YNB) were prepared as described elsewhere [33]. The YPD medium contained 10 g/L of yeast extract (Difco, Paris, France), 10 g/L of Bacto Peptone (Difco, Paris, France), and 10 g/L of glucose (Sigma Aldrich, Saint-Quentin Fallavier, France). The YNB medium contained 1.7 g/L of yeast nitrogen base without amino acids and ammonium sulfate (YNBww; Difco, Paris, France), 10 g/L of glucose (Sigma), 5.3 g/L of NH_4Cl, and 50 mM phosphate buffer (pH 6.8). This minimal medium was supplemented with uracil (0.1 g/L) and/or leucine (0.1 g/L) as necessary. YP_2D_4 medium contained 10 g/L of yeast extract (Difco, Paris, France), 20 g/L of Bacto Peptone (Difco, Paris, France), and 40 g/L of glucose (Sigma Aldrich, Saint-Quentin Fallavier, France). Solid media were created by adding 1.6% agar.

Plasmid and strain construction

The structure of the plasmids constructed in this study was typical of that of the expression vector JMP62 [6] (Fig. 1a). The plasmids contained an excisable marker (the *I-sceI* fragment flanked by LoxP/LoxR [37]), and the promoter and gene of interest were carried in the *ClaI*-*BamHI* and *BamHI*-*AvrII* fragments, respectively. The zeta region for expression cassette integration was flanked by the *NotI* site, which is involved in the release of the expression cassette prior to transformation. Plasmid and strain construction are described in Additional file 2: Figure S1. In most cases, the genes of interest were introduced by digesting the corresponding donor plasmid using *BamHI*-*AvrII* (Additional file 2: Figure S1a). Promoter exchange was performed by digesting the donor plasmid using *ClaI*-*BamHI*; *ClaI* was used to insert the modified promoter (Additional file 2: Figure S1b).

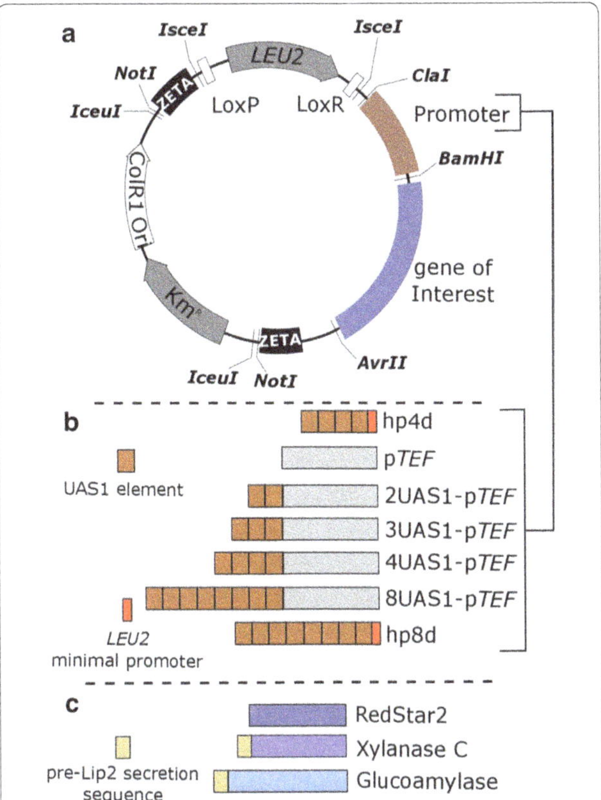

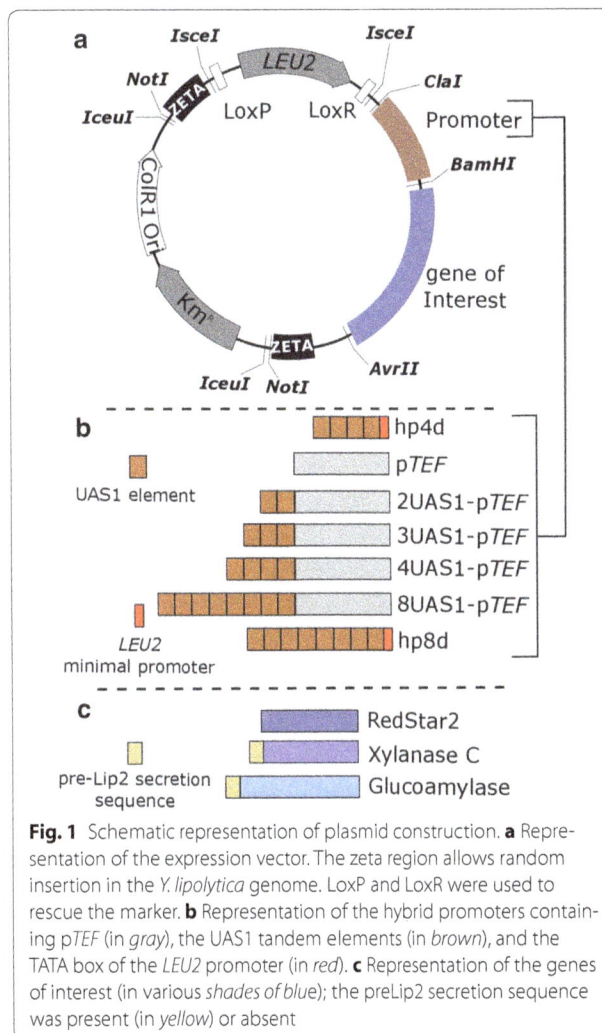

Fig. 1 Schematic representation of plasmid construction. **a** Representation of the expression vector. The zeta region allows random insertion in the *Y. lipolytica* genome. LoxP and LoxR were used to rescue the marker. **b** Representation of the hybrid promoters containing p*TEF* (in *gray*), the UAS1 tandem elements (in *brown*), and the TATA box of the *LEU2* promoter (in *red*). **c** Representation of the genes of interest (in various *shades of blue*); the preLip2 secretion sequence was present (in *yellow*) or absent

The two, three, or four UAS1 tandem element fragments were amplified by PCR using HYB-*ClaI*3′Hp4d5′ and HYB-*BstbI*5′Hp4d3′ as primers (Table 1; Additional file 2: Figure S1). The corresponding fragments were ligated into pCR4Blunt-TOPO® in accordance with the manufacturer's instructions (Invitrogen, Saint-Aubin, France).

GA was cloned into the JMP2482, JMP2484, JMP2397, JMP2607, JMP2471, and JMP2473 plasmids at the *BamHI* and *AvrII* restriction sites, yielding JMP3781 (*LEU2*ex 2UAS1-p*TEF-GA*), JMP3782 (*LEU2*ex 3UAS1-p*TEF-GA*), JMP3783 (*LEU2*ex 4UAS1-p*TEF-GA*), JMP3784 (*LEU2*ex 8UAS1-p*TEF-GA*), JMP3785 (*LEU2*ex hp4d-*GA*), and JMP3786 (*LEU2*ex hp8d-*GA*), respectively.

XlnC was cloned into the JMP2482, JMP2484, JMP2397, JMP2607, JMP2471, and JMP2473 plasmids at the *BamHI* and *AvrII* restriction sites, yielding the JMP3096 (*LEU2*ex 2UAS1-p*TEF-XlnC*), JMP3097 (*LEU2*ex 3UAS1-p*TEF-XlnC*), JMP3098 (*LEU2*ex 4UAS1-p*TEF-XlnC*), JMP3099 (*LEU2*ex 8UAS1-p*TEF-XlnC*), JMP3100 (*LEU2*ex hp4d-*XlnC*), and JMP3101 (*LEU2*ex hp8d-*XlnC*) plasmids, respectively.

The sequences of the genes encoding YFP and α-amylase are provided in Additional file 3: Data S1. These genes were inserted into pENTR™/D-TOPO® in accordance with the manufacturer's instructions using the primers listed in Table 1.

The overexpression cassettes, obtained by digesting the plasmids with *NotI*, were used to transform individual strains via the lithium-acetate method [18]. Transformants were selected utilizing YNB Ura, YNB Leu, or YNB medium, depending on their genotype, and their genomic DNA was prepared as described by Querol and colleagues [34]. The primers used to verify expression cassette insertion are given in Table 1.

Restriction enzymes were obtained from OZYME (Saint-Quentin-en-Yvelines, France). PCR was performed

Table 1 List of primers used in this study

Primer	Sequence	Use
HYB-ClaI3′Hp4d5′	CCCTACATCGATACGCGTGC	Hybrid promoter construction
HYB-BstbI5′Hp4d3′	CCTTCGAACGCACTTTTGCCCGTGATCAG	
GATO_Amont_ClaI_for	CCCTGTTATCCCTAGAATCGAT	Verification of plasmid construction and insertion into the *Y. lipolytica* genome
GATO_AvaI_AvrII_rev	TTAGATACCACAGACACCCTAG	
GATO_pTEF_BamHI_for	AACTCACACCCGAAGGATCC	
GATO_HP4d_BamHI_for	GAACCCGAAACTAAGGATCC	
YFP-pool-Fw	CACCATGGTGAGCAAGGGCGAGGAGC	Insertion of YFP gene into pENTR™/D-TOPO®
YFP-pool-Rv	TTACTTGTACAGCTCGTCCATGCC	
Amy-pool-Fw	CACCATGAAGCTGTCTACCATTCTG	Insertion of α-amylase gene into pENTR™/D-TOPO®
Amy-pool-Rv	TCAAATCTTCTCCCAAATAGCG	
1529BamHIcorrigéF	CCTTGTCAACTCACACCCGAAGGATCCATCACAAGTTTGTAC	Addition of a *Bam*HI site close to the promoter in JMP1529 to obtain JMP3030
1529BglIIcorrigéR	TCTGGCTTTTAGTAAGCCAGATCTACGCGTTTACGCCCCGCC	
1529BamHIcorrigéR	GTACAAACTTGTGATGGATCCTTCGGGTGTGAGTTGACAAGG	
qPCR_XlnCF	CGAGCTGCCGATCCCAATGCC	qPCR related to the XlnC gene
qPCR_XlnCR	GCTCCACCGCCTGCAGACA	
qPCR_YALI0D08272F	AGGCCCAGTCCAAGCGAGGT	qPCR related to the actin gene
qPCR_YALI0D08272R	TCGGTGAGCAGGACGGGGTG	

using an Eppendorf 2720 thermal cycler; GoTaq DNA polymerases (Promega, Madison, WI, USA) were employed to verify the results and PyroBest DNA polymerases (Takara, Saint-Germain-en-Laye, France) were employed to carry out cloning. PCR and DNA fragment purification were performed as previously described [35]. The amounts of DNA obtained were measured using MySpec (VWR, Fontenay-sous-Bois, France). All the reactions were performed in accordance with the manufacturer's instructions. The sequencing of the cloned fragments was performed by GATC Biotech (Konstanz, Germany). Clone Manager software was used for the gene sequence analysis (Sci-Ed Software, Morrisville, NC, USA).

Plasmid pool

Forty ng/μL of each of the recipient plasmids was mixed with pENTR™/D-TOPO® containing the YFP or α-amylase gene. The transfer of the genes of interest was performed using LR Clonase® in accordance with the manufacturer's instructions (Invitrogen, Saint-Aubin, France). The mixture was used to transform *E. coli* strain DB3.1. The resulting transformants were then pooled, and their DNA was extracted and digested before *Y. lipolytica* was transformed in turn.

Sds page

Supernatant was obtained from cultures grown for 72 h in YNB, YPD, or YP_2D_4 media and was concentrated tenfold in 30 mM Tris (pH 8.0) and 50 mM NaCl using Amicon Ultra-0.5 10 K centrifugal filters (Merck Millipore Ltd, Ireland). Protein production was analyzed via polyacrylamide gel electrophoresis (SDS-PAGE); 4–12% Tris–Glycine gels and an XCell SureLock™ Mini-Cell electrophoresis system (Novex, Life Technologies, Saint-Aubin, France) were used. Prism (MW1; 19–130 kDa) and wide-range (MW2; 14–212 kDa) protein molecular weight markers were used as standards (VWR Chemicals, Fontenay-sous-Bois, France). The gels were stained with 0.2% Coomassie Brilliant Blue R dye (Thermo Fisher Scientific, Villebon-sur-Yvette, France).

Protein content

Twenty-μL samples were analyzed for protein content using the Coomassie (Bradford) Protein Assay Kit (Thermo Fisher Scientific, Villebon-sur-Yvette, France) in accordance with the manufacturer's instructions.

Glucoamylase activity

GA activity was measured as previously described [36], with the following modifications. Samples containing 40 μL of supernatant were incubated for 2–10 min with 1.8 mL of a 0.2% soluble cornstarch solution (30 °C, pH 5). The resulting glucose concentration was determined via high-performance liquid chromatography: an UltiMate® 3000 system (Dionex-Thermo Fisher Scientific, UK) with an Aminex HPX87H column coupled to an RI detector was used. The column was eluted with 0.01 N H_2SO_4 at room temperature and a flow rate of 0.6 mL/min. Identification and quantification were achieved via comparison to standards. Enzyme activity was expressed in U mL/L of supernatant, where one unit of GA activity (1 U) was defined as the amount of GA required to release 1 μmol of glucose per minute.

Xylanase activity

XlnC activity was determined using the EnzChek® Ultra Xylanase Assay Kit (Molecular Probes Invitrogen Ltd., Paisley, UK) in 30 mM Tris (pH 8.0) and 50 mM NaCl at 25 °C in a BioLector® (Biotek, Colmar, France). Prior to the assays, supernatant from cultures grown in YNB medium was diluted 50- and 100-fold, and supernatant from cultures grown in YPD or YP_2D_4 was diluted 500- and 1000-fold. As in the case of GA, one unit of XlnC activity (1 U) was defined as the amount of XlnC required to release 1 μmol of xylose per minute.

Growth analysis

The growth of the *Y. lipolytica* strains was analyzed using a microtiter plate reader, as previously described [37]. RedStar2 fluorescence and YFP fluorescence were analyzed at emission wavelength settings of 558 and 586 nm, respectively; the reception wavelength settings were 505 and 530 nm, respectively.

Microscopic analysis

Images were acquired using a Zeiss Axio Imager M2 microscope (Zeiss, Le Pecq, France) and Axiovision v. 4.8 software (Zeiss, Le Pecq, France).

qPCR analysis

RNA extraction was performed using the RNeasy Mini Kit (Qiagen, Courtaboeuf, France) followed by DNA digestion with DNase I (RNase-free; New England BioLabs, Evry, France). cDNA synthesis was performed with the Maxima First Strand cDNA Synthesis Kit with dsDNase (Thermofisher Scientific, Courtaboeuf, France). PCR quantification was performed with CFX Connect™ Real-Time PCR Detection System (Bio-Rad, Marnes-la-Coquette, France) using the SsoAdvanced™ Universal SYBR® Green Supermix Kit (Bio-Rad, Marnes-la-Coquette, France). The number of XlnC mRNA copies was determined using the cycle threshold (Ct) values, which were standardized using results for the expression of the actin gene (YALI0D08272g); the number of XlnC mRNA copies found in the strain containing p*TEF-XlnC* was employed as a reference.

Results and discussion

RedStar2 expression varies with promoter strength

To examine how protein expression varied with promoter strength, we constructed seven promoters (see diagram in Fig. 1b). Two were conventional promoters: p*TEF* and hp4d. Four new hybrid promoters were generated by combining two, three, four, or eight UAS1 tandem elements taken from hp4d with the *TEF* promoter, yielding 2UAS1-p*TEF*, 3UAS1-p*TEF*, 4UAS1-p*TEF*, and 8UAS1-p*TEF*, respectively (Fig. 1b). We also created a derivative of the hp4d promoter by adding four supplementary UAS1 tandem elements, thus generating hp8d (Fig. 1b).

Based on previous studies, hp4d and p*TEF* should be the weakest promoters, while hp8d and 8UAS1-p*TEF* should be the strongest. All of these promoters were ligated into a JMP62-*LEU2* plasmid containing the *LEU2* marker and a long-terminal-repeat zeta element that allows random insertion in *Y. lipolytica* (Fig. 1a) [38]. RedStar2 was used as a reporter system to measure promoter strength (Fig. 1c); it was chosen because the protein's fluorescence is easy to detect and quantify in *Y. lipolytica* [35, 39, 40]. RedStar2 fluorescence was analyzed by microscopy (Fig. 2a) and using a Biotek BioLector® (Fig. 2b, c). Since all the strains showed similar growth patterns (Fig. 2c), their fluorescence levels could be compared. As expected, there was a correlation between putative promoter strength and strain fluorescence (Fig. 2a, b): the stronger the promoter, the greater the fluorescence. Therefore, the strains containing hp4d and p*TEF* had the weakest fluorescence, while the strains containing hp8d and 8UAS1-p*TEF* had the strongest fluorescence. Over time, the fluorescence of strains containing hp8d and 8UAS1-p*TEF* increased 2.3- and 5.3-fold compared to their respective controls, the strains containing hp4d and p*TEF*. Therefore, our results show that increasing the number of UAS1 tandem elements in hybrid promoters resulted in a gradual increase in RedStar2 expression levels (Fig. 2b), confirming the previous findings of Blazeck and colleagues [27, 28]. Thus, our seven promoters varied greatly in strength: there was a 29-fold difference between the weakest (p*TEF*: 4000 AU) and strongest promoter (php8d: 115,000 AU) (Fig. 2b).

Promoter strength affects xylanase C production but not glucoamylase production

We used GA and XlnC to examine how our promoters could be used to produce proteins of industrial interest. GA is used to degrade lignocellulosic materials, the starch in oligosaccharides, or glucose, and it can thus be used by microorganisms to produce biolipids, bioethanol, and other bioindustrial materials [30, 31, 41]. XlnC is a commonly used enzyme in bioprocesses in the paper, textile, and pet-food industries. Therefore, enhancing its production could be of great interest. GA and XlnC activity are also easy to measure (see refs. [42, 43] for GA and "Methods" section for XlnC), making them good candidates for examining the relationship between protein production and promoter strength. To facilitate

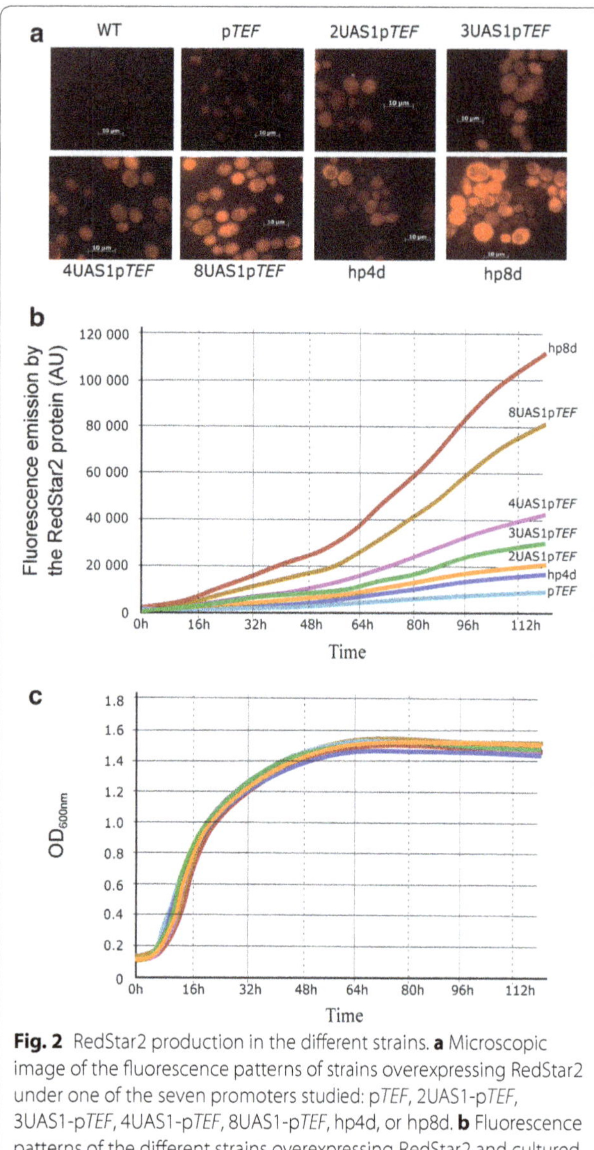

Fig. 2 RedStar2 production in the different strains. **a** Microscopic image of the fluorescence patterns of strains overexpressing RedStar2 under one of the seven promoters studied: p*TEF*, 2UAS1-p*TEF*, 3UAS1-p*TEF*, 4UAS1-p*TEF*, 8UAS1-p*TEF*, hp4d, or hp8d. **b** Fluorescence patterns of the different strains overexpressing RedStar2 and cultured in YNB. **c** Growth patterns of the different strains overexpressing RedStar2 and cultured in YNB

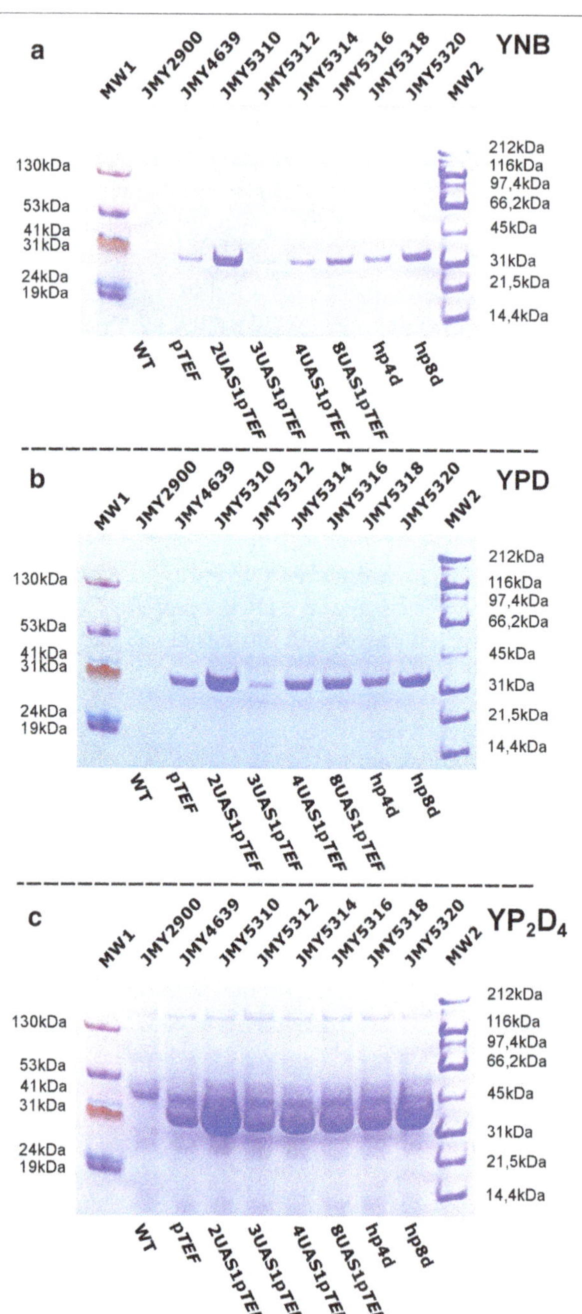

Fig. 3 Production of secreted xylanase by the different strains in the different media. SDS-PAGE gel showing xylanase C production by the different strains. **a** YNB medium, **b** YPD medium, and **c** YP_2D_4 medium. MW1 and MW2 represent the prism and the wide-range protein molecular weight markers, respectively

our analyses (i.e., the visualization of the electrophoresis results and the interpretation of the enzyme assays), the preLip2 secretion sequence was added to the *GA* and *XlnC* genes. This sequence allows proteins to be secreted into the growth medium [6, 14, 30]. Both genes were cloned into different vectors containing the seven different promoters, which were subsequently used to transform *Y. lipolytica*. Cultures were then grown in three media—a defined medium, YNB; a rich medium, YPD; and a very rich medium, YP_2D_4—and the levels of secreted GA and XlnC were analyzed (Fig. 3; Additional file 4: Figure S2, Additional file 5: Table S2).

As expected, GA production varied with promoter strength and increased with medium richness (Additional file 4: Figure S2a–d). However, high production levels may or may not translate into high activity levels. To determine if there was a correlation between the two variables, GA activity was estimated by measuring the disappearance of starch and the appearance of glucose. Activity was found to be positively associated with production (Additional file 4: Figure S2e).

In contrast, XlnC production was not associated with promoter strength. Indeed, across all media, thicker bands were observed for strains containing 2UAS1-p*TEF* and, to a lesser extent, hp8d, whereas band thickness was equivalent for strains containing 3UAS1-p*TEF*, 4UAS1-p*TEF*, 8UAS1-p*TEF*, and hp4d (Figs. 3, 4a; Additional file 5: Table S2). The results were consistent when additional transformants were analyzed. Semi-quantitative PCR confirmed that only one copy of *XlnC* was inserted into the genome of the strain containing 2UAS1-p*TEF* (data not shown). Interestingly, we found that XlnC production was 2–4 times higher in the strain containing 2UAS1-p*TEF* than in the strains containing p*TEF*, 8UAS1-p*TEF*, and hp4d. In our experiment, in YP_2D_4, maximum XlnC production was about 153 mg/L. The strain containing 8UAS1-p*TEF* produced slightly more XlnC than the strains containing 3UAS1-p*TEF* and 4UAS1-p*TEF* when the yeasts were cultured in YNB. However, its levels of production were similar or lower when the yeasts were cultured in YPD or YP_2D_4. In various microorganisms, several bottlenecks in heterologous protein production have been identified; they include transcription, protein folding and glycosylation, translocation, signal peptide processing, and proteolysis [41–43]. Therefore, several hypotheses could explain why 2UAS1-p*TEF* was the best promoter for XlnC production. To evaluate if this result could be attributed to the 2UAS1-p*TEF* promoter resulting in higher levels of *XlnC* transcription, XlnC mRNA levels were analyzed using qRT-PCR (Fig. 5). However, mRNA levels were positively correlated with promoter strength. This result is consistent with those of a previous study [44], in which researchers observed that the production of an insulin precursor and of amylase was lower under the *TEF1* promoter than under the *TPI* promoter even though their

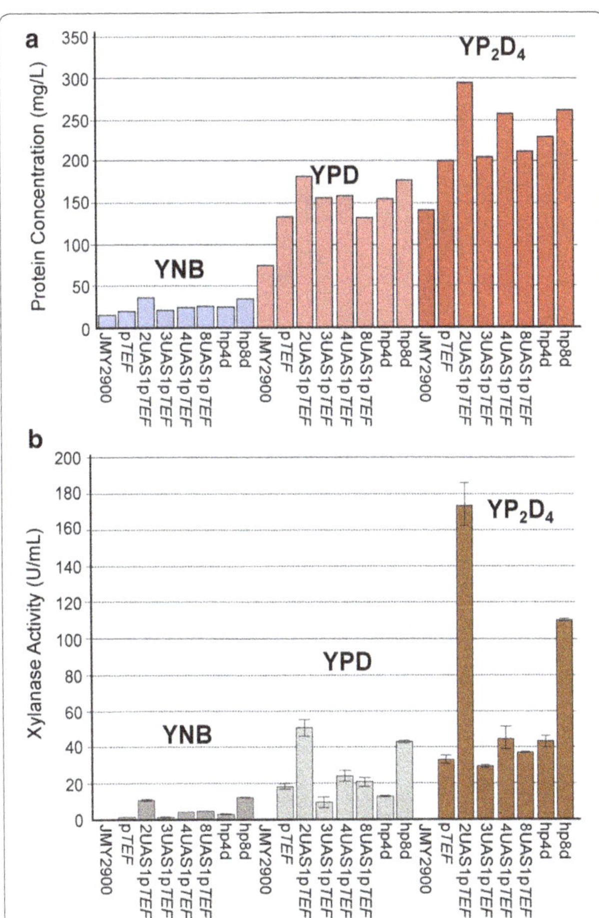

Fig. 4 Supernatant protein content and xylanase C activity levels in the different strains in the different media. **a** Total protein content of the supernatant samples containing xylanase C, as assessed by the Bradford assay, for the different strains across the different media (YNB: *blue*; YPD: *pink*; and YP_2D_4: *red*). **b** Xylanase C activity in the different strains in the different media (YNB: *dark gray*; YPD: *light gray*; and YP_2D_4: *brown*)

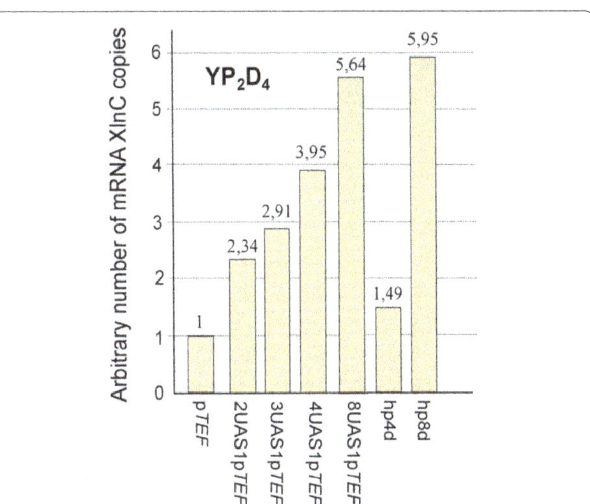

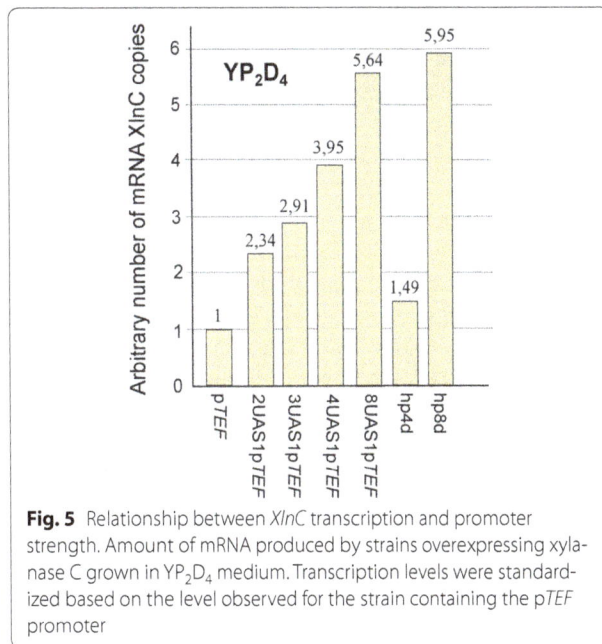

Fig. 5 Relationship between *XlnC* transcription and promoter strength. Amount of mRNA produced by strains overexpressing xylanase C grown in YP_2D_4 medium. Transcription levels were standardized based on the level observed for the strain containing the p*TEF* promoter

transcription was greater under the *TEF1* promoter. However, it is possible that the use of promoters stronger than 2UAS1-p*TEF* could have resulted in excessive protein production, which could have negatively affected protein folding because of the titration of chaperon proteins and the saturation of secretion machinery, as found previously [43].

As for GA, we examined the correlation between XlnC production and activity (Fig. 4b). As expected, the WT strain, JMY2900, demonstrated no XlnC activity. Surprisingly, activity levels were not always associated with production levels, which could suggest that there was co-secretion of non-active or less-active forms of the enzyme. Although the two variables were correlated when the strains were grown in YNB, the correlation was weak or completely absent when the strains were grown in YP_2D_4 or YPD, respectively (Fig. 4a, b). For instance, the strain containing 3UAS1-p*TEF* had a production level similar to that of the strain containing 4UAS1-p*TEF*, but the former's activity level was much lower. Indeed, its activity level resembled that of the strain containing p*TEF*. Interestingly, activity levels were 1.5–2 times higher than expected for the strains containing 2UAS1-p*TEF* and hp8d (Fig. 4b). Oddly, although these promoters increased protein production two to fourfold, compared to the strain containing p*TEF*, activity increased three to sixfold (Fig. 4a, b; Additional file 5: Table S2, Additional file 6: Table S3). These results underscore that enzyme expression, production, and activity are not always linearly related to promoter strength. Indeed, these relationships may vary and depend on the specific enzyme and growth medium used.

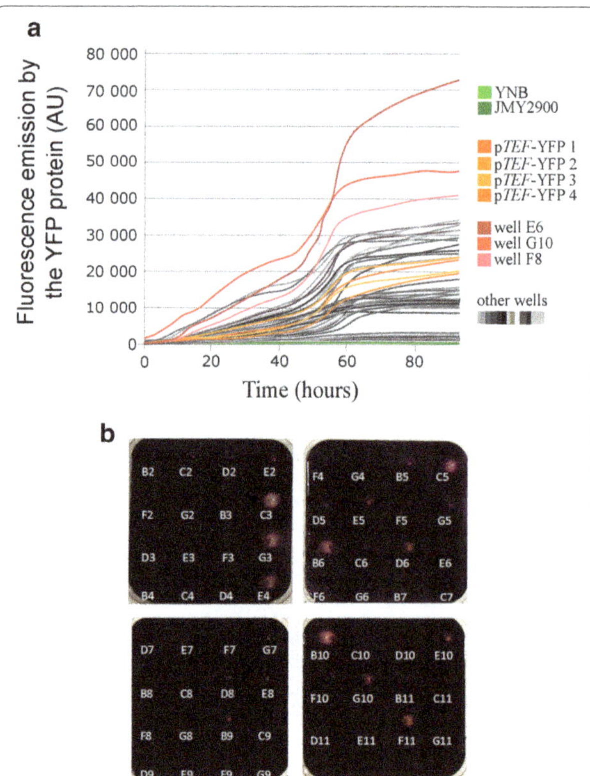

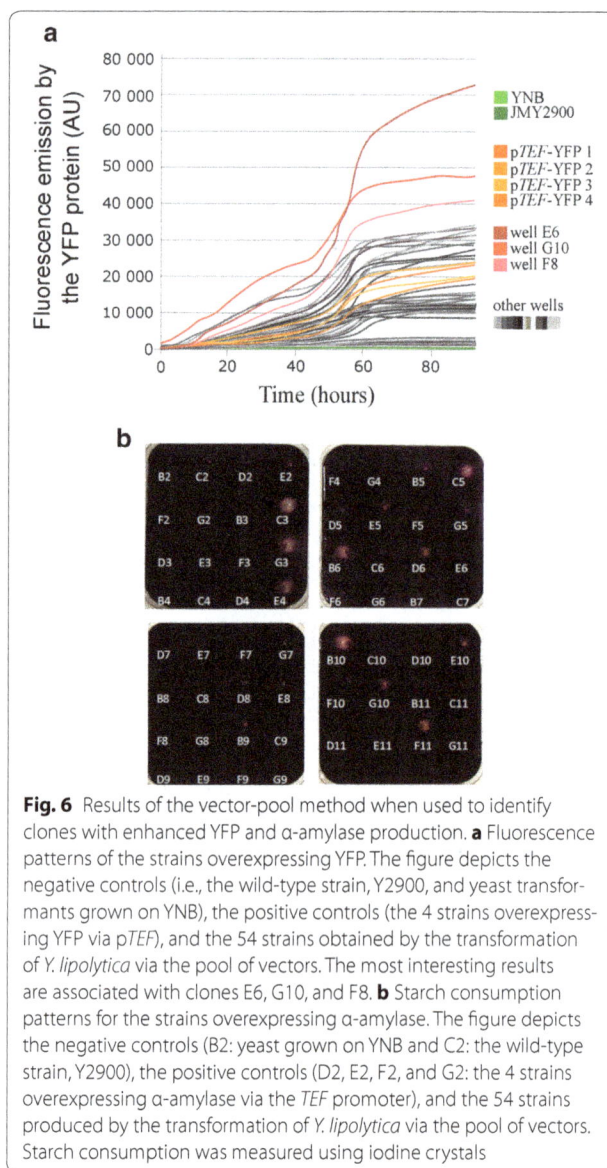

Fig. 6 Results of the vector-pool method when used to identify clones with enhanced YFP and α-amylase production. **a** Fluorescence patterns of the strains overexpressing YFP. The figure depicts the negative controls (i.e., the wild-type strain, Y2900, and yeast transformants grown on YNB), the positive controls (the 4 strains overexpressing YFP via p*TEF*), and the 54 strains obtained by the transformation of *Y. lipolytica* via the pool of vectors. The most interesting results are associated with clones E6, G10, and F8. **b** Starch consumption patterns for the strains overexpressing α-amylase. The figure depicts the negative controls (B2: yeast grown on YNB and C2: the wild-type strain, Y2900), the positive controls (D2, E2, F2, and G2: the 4 strains overexpressing α-amylase via the *TEF* promoter), and the 54 strains produced by the transformation of *Y. lipolytica* via the pool of vectors. Starch consumption was measured using iodine crystals

A gateway vector pool for selecting the best protein producer

Since promoter strength was not necessarily correlated with heterologous protein production, we decided to develop a method for rapidly identifying transformants with optimized production; we used a pool of vectors containing promoters that varied in strength. To simplify the approach, we employed a gateway system that allowed in vitro cloning and the counter-selection of the correct clone using CcdB toxicity. We constructed a derivative of the gateway plasmid JMP1529 described in Leplat et al. [39]: JMP3030 (gateway-*Cla*I-p*TEF*-*Bam*HI). Derivatives were constructed using *Cla*I-*Bam*HI-based promoter exchange (Additional file 1: Table S1).

We analyzed the expression of YFP and secreted α-amylase (Fig. 6). Briefly, we first inserted the genes encoding YFP and α-amylase into pENTR™/D-TOPO®. We then transferred these genes into a pool of vectors using LR Clonase® (Additional file 7: Figure S3). After transforming *Y. lipolytica*, we analyzed 54 clones for YFP and α-amylase expression (Fig. 6). We found that some clones displayed higher activity levels than others—YFP activity was especially high for the E6, G10, and F8 clones (72,000 U; 48,000 U; and 41,000 U, respectively), and α-amylase activity was especially high for the C3, G3, E4, B6, B10, and F11 clones. Analysis of the promoters involved in the expression of these genes revealed that most of the clones contained a promoter that was stronger than p*TEF* (Table 2). Indeed, with the exception of G10, which contained p*TEF*, the clones contained hp4d, hp8d, 4UAS-p*TEF*, or 8UAS-p*TEF*. However, in some cases, it was difficult to identify the promoter since sequencing was impaired by the multiple UAS1 tandem elements and there was not enough differentiation among fragment sizes to use a PCR-based approach. Therefore, we have proposed two candidate promoters for B6 (α-amylase) and E6 (YFP). Using this method, we have identified several good producers for both enzymes, which shows that it could be very helpful to use a pool of plasmid vectors containing variable-strength promoters to obtain strains that have optimal activity levels.

Conclusions

Blazeck and colleagues [27, 28] developed very strong promoters to optimize protein expression in *Y. lipolytica*; however, these promoters were only used to produce intracellular proteins, such as fluorescent proteins, or in β-galactosidase assays. We constructed similar versions of these promoters (p*TEF*, 2UAS1-p*TEF*, 3UAS1-p*TEF*, 4UAS1-p*TEF*, 8UAS1-p*TEF*, hp4d, and hp8d) and analyzed their impact on the production of intracellular proteins, namely RedStar2 and YFP, as well as extracellular proteins, namely glucoamylase, xylanase C, and α-amylase (see summary in Table 3). We found that, most of the time, having the strongest promoter (8UAS-p*TEF*) resulted in the highest levels of protein production and activity (i.e., in the cases of RedStar2, glucoamylase, YFP, and α-amylase). However, the best promoters for xylanase C were 2UAS1-p*TEF* and hp8d. Our results show that stronger promoters do not always optimize protein production and activity. It could be that either transcriptional or post-translational regulation, such as RNA processing and stability, translation efficiency, or protein stability and modification [45, 46], places limits on this relationship. As a result, multiple promoters should always be tested. To limit clone number and keep the process simple, we developed a straightforward strategy for accomplishing this aim: exploiting a pool of vectors containing promoters of different strengths. Cloning was facilitated by using the gateway system and LR Clonase®. Indeed, in a single step, it was possible to obtain a collection of vectors containing variable-strength promoters upstream from the gene of interest. Once the study organism has been transformed, screening tests can be used to select the best strains. This approach could be very helpful to those seeking to improve protein production, whether in a research or an industrial setting. The pool should contain a decent number of promoters and include inducible promoters, which could be particularly important when dealing with toxic proteins.

Table 2 Promoters upstream of the α-amylase gene and YFP gene in the different clones

	Promoter
α-Amylase	
C3	4UAS1-p*TEF*
G3	hp4d
E4	hp8d
B6	3UAS1-p*TEF* or 4UAS1-p*TEF*
B10	8UAS1-p*TEF*
F11	8UAS1-p*TEF*
YFP	
E6	4UAS1-p*TEF* or hp8d
F8	8UAS1-p*TEF*
G10	p*TEF*

The clone names are the same as in Fig. 6

Table 3 Relative results for the experiments examining RedStar2, glucoamylase, and xylanase C expression under the seven different promoters studied

	Redstar2	GA		XlnC		
	Activity	Production	Activity	RNA	Production	Activity
p*TEF*	+	+	+	+	+	+
2UAS-p*TEF*	++	++	++	++	+++++	+++++
3UAS-p*TEF*	+++	++	++	+++	±	+
4UAS-p*TEF*	++++	+++	+++	++++	+	+
8UAS-p*TEF*	+++++	++++	++++	+++++	+++	+
hp4d	++	+	++	+	++	+
hp8d	+++++	+++	++++	+++++	++++	+++

The number of crosses indicate very low (±), low (+), medium (++), high (+++), very high (++++) and extremely high (+++++) levels

Additional files

Additional file 1: Table S1. Details on the strains and plasmids used in this study.

Additional file 2: Figure S1. Schematic representation of plasmid and strain construction.

Additional file 3: Data S1. Sequences of the genes used in this study.

Additional file 4: Figure S2. Production and activity of secreted glucoamylase for the different strains in the different media.

Additional file 5: Table S2. Glucoamylase and xylanase C concentrations.

Additional file 6: Table S3. Xylanase activity levels.

Additional file 7: Figure S3. Schematic representation of the construction of the vector pool.

Abbreviations

GA: glucoamylase; XlnC: xylanase C; YFP: yellow fluorescent protein; PCR: polymerase chain reaction.

Authors' contributions

RD: wrote the project proposal, designed the experiments, built some of the plasmids and strains, performed the RedStar2 experiment, performed the vector-pool experiment, analyzed the results, and wrote the manuscript; FB: built the different promoters as well as some of the plasmids and strains and performed the RedStar2 experiment; TD: wrote the project proposal; RLA: performed the test for GA activity; JV: built some of the strains and performed the vector-pool experiment; MT: quantified protein production and ran the related gels; ST: built some of the plasmids and strains; JM: wrote the project proposal and designed the experiments; CL: wrote the project proposal, designed the experiments, built some of the plasmids and strains, performed the RedStar2 experiment, quantified protein production, ran the related gels, performed the test for XlnC activity, carried out quantitative PCR, analyzed the results, and wrote the manuscript. All authors read and approved the final manuscript.

Acknowledgements

We thank the French government for giving us IDEX funding (Grant No 2015-0445I). R. Ledesma-Amaro received financial support from the European Union (Marie-Curie FP7 COFUND People Program: AgreenSkills Fellowship). We thank Jessica Pearce and Lindsay Higgins for their language editing services.

Competing interests

The authors declare that they have no competing interests.

Funding

C. Leplat received IDEX funding (Grant No 2015-0445I), which paid for all the experiments and J. Vion's salary. R. Ledesma-Amaro received financial support from the European Union (Marie-Curie FP7 COFUND People Program: AgreenSkills Fellowship).

References

1. Domínguez A, Fermiñán E, Sánchez M, González FJ, Pérez-Campo FM, García S, Herrero AB, San Vicente A, Cabello J, Prado M, Iglesias FJ, Choupina A, Burguillo FJ, Fernández-Lago L, López MC. Non-conventional yeasts as hosts for heterologous protein production. Int Microbiol. 1998;1:131–42.
2. Madhavan A, Sukumaran RK. Promoter and signal sequence from filamentous fungus can drive recombinant protein production in the yeast *Kluyveromyces lactis*. Bioresour Technol. 2014;165:302–8.
3. Bragança CR, Colombo LT, Roberti AS, Alvim MC, Cardoso SA, Reis KC, de Paula SO, da Silveira WB, Passos FM. Construction of recombinant *Kluyveromyces marxianus* UFV-3 to express dengue virus type 1 nonstructural protein 1 (NS1). Appl Microbiol Biotechnol. 2015;99:1191–203.
4. Spohner SC, Schaum V, Quitmann H, Czermak P. *Kluyveromyces lactis*: an emerging tool in biotechnology. J Biotechnol. 2016;222:104–16.
5. Wagner JM, Alper HS. Synthetic biology and molecular genetics in non-conventional yeasts: current tools and future advances. Fungal Genet Biol. 2016;89:126–36.
6. Nicaud JM, Madzak C, van den Broek P, Gysler C, Duboc P, Niederberger P, Gaillardin C. Protein expression and secretion in the yeast *Yarrowia lipolytica*. FEMS Yeast Res. 2002;2:371–9.
7. Madzak C, Gaillardin C, Beckerich JM. Heterologous protein expression and secretion in the non-conventional yeast *Yarrowia lipolytica*: a review. J Biotechnol. 2004;109:63–81.
8. Nicaud JM. *Yarrowia lipolytica*. Yeast. 2012;29:409–18.
9. Madzak C, Beckerich JM. Heterologous protein expression and secretion in *Yarrowia lipolytica*. In: *Yarrowia lipolytica*, vol. 25. Berlin: Springer; 2013. p. 1–76.
10. Müller S, Sandal T, Kamp-Hansen P, Dalbøge H. Comparison of expression systems in the yeasts *Saccharomyces cerevisiae, Hansenula polymorpha, Klyveromyces lactis, Schizosaccharomyces pombe* and *Yarrowia lipolytica*. Cloning of two novel promoters from *Yarrowia lipolytica*. Yeast. 1998;14:1267–83.
11. Madzak C, Treton B, Blanchin-Roland S. Strong hybrid promoters and integrative expression/secretion vectors for quasiconstitutive expression of heterologous proteins in the yeast *Yarrowia lipolytica*. Mol Microbiol Biotechnol. 2000;2:207–16.
12. Nthangeni MB, Urban P, Pompon D, Smit MS, Nicaud JM. The use of *Yarrowia lipolytica* for the expression of human cytochrome P450 CYP1A1. Yeast. 2004;21:583–92.
13. Sassi H, Delvigne F, Kar T, Nicaud JM, Coq AM, Steels S, Fickers P. Deciphering how *LIP2* and *POX2* promoters can optimally regulate recombinant protein production in the yeast *Yarrowia lipolytica*. Microb Cell Fact. 2016;15:159.
14. Gasmi N, Fudalej F, Kallel H, Nicaud JM. A molecular approach to optimize hIFN α2b expression and secretion in *Yarrowia lipolytica*. Appl Microbiol Biotechnol. 2011;89:109–19.
15. Barth G, Gaillardin C. *Yarrowia lipolytica*. In: Wolf K, editor. Non conventional yeasts in biotechnology, vol. 1. Springer: Germany; 1996. p. 313–88.
16. Chen DC, Beckerich JM, Gaillardin C. One-step transformation of the dimorphic yeast *Yarrowia lipolytica*. Appl Microbiol Biotechnol. 1997;48:232–5.
17. Xuan JW, Fournier P, Gaillardin C. Cloning of the LYS5 gene encoding saccharopine dehydrogenase from the yeast *Yarrowia lipolytica* by target integration. Curr Genet. 1988;14:15–21.
18. Le Dall MT, Nicaud JM, Gaillardin C. Multiple-copy integration in the yeast *Yarrowia lipolytica*. Curr Genet. 1994;26:38–44.
19. Boonvitthya N, Bozonnet S, Burapatana V, O'Donohue MJ, Chulalaksananukul W. Comparison of the heterologous expression of *Trichoderma reesei* endoglucanase II and cellobiohydrolase II in the yeasts *Pichia pastoris* and *Yarrowia lipolytica*. Mol Biotechnol. 2013;54:158–69.
20. Wang W, Wei H, Alahuhta M, Chen X, Hyman D, Johnson DK, Zhang M, Himmel ME. Heterologous expression of xylanase enzymes in lipogenic yeast *Yarrowia lipolytica*. PLoS ONE. 2014;9:e111443.
21. Hong SP, Seip J, Walters-Pollak D, Rupert R, Jackson R, Xue Z, Zhu Q. Engineering *Yarrowia lipolytica* to express secretory invertase with strong $FBA1_{IN}$ promoter. Yeast. 2012;29:59–72.
22. De Pourcq K, Vervecken W, Dewerte I, Valevska A, Van Hecke A, Callewaert N. Engineering the yeast *Yarrowia lipolytica* for the production of therapeutic proteins homogeneously glycosylated with Man 8 GlcNAc 2 and Man5 GlcNAc2. Microb Cell Fact. 2012;11:53.
23. Lazar Z, Rossignol T, Verbeke J, Crutz-Le Coq AM, Nicaud JM, Robak M. Optimized invertase expression and secretion cassette for improving *Yarrowia lipolytica* growth on sucrose for industrial applications. J Ind Microbiol Biotechnol. 2013;40:1273–83.

24. Moon HY, Van TL, Cheon SA, Choo J, Kim JY, Kang HA. Cell-surface expression of *Aspergillus saitoi*-derived functional α-1,2-mannosidase on *Yarrowia lipolytica* for glycan remodeling. J Microbiol. 2013;51:506–14.
25. Nars G, Saurel O, Bordes F, Saves I, Remaud-Siméon M, André I, Milon A, Marty A. Production of stable isotope labelled lipase Lip2 from *Yarrowia lipolytica* for NMR: investigation of several expression systems. Protein Expr Purif. 2014;101:14–20.
26. Madzak C, Blanchin-Roland S, Cordero Otero RR, Gaillardin C. Functional analysis of upstream regulating regions from the *Yarrowia lipolytica XPR2* promoter. Microbiology. 1999;145:75–87.
27. Blazeck J, Liu L, Redden H, Alper H. Tuning gene expression in *Yarrowia lipolytica* by a hybrid promoter approach. Appl Environ Microbiol. 2011;77:7905–14.
28. Blazeck J, Reed B, Garg R, Gerstner R, Pan A, Agarwala V, Alper HS. Generalizing a hybrid synthetic promoter approach in *Yarrowia lipolytica*. Appl Microbiol Biotechnol. 2013;97:3037–52.
29. Shabbir Hussain M, Gambill L, Smith S, Blenner MA. Engineering promoter architecture in oleaginous yeast *Yarrowia lipolytica*. ACS Synth Biol. 2016;5:213–23.
30. Ledesma-Amaro R, Dulermo T, Nicaud JM. Engineering *Yarrowia lipolytica* to produce biodiesel from raw starch. Biotechnol Biofuels. 2015;8:148.
31. Mehmood N, Husson E, Jacquard C, Wewetzer S, Büchs J, Sarazin C, Gosselin I. Impact of two ionic liquids, 1-ethyl-3-methylimidazolium acetate and 1-ethyl-3-methylimidazolium methylphosphonate, on *Saccharomyces cerevisiae*: metabolic, physiologic, and morphological investigations. Biotechnol Biofuels. 2015;8:17.
32. Sambrook J, Maniatis T, Fritsch EF. Molecular cloning: a laboratory manual. 2nd ed. Cold Spring Harbor: Cold Spring Harbor Laboratory Press; 1989.
33. Mlícková K, Roux E, Athenstaedt K, d'Andrea S, Daum G, Chardot T, Nicaud JM. Lipid accumulation, lipid body formation, and acyl coenzyme A oxidases of the yeast *Yarrowia lipolytica*. Appl Environ Microbiol. 2004;70:3918–24.
34. Dulermo R, Gamboa-Meléndez H, Michely S, Thevenieau F, Neuvéglise C, Nicaud JM. The evolution of Jen3 proteins and their role in dicarboxylic acid transport in *Yarrowia*. Microbiologyopen. 2015;4:100–20.
35. Querol A, Barrio E, Huerta T, Ramón D. Molecular monitoring of wine fermentations conducted by active dry yeast strains. Appl Environ Microbiol. 1992;58:2948–53.
36. Dulermo R, Gamboa-Meléndez H, Ledesma-Amaro R, Thévenieau F, Nicaud JM. Unraveling fatty acid transport and activation mechanisms in *Yarrowia lipolytica*. Biochim Biophys Acta. 2015;1851:1202–17.
37. Viktor MJ, Rose SH, van Zyl WH, Viljoen-Bloom M. Raw starch conversion by Saccharomyces cerevisiae expressing *Aspergillus tubingensis* amylases. Biotechnol Biofuels. 2013;6:167.
38. Pignède G, Wang H, Fudalej F, Gaillardin C, Seman M, Nicaud JM. Autocloning vectors for gene expression and amplification for the yeast *Y. lipolytica*. Appl Environ Microbiol. 2000;66:3283–9.
39. Leplat C, Nicaud JM, Rossignol T. High-throughput transformation method for *Yarrowia lipolytica* mutant library screening. FEMS Yeast Res. 2015;15:fov052.
40. Dulermo R, Gamboa-Meléndez H, Dulermo T, Thevenieau F, Nicaud JM. The fatty acid transport protein Fat1p is involved in the export of fatty acids from lipid bodies in *Yarrowia lipolytica*. FEMS Yeast Res. 2014;14:883–96.
41. Idiris A, Tohda H, Kumagai H, Takegawa K. Engineering of protein secretion in yeast: strategies and impact on protein production. Appl Microbiol Biotechnol. 2010;86:403–17.
42. Li W, Zhou X, Lu P. Bottlenecks in the expression and secretion of heterologous proteins in *Bacillus subtilis*. Res Microbiol. 2004;155:605–10.
43. Ahmad M, Hirz M, Pichler H, Schwab H. Protein expression in *Pichia pastoris*: recent achievements and perspectives for heterologous protein production. Appl Microbiol Biotechnol. 2014;98:5301–17.
44. Liu Z, Tyo KE, Martínez JL, Petranovic D, Nielsen J. Different expression systems for production of recombinant proteins in *Saccharomyces cerevisiae*. Biotechnol Bioeng. 2012;109:1259–68.
45. Greenbaum D, Colangelo C, Williams K, Gerstein M. Comparing protein abundance and mRNA expression levels on a genomic scale. Genome Biol. 2003;4:117.
46. Vogel C, Marcotte EM. Insights into the regulation of protein abundance from proteomic and transcriptomic analyses. Nat Rev Genet. 2012;13:227–32.

7

Inheritance of brewing-relevant phenotypes in constructed *Saccharomyces cerevisiae* × *Saccharomyces eubayanus* hybrids

Kristoffer Krogerus[1,2*], Tuulikki Seppänen-Laakso[1], Sandra Castillo[1] and Brian Gibson[1]

Abstract

Background: Interspecific hybridization has proven to be a potentially valuable technique for generating de novo lager yeast strains that possess diverse and improved traits compared to their parent strains. To further enhance the value of hybridization for strain development, it would be desirable to combine phenotypic traits from more than two parent strains, as well as remove unwanted traits from hybrids. One such trait, that has limited the industrial use of de novo lager yeast hybrids, is their inherent tendency to produce phenolic off-flavours; an undesirable trait inherited from the *Saccharomyces eubayanus* parent. Trait removal and the addition of traits from a third strain could be achieved through sporulation and meiotic recombination or further mating. However, interspecies hybrids tend to be sterile, which impedes this opportunity.

Results: Here we generated a set of five hybrids from three different parent strains, two of which contained DNA from all three parent strains. These hybrids were constructed with fertile allotetraploid intermediates, which were capable of efficient sporulation. We used these eight brewing strains to examine two brewing-relevant phenotypes: stress tolerance and phenolic off-flavour formation. Lipidomics and multivariate analysis revealed links between several lipid species and the ability to ferment in low temperatures and high ethanol concentrations. Unsaturated fatty acids, such as oleic acid, and ergosterol were shown to positively influence growth at high ethanol concentrations. The ability to produce phenolic off-flavours was also successfully removed from one of the hybrids, Hybrid T2, through meiotic segregation. The potential application of these strains in industrial fermentations was demonstrated in wort fermentations, which revealed that the meiotic segregant Hybrid T2 not only didn't produce any phenolic off-flavours, but also reached the highest ethanol concentration and consumed the most maltotriose.

Conclusions: Our study demonstrates the possibility of constructing complex yeast hybrids that possess traits that are relevant to industrial lager beer fermentation and that are derived from several parent strains. Yeast lipid composition was also shown to have a central role in determining ethanol and cold tolerance in brewing strains.

Keywords: Yeast, Beer, Rare mating, Lipid, Fatty acid, Phenolic off-flavour, Aroma

Background

Yeast hybrids have been extensively used for centuries in the brewing and winemaking industries [1]. Lager yeast in particular, a natural interspecies hybrid between *Saccharomyces cerevisiae* and *Saccharomyces eubayanus*, is used for the majority of global industrial beer production. It possesses a range of desirable phenotypes that are relevant for the production of lager beer: cold tolerance, efficient use of wort sugars, and low formation of undesirable off-flavours. Recent studies have revealed that generating new lager yeast hybrids is a powerful strain-development tool, as hybrid strains have exhibited various improved traits including faster fermentation rates, more complete sugar use, and increases in aroma compound production [2–4]. Hybridization enables the combination and enhancement of phenotypic features

*Correspondence: kristoffer.krogerus@aalto.fi
[1] VTT Technical Research Centre of Finland, Tietotie 2, P.O. Box 1000, 02044 Espoo, Finland
Full list of author information is available at the end of the article

from two different parent strains [5]. In order to further improve the potential of hybridization for strain development, it would be desirable to combine phenotypic traits from more than two parent strains, as well as remove unwanted traits from the hybrid. This could be achieved through sporulation, meiotic recombination and further mating of the hybrid. However, interspecies yeast hybrids, such as lager yeast, tend to be sterile, and therefore their sporulation efficiencies and spore viabilities are usually poor [6–9]. Studies have revealed that allotetraploid hybrids are usually not constrained by sterility [6, 8], and these tend to be capable of producing viable diploid spores. Hence, allotetraploid interspecific hybrids may undergo meiosis, during which recombination may give rise to crossovers and gene conversions [10]. This in turn causes phenotypic variation, as traits may get strengthened, weakened or even removed [11].

Many strains of *Saccharomyces* produce vinyl phenols (POF; phenolic off-flavours) from hydroxycinnamic acids, and these phenolic compounds are considered undesirable in lager beer. The most well-studied of these vinyl phenols is 4-vinyl guaiacol, which is formed from ferulic acid. The ability of brewing yeast to produce volatile phenols has been attributed to the adjacent *PAD1* and *FDC1* genes, both of which are essential for the POF+ phenotype [12]. Wild yeast strains, such as the *S. eubayanus* strains that are available for de novo lager yeast creation, tend to have functional *PAD1* and *FDC1* genes, while domesticated POF− brewing yeast have nonsense or frameshift mutations in these genes, rendering them non-functional [13–15]. As the functional genes from *S. eubayanus* are passed on to any lager hybrids created from it, these hybrids have all been afflicted with the POF+ phenotype [2–4]. However, if any non-functional alleles of *PAD1* or *FDC1* are present in the hybrid genome, it may be possible to remove the POF+ phenotype through meiotic recombination as demonstrated in studies with intraspecific *S. cerevisiae* hybrids [13, 16].

Besides the ability to produce clean flavour profiles (i.e. no off-aromas such as 4-vinyl guaiacol), one of the main traits of lager yeast is its ability to stay metabolically active and ferment in the cold [1]. Recent studies have revealed that the *S. eubayanus* parent strain contributes this cold tolerance to lager yeast [4, 17]. However, unlike the POF+ phenotype, the mechanisms that contribute to cold tolerance in brewing yeast are not fully understood. It has been revealed that low temperature affects protein translation and folding efficiencies, mRNA stabilities and the product activity and expression of central metabolic genes [18–24]. Recent studies have also suggested that differences in the lipid composition of the plasma membrane play a vital role in temperature tolerance [25–27]. The fluidity of the plasma membrane is affected by its lipid composition and the temperature [26, 28], and a decrease in membrane fluidity caused by a low fermentation temperature can, in turn, result in impaired transporter function and lower nutrient uptake [27, 29]. Relatively high levels of ergosterol and low levels of unsaturated fatty acids could, for example, result in a greater tendency of cell membranes to freeze at lower temperatures, thereby reducing functionality [26, 30, 31]. The lipid composition of yeast has also been shown to influence ethanol tolerance [32]. Similarly to cold tolerance, ethanol tolerance has been shown to be dependent on unsaturated fatty acid and ergosterol concentrations [31, 33]. They have been hypothesized to function by maintaining optimum membrane thickness and fluidity by counteracting the fluidizing effects of ethanol. Here, we wanted to investigate what lipid species correlated with good fermentation performance at low temperatures and high alcohol levels, by examining how the lipidomes of brewing yeasts react to changes in temperature and ethanol concentration.

In this study we wanted to both demonstrate the possibility of using an allotetraploid interspecific hybrid as an intermediate to create a POF− lager hybrid with DNA from three parent strains, and use these hybrids to elucidate relationships between different lipid species and tolerance towards low temperatures and high ethanol concentrations. This was accomplished by first rare mating a set of three parent strains, all with different desirable properties, to produce five different hybrid strains. Their fermentation kinetics and lipid compositions were compared in small-scale fermentations performed at low temperatures and in the presence of ethanol. Lipids were analysed using both a targeted GC/MS and non-targeted UPLC/MS approach. Using multivariate analysis, it was revealed that high levels of unsaturated fatty acids and especially phospholipids containing palmitoleic and oleic acid were associated with good fermentation performance at low temperatures and high ethanol concentrations. Furthermore, ergosterol was associated with good fermentation performance at high ethanol concentrations, but had a negative effect at low temperatures. These observations were supported with growth data of laboratory strains lacking *OLE1* and *ERG4* genes in media containing ethanol and oleic acid. The potential application of these strains in industrial fermentations was finally demonstrated in 2-L wort fermentations. These revealed that Hybrid T2, a meiotic segregant of Hybrid T1 (containing DNA from all three parent strains), reached the highest ethanol concentration, consumed the most maltotriose, and did not produce any 4-vinyl guaiacol, therefore making it a suitable candidate for industrial lager beer fermentations.

Results

Generation of inter- and intra-specific hybrids

The set of 8 brewing strains that were used in this study consisted of 3 parent strains and 5 hybrid strains (Table 1). The three parent strains, P1 and P2 being industrial ale strains and P3 being the type strain of *S. eubayanus*, were chosen for their varying phenotypic properties. Of the three, P1 is the only strain that is able to use maltotriose during fermentation, P2 is the only strain that does not produce 4-vinyl guaiacol (i.e. it is POF−), while P3 is the cold-tolerant parent strain of lager yeast. From the three parent strains, the three hybrid strains H1–H3 were first successfully constructed through rare mating according to the schematic in Fig. 1. The hybrid status of these strains was confirmed using PCR (ITS, species-specific, and interdelta primers), which showed that they contained DNA from all their parent strains (Additional file 1: Figure S1). Flow cytometry also revealed that these hybrid strains were all polyploid (Additional file 1: Figure S2), containing higher amounts of DNA than any of the 3 parent strains P1–P3.

The tetraploid nature of H1 allowed it to form viable spores (47% viability) despite being an interspecific hybrid, and these spores could subsequently be mated with P2 to form the triple hybrid T1 (Fig. 1). PCR was again used to confirm that the strain contained DNA from all three parent strains (Additional file 1: Figure S1). Whole genome sequencing of the strains also revealed that Hybrid T1 contained a higher ratio of *S. cerevisiae*-derived to *S. eubayanus*-derived chromosomes (approximately 3:1; Additional file 1: Figure S3E). Additionally, the strain appeared to contain a chimeric chromosome consisting of approximately 355 kbp of the *S. cerevisiae* P1-derived chromosome II and 437 kbp of the *S. eubayanus* P3-derived chromosome IV, apparently formed during sporulation of Hybrid H1. Like Hybrid H1, Hybrid T1 was also able to form viable spores (38% viability). The spore clones of T1 were screened for the POF phenotype in media containing ferulic acid, and out of 12 spores clones that were assayed, only 3 were POF−. The best growing of these was given the name Hybrid T2, i.e. a POF− meiotic segregant of Hybrid T1. Whole genome sequencing of the strains revealed that Hybrid T2 contained at least one copy of the *S. cerevisiae*-derived chromosomes, but had lost several chromosomes derived from *S. eubayanus* (Additional file 1: Figure S3F).

Effects of temperature and ethanol on fermentation rate and lipid composition

To investigate how the fermentation rate and the lipid composition of the 8 brewing strains was affected by the fermentation temperature and ethanol content of the growth media, a set of 100 mL flask fermentations were carried out. The fermentations were performed at two temperatures, 10 and 20 °C, and with two initial ethanol concentrations, 0 and 8% (v/v). All eight strains were able to finish fermentation (a decrease in approx. 7 °brix) at both 10 and 20 °C in the media without any added ethanol (Fig. 2a, b; Table 2). While, *S. eubayanus* (P3) had the shortest lag time (λ) at 10 °C as we had expected, *S. cerevisiae* P1 and P2 rather surprisingly showed the highest maximum fermentation rate (μ) at this temperature. Despite the fast fermentation rate, the amount of biomass produced at 10 °C without any added ethanol by *S. cerevisiae* P1 and P2, and the intraspecific Hybrid H3 all showed the lowest level of accumulated biomass ($p < 0.05$) when compared to the fermentations at 20 °C without any added ethanol.

Fermentations were not as efficient in the growth media supplemented with 8% (v/v) ethanol. At 10 °C, it

Table 1 Yeast strains used in the study

ID	Species	Information	Source
Brewing strains			
P1	*S. cerevisiae*	VTT-A81062. Maltotriose fermentation, POF+	VTT culture collection
P2	*S. cerevisiae*	WLP099. No maltotriose fermentation, POF−	White Labs Inc
P3	*S. eubayanus*	VTT-C12902. No maltotriose fermentation, POF+	VTT culture collection
H1	Hybrid	P1 × P3 interspecific hybrid (VTT-A15225)	Created in this study
H2	Hybrid	P2 × P3 interspecific hybrid	Created in this study
H3	Hybrid	P1 × P2 intraspecific hybrid	Created in this study
T1	Hybrid	H1 × P2 interspecific triple hybrid	Created in this study
T2	Hybrid	Meiotic segregant of T1	Created in this study
Laboratory strains			
WT	*S. cerevisiae*	BY4741 *MAT***a**; *his3Δ1*; *leu2Δ0*; *met15Δ0*; *ura3Δ0*	EUROSCARF (Y00000)
ole1Δ	*S. cerevisiae*	BY4741 *MAT***a**; *his3Δ1*; *leu2Δ0*; *met15Δ0*; *ura3Δ0*; *ole1-m2:kanMX*	EUROSCARF (Y40963)
erg4Δ	*S. cerevisiae*	BY4741 *MAT***a**; *his3Δ1*; *leu2Δ0*; *met15Δ0*; *ura3Δ0*; *YGL012w::kanMX4*	EUROSCARF (Y04380)

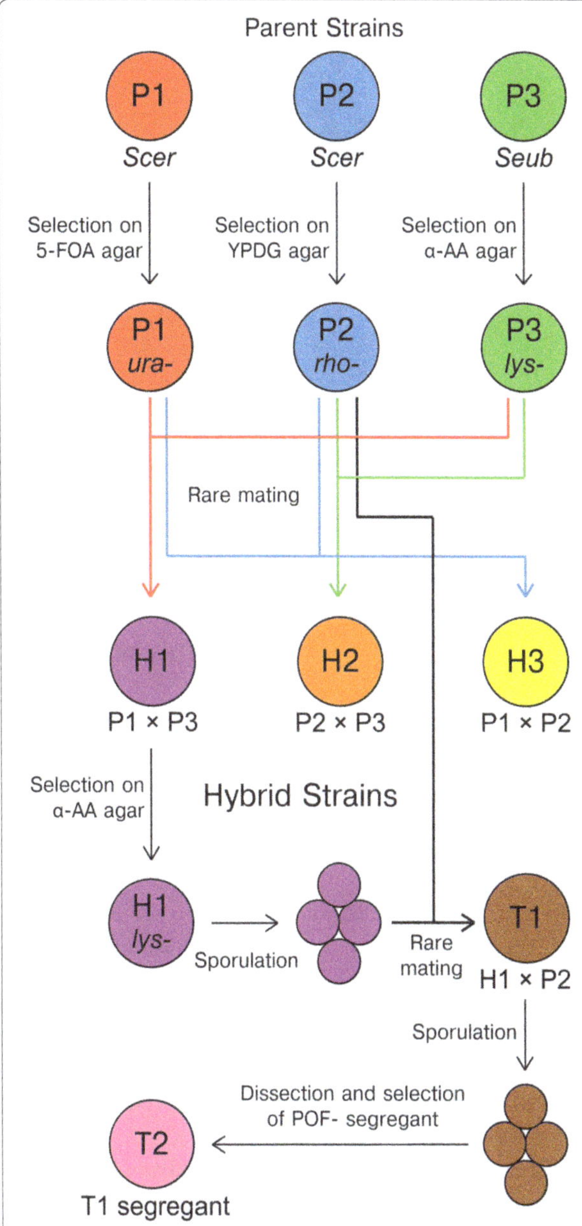

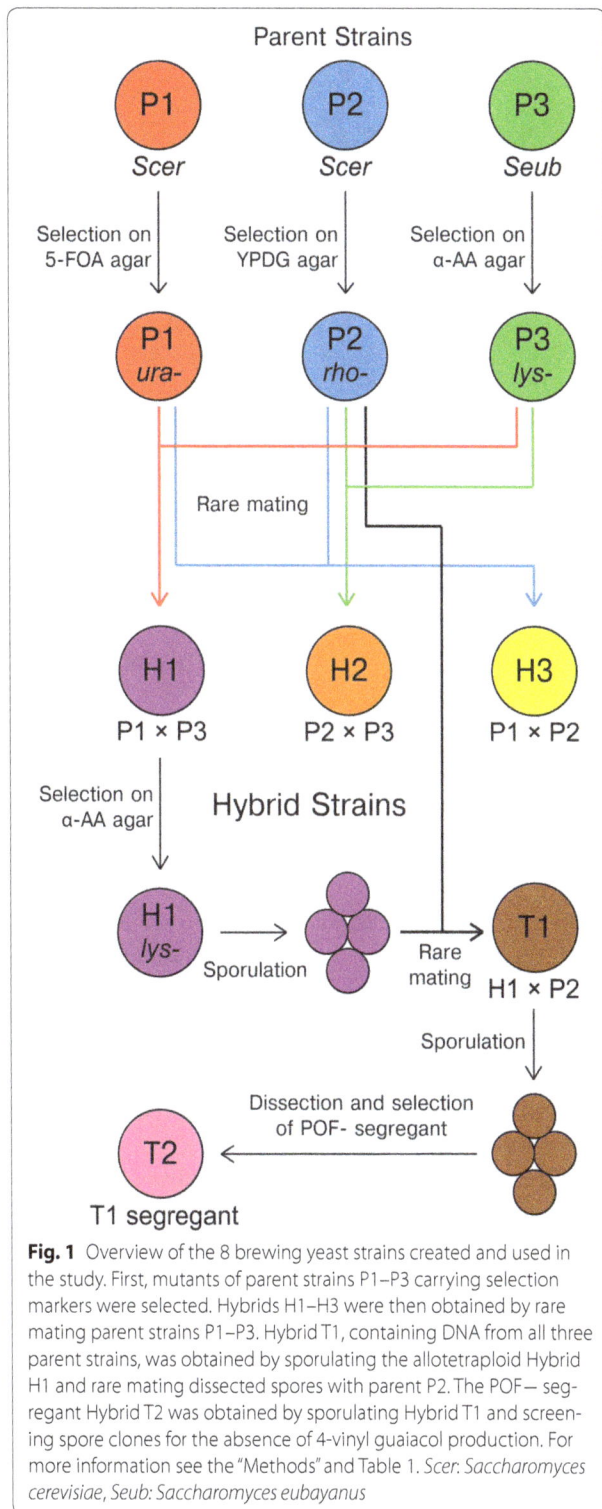

Fig. 1 Overview of the 8 brewing yeast strains created and used in the study. First, mutants of parent strains P1–P3 carrying selection markers were selected. Hybrids H1–H3 were then obtained by rare mating parent strains P1–P3. Hybrid T1, containing DNA from all three parent strains, was obtained by sporulating the allotetraploid Hybrid H1 and rare mating dissected spores with parent P2. The POF− segregant Hybrid T2 was obtained by sporulating Hybrid T1 and screening spore clones for the absence of 4-vinyl guaiacol production. For more information see the "Methods" and Table 1. *Scer: Saccharomyces cerevisiae*, *Seub: Saccharomyces eubayanus*

was only *S. eubayanus* P3 and Hybrid H1 which managed to reach the same maximum fermentation level (A) as in the unsupplemented growth media. However, at 20 °C P3 both fermented and grew the worst of the 8 strains, suggesting it is ethanol-tolerant at 10 °C but not 20 °C. *S. cerevisiae* P1 on the other hand, performed relatively well in the presence of ethanol at 20 °C, but not at 10 °C. The interspecific hybrid between the two, Hybrid H1, was able to ferment in the presence of ethanol at both 10 and 20 °C, and even outperforming P1 with regards to maximum fermentation rate (μ) at 20 °C. The two triple hybrids T1 and T2, as well as the *S. cerevisiae* P2 (the genome of which dominates in T1 and T2), performed poorly in the presence of ethanol at both 10 and 20 °C. Hybrid T2 did not grow or ferment at 10 °C and with 8% supplemented ethanol, and samples of it at this condition were thus excluded from subsequent lipid analysis.

Analysis of the fatty acid composition, squalene content, and ergosterol content of the 8 brewing yeast strains sampled at the end of the exponential phase during the flask fermentations revealed significant differences between the different strains and conditions (Fig. 3; Additional file 1: Table S1). In almost all samples, the unsaturated fatty acids palmitoleic acid (C16:1) and oleic acid (C18:1) were the most prevalent of the fatty acids. Of the strains, *S. eubayanus* P3 was particularly high in palmitoleic acid content, while *S. cerevisiae* P2 was particularly high in oleic acid content. When conditions were changed from the control (20 °C and no supplemented ethanol), there was a slight increase in both the ratio of unsaturated to saturated fatty acids (from 2.0 up to 4.9) and average chain length (from 16.8 up to 17.0). An exception was the fermentations at 10 °C and 8% supplemented ethanol, where the ratio of unsaturated to saturated fatty acids instead decreased slightly (from 2.0 to 1.7). The concentrations of ergosterol were in general higher at the fermentations performed at 20 °C compared to those at 10 °C. Of the strains, *S. cerevisiae* P1 tended to have the highest levels of ergosterol, while *S. eubayanus* P3 tended to have the lowest levels of ergosterol. No clear patterns were observed between the strains and the conditions for the squalene concentrations, other than that the highest concentrations of squalene were measured for the fermentations performed at 10 °C and 8% supplemented ethanol.

The UPLC/MS lipidomics analysis of the 8 brewing yeast strains sampled at the end of the exponential phase during the flask fermentations also revealed differences between the different strains and conditions (Fig. 4). Based on retention times and m/z spectra we identified 60 lipid species from the samples (Additional file 2). The most prevalent lipid groups in the samples were the phosphatidylethanolamines (PE) and phosphatidylinositols (PI). Of the individual lipid species, PE(32:2), PE(34:2), PI(34:1), PI(34:2) and PI(36:1) were present in the highest concentrations (Fig. 4a). As the growth temperature was lowered from 20 to 10 °C and ethanol was

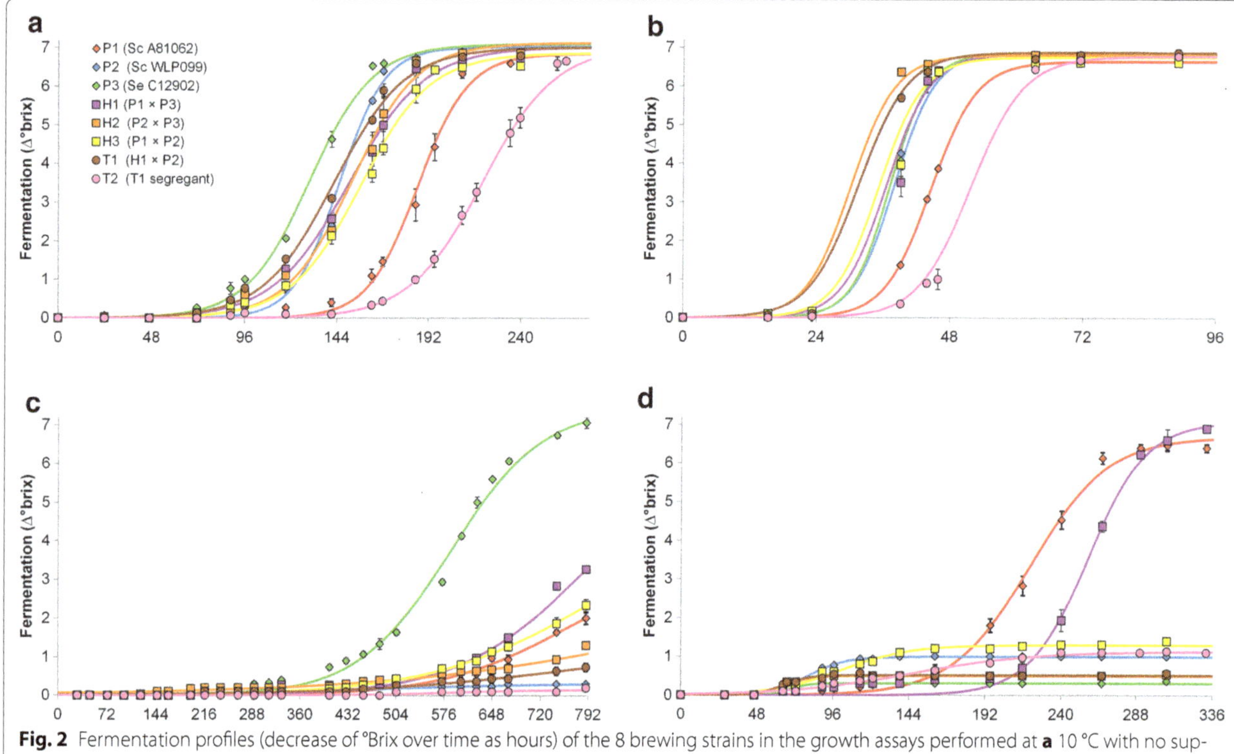

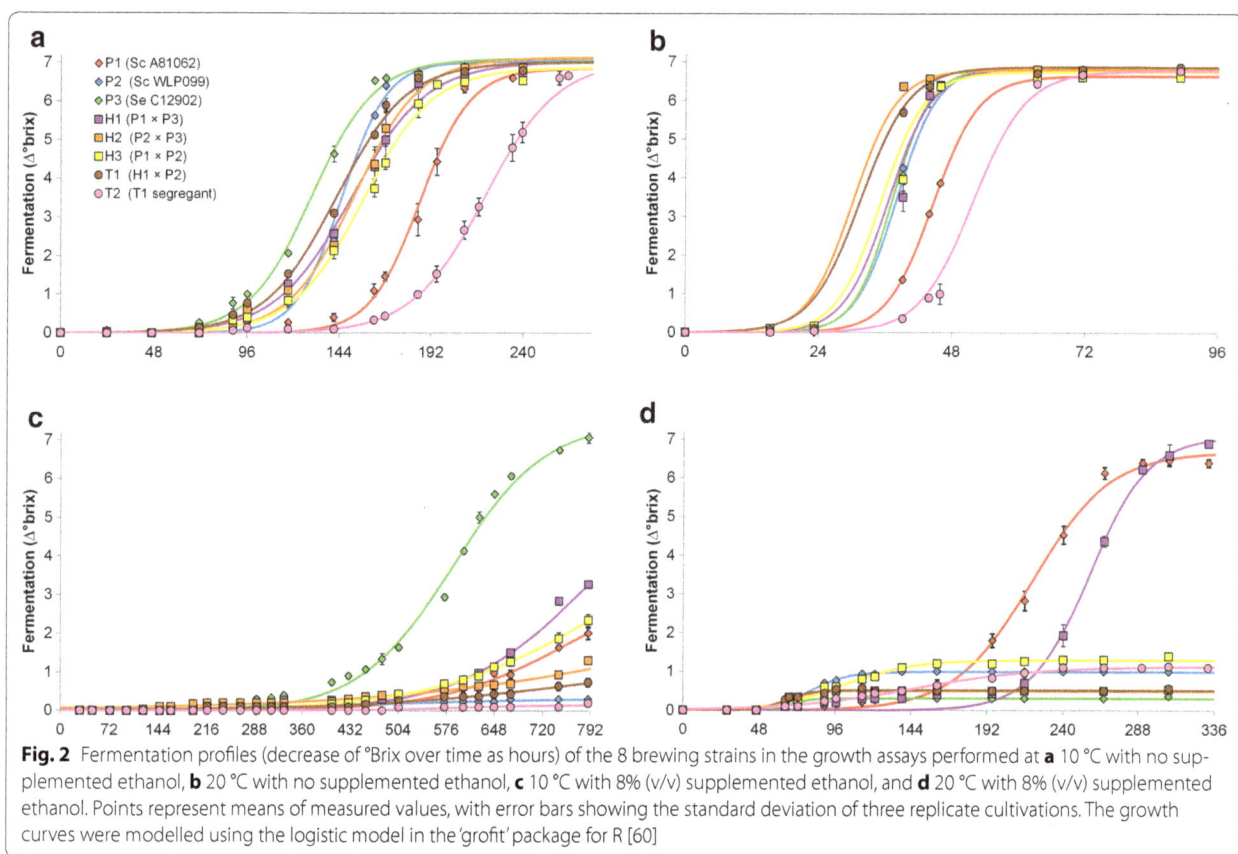

Fig. 2 Fermentation profiles (decrease of °Brix over time as hours) of the 8 brewing strains in the growth assays performed at **a** 10 °C with no supplemented ethanol, **b** 20 °C with no supplemented ethanol, **c** 10 °C with 8% (v/v) supplemented ethanol, and **d** 20 °C with 8% (v/v) supplemented ethanol. Points represent means of measured values, with error bars showing the standard deviation of three replicate cultivations. The growth curves were modelled using the logistic model in the 'grofit' package for R [60]

supplemented to the growth media, significant changes were observed for 55% of the lipid species (Fig. 4b). Similarly to what was revealed by the fatty acid analysis, there was an increase in unsaturation ratio and chain length among the lipid species when the temperature was lowered and the ethanol content was increased.

Using Partial Least Squares Discriminant Analysis (PLS-DA), it was possible to separate the four different fermentation conditions based on the lipid data. A PLS-DA model was first constructed based on the measured fatty acid, squalene and ergosterol compositions (Fig. 5a, b). The PLS-DA model was cross-validated and can be considered significant ($Q^2 > 0.5$) [34]. As indicated by the compositions in Fig. 3, there is considerable overlap between the fermentations at 10 °C and no supplemented ethanol and 20 °C with 8% supplemented in the PLS-DA model, suggesting that the fatty acid composition of these strains respond similarly in regards to cold and ethanol stress. A cross-validated and significant PLS-DA model could also separate the four different fermentation conditions based on the compositions of the 60 lipid species determined by UPLC/MS (Fig. 5c, d). The four groups were separated by initial ethanol content along the first model component (t1) and the temperature along the second model component (t2). Like the PLS-DA model produced from the fatty acid, squalene and ergosterol data (Fig. 5a, b), there was considerable overlap between the lipid profiles of the fermentations at lower temperature and with supplemented ethanol, again suggesting that the lipid composition of these strains respond similarly in regards to cold and ethanol stress.

In order to elucidate what lipid species were associated with the yeast strains showing good fermentation performance at 10 °C and in the presence of ethanol, multivariate analysis was performed on both the lipid data (GC/MS and UPLC/MS) and the fermentation parameters (Table 2). Fermentation performance was quantitated based on the ratio (A/λ) of the maximum fermentation level (A in Table 2) and the lag time (λ in Table 2). Hence, this value increases with an increased fermentation level or a shorter lag time. It was attempted to construct PLS regression models for the different fermentation conditions using this ratio (A/λ) as the *Y* response variable and the lipid data as the *X* predictor variables (Fig. 6). Only the models produced using the data from the fermentation at 10 °C and no supplemented ethanol (Fig. 6a, b) and 20 °C and 8% supplemented ethanol (Fig. 6e, f) were successfully cross-validated ($Q^2 > 0.5$ and significant at $p < 0.05$ by random permutation test [34]). The model produced using the data from the fermentation at 10 °C

Table 2 Modelled (*A*, μ, λ) and measured (dry mass) fermentation and growth parameters of the fermentation assays that were performed at different temperatures (10 and 20 °C) and supplemented ethanol levels (0 and 8% (v/v) EtOH)

Strain and condition		*A* (°brix)	μ (°brix h^{-1})	λ (hours)	Dry mass (g L^{-1})
10 °C, 0% EtOH	P1	6.85 (±0.18)^{ab}	0.132 (±0.0107)^{ab}	160 (±2.3)^{b}	9.2 (±0.65)^{cd}
	P2	6.63 (±0.13)^{b}	0.140 (±0.0172)^{a}	122 (±2.3)^{c}	9.8 (±0.05)^{bcd}
	P3	7.10 (±0.15)^{a}	0.108 (±0.0064)^{bc}	97 (±2.2)^{e}	11.0 (±0.33)^{abc}
	H1 (P1 × P3)	7.02 (±0.18)^{ab}	0.089 (±0.0059)^{c}	111 (±3.1)^{d}	12.2 (±1.62)^{ab}
	H2 (P2 × P3)	7.13 (±0.19)^{a}	0.106 (±0.0083)^{c}	118 (±3.2)^{c}	12.6 (±0.14)^{a}
	H3 (P1 × P2)	6.87 (±0.16)^{ab}	0.095 (±0.0058)^{c}	120 (±2.6)^{c}	7.5 (±1.75)^{d}
	T1 (H1 × P2)	7.01 (±0.14)^{ab}	0.095 (±0.0054)^{c}	106 (±2.5)^{d}	12.7 (±0.39)^{a}
	T2 (T1 segregant)	7.05 (±0.13)^{ab}	0.093 (±0.0015)^{c}	182 (±0.7)^{a}	10.9 (±0.16)^{abc}
20 °C, 0% EtOH	P1	6.63 (±0.02)^{b}	0.403 (±0.0069)^{a}	36 (±0.1)^{b}	12.5 (±3.12)^{a}
	P2	6.86 (±0.05)^{a}	0.470 (±0.0630)^{a}	31 (±0.6)^{c}	11.6 (±0.16)^{a}
	P3	6.83 (±0.07)^{a}	0.501 (±0.1218)^{a}	30 (±0.8)^{cd}	10.6 (±0.58)^{a}
	H1 (P1 × P3)	6.83 (±0.04)^{a}	0.435 (±0.0406)^{a}	29 (±0.3)^{de}	13.9 (±1.55)^{a}
	H2 (P2 × P3)	6.80 (±0.03)^{a}	0.426 (±0.0470)^{a}	22 (±0.7)^{f}	10.6 (±0.18)^{a}
	H3 (P1 × P2)	6.74 (±0.06)^{ab}	0.428 (±0.1647)^{a}	27 (±0.7)^{e}	13.3 (±1.88)^{a}
	T1 (H1 × P2)	6.87 (±0.05)^{a}	0.392 (±0.0494)^{a}	23 (±1.0)^{f}	13.1 (±0.80)^{a}
	T2 (T1 segregant)	6.77 (±0.08)^{ab}	0.357 (±0.0292)^{a}	42 (±0.4)^{a}	11.5 (±0.22)^{a}
10 °C, 8% EtOH	P1	3.45 (±0.53)^{d}	0.009 (±0.0007)^{bc}	569 (±14.9)^{ab}	4.3 (±0.4)^{bc}
	P2	0.32 (±0.03)^{e}	0.001 (±0.0001)^{d}	186 (±28.3)^{d}	0.3 (±0.04)^{bc}
	P3	7.50 (±0.20)^{a}	0.027 (±0.0010)^{a}	446 (±5.7)^{c}	15.7 (±1.97)^{a}
	H1 (P1 × P3)	6.45 (±0.96)^{ab}	0.017 (±0.0082)^{b}	596 (±58.9)^{a}	8.1 (±5.26)^{b}
	H2 (P2 × P3)	4.22 (±0.63)^{cd}	0.004 (±0.0008)^{cd}	559 (±43.5)^{ab}	6.4 (±0.24)^{b}
	H3 (P1 × P2)	5.14 (±0.60)^{bc}	0.010 (±0.0008)^{bc}	561 (±16.6)^{ab}	3.3 (±0.09)^{bc}
	T1 (H1 × P2)	1.53 (±0.51)^{e}	0.002 (±0.0005)^{cd}	464 (±66.3)^{bc}	3.3 (±0.70)^{bc}
	T2 (T1 segregant)	NA	NA	NA	NA
20 °C, 8% EtOH	P1	6.69 (±0.11)^{b}	0.068 (±0.0036)^{b}	171 (±2.8)^{b}	7.5 (±1.53)^{b}
	P2	1.00 (±0.01)^{d}	0.020 (±0.0011)^{c}	56 (±1.3)^{d}	1.5 (±0.08)^{c}
	P3	0.31 (±0.01)^{e}	0.013 (±0.0031)^{cd}	57 (±2.9)^{d}	0.8 (±0.10)^{c}
	H1 (P1 × P3)	7.09 (±0.19)^{a}	0.095 (±0.0068)^{a}	220 (±2.9)^{a}	14.9 (±0.10)^{a}
	H2 (P2 × P3)	0.50 (±0.01)^{e}	0.015 (±0.0033)^{cd}	49 (±4.1)^{d}	1.5 (±0.06)^{c}
	H3 (P1 × P2)	1.29 (±0.03)^{c}	0.015 (±0.0014)^{cd}	58 (±4.4)^{d}	1.5 (±0.09)^{c}
	T1 (H1 × P2)	0.51 (±0.01)^{e}	0.016 (±0.0037)^{cd}	49 (±4.4)^{d}	1.1 (±0.06)^{c}
	T2 (T1 segregant)	1.14 (±0.03)^{cd}	0.007 (±0.0005)^{d}	70 (±4.5)^{c}	1.3 (±0.04)^{c}

NA not available

The growth curves were modelled using the logistic model in the 'grofit' package for R. Values were determined from three independent cultivations (standard deviation in parentheses). Values in the same group in the same column with different superscript letters (a–f) differ significantly ($p < 0.05$ as determined with one-way ANOVA with Tukey's post hoc test)

and 8% supplemented ethanol (Fig. 6c, d) had a Q^2 value of 0.49 ($p < 0.05$), i.e. just below the threshold of 0.5, and will still be presented.

The three models all showed an increase in fermentation performance along the first model component (t1), with the highest *Y* values (i.e. A/λ) in the upper right quadrant (Fig. 6a, c, e). The 10 individual lipid species which had the largest Variable Importance in Projection (VIP) scores and regression coefficients for each PLS model are shown in Table 3. These are the lipid species that had either the largest positive or negative effect on the *Y* response variable, i.e. the fermentation performance. At 10 °C and no supplemented ethanol, phosphatidylcholines (PC), phosphatidylinositols (PI) and sterol esters containing unsaturated fatty acids seem to positively correlate with fermentation performance, while longer-chain phosphatidylinositols and phosphatidylserines (e.g. PI(36:0), PI(36:1), PI(36:2) and PS(36:1)) and saturated fatty acids (e.g. C14:0 and C18:0) were negatively correlated with fermentation performance (Fig. 6a, b; Table 3). Similarly at 10 °C and 8% supplemented ethanol, phosphatidylcholines (PC) and phosphatidylethanolamines (PE) containing palmitoleic acid [e.g. PC(30:1), PC(30:2) and PE(32:2)] seem

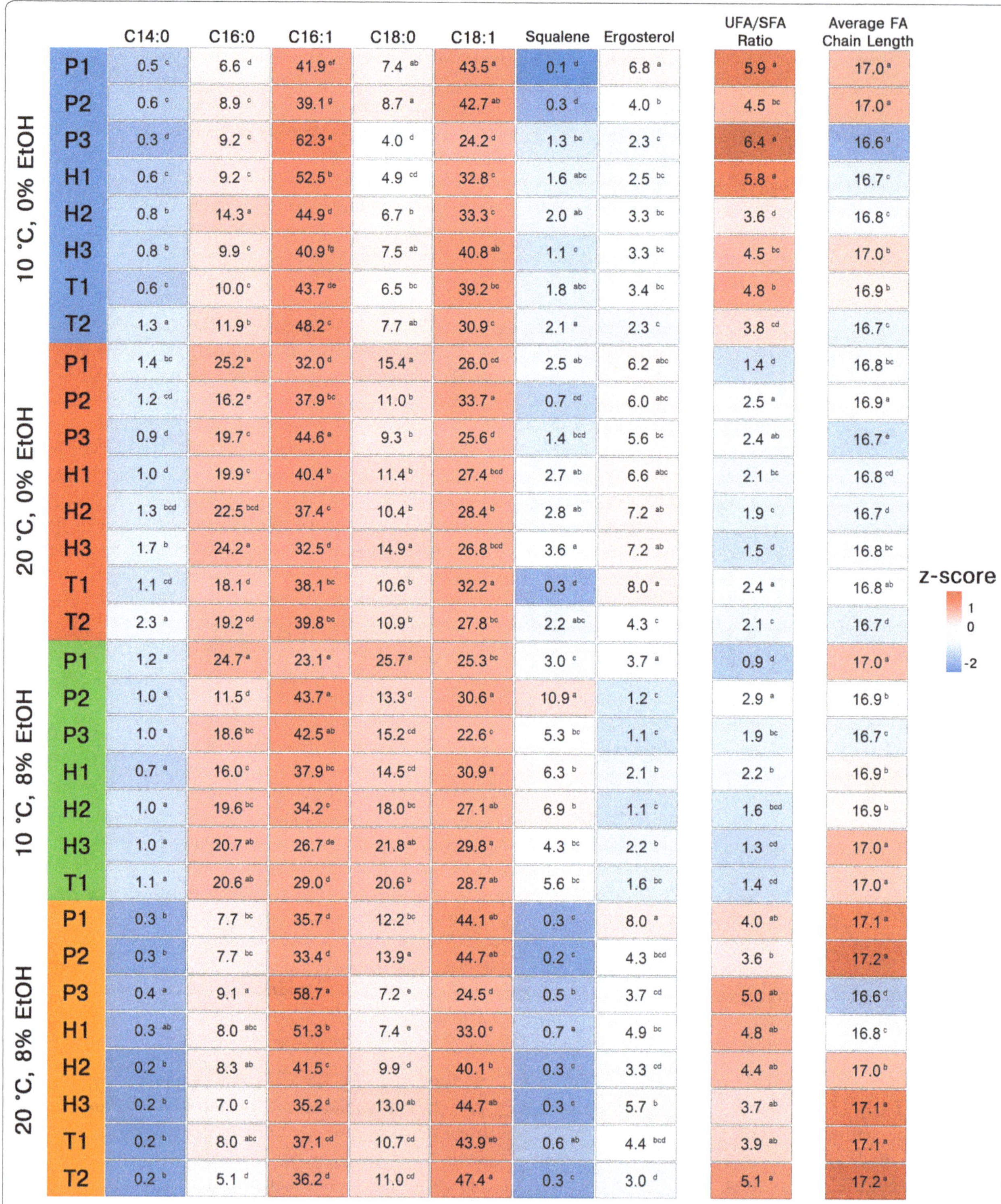

Condition	Strain	C14:0	C16:0	C16:1	C18:0	C18:1	Squalene	Ergosterol	UFA/SFA Ratio	Average FA Chain Length
10 °C, 0% EtOH	P1	0.5^{c}	6.6^{d}	41.9^{ef}	7.4^{ab}	43.5^{a}	0.1^{d}	6.8^{a}	5.9^{a}	17.0^{a}
	P2	0.6^{c}	8.9^{c}	39.1^{g}	8.7^{a}	42.7^{ab}	0.3^{d}	4.0^{b}	4.5^{bc}	17.0^{a}
	P3	0.3^{d}	9.2^{c}	62.3^{a}	4.0^{d}	24.2^{d}	1.3^{bc}	2.3^{c}	6.4^{a}	16.6^{d}
	H1	0.6^{c}	9.2^{c}	52.5^{b}	4.9^{cd}	32.8^{c}	1.6^{abc}	2.5^{bc}	5.8^{a}	16.7^{c}
	H2	0.8^{b}	14.3^{a}	44.9^{d}	6.7^{b}	33.3^{c}	2.0^{ab}	3.3^{bc}	3.6^{d}	16.8^{c}
	H3	0.8^{b}	9.9^{c}	40.9^{fg}	7.5^{ab}	40.8^{ab}	1.1^{c}	3.3^{bc}	4.5^{bc}	17.0^{b}
	T1	0.6^{c}	10.0^{c}	43.7^{de}	6.5^{bc}	39.2^{bc}	1.8^{abc}	3.4^{bc}	4.8^{b}	16.9^{b}
	T2	1.3^{a}	11.9^{b}	48.2^{c}	7.7^{ab}	30.9^{c}	2.1^{a}	2.3^{c}	3.8^{cd}	16.7^{c}
20 °C, 0% EtOH	P1	1.4^{bc}	25.2^{a}	32.0^{d}	15.4^{a}	26.0^{cd}	2.5^{ab}	6.2^{abc}	1.4^{d}	16.8^{bc}
	P2	1.2^{cd}	16.2^{e}	37.9^{bc}	11.0^{b}	33.7^{a}	0.7^{cd}	6.0^{abc}	2.5^{a}	16.9^{a}
	P3	0.9^{d}	19.7^{c}	44.6^{a}	9.3^{b}	25.6^{d}	1.4^{bcd}	5.6^{bc}	2.4^{ab}	16.7^{e}
	H1	1.0^{d}	19.9^{c}	40.4^{b}	11.4^{b}	27.4^{bcd}	2.7^{ab}	6.6^{abc}	2.1^{bc}	16.8^{cd}
	H2	1.3^{bcd}	22.5^{bcd}	37.4^{c}	10.4^{b}	28.4^{b}	2.8^{ab}	7.2^{ab}	1.9^{c}	16.7^{d}
	H3	1.7^{b}	24.2^{a}	32.5^{d}	14.9^{a}	26.8^{bcd}	3.6^{a}	7.2^{ab}	1.5^{d}	16.8^{bc}
	T1	1.1^{cd}	18.1^{d}	38.1^{bc}	10.6^{b}	32.2^{a}	0.3^{d}	8.0^{a}	2.4^{a}	16.8^{ab}
	T2	2.3^{a}	19.2^{cd}	39.8^{bc}	10.9^{b}	27.8^{bc}	2.2^{abc}	4.3^{c}	2.1^{c}	16.7^{d}
10 °C, 8% EtOH	P1	1.2^{a}	24.7^{a}	23.1^{e}	25.7^{a}	25.3^{bc}	3.0^{c}	3.7^{a}	0.9^{d}	17.0^{a}
	P2	1.0^{a}	11.5^{d}	43.7^{a}	13.3^{d}	30.6^{a}	10.9^{a}	1.2^{c}	2.9^{a}	16.9^{b}
	P3	1.0^{a}	18.6^{bc}	42.5^{ab}	15.2^{cd}	22.6^{c}	5.3^{bc}	1.1^{c}	1.9^{bc}	16.7^{c}
	H1	0.7^{a}	16.0^{c}	37.9^{bc}	14.5^{cd}	30.9^{a}	6.3^{b}	2.1^{b}	2.2^{b}	16.9^{b}
	H2	1.0^{a}	19.6^{bc}	34.2^{c}	18.0^{bc}	27.1^{ab}	6.9^{b}	1.1^{c}	1.6^{bcd}	16.9^{b}
	H3	1.0^{a}	20.7^{ab}	26.7^{de}	21.8^{ab}	29.8^{a}	4.3^{bc}	2.2^{b}	1.3^{cd}	17.0^{a}
	T1	1.1^{a}	20.6^{ab}	29.0^{d}	20.6^{b}	28.7^{ab}	5.6^{bc}	1.6^{bc}	1.4^{cd}	17.0^{a}
20 °C, 8% EtOH	P1	0.3^{b}	7.7^{bc}	35.7^{d}	12.2^{bc}	44.1^{ab}	0.3^{c}	8.0^{a}	4.0^{ab}	17.1^{a}
	P2	0.3^{b}	7.7^{bc}	33.4^{d}	13.9^{a}	44.7^{ab}	0.2^{c}	4.3^{bcd}	3.6^{b}	17.2^{a}
	P3	0.4^{a}	9.1^{a}	58.7^{a}	7.2^{e}	24.5^{d}	0.5^{b}	3.7^{cd}	5.0^{ab}	16.6^{d}
	H1	0.3^{ab}	8.0^{abc}	51.3^{b}	7.4^{e}	33.0^{c}	0.7^{a}	4.9^{bc}	4.8^{ab}	16.8^{c}
	H2	0.2^{b}	8.3^{ab}	41.5^{c}	9.9^{d}	40.1^{b}	0.3^{c}	3.3^{cd}	4.4^{ab}	17.0^{b}
	H3	0.2^{b}	7.0^{c}	35.2^{d}	13.0^{ab}	44.7^{ab}	0.3^{c}	5.7^{b}	3.7^{ab}	17.1^{a}
	T1	0.2^{b}	8.0^{abc}	37.1^{cd}	10.7^{cd}	43.9^{ab}	0.6^{ab}	4.4^{bcd}	3.9^{ab}	17.1^{a}
	T2	0.2^{b}	5.1^{d}	36.2^{d}	11.0^{cd}	47.4^{a}	0.3^{c}	3.0^{d}	5.1^{a}	17.2^{a}

Fig. 3 The relative concentrations (% of total lipid content) of fatty acids, squalene and ergosterol, unsaturated (UFA) to saturated (SFA) fatty acid ratio, and average fatty acid (FA) chain length in the lipids extracted from cells in late exponential phase during the growth assays. Values are means from three independent cultivations. Values in the same group in the same column with different superscript letters (*a–g*) differ significantly ($p < 0.05$ as determined with one-way ANOVA with Tukey's post hoc test). The heatmap was generated based on the z-scores (a *blue color* indicates a negative z-score and lower concentration, while a *red color* indicates a positive z-score and a higher concentration)

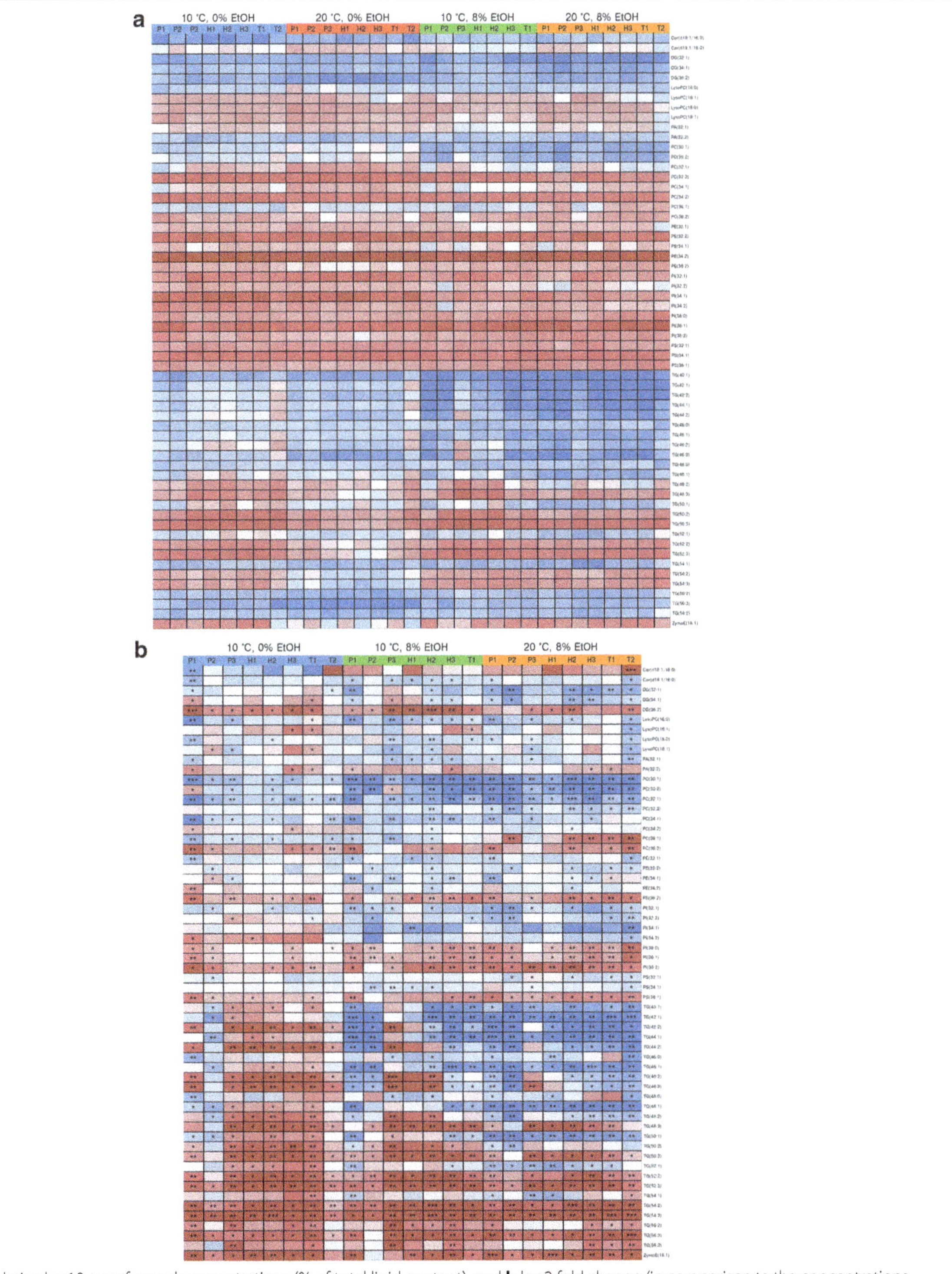

Fig. 4 The **a** relative log10-transformed concentrations (% of total lipid content), and **b** log2 fold change (in comparison to the concentrations measured for the cultivations performed at 20 °C and 0% supplemented EtOH) of the relative concentrations (% of total lipid content), of the 60 lipid species determined by UPLC/MS lipidomics in the lipids extracted from cells in late exponential phase during the growth assays. Values are means from three independent cultivations. **a** The heatmap was generated based on the z-scores (a *blue color* indicates a negative z-score and lower concentration, while a *red color* indicates a positive z-score and a higher concentration). **b** A *blue color* indicates a decrease in concentration, while a *red color* indicates an increase in concentration compared to the cultivation at 20 °C and 0% supplemented EtOH. The *asterisks* indicate whether the change was significant as determined by Student's t test (unpaired, two-tailed and unequal variance) corrected with the Benjamini-Hochberg procedure (*$p < 0.05$; **$p < 0.01$; ***$p < 0.001$)

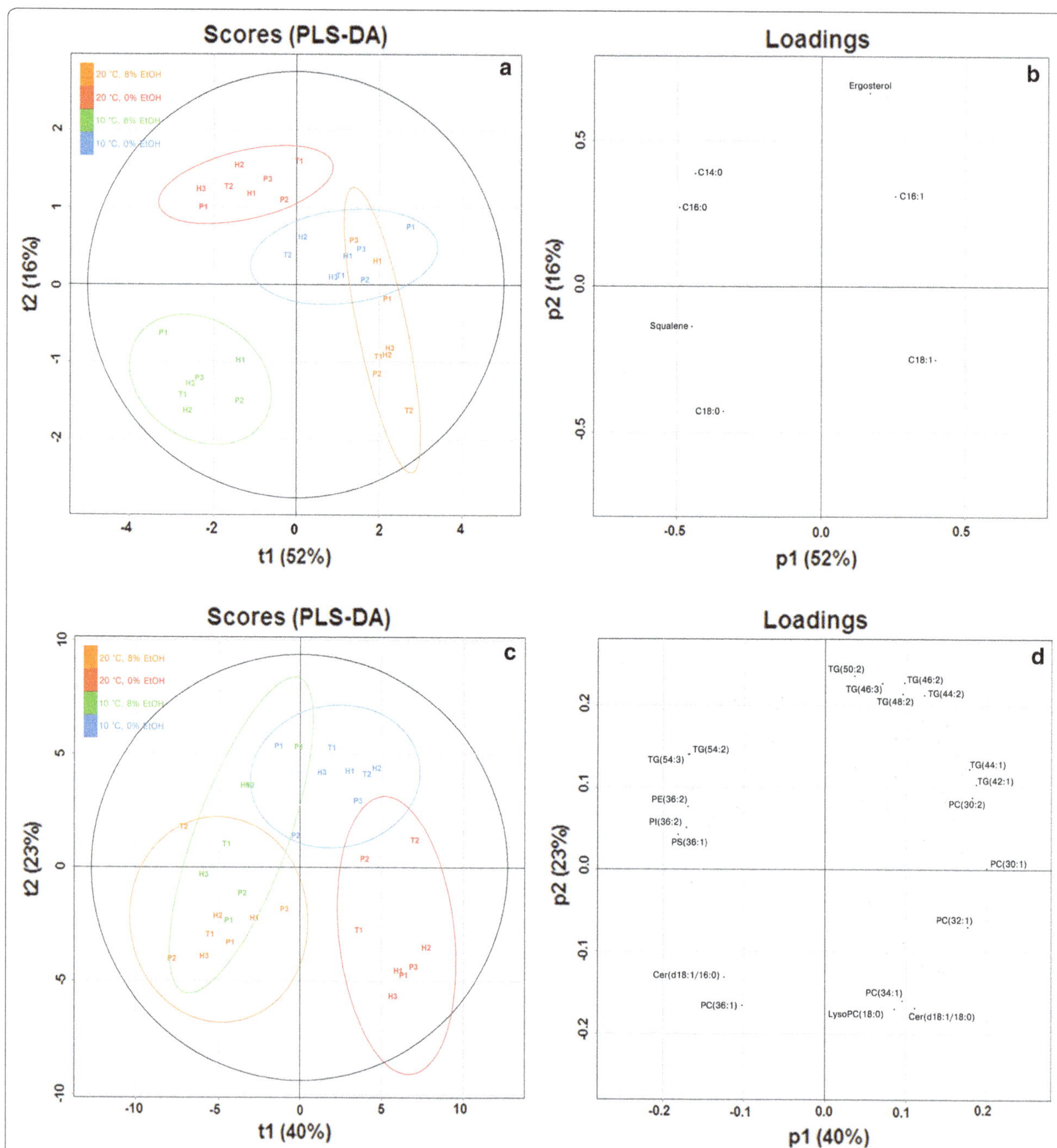

Fig. 5 Score and loading plots of the PLS-DA models built on **a**, **b** the fatty acid, squalene and ergosterol compositions obtained from GC/MS analysis, and **c**, **d** the compositions of the 60 lipid species obtained from UPLC/MS lipidomics analysis. Model properties: **a**, **b** $R^2 = 0.69$, $Q^2 = 0.59$ ($p < 0.05$ by random permutation test), and model uses 3 components; **c**, **d** $R^2 = 0.64$, $Q^2 = 0.54$ ($p < 0.05$ by random permutation test), and model uses 3 components

to positively correlate with fermentation performance, while longer-chain phosphatidylcholines and phosphatidylethanolamines [e.g. PC(34:1), PC(36:1), and PE(34:1)] were negatively correlated with fermentation performance (Fig. 6c, d; Table 3). Finally, at 20 °C and 8% supplemented ethanol, phosphatidylethanolamines (PE), phosphatidylinositols (PI), and phosphatidylserines (PS) containing either or both oleic and palmitoleic acid [e.g. PE(34:2), PI(34:1), PI(34:2), and PS(34:1)] seem to positively correlate with fermentation performance (Fig. 6e,

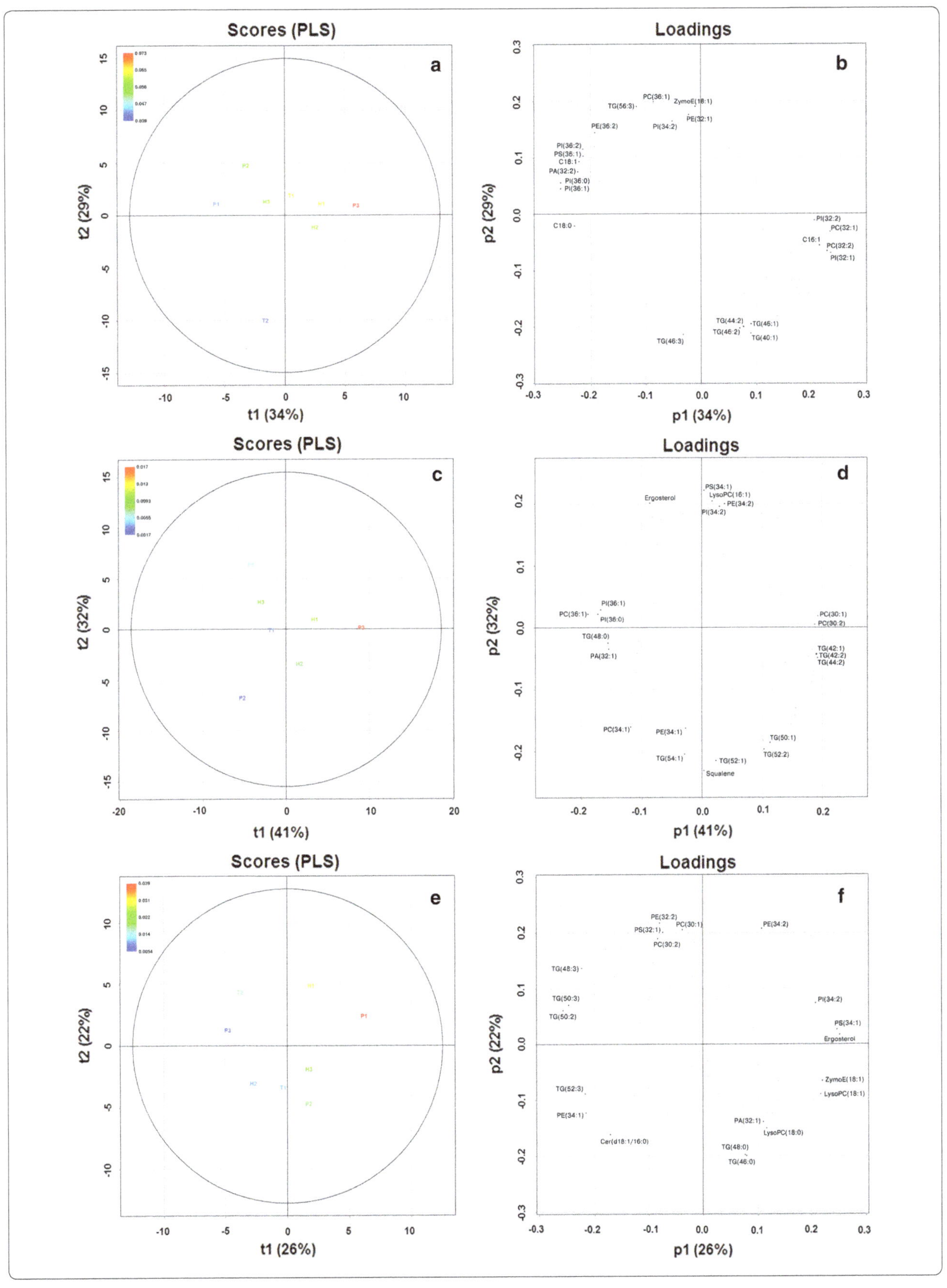

Scores (PLS)
a
t1 (34%)
t2 (29%)
Loadings
b
p1 (34%)
p2 (29%)
PC(36:1)
ZymoE(18:1)
TG(56:3)
PE(32:1)
PE(36:2)
PI(34:2)
PI(36:2)
PS(36:1)
C18:1
PA(32:2)
PI(36:0)
PI(36:1)
C18:0
PI(32:2)
PC(32:1)
C16:1
PC(32:2)
PI(32:1)
TG(44:2)
TG(46:1)
TG(46:2)
TG(40:1)
TG(46:3)
Scores (PLS)
c
t1 (41%)
t2 (32%)
Loadings
d
p1 (41%)
p2 (32%)
PS(34:1)
Ergosterol
LysoPC(16:1)
PE(34:2)
PI(34:2)
PI(36:1)
PC(36:1)
PI(36:0)
PC(30:1)
PC(30:2)
TG(48:0)
PA(32:1)
TG(42:1)
TG(42:2)
TG(44:2)
PC(34:1)
PE(34:1)
TG(50:1)
TG(52:2)
TG(54:1)
TG(52:1)
Squalene
Scores (PLS)
e
t1 (26%)
t2 (22%)
Loadings
f
p1 (26%)
p2 (22%)
PE(32:2)
PC(30:1)
PS(32:1)
PE(34:2)
PC(30:2)
TG(48:3)
TG(50:3)
PI(34:2)
TG(50:2)
PS(34:1)
Ergosterol
ZymoE(18:1)
TG(52:3)
LysoPC(18:1)
PE(34:1)
PA(32:1)
LysoPC(18:0)
Cer(d18:1/16:0)
TG(48:0)
TG(46:0)

(See figure on previous page.)

Fig. 6 Score and loading plots of the PLS models built from the fermentation and lipid data obtained at different temperatures and supplemented ethanol levels (**a, b** 10 °C, 0% EtOH; **c, d** 10 °C, 8% EtOH; **e, f** 20 °C, 8% EtOH). The *Y* variable of the models was the maximum fermentation level divided by the lag time (*A* and λ from Table 2, respectively), while the *X* variables were the combined dataset of the compositions of fatty acids, squalene and ergosterol obtained from GC/MS analysis and the compositions of the 60 lipid species obtained from UPLC/MS lipidomics analysis. Model properties: **a, b** $R^2 = 0.95$, $Q^2 = 0.76$ ($p < 0.05$ by random permutation test), and model uses 2 components; **c, d** $R^2 = 0.87$, $Q^2 = 0.49$ ($p < 0.05$ by random permutation test), and model uses 2 components; **e, f** $R^2 = 0.97$, $Q^2 = 0.51$ ($p < 0.05$ by random permutation test), and model uses 2 components

f; Table 3). Ergosterol was also positively correlated with fermentation performance. These results would suggest that good fermentation performance at both lower temperatures and in higher ethanol concentrations are associated with increased amounts of unsaturated fatty acids in the yeast. Increased ergosterol concentrations seem to have a positive effect on fermentation performance in higher ethanol concentrations, but this effect is negated at lower temperatures.

Supplementation of oleic acid enhances growth in the presence of ethanol

It has been shown that *Saccharomyces* strains are able to incorporate exogenous fatty acids from the growth media and these fatty acids can constitute a considerable fraction of the total fatty acids in the cell [35]. To test the effects that ergosterol and unsaturated fatty acids have on the ability to grow in the presence of ethanol and at sub-optimal temperatures, microplate cultivations of a laboratory strain WT and two knockout strains, *ole1Δ* and *erg4Δ*, unable to synthesize unsaturated fatty acids and ergosterol respectively (Table 1), were carried out in media containing varying concentrations of supplemented ethanol (1, 5 or 10%) and oleic acid (0 or 0.8 mM) at two different temperatures (20 and 15 °C). At 20 °C in 1% ethanol and no added oleic acid, the three strains performed similarly (Fig. 7a; Additional file 1: Table S2). As the ethanol concentration was increased to 5 and 10%, the *erg4Δ* strain grew slightly slower and had longer lag times than the other strains, suggesting that ergosterol enhances ethanol tolerance (Fig. 7b, c; Additional file 1: Table S2). At 15 °C, this effect wasn't observed, as the *erg4Δ* strain performed similarly to the WT strain when no oleic acid was added. While ergosterol may enhance ethanol tolerance, results suggest that it may decrease the ability to grow in sub-optimal temperatures. The addition of oleic acid to the growth media resulted in longer lag times for all strains in all conditions (Fig. 7; Additional file 1: Table S2). However, the presence of oleic acid considerably increased the growth rate of all the strains. This was particularly apparent at 20 °C and 10% ethanol, as well as at 15 °C with 1 and 5% ethanol, where e.g. the *ole1Δ* strain reached the maximum OD600 level several hours earlier when grown in the presence of oleic acid (Fig. 7c–e; Additional file 1: Table S2). This suggests that exogenous oleic acid enhances growth when ethanol concentration is increased in the media and at sub-optimal growth temperatures.

Fermentations in wort confirm POF− phenotype

To examine the performance of the 8 brewing strains in a brewery environment, fermentations were carried out

Table 3 The 10 most significant Variable Importance in Projection (VIP) scores and regressions coefficients of the three PLS models presented in Fig. 6

10 °C, 0% supplemented EtOH				10 °C, 8% supplemented EtOH				20 °C, 8% supplemented EtOH			
VIP scores		Regression coefficients		VIP scores		Regression coefficients		VIP scores		Regression coefficients	
PC(30:2)	1.99	PC(30:2)	−0.1029	PC(30:1)	1.62	TG(58:2)	0.0524	PE(34:1)	2.01	Cer(d18:1/16:0)	−0.1003
PC(34:1)	1.71	TG(46:3)	−0.0882	TG(58:2)	1.58	TG(56:3)	0.0477	Cer(d18:1/16:0)	1.99	PI(34:1)	0.0967
TG(46:3)	1.66	C14:0 (FA)	−0.0876	PC(30:2)	1.47	PC(30:1)	0.0450	PI(34:1)	1.96	PE(34:1)	−0.0963
C14:0 (FA)	1.65	PC(34:1)	0.0859	TG(56:3)	1.45	TG(40:1)	0.0421	TG(52:3)	1.87	PE(34:2)	0.0963
C18:0 (FA)	1.60	C18:0 (FA)	−0.0741	TG(40:1)	1.44	PC(34:1)	−0.0418	PE(34:2)	1.83	TG(52:3)	−0.0869
PC(32:1)	1.49	ZymoE(18:1)	0.0726	TG(56:2)	1.42	PA(32:2)	−0.0416	PS(34:1)	1.80	PI(34:2)	0.0779
PI(36:1)	1.47	PC(32:1)	0.0652	PA(32:2)	1.40	PE(32:2)	0.0399	Ergosterol	1.80	PS(34:1)	0.0765
PI(36:0)	1.41	PI(36:1)	−0.0596	TG(44:2)	1.38	TG(54:3)	0.0397	PI(34:2)	1.71	Ergosterol	0.0752
ZymoE(18:1)	1.39	PI(34:1)	0.0587	TG(42:1)	1.38	PE(34:1)	−0.0397	TG(50:2)	1.58	TG(50:3)	−0.0566
PC(32:2)	1.30	PE(32:1)	0.0581	PC(36:1)	1.38	PC(30:2)	0.0381	TG(50:3)	1.55	TG(50:2)	−0.0559

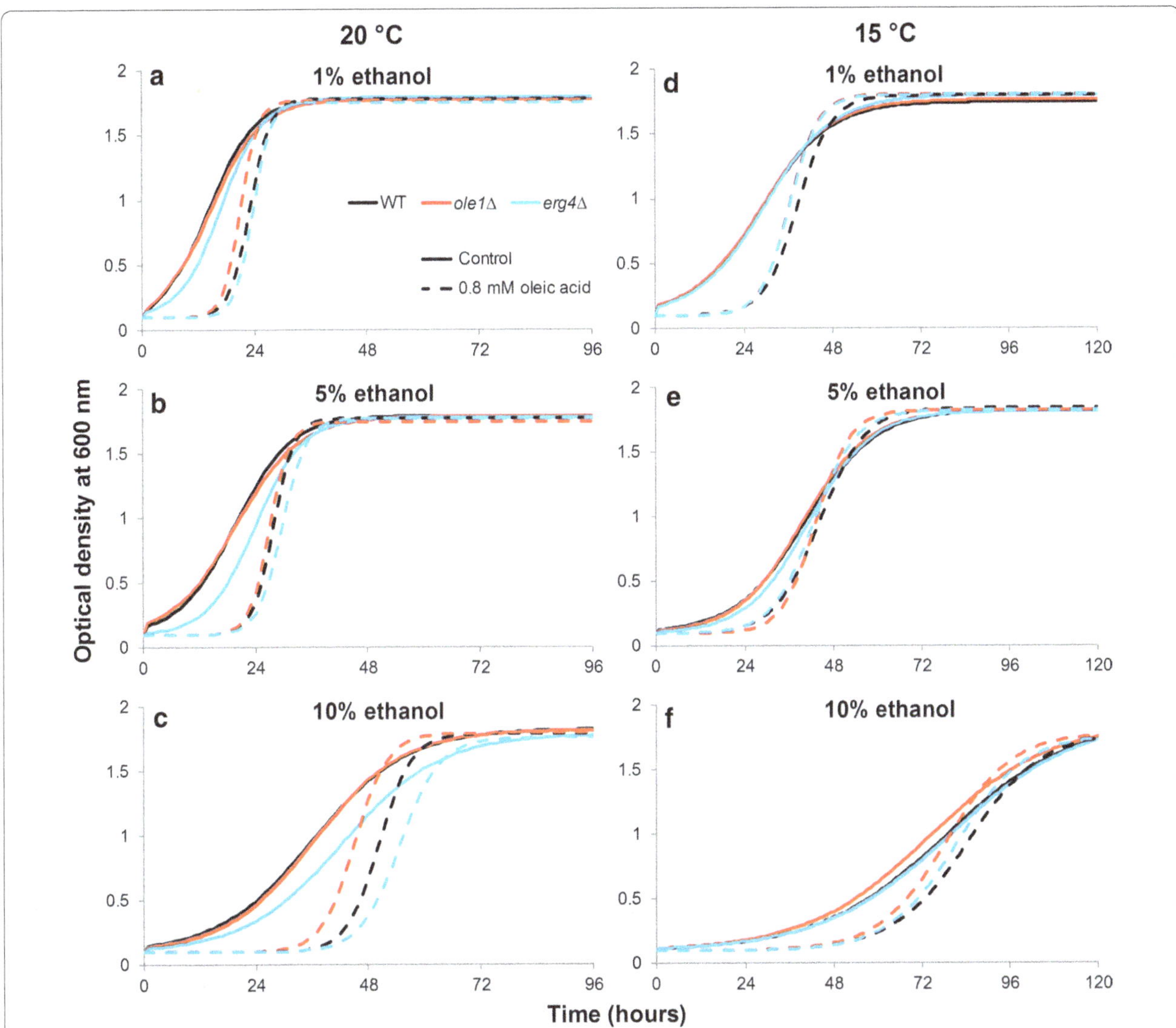

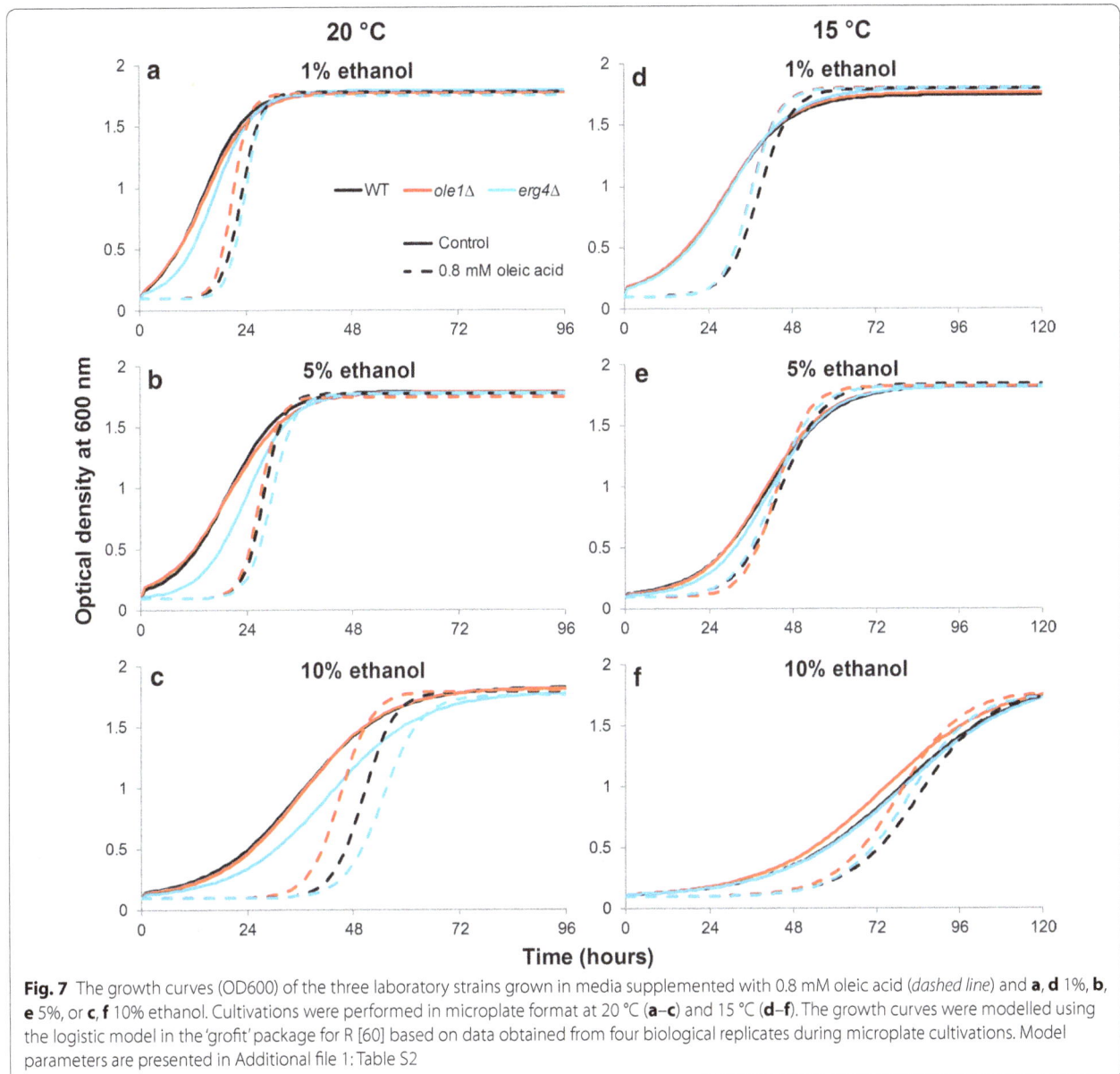

Fig. 7 The growth curves (OD600) of the three laboratory strains grown in media supplemented with 0.8 mM oleic acid (*dashed line*) and **a**, **d** 1%, **b**, **e** 5%, or **c**, **f** 10% ethanol. Cultivations were performed in microplate format at 20 °C (**a**–**c**) and 15 °C (**d**–**f**). The growth curves were modelled using the logistic model in the 'grofit' package for R [60] based on data obtained from four biological replicates during microplate cultivations. Model parameters are presented in Additional file 1: Table S2

in high-gravity 15 °P all-malt wort at 15 °C. These conditions were chosen to replicate those of industrial lager fermentations. There was considerable variation in fermentation performance between the eight strains, as is revealed by the wort alcohol content over time (Fig. 8a). Of the 8 strains, Hybrid T1 and *S. cerevisiae* P2 had the highest overall fermentation rates, but these slowed down considerably after reaching 5.8% (v/v) alcohol. Analysis of the sugar concentrations in the beer revealed that *S. cerevisiae* P2 was unable to ferment the maltotriose in the wort, while Hybrid T1 had only consumed approx. 15% of the initial maltotriose present in the wort (Fig. 8d). Of the 8 strains, Hybrid T2 reached the highest beer alcohol content [7.3% (v/v)], followed closely by Hybrids H1 and H3 (Fig. 8a). The maltotriose concentrations in the beers revealed that maltotriose had been efficiently used by these three hybrids (Fig. 8d). *S. cerevisiae* P1 had also used maltotriose efficiently, but it was only able to produce 5.1% (v/v) alcohol after 14 days of fermentation. *S. eubayanus* P3 on the other hand, was also unable to use maltotriose and the strain performed poorly in the wort reaching only 1.6% (v/v) alcohol.

The beers produced with the 8 brewing strains also varied considerably in concentrations of aroma-active compounds (Fig. 8b). The most ester-rich beers were produced with Hybrids H1, H3 and T1, and these contained as high or higher concentrations of most esters compared to either of the parent strains. Comparing the

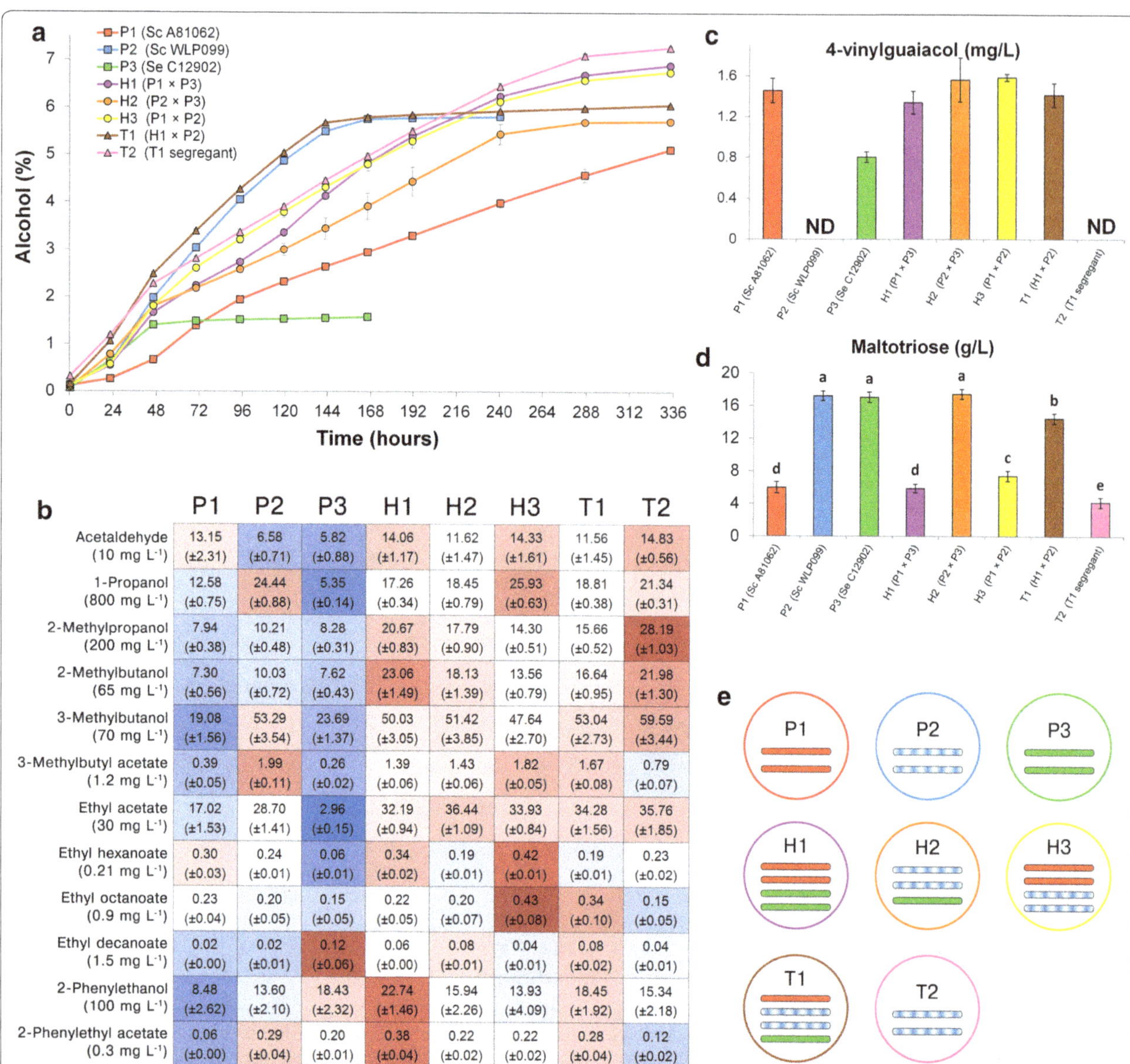

b

	P1	P2	P3	H1	H2	H3	T1	T2
Acetaldehyde (10 mg L-1)	13.15 (±2.31)	6.58 (±0.71)	5.82 (±0.88)	14.06 (±1.17)	11.62 (±1.47)	14.33 (±1.61)	11.56 (±1.45)	14.83 (±0.56)
1-Propanol (800 mg L-1)	12.58 (±0.75)	24.44 (±0.88)	5.35 (±0.14)	17.26 (±0.34)	18.45 (±0.79)	25.93 (±0.63)	18.81 (±0.38)	21.34 (±0.31)
2-Methylpropanol (200 mg L-1)	7.94 (±0.38)	10.21 (±0.48)	8.28 (±0.31)	20.67 (±0.83)	17.79 (±0.90)	14.30 (±0.51)	15.66 (±0.52)	28.19 (±1.03)
2-Methylbutanol (65 mg L-1)	7.30 (±0.56)	10.03 (±0.72)	7.62 (±0.43)	23.06 (±1.49)	18.13 (±1.39)	13.56 (±0.79)	16.64 (±0.95)	21.98 (±1.30)
3-Methylbutanol (70 mg L-1)	19.08 (±1.56)	53.29 (±3.54)	23.69 (±1.37)	50.03 (±3.05)	51.42 (±3.85)	47.64 (±2.70)	53.04 (±2.73)	59.59 (±3.44)
3-Methylbutyl acetate (1.2 mg L-1)	0.39 (±0.05)	1.99 (±0.11)	0.26 (±0.02)	1.39 (±0.06)	1.43 (±0.06)	1.82 (±0.05)	1.67 (±0.08)	0.79 (±0.07)
Ethyl acetate (30 mg L-1)	17.02 (±1.53)	28.70 (±1.41)	2.96 (±0.15)	32.19 (±0.94)	36.44 (±1.09)	33.93 (±0.84)	34.28 (±1.56)	35.76 (±1.85)
Ethyl hexanoate (0.21 mg L-1)	0.30 (±0.03)	0.24 (±0.01)	0.06 (±0.01)	0.34 (±0.02)	0.19 (±0.01)	0.42 (±0.01)	0.19 (±0.01)	0.23 (±0.02)
Ethyl octanoate (0.9 mg L-1)	0.23 (±0.04)	0.20 (±0.05)	0.15 (±0.05)	0.22 (±0.05)	0.20 (±0.07)	0.43 (±0.08)	0.34 (±0.10)	0.15 (±0.05)
Ethyl decanoate (1.5 mg L-1)	0.02 (±0.00)	0.02 (±0.01)	0.12 (±0.06)	0.06 (±0.00)	0.08 (±0.01)	0.04 (±0.01)	0.08 (±0.02)	0.04 (±0.01)
2-Phenylethanol (100 mg L-1)	8.48 (±2.62)	13.60 (±2.10)	18.43 (±2.32)	22.74 (±1.46)	15.94 (±2.26)	13.93 (±4.09)	18.45 (±1.92)	15.34 (±2.18)
2-Phenylethyl acetate (0.3 mg L-1)	0.06 (±0.00)	0.29 (±0.04)	0.20 (±0.01)	0.38 (±0.04)	0.22 (±0.02)	0.22 (±0.02)	0.28 (±0.04)	0.12 (±0.02)

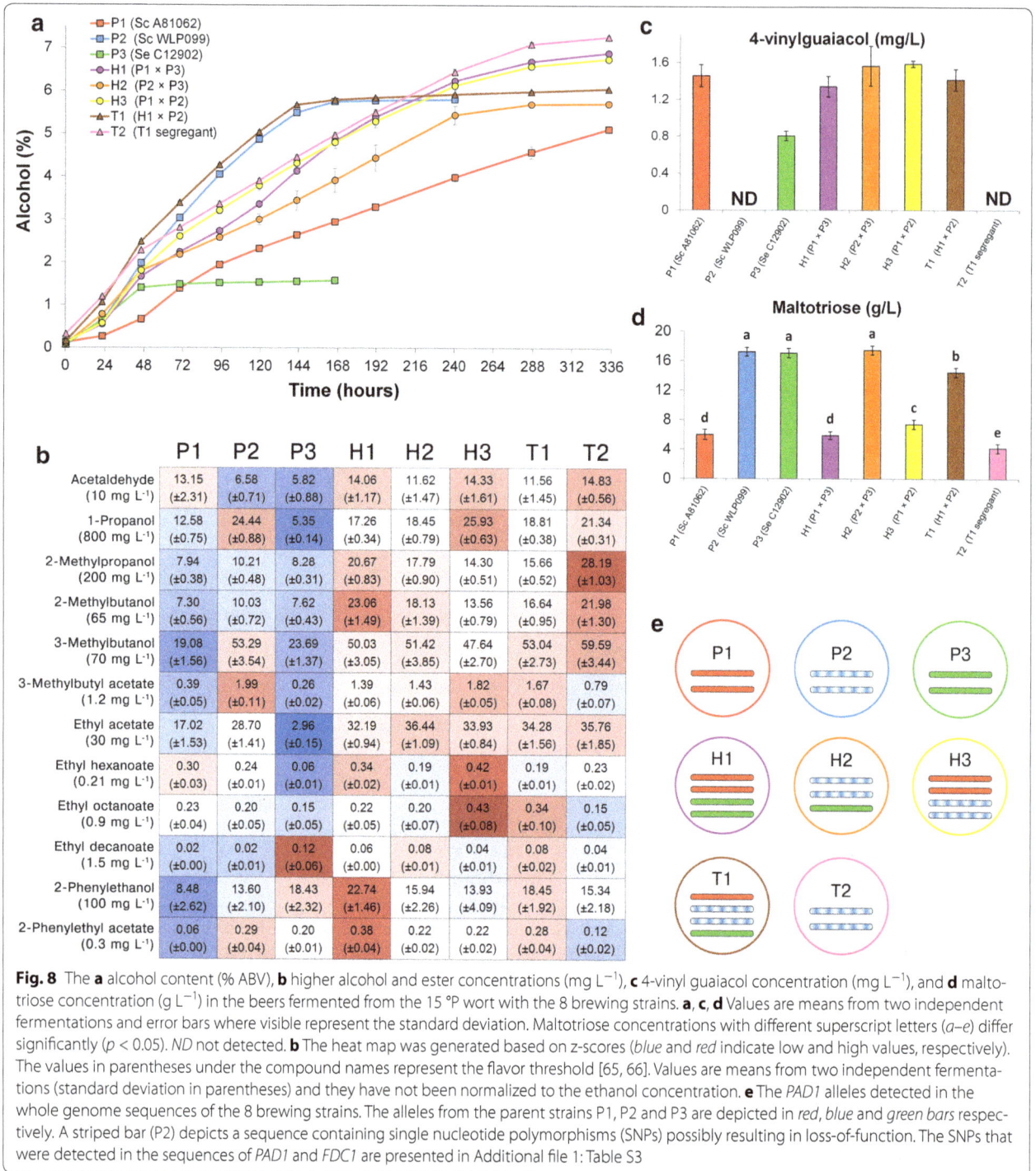

Fig. 8 The **a** alcohol content (% ABV), **b** higher alcohol and ester concentrations (mg L^{-1}), **c** 4-vinyl guaiacol concentration (mg L^{-1}), and **d** maltotriose concentration (g L^{-1}) in the beers fermented from the 15 °P wort with the 8 brewing strains. **a**, **c**, **d** Values are means from two independent fermentations and error bars where visible represent the standard deviation. Maltotriose concentrations with different superscript letters (*a–e*) differ significantly ($p < 0.05$). *ND* not detected. **b** The heat map was generated based on z-scores (*blue* and *red* indicate low and high values, respectively). The values in parentheses under the compound names represent the flavor threshold [65, 66]. Values are means from two independent fermentations (standard deviation in parentheses) and they have not been normalized to the ethanol concentration. **e** The *PAD1* alleles detected in the whole genome sequences of the 8 brewing strains. The alleles from the parent strains P1, P2 and P3 are depicted in *red*, *blue* and *green bars* respectively. A striped bar (P2) depicts a sequence containing single nucleotide polymorphisms (SNPs) possibly resulting in loss-of-function. The SNPs that were detected in the sequences of *PAD1* and *FDC1* are presented in Additional file 1: Table S3

aroma profiles of the beers made with Hybrids T1 and T2, it is revealed that the meiotic segregant T2 produced lower concentrations of most esters, while its beer contained higher concentrations of most higher alcohols. This would suggest lower activities or expression of alcohol acetyl transferases in Hybrid T2. Of the 8 brewing strains, the POF− *S. cerevisiae* P2 and Hybrid T2 strains were the only ones that did not produce any detectable amounts of 4-vinyl guaiacol (detection limit 0.2 mg L^{-1}), thus confirming their POF− phenotype (Fig. 8c). All other strains produced 4-vinyl guaiacol in concentrations above the flavour threshold of 0.3–0.5 mg L^{-1} [36]. Yeasts produce 4-vinyl guaiacol from ferulic acid with the aid of the *PAD1*- and *FDC1*-encoded enzymes. By comparing

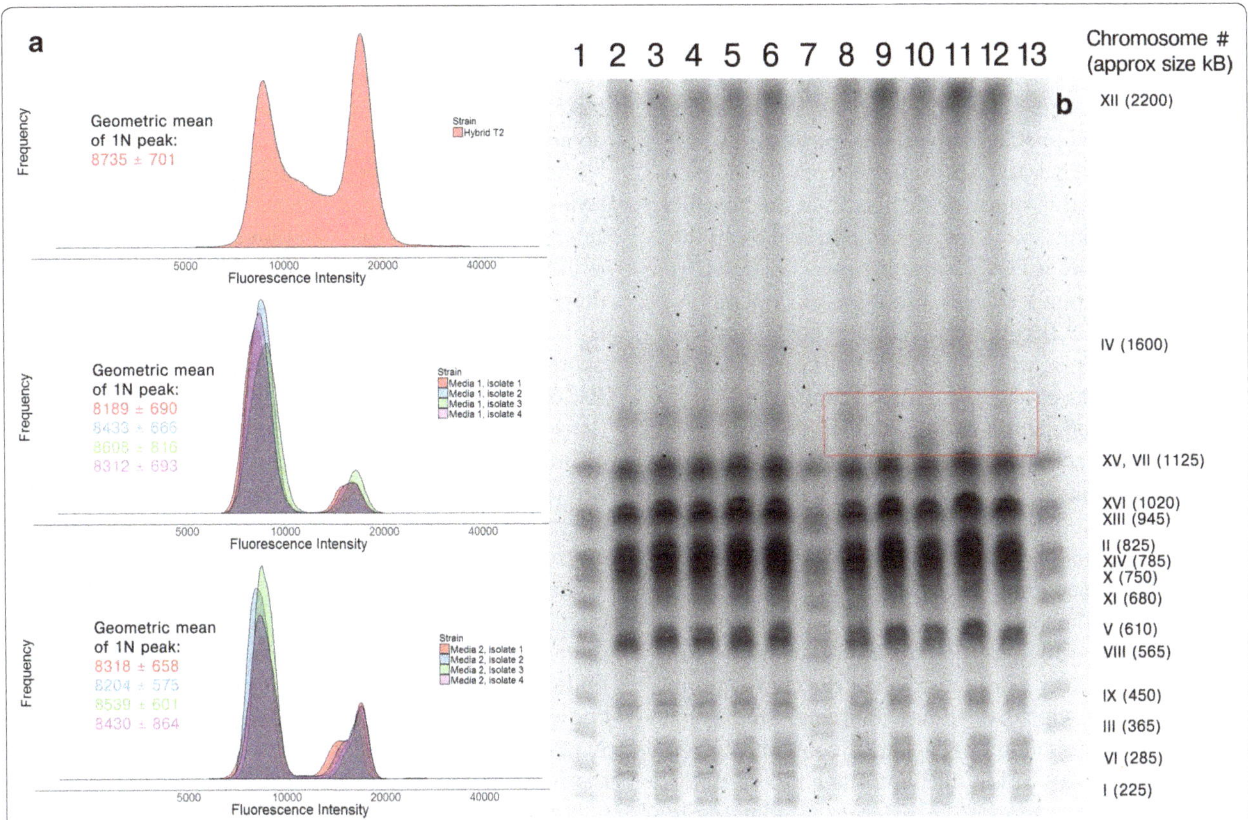

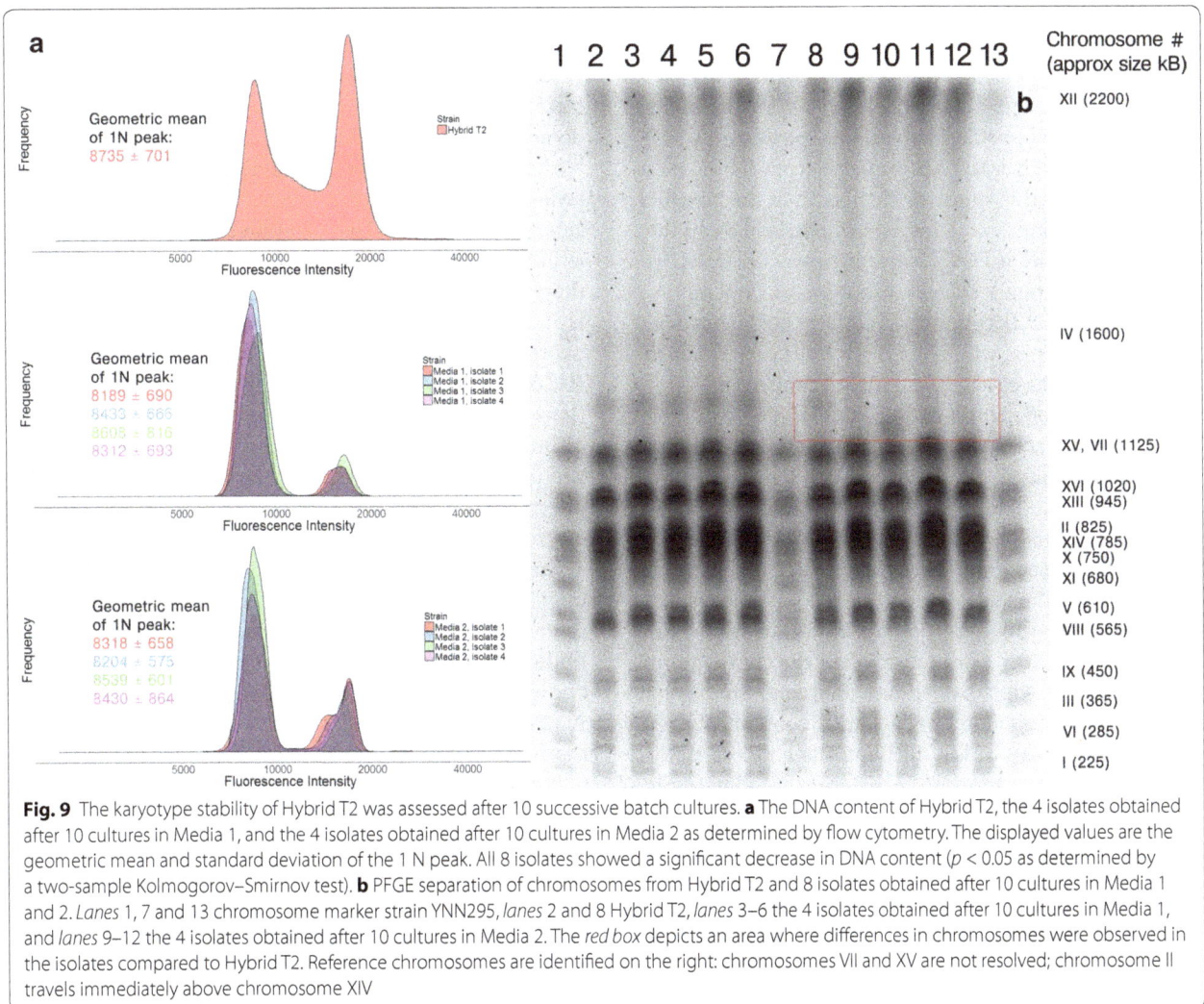

Fig. 9 The karyotype stability of Hybrid T2 was assessed after 10 successive batch cultures. **a** The DNA content of Hybrid T2, the 4 isolates obtained after 10 cultures in Media 1, and the 4 isolates obtained after 10 cultures in Media 2 as determined by flow cytometry. The displayed values are the geometric mean and standard deviation of the 1 N peak. All 8 isolates showed a significant decrease in DNA content ($p < 0.05$ as determined by a two-sample Kolmogorov–Smirnov test). **b** PFGE separation of chromosomes from Hybrid T2 and 8 isolates obtained after 10 cultures in Media 1 and 2. *Lanes* 1, 7 and 13 chromosome marker strain YNN295, *lanes* 2 and 8 Hybrid T2, *lanes* 3–6 the 4 isolates obtained after 10 cultures in Media 1, and *lanes* 9–12 the 4 isolates obtained after 10 cultures in Media 2. The *red box* depicts an area where differences in chromosomes were observed in the isolates compared to Hybrid T2. Reference chromosomes are identified on the right: chromosomes VII and XV are not resolved; chromosome II travels immediately above chromosome XIV

the sequences of these two genes in the 8 brewing strains, we found that Hybrid T2 only carried the *PAD1* allele that was derived from *S. cerevisiae* P2 (Fig. 8e; Additional file 1: Table S3). This particular allele, contained a possible loss-of-function SNP at position 638 (A>G, resulting in an amino acid substitution of aspartate to glycine). The other strains, including Hybrid T1, which Hybrid T2 was derived from, carried either or both of the functional *PAD1* alleles derived from *S. cerevisiae* P1 or *S. eubayanus* P3. While the alleles of *FDC1* in the parent strains contained different SNPs (Additional file 1: Table S3), none of them appeared to cause loss-of-function.

During industrial beer fermentations, the yeast is commonly reused for multiple consecutive fermentations (typically up to 10 times depending on brewery), and it is therefore essential that the yeast genome maintains stable. In order to further evaluate the industrial applicability of Hybrid T2, its karyotype stability was assessed. After 10 successive batch cultures in two different growth media (Media 1 containing 1% yeast extract, 2% peptone, 1% maltose and 1% maltotriose; and Media 2 containing 1% yeast extract, 2% peptone, 2% maltose and 14% sorbitol), there was a significant decrease ($p < 0.05$) in DNA content for all eight 10-week isolates compared to the original strain (Fig. 9a). Pulsed-field gel electrophoresis (PFGE) further revealed that the 4 isolates obtained from Media 1 showed identical chromosome profiles as Hybrid T2 (Fig. 9b), suggesting that differences in DNA content could be a result of differences in chromosome copy number rather than complete loss of a chromosome. The 4 isolates obtained from Media 2 on the other hand, showed a loss of a whole or part of a chromosome (size between 1125 and 1600 kB) compared to Hybrid T2 (area highlighted with a red box in Fig. 9b). These results highlight the importance of stabilisation before any hybrids can be seen as viable candidates for industrial beer fermentations.

Discussion

Interspecific hybridization has been shown to be a promising tool for increasing the diversity of lager yeasts available to the brewing industry. However, de novo lager yeast hybrids have so far inadvertently possessed a POF+ phenotype; a trait which they inherit from the *S. eubayanus* parent. In this study, we demonstrate the possibility of using an allotetraploid interspecific hybrid as an intermediate to create a POF− lager hybrid with DNA from three parent strains, and we use a set of brewing yeast hybrids to reveal that unsaturated fatty acids and ergosterol are associated with good fermentation performance at low temperatures and high ethanol concentrations.

Sporulation efficiency and spore viability tend to be poor in interspecies yeast hybrids, which limits the possibility of introducing variation to such a hybrid through meiotic recombination and chromosomal cross-over. The mechanisms contributing to hybrid sterility are not completely understood, but studies suggests that large sequence divergence, abnormal chromosome segregation and reciprocal gene loss contribute to it [9]. However, hybrid fertility can be recovered through a number of ways, one of which is genome doubling. Studies have revealed that doubling the genome content of sterile allodiploid hybrids result in allotetraploids capable of producing viable spores [6, 8]. Here, the allotetraploid interspecies hybrids H1 and T1 showed spore viabilities of 47 and 38%, respectively, and sporulation allowed us to not only cross Hybrid H1 with a third parent strain (P2), but also remove the POF+ phenotype from Hybrid T1. As was revealed from the genome sequences of Hybrids T1 and T2, chromosome losses may occur during spore formation (e.g. the *S. eubayanus*-derived chromosome XIII carrying *PAD1* and *FDC1*), suggesting that genetically diverse strains can be obtained. Hence, the use of fertile intermediates allows for the construction of complex interspecific hybrid strains, which can be screened and selected to contain desirable traits. While not investigated here, one may expect that meiotic segregants of interspecies hybrids vary considerably phenotypically as a result of orthologous gene segregation and the creation of unique biochemical pathways and regulatory mechanisms [37, 38]. We believe this would be particularly beneficial for generating novel and diverse lager yeast strains, as there currently exist limited genetic and phenotypic diversity between natural lager yeast hybrids [1, 39]. As the stability test of Hybrid T2 revealed that some genetic changes occurred after 10 successive batch cultures in relatively non-stressful media, it is vital that any hybrids are stabilised and phenotypically reassessed before they are viable candidates for industrial beer fermentations. Previous studies have shown that sufficient stabilisation can be achieved by growing the hybrids for 30–70 generations under fermentative conditions in high-sugar media [4, 40]. The inherent instability of interspecific yeast hybrids could also be exploited for further strain development through adaptive evolution, as was recently demonstrated for biofuel applications [41].

Phenolic off-flavours are undesirable in many beer styles, and in lager beer especially their presence is considered a flaw [36]. Hybridization and subsequent sporulation has been proposed as one technique of removing the POF+ phenotype from crossed *S. cerevisiae* strains [16]. Gallone et al. [13] have also shown that a POF− hybrid can be constructed when both parent strains carry *PAD1* or *FDC1* alleles containing loss-of-function mutations. *S. eubayanus* however, contains functional alleles of both *PAD1* and *FDC1*, so any hybrids made from it will initially be POF+ as well. Here we obtained a POF− interspecies hybrid through the use of a fertile tetraploid intermediate, and demonstrated that it didn't produce 4-vinyl guaiacol in wort fermentations and only contained *PAD1* and *FDC1* alleles derived from the POF− *S. cerevisiae* P2 parent. No nonsense or frameshift mutations were detected in the *PAD1* and *FDC1* alleles of P2, but a SNP at position 638 (A>G) of *PAD1* results in an Asp213Gly amino acid substitution, possibly affecting the functionality of the enzyme. This same SNP (*PAD1*: 638 A>G) was present in the POF− K1V-1116 strain that was studied by Mukai et al. [14] and the POF− strains Beer024, Beer033, Beer088, Wine001, Wine009 and Spirits002 that were studied by Gallone et al. [13]. However, it may be possible that this SNP is not the cause for the POF− phenotype in P2 and T2, as well as the POF− strains from the previously mentioned studies, and rather another unknown mechanism is responsible.

Among the defining characteristics of lager yeast is their ability to grow and ferment at low temperatures and high ethanol concentrations. Unlike the POF+ phenotype, the mechanisms that contribute to cold and ethanol tolerance in brewing yeast are more complex and not fully understood. Differences in the lipid composition of the plasma membrane have however been shown to play a vital role in determining both temperature and ethanol tolerance [25–27, 33]. Here, we observed variations in cellular lipid composition both between yeast strains and between environmental conditions. When temperatures were lowered or ethanol content was increased, the degree of unsaturation tended to increase. Similar conclusions have also been reached in recent studies in wine-making conditions [25, 26, 32]. It is assumed that these changes in response to a lowered temperature or an increase in ethanol concentration are vital for maintaining membrane fluidity and functionality [25, 33, 42]. The PLS models that were constructed between the lipid dataset and fermentation parameters at the different conditions revealed that yeast strains which performed well at low temperatures and/or high ethanol concentrations were associated with increased concentrations of phospholipids and triacylglycerides containing unsaturated fatty acids, such as palmitoleic acid (C16:1) and oleic acid (C18:1). Ergosterol was also shown to be positively correlated with ethanol tolerance, particularly at 20 °C. Studies have revealed that ergosterol plays an important role in maintaining membrane rigidity and protecting against ethanol toxicity when ethanol concentrations are increased [30, 31]. Results from microplate cultivations of *ole1*Δ and *erg4*Δ knockout strains supplemented with oleic acid and ethanol were in agreement with these observations. Redón et al. [35] also observed that wine fermentations at low temperatures proceeded faster when unsaturated fatty acids such as palmitoleic and linolenic acid were supplemented to the must. Despite the PLS models revealing that lipid composition does seem to influence cold and ethanol tolerance in brewing yeast, we feel that the lipidomics data generated here did not reveal any obvious patterns between the two. It is thus plausible that a combination of other factors, e.g. protein translation and folding efficiencies, mRNA stabilities and the product activity and expression of central metabolic genes [18–24], also contribute to determining cold and ethanol tolerance, and suggest that this topic should be addressed in future studies.

Conclusions

Recent studies have revealed that the creation of de novo lager yeast hybrids has the potential to considerably increase the genetic and phenotypic diversity of lager yeast strains available to the brewing industry. During interspecific hybridization, one is typically limited to combining traits from two different strains. Here we demonstrated the possibility of constructing complex yeast hybrids, through the use of a fertile allotetraploid intermediate, that possess traits that are relevant to industrial lager beer fermentation and that are derived from several parent strains. Yeast lipid composition was also shown to have a central role in determining ethanol and cold tolerance in brewing strains. The presence of unsaturated fatty acids and ergosterol in particular, were shown to benefit growth in the presence of ethanol and at lower temperatures.

Methods

Yeast strains and hybrid generation

The yeast strains used in the study are listed in Table 1. The interspecific yeast hybrids H1 (P1 × P2) and H2 (P2 × P3) and intraspecific yeast hybrid H3 (P1 × P2) were generated using rare mating according to the method described in Krogerus et al. [3]. Prior to mating, a uracil auxotroph (*ura-*) of P1 was selected on 5-fluoroorotic acid (5-FOA) agar [43], a respiratory-deficient mutant (*rho-*) of P2 was selected on YPDG agar containing 3% glycerol and 0.1% glucose, and a lysine auxotroph (*lys-*) of P3 was selected on α-aminoadipic (α-AA) agar [44] to allow for the selection of hybrids on minimal selection agar medium (0.67% Yeast Nitrogen Base without amino acids, 3% glycerol, 3% ethanol and 2% agar). The interspecific hybrid H1 was further mated with P2 (*rho-*) to form the interspecific triple hybrid T1 [(P1 × P3) × P2]. Prior to mating, a lysine auxotroph (*lys-*) of H1 was first selected as described above, after which ascospores of it were generated on 1% potassium acetate agar. Spore viability was calculated based on the amount of colonies formed from the dissection of up to 20 tetrads. Ascus walls were digested with Zymolyase 100T, after which spores of H1 (*lys-*) were mixed with P2 (*rho-*). Hybrid T1 was selected on minimal selection agar. Hybrid T2, a meiotic segregant of T1, was created by generating ascospores of Hybrid T1 on 1% potassium acetate agar and then dissecting the spores on YPD agar. All spore dissections were carried out using the Singer MSM 400 dissecting microscope (Singer Instruments, UK). The viable spore clones were then screened for the POF phenotype in a small-scale assay. 1 ml of YPM supplemented with 100 mg L^{-1} of *trans*-ferulic acid was inoculated with a colony of the spore clones, and they were allowed to incubate for 48 h at 25 °C. The POF phenotype was determined sensorially by examining for the presence (POF+) or absence (POF−) of the distinct clove-like aroma of 4-vinyl guaiacol. Hybrid T2 was a spore clone of T1 which did not produce 4-vinyl guaiacol in this assay. An overview of these brewing strains and their creation is presented in Fig. 1.

The hybrid status of isolates was confirmed by PCR as described in Krogerus et al. [2]. Briefly, the rDNA ITS region was amplified using the primers ITS1 (5′-TCCGTAGGTGAACCTGCGG-3′) and ITS4 (5′-TCCTCCGCTTATTGATATGC-3′) and amplicons were digested using the HaeIII restriction enzyme (New England BioLabs, USA) as described previously [45]. Amplification of the *S. eubayanus*-specific *FSY1* gene (amplicon size 228 bp) and the *S. cerevisiae*-specific *MEX67* gene (amplicon size 150 bp) was also performed using the primers SeubF3 (5′-GTCCCTGTACCAATTTAATATTGCGC-3′), SeubR2 (5′-TTTCACATCTCTTAGTCTTTTCC-AGACG-3′), ScerF2 (5′-GCGCTTTACATTCAGATCCCGAG-3′), and ScerR2 (5′-TAAGTTGGTTGTCAGCAAGATTG-3′) as described by Muir et al. [46] and Pengelly and Wheals [47]. Additionally, the hybrid statuses of the intraspecific hybrid H3 and the triple hybrids T1 and T2 were confirmed by amplifying interdelta sequences using the delta12 (5′-TCAACAATGGAATCCCAAC-3′) and delta21 (5′-CATCTTAACACCGTATATGA-3′) primers as described by Legras and Karst [48].

Flow cytometry was also performed on the brewing yeast strains to estimate ploidy essentially as described by Haase and Reed [49]. Cells were grown overnight in YPD medium (1% yeast extract, 2% peptone, 2% glucose), and approximately 1 × 10^7 cells were washed with 1 mL of 50 mM citrate buffer. Cells were then fixed with cold 70% ethanol, and incubated at room temperature for 1 h. Cells were then washed with 50 mM citrate buffer (pH 7.2), resuspended in 50 mM citrate buffer containing 0.25 mg mL^{-1} RNAse A and incubated overnight at 37 °C. 1 mg mL^{-1} of Proteinase K was then added, and cells were incubated for 1 h at 50 °C. Cells were then stained with SYTOX Green (2 μM; Life Technologies, USA), and their DNA content was determined using a FACSAria IIu cytometer (Becton–Dickinson, USA). DNA contents were estimated by comparing fluorescence intensities with those of *S. cerevisiae* haploid (CEN.PK113-1A) and diploid (CEN.PK) reference strains. Measurements were performed on duplicate independent yeast cultures, and 100,000 events were collected per sample during flow cytometry.

Genome sequencing

Whole genome sequences of the brewing strains P1 and P3 have been published previously [3, 50]. The other 6 brewing strains were sequenced by Biomedicum Genomics (Helsinki, Finland). In brief, an Illumina NexteraXT pair-end 150 bp library was prepared for each hybrid and sequencing was carried out with a NextSeq500 instrument. Pair-end reads from the NextSeq500 sequencing were quality-analysed with FastQC [51] and trimmed and filtered with Skewer [52]. Alignment, realignment and variant analysis was carried out using SpeedSeq [53] and its FreeBayes SNP prediction and CNVnator copy number variation prediction modules [54, 55]. Reads of *S. cerevisiae* P2 were aligned to that of *S. cerevisiae* S288c [56], while reads of hybrid strains were aligned to concatenated reference sequences of strains P1 and P3 [3, 50] as described previously [3]. SNPs predicted by FreeBayes with less than five left and right aligning reads were discarded. Prior to SpeedSeq variant analysis, alignments were filtered to a minimum MAPQ of 50 with SAMtools [57]. Quality of alignments was assessed with QualiMap [58]. The median coverage over 10,000 bp windows was calculated with BEDTools [59] and visualized with R (http://www.r-project.org/). The coverage of *S. cerevisiae* P2 and the 5 hybrid strains (H1-H3 and T1-T2) across the *S. cerevisiae* and *S. eubayanus* reference genomes are displayed in Additional file 1: Figure S3.

Fermentation assay

The fermentation kinetics and lipid compositions of the 8 brewing strains at two different temperatures (10 and 20 °C) and two different initial ethanol concentrations [0 and 8% (v/v)] were assayed in 100 mL shake-flask fermentations. The strains were grown in media containing 1% yeast extract, 2% peptone, 8% maltose, and up to 8% ethanol. Prior to the assay, the strains were pre-cultivated in media containing 1% yeast extract, 2% peptone and 4% maltose for 24 h at 20 °C. The OD600 of the pre-cultivations was measured, and the growth assays were started by inoculating 100 mL of media to a starting OD600 of 0.01 in triplicate flasks. Flasks were capped and then incubated at either 10 or 20 °C with light agitation (80 RPM) for up to 33 days. Fermentation was monitored (up to twice daily) by drawing 100 μL samples and measuring the refractive index (°brix) with a Quick-Brix 90 digital refractometer (Mettler-Toledo AG, Switzerland). Samples were drawn for lipid analysis at the end of the exponential fermentation phase. 15 mL samples of fermentation media were centrifuged for 5 min at 9000×*g*, after which the yeast pellet was washed twice in 15 mL of ice-cold deionized water. The washed yeast pellet was then transferred to a cryotube, and flash-frozen in liquid nitrogen. The samples were stored at −80 °C prior to lipid analysis. After fermentations were complete, the biomass concentration determined by drawing and centrifuging 30 mL samples of the fermentation media (10 min at 9000×*g*), washing the yeast pellets gained from centrifugation twice with 25 mL deionized H_2O and then suspending the washed yeast in a total of 6 mL deionized H_2O. The suspension was then transferred to a pre-weighed porcelain crucible, and was dried overnight at 105 °C and allowed to cool in a desiccator before the change of mass was measured. Fermentation curves for the fermentations were modelled based on the decrease in °brix over time using the 'grofit'-package for R [60]. The fermentation parameters were determined using the logistic model in 'grofit'.

Lipid analysis

Prior to lipid extraction, the frozen cell samples were freeze-dried at −55 °C overnight (Martin Christ Alpha 1-4 LDplus, Germany). For fatty acid (free and bound) and sterol analysis, 10 mg of freeze-dried sample was rehydrated into 200 μL of 0.9% sodium chloride solution and spiked with heptadecanoic acid (FFA 17:0; 5.36 μg) and glyceryl triheptadecanoate [TG(17:0/17:0/17:0); 21.11 μg]. Lipids were extracted with chloroform:methanol (2:1 v/v; 800 μL) by homogenizing the samples with grinding balls in a Retsch mixer mill MM400 homogenizer (Retsch GmbH, Haan, Germany) at 20 Hz for 2 min. After 30 min standing at room temperature, the samples were centrifuged at 10,620×*g* for 5 min and the lower layer was separated into glass tubes and evaporated to dryness under nitrogen flow.

The evaporation residues from lipid extractions were dissolved into petroleum ether (b.p. 40–60 °C; 700 μL). Fatty acids were transesterified with sodium methoxide (NaOMe; 0.5 M; 250 μL) in dry methanol by boiling at 45 °C for 5 min. The mixture was acidified with 15% sodium hydrogen sulphate ($NaHSO_4$; 500 μL). The petroleum ether phase containing the fatty acid methyl esters (FAME) as well as free fatty acids (FFA) was collected into an Eppendorf tube and centrifuged (10,620×*g*; 5 min). Half of the petroleum ether layer was separated into a GC vial and evaporated, after which the residue was dissolved into hexane (100 μL) and taken into a vial insert. FAMEs were analysed on an Agilent 7890A GC combined with an Agilent 5975C mass selective detector controlled by MSD ChemStation software (Agilent Technologies Inc., Santa Clara, CA, USA). The column was an Agilent FFAP silica capillary column (25 m × 0.2 mm × 0.3 μm). Helium was used as carrier gas and the samples were injected in splitless mode. The oven temperature programme was from 70 °C (2 min) to 240 °C at a rate of 15 °C min^{-1}, total run time was 39 min. The temperatures of the injector and MS source were 260 and 230 °C, respectively. The samples (1 μL) were injected by a Gerstel MPS injection system (Gerstel GmbH & Co. KG, Mülheim an der Ruhr, Germany) and the data were collected in EI mode (70 eV) at a mass range of m/z 40–600.

The other half of the petroleum ether layer was transferred to a GC vial and evaporated into dryness for the determination of free fatty acid and sterols. The residue was dissolved into 50 μL of DCM and derivatized with *N*-Methyl-*N*-(trimethylsilyl)trifluoroacetamide (MSTFA; 40 μL) and trimethylchlorosilane (TMCS; 10 μL) by incubating at 80 °C for 20 min. The samples (1 μL aliquots) were injected in splitless mode at 300 °C, and analysed on an Agilent DB-5MS column (30 m × 0.2 mm × 0.25 μm). The oven temperature programme was from 50 °C (1.5 min) to 325 °C at a rate of 10 °C min^{-1}, total run time was 49 min.

For lipidomics analyses, approx. 5 mg of freeze-dried cell samples were rehydrated in 50 μL of 0.9% sodium chloride in Eppendorf tubes, mixed with 400 μL chloroform:methanol (2:1) and spiked with 10 μL of an internal standard mixture 1 [IS1; containing LysoPC(17:0), MG(17:0), DG(17:0/17:0), TG(17:0/17:0/17:0), PG(17:0/17:0), Cer(d18:1/17:0), PC(17:0/17:0), PE(17:0/17:0), CE(19:0), CA(14:0), C12(β)-D-GluCer and C8-L-threo-LacCer] (Avanti Polar Lipids, Alabaster, AL, USA; Larodan Fine Chemicals AB, Malmö, Sweden; Nu-Chek Prep, Inc., Elysian, MN, USA) at concentration levels of 0.4–3.2 μg/sample. The samples were homogenized with grinding balls in a Retsch mixer mill MM400 homogenizer at 25 Hz for 2 min and after 30 min standing were centrifuged at 10,620×*g* for 3 min. Before UPLC-MS analysis, a 20 μL aliquot of a labelled lipid standard mixture [IS2; containing LysoPC(16:0-D_3), PC(16:0/16:0-D_6) and TG(16:0/16:0/16:0-^{13}C3)] (Avanti Polar Lipids, Alabaster, AL, USA) was added into the separated lipid extracts.

Lipid extracts were analyzed on a Waters Q-TOF Premier mass spectrometer combined with an Acquity Ultra Performance LC™ (UPLC) under the control of MassLynx software (v 4.1; Waters Inc., Milford, MA, USA). The column (at 50 °C) was an Acquity UPLC™ BEH C18 (2.1 × 100 mm with 1.7 μm particles). The solvent system included (A) ultrapure water (1% 1 M NH_4Ac, 0.1% HCOOH) and (B) LC/MS grade acetonitrile/isopropanol (1:1, 1% 1 M NH_4Ac, 0.1% HCOOH). In ESI− mode, the same solvent system but without acid was used. The gradient started from 65% A/35% B, reached 80% B in 2 min, 100% B in 7 min and remained there for the next 7 min. There was a 5 min re-equilibration step before the next run. The flow rate was 0.400 mL min^{-1} and the injected amount 1.0 μL (Acquity Sample Organizer at 10 °C). The ESI source was at 120 °C and the capillary voltage 3.0 and 2.5 kV in positive and negative mode, respectively. N_2 was used as desolvation gas (800 L h^{-1}) at 270 °C.

The data were collected at a mass range of m/z 300–1200 with a scan duration of 0.2 s. Reserpine was used as the lock spray reference compound in ESI+ mode and leucine enkephaline in ESI− mode. Data processing was carried out with MZmine software (version 2.20) [61] enabling peak integration and alignment of the peaks. An internal spectral MS/MS library was used for identification of the compounds.

Quantification of lipid subspecies was based on peak heights of internal standards. All monoacyl lipids except cholesterol esters, such as monoacylglycerols and monoacylglycerophospholipids, were normalized with LysoPC(17:0), all diacyl lipids except ethanolamine phospholipids were normalized with PC(17:0/17:0), all ceramides with Cer(d18:1/17:0), all diacyl ethanolamine phospholipids with PE(17:0/17:0), and TGs and sterylesters were normalized with TG(17:0/17:0/17:0) and CE(19:0), respectively. Other (unidentified) molecular species were normalized with LysoPC(17:0) for retention time <300 s, PC(17:0/17:0) for retention time between 300 and 410 s, and TG(17:0/17:0/17:0) for higher retention times.

Microplate cultivations

To assess the roles of oleic acid (C18:1) on cold and ethanol tolerance in yeast, microcultures were carried out in media containing various concentrations of supplemented oleic acid (0 or 0.8 mM) and ethanol (1, 5 or 10% v/v). This concentration of oleic acid was chosen based on values found previously in literature [35]. The microcultures were carried out in 100-well honeycomb microtiter plates at 15 and 20 °C (with continuous shaking), and their growth dynamics were monitored with a Bioscreen C MBR incubator and plate reader (Oy Growth Curves Ab, Finland). The wells of the microtiter plates were filled with 300 μL of YPDt medium (1% yeast extract, 2% peptone, 2% glucose, and 1% Tergitol NP-40) supplemented with oleic acid (0 or 0.8 mM) and ethanol (1, 5 or 10% v/v). Oleic acid was added to the media from a 100 × stock solution (80 mM oleic acid) prepared in 50% ethanol and 35% Tergitol NP-40. Precultures of the laboratory strains WT, *ole1*Δ, and *erg4*Δ (Table 1) were started in 10 mL YPD medium (1% yeast extract, 2% peptone, and 2% glucose) and incubated at 25 °C with shaking at 120 rpm. The cultures were centrifuged and the yeast pellets were washed once with sterile deionized water. The yeast was then resuspended in YPDt medium to an OD600 value of 10. The microcultures were started by inoculating the microtiter plates with 3 μL of cell suspension per well (for an initial OD600 value of 0.1) and placing the plates in the Bioscreen C MBR. The optical density of the microcultures at 600 nm was automatically read every 30 min. Four replicates were performed for each strain in each medium. Growth curves for the microcultures were modelled based on the OD600 values over time using the 'grofit'-package for R [60].

Fermentation and analysis

The set of eight brewing strains were characterized in fermentations performed in a 15 °Plato high gravity wort at 15 °C. Yeast was propagated essentially as described previously [3], with the use of a 'Generation 0' fermentation prior to the actual experimental fermentations. The experimental fermentations were carried out in duplicate, in 2-L cylindroconical stainless steel fermenting vessels, containing 1.5 L of wort medium. The 15 °Plato wort (69 g maltose, 17.4 g maltotriose, 15.1 g glucose, and 5.0 g fructose per litre) was produced at the VTT Pilot Brewery from barley malt. Yeast was inoculated at a rate of 15 × 10^6 viable cells mL^{-1}. The wort was oxygenated to 15 mg L^{-1} prior to pitching (Oxygen Indicator Model 26073 and Sensor 21158, Orbisphere Laboratories, Switzerland). The fermentations were carried out at 15 °C until an apparent attenuation of 80% (corresponding to approx 7% ABV) was reached, until no change in residual extract was observed for 24 h, or for a maximum of 14 days.

Wort samples were drawn regularly from the fermentation vessels aseptically, and placed directly on ice, after which the yeast was separated from the fermenting wort by centrifugation (9000×*g*, 10 min, 1 °C). Samples for yeast-derived flavour compounds, fermentable sugars and 4-vinyl guaiacol analysis were drawn from the beer when fermentations were ended. Yeast viability was measured from the yeast that was collected at the end of the fermentations using a Nucleocounter® YC-100™ (ChemoMetec, Denmark).

The alcohol level (% v/v) and pH of samples was determined from the centrifuged and degassed fermentation samples using an Anton Paar Density Meter DMA 5000 M with Alcolyzer Beer ME and pH ME modules (Anton Paar GmbH, Austria). The yeast dry mass content of the samples (i.e. yeast in suspension) was determined by washing the yeast pellets gained from centrifugation twice with 25 mL deionized H_2O and then suspending the washed yeast in a total of 6 mL deionized H_2O. The suspension was then transferred to a pre-weighed porcelain crucible, and was dried overnight at 105 °C and allowed to cool in a desiccator before the change of mass was measured. The measured pH values and suspended dry mass are presented in Additional file 1: Figure S4.

Concentrations of fermentable sugars (glucose, fructose, maltose and maltotriose) were measured by HPLC using a Waters 2695 Separation Module and Waters System Interphase Module liquid chromatograph coupled with a Waters 2414 differential refractometer (Waters Co., Milford, MA, USA). An Aminex HPX-87H Organic Acid Analysis Column (300 × 7.8 mm, Bio-Rad, USA) was equilibrated with 5 mM H_2SO_4 (Titrisol, Merck, Germany) in water at 55 °C and samples were eluted with 5 mM H_2SO_4 in water at a 0.3 mL min^{-1} flow rate.

Yeast-derived flavour compounds were determined by headspace gas chromatography with flame ionization detector (HS-GC-FID) analysis. 4 mL samples were filtered (0.45 μm), incubated at 60 °C for 30 min and then 1 mL of gas phase was injected (split mode; 225 °C; split flow of 30 mL min^{-1}) into a gas chromatograph equipped with an FID detector and headspace autosampler (Agilent 7890 Series; Palo Alto, CA, USA). Analytes were separated on a HP-5 capillary column (50 m × 320 μm × 1.05 μm column, Agilent, USA). The carrier gas was helium (constant flow of 1.4 mL min^{-1}). The temperature program was 50 °C for 3 min, 10 °C min^{-1} to 100 °C, 5 °C min^{-1} to 140 °C, 15 °C min^{-1} to 260 °C and then isothermal for 1 min. Compounds were identified by comparison with authentic standards and were quantified using standard curves. 1-Butanol was used as internal standard.

4-Vinyl guaiacol was analyzed using HPLC-PAD based on methods described by Coghe et al. [62] and McMurrough et al. [63]. The chromatography was carried out using a Waters Alliance HPLC system consisting of a Waters e2695 Separations Module equipped with a XTerra® MS C18 column (5 μm, 4.6 × 150 mm) and a Waters 2996 Photodiode Array Detector. The mobile phase consisted of $H_2O/CH_3OH/H_3PO_4$ (64:35:1, v/v) and flow rate was 0.5 mL min^{-1}. The diode array detector was used at 190–400 nm. 4-Vinyl guaiacol was quantified at 260 nm using standard curves of the pure compound (0.3–30 mg L^{-1}).

Stability of Hybrid T2

The karyotype stability of Hybrid T2 was evaluated after 10 successive batch cultures (corresponding to approximately 65 cell generations) in two different media. Media 1 consisted of 1% yeast extract, 2% peptone, 1% maltose and 1% maltotriose, while Media 2 consisted of 1% yeast extract, 2% peptone, 2% maltose and 14% sorbitol. Media 2 was used to mimic the osmotic stress occurring during high-gravity wort fermentations. Hybrid T2 was first grown overnight in YPM at 25 °C. This culture was used to inoculate 1 mL of Media 1 or Media 2 in duplicate to a starting OD600 of 0.1. The cultures were allowed to grow for 7 days at 18 °C, after which they were used to inoculate a fresh 1 mL aliquot of Media 1 or Media 2 to a starting OD600 of 0.1. This was repeated for a total of 10 successive cultures (10 weeks). After this, 20 µL aliquots of the cultures were spread out on YPM agar, and a total of 8 colonies were randomly selected and isolated for further analysis (2 isolates per duplicate per media).

The karyotype stability of the 10-week isolates was assessed by determining their DNA content by flow cytometry as described above, and their karyotypes by pulsed-field gel electrophoresis (PFGE). PFGE was carried out essentially as described previously [2]. Sample plugs were prepared with the CHEF Genomic DNA Plug Kit for Yeast (Bio-Rad) according to the manufacturer's instructions with minor modifications. Instead of lyticase treatment, the plugs were treated with 0.1 mg mL^{-1} Zymolyase 100T in buffer containing 1 mM dithiothreitol. The sample plugs were loaded into the wells of a 1.0% pulse field certified agarose (Bio-Rad) gel. PFGE was performed at 14 °C in 0.5 × TBE buffer [89 mM Tris, 89 mM boric acid, 2 mM EDTA (pH 8)]. A CHEF Mapper XA pulsed field electrophoresis system (Bio-Rad) was used with the following settings: 6 V cm^{-1} in a 120° angle, pulse length increasing linearly from 26 to 228 s, and total running time of 40 h. A commercial chromosome marker preparation from *S. cerevisiae* strain YNN295 (Bio-Rad) was used for molecular mass calibration. After electrophoresis, the gels were stained with ethidium bromide and scanned with Gel Doc XR+ imaging system (Bio-Rad).

Data analysis

Data and statistical analyses were performed with R (http://www.r-project.org/). The distributions of the lipidomic data were estimated by histograms and the Shapiro–Wilk test, and the lipidomic data was consequently log10-transformed to correct for skewed distributions. The change in lipid composition compared to the control cultivation at 20 °C and 0% supplemented EtOH was tested by Student's t test (two-tailed, unpaired, and unequal variances). To control for multiple testing, the *p* values were further adjusted for Benjamini–Hochberg false discovery rate (FDR). Strain-specific differences in fatty acid, squalene and ergosterol concentrations were tested with one-way ANOVA with Tukey's post hoc test. Multivariate analysis was performed with Partial Least Squares (PLS) and PLS-Discriminant Analysis (PLS-DA) using the 'ropls' package in R [64]. PLS-DA was initially performed on the lipid data of all samples in order to determine whether the lipid compositions of the yeast in low temperatures or high alcohol levels could be distinguished from those at control conditions (20 °C, 0% supplemented EtOH). PLS models were constructed from the fermentation and lipid data obtained at the different temperatures and supplemented ethanol levels in order to elucidate which lipid species contributed positively and negatively to fermentation performance at those conditions. The *Y* response variable of the models was the maximum fermentation level divided by the lag time (*A* and λ from Table 2, respectively), while the *X* predictor variables were the combined dataset of the compositions of fatty acids, squalene and ergosterol obtained from GC/MS analysis and the compositions of the 60 lipid species obtained from UPLC/MS lipidomics analysis. PLS(-DA) models were cross-validated ($Q^2 > 0.5$ was considered significant [34]), and the significance of the Q^2 value was tested with 200 random permutations of the *X* dataset.

Additional files

Additional file 1: Table S1. The relative concentrations (% of total lipid content) of fatty acids, squalene and ergosterol, unsaturated to saturated fatty acid ratio, and average fatty acid chain length in the lipids extracted from cells in late exponential phase during the growth assays. **Table S2.** Modelled (*A*, µ, λ) growth parameters of the microplate cultivations performed with the three laboratory strains grown in media supplemented with 0.8 mM oleic acid and various concentrations of ethanol (growth curves are presented in Fig. 7 in the main article). **Table S3.** The alleles of the *PAD1* and *FDC1* genes (responsible for the 'phenolic off-flavour'-phenotype) that were detected in the 8 brewing strains based on single nucleotide polymorphisms. **Figure S1.** Confirmation of hybridization by (**A**) interdelta PCR, (**B**) rDNA ITS PCR and RFLP, and (**C**) amplification of *FSY1* and *MEX67* genes using species-specific primers. **Figure S2.** DNA content of the (**A**) *S. cerevisiae* haploid (CEN.PK113-1A) and diploid (CEN.PK) reference strains, (**B**) P1-P3 parent strains (all diploid), (**C**) H1-H3 hybrid strains (allotetraploid, allotriploid and allotetraploid, respectively), and (**D**) T1-T2 hybrids strains (allotetraploid and allodiploid, respectively) by flow cytometry. **Figure S3.** The sequencing coverage (median in 10 kbp windows) over the *S. cerevisiae*- (black and red) and *S. eubayanus*-derived (black and blue) chromosomes of parent and hybrid strains (**A**) P2 (265x), (**B**) H1 (87x), (**C**) H2 (304x), (**D**) H3 (275x), (**E**) T1 (295x), and (**F**) T2 (317x). **Figure S4** The (**A**) suspended yeast dry mass (g L^{-1}) and (**B**) pH in the beers fermented from the 15 °P wort with the 8 brewing strains.

Additional file 2. The masses (m/z), retention times (s), and identifications of 60 lipid subspecies in the 8 brewing strains at the 4 different growth conditions. The values represent relative amounts (%) of the 60 identified compounds out of the total 488 obtained in lipidomics analyses. The identified lipid species constitute an average of 43.5% of the total amount of lipids.

Abbreviations

CA: cardiolipin; CE: cholesterol ester; Cer: ceramide; DG: diacylglyceride; FA: fatty acid; GluCer: glucosyl ceramide; LacCer: lactosyl ceramide; MG: monacylglyceride; PA: phosphatidic acid; PC: phosphatidylcholine; PE: phosphatidylethanolamine; PG: phosphatidylglycerol; PI: phosphatidylinositol; PLS: partial least squares; PLS-DA: partial least squares discriminant analysis; POF: phenolic off-flavour; PS: phosphatidylserine; SFA: saturated fatty acid; SNP: single nucleotide polymorphism; TG: triacylglyceride; UFA: unsaturated fatty acid; ZymoE: zymosterol ester.

Authors' contributions

Conceived and designed the experiments: KK TSL BG. Performed the experiments: KK. Analyzed the data: KK SC TSL. Supervised the work: BG. Wrote the manuscript: KK TSL BG. All authors read and approved the final manuscript.

Author details

[1] VTT Technical Research Centre of Finland, Tietotie 2, P.O. Box 1000, 02044 Espoo, Finland. [2] Department of Biotechnology and Chemical Technology, Aalto University, School of Chemical Technology, Kemistintie 1, Aalto, P.O. Box 16100, 00076 Espoo, Finland.

Acknowledgements

We thank Annika Wilhelmson for her support throughout, Ulla Lahtinen, Anna-Liisa Ruskeepää, and Airi Hyrkäs for assistance with lipid analysis, Eero Mattila for wort preparation and other assistance in the VTT Pilot Brewery, and Aila Siltala for skilled technical assistance.

Competing interests

The authors declare that they have no competing interests.

Funding

This work was supported by the Alfred Kordelin Foundation, Svenska Kulturfonden - The Swedish Cultural Foundation in Finland, PBL Brewing Laboratory, and the Academy of Finland (Academy Project 276480).

References

1. Gibson B, Liti G. *Saccharomyces pastorianus*: genomic insights inspiring innovation for industry. Yeast. 2015;32:17–27.
2. Krogerus K, Magalhães F, Vidgren V, Gibson B. New lager yeast strains generated by interspecific hybridization. J Ind Microbiol Biotechnol. 2015;42:769–78.
3. Krogerus K, Arvas M, De Chiara M, Magalhães F, Mattinen L, Oja M, et al. Ploidy influences the functional attributes of de novo lager yeast hybrids. Appl Microbiol Biotechnol. 2016;100:7203–22.
4. Mertens S, Steensels J, Saels V, De Rouck G, Aerts G, Verstrepen KJ. A large set of newly created interspecific *Saccharomyces* hybrids increases aromatic diversity in lager beers. Appl Environ Microbiol. 2015;81:8202–14.
5. Krogerus K, Magalhães F, Vidgren V, Gibson B. Novel brewing yeast hybrids: creation and application. Appl Microbiol Biotechnol. 2017;101:65–78.
6. Greig D, Borts RH, Louis EJ, Travisano M. Epistasis and hybrid sterility in *Saccharomyces*. Proc Biol Sci. 2002;269:1167–71.
7. Marinoni G, Manuel M, Petersen RF, Hvidtfeldt J, Sulo P, Piskur J. Horizontal transfer of genetic material among *Saccharomyces* yeasts horizontal transfer of genetic material among saccharomyces yeasts. J Bacteriol. 1999;181:6488–96.
8. Sebastiani F, Barberio C, Casalone E, Cavalieri D, Polsinelli M. Crosses between *Saccharomyces cerevisiae* and *Saccharomyces bayanus* generate fertile hybrids. Res Microbiol. 2002;153:53–8.
9. Morales L, Dujon B. Evolutionary role of interspecies hybridization and genetic exchanges in yeasts. Microbiol Mol Biol Rev. 2012;76:721–39.
10. Mancera E, Bourgon R, Brozzi A, Huber W, Steinmetz LM. High-resolution mapping of meiotic crossovers and non-crossovers in yeast. Nature. 2008;454:479–85.
11. Marullo P, Aigle M, Bely M, Masneuf-Pomarède I, Durrens P, Dubourdieu D, et al. Single QTL mapping and nucleotide-level resolution of a physiologic trait in wine *Saccharomyces cerevisiae* strains. FEMS Yeast Res. 2007;7:941–52.
12. Mukai N, Masaki K, Fujii T, Kawamukai M, Iefuji H. PAD1 and FDC1 are essential for the decarboxylation of phenylacrylic acids in *Saccharomyces cerevisiae*. J Biosci Bioeng. 2010;109:564–9.
13. Gallone B, Steensels J, Baele G, Maere S, Verstrepen KJ, Prahl T, et al. Domestication and Divergence of *Saccharomyces cerevisiae* Beer Yeasts. Cell. 2016;166(1397–1410):e16.
14. Mukai N, Masaki K, Fujii T, Iefuji H. Single nucleotide polymorphisms of PAD1 and FDC1 show a positive relationship with ferulic acid decarboxylation ability among industrial yeasts used in alcoholic beverage production. J Biosci Bioeng. 2014;118:50–5.
15. Gonçalves M, Pontes A, Almeida P, Barbosa R, Serra M, Libkind D, et al. Distinct domestication trajectories in top-fermenting beer yeasts and wine yeasts. Curr Biol. 2016;26:2750–61.
16. Tubb RS, Searle BA, Goodey AR, Brown AJP. Rare mating and transformation for construction of novel brewing yeasts. In: Proceedings of 18th Congress European Brewery Convention. 1981; p. 487–96.
17. Gibson BR, Storgårds E, Krogerus K, Vidgren V. Comparative physiology and fermentation performance of Saaz and Frohberg lager yeast strains and the parental species *Saccharomyces eubayanus*. Yeast. 2013;30:255–66.
18. Aguilera J, Randez-Gil F, Prieto JA. Cold response in *Saccharomyces cerevisiae*: new functions for old mechanisms. FEMS Microbiol Rev. 2007;31:327–41.
19. Deed RC, Deed NK, Gardner RC. Transcriptional response of *Saccharomyces cerevisiae* to low temperature during wine fermentation. Antonie Van Leeuwenhoek. 2015;107:1029–48.
20. López-Malo M, Querol A, Guillamon JM. Metabolomic comparison of *Saccharomyces cerevisiae* and the cryotolerant species *S. bayanus* var. *uvarum* and *S. kudriavzevii* during wine fermentation at low temperature. PLoS ONE. 2013;8:e60135.
21. Paget CM, Schwartz JM, Delneri D. Environmental systems biology of cold-tolerant phenotype in *Saccharomyces* species adapted to grow at different temperatures. Mol Ecol. 2014;23:5241–57.
22. Sahara T, Goda T, Ohgiya S. Comprehensive expression analysis of time-dependent genetic responses in yeast cells to low temperature. J Biol Chem. 2002;277:50015–21.
23. Schade B, Jansen G, Whiteway M, Entian KD, Thomas DY, Goethe-university JW, et al. Cold adaptation in budding yeast. Mol Biol Cell. 2004;15:5492–502.
24. Tai SL, Daran-Lapujade P, Walsh MC, Pronk JT, Daran J-M. Acclimation of *Saccharomyces cerevisiae* to low temperature: a chemostat-based transcriptome analysis. Mol Biol Cell. 2007;18:5100–12.
25. Henderson CM, Zeno WF, Lerno LA, Longo ML, Block DE. Fermentation temperature modulates phosphatidylethanolamine and phosphatidylinositol levels in the cell membrane of *Saccharomyces cerevisiae*. Appl Environ Microbiol. 2013;79:5345–56.
26. Redón M, Guillamón JM, Mas A, Rozès N. Effect of growth temperature on yeast lipid composition and alcoholic fermentation at low temperature. Eur Food Res Technol. 2011;232:517–27.
27. Vicent I, Navarro A, Mulet JM, Sharma S, Serrano R. Uptake of inorganic phosphate is a limiting factor for *Saccharomyces cerevisiae* during growth at low temperatures. FEMS Yeast Res. 2015;15:1–13.
28. Beltran G, Novo M, Guillamón JM, Mas A, Rozès N. Effect of fermentation temperature and culture media on the yeast lipid composition and wine volatile compounds. Int J Food Microbiol. 2008;121:169–77.

29. Abe F, Horikoshi K. Tryptophan permease gene TAT2 confers high-pressure growth in Saccharomyces cerevisiae. Mol Cell Biol. 2000;20:8093–102.
30. Abe F, Hiraki T. Mechanistic role of ergosterol in membrane rigidity and cycloheximide resistance in *Saccharomyces cerevisiae*. Biochim Biophys Acta Biomembr. 2009;1788:743–52.
31. Vanegas JM, Contreras MF, Faller R, Longo ML. Role of unsaturated lipid and ergosterol in ethanol tolerance of model yeast biomembranes. Biophys J. 2012;102:507–16.
32. Henderson CM, Lozada-Contreras M, Jiranek V, Longo ML, Block DE. Ethanol production and maximum cell growth are highly correlated with membrane lipid composition during fermentation as determined by lipidomic analysis of 22 *Saccharomyces cerevisiae* strains. Appl Environ Microbiol. 2013;79:91–104.
33. You KM, Rosenfield C, Knipple DC. Ethanol tolerance in the yeast *Saccharomyces cerevisiae* is dependent on cellular oleic acid content. Appl Environ Microbiol. 2003;69:1499.
34. Triba MN, Le Moyec L, Amathieu R, Goossens C, Bouchemal N, Nahon P, et al. PLS/OPLS models in metabolomics: the impact of permutation of dataset rows on the K-fold cross-validation quality parameters. Mol BioSyst. 2014;11:13–9.
35. Redón M, Guillamón JM, Mas A, Rozès N. Effect of lipid supplementation upon *Saccharomyces cerevisiae* lipid composition and fermentation performance at low temperature. Eur Food Res Technol. 2009;228:833–40.
36. Vanbeneden N, Gils F, Delvaux F, Delvaux FR. Formation of 4-vinyl and 4-ethyl derivatives from hydroxycinnamic acids: occurrence of volatile phenolic flavour compounds in beer and distribution of Pad1-activity among brewing yeasts. Food Chem. 2008;107:221–30.
37. Landry CR, Lemos B, Rifkin SA, Dickinson WJ, Hartl DL. Genetic properties influencing the evolvability of gene expression. Science. 2007;317:118–21.
38. Tirosh I, Reikhav S, Levy AA, Barkai N. A yeast hybrid provides insight into the evolution of gene expression regulation. Science. 2009;324:659–62.
39. Dunn B, Sherlock G. Reconstruction of the genome origins and evolution of the hybrid lager yeast *Saccharomyces pastorianus*. Genome Res. 2008;18:1610–23.
40. Pérez-Través L, Lopes CA, Barrio E, Querol A. Stabilization process in *Saccharomyces* intra and interspecific hybrids in fermentative conditions. Int Microbiol. 2014;17:213–24.
41. Peris D, Moriarty RV, Alexander WG, Baker E, Sylvester K, Sardi M, et al. Hybridization and adaptive evolution of diverse *Saccharomyces* species for cellulosic biofuel production. Biotechnol Biofuels. 2017;154:78.
42. Alexandre H, Rousseaux I, Charpentier C. Relationship between ethanol tolerance, lipid composition and plasma membrane fluidity in *Saccharomyces cerevisiae* and *Kloeckera apiculata*. FEMS Microbiol Lett. 1994;124:17–22.
43. Boeke JD, Trueheart J, Natsoulis G, Fink GR. 5-Fluoroorotic acid as a selective agent in yeast molecular genetic. Methods Enzymol. 1987;154:164–75.
44. Zaret KS, Sherman F. α-Aminoadipate as a primary nitrogen source for *Saccharomyces cerevisiae* mutants. J Bacteriol. 1985;162:579–83.
45. Pham T, Wimalasena T, Box WG, Koivuranta K, Storgårds E, Smart KA, et al. Evaluation of ITS PCR and RFLP for differentiation and identification of brewing yeast and brewery "wild" yeast contaminants. J Inst Brew. 2011;117:556–68.
46. Muir A, Harrison E, Wheals A. A multiplex set of species-specific primers for rapid identification of members of the genus *Saccharomyces*. FEMS Yeast Res. 2011;11:552–63.
47. Pengelly RJ, Wheals AE. Rapid identification of *Saccharomyces eubayanus* and its hybrids. FEMS Yeast Res. 2013;13:156–61.
48. Legras JL, Karst F. Optimisation of interdelta analysis for *Saccharomyces cerevisiae* strain characterisation. FEMS Microbiol Lett. 2003;221:249–55.
49. Haase SB, Reed SI. Improved flow cytometric analysis of the budding yeast cell cycle. Cell Cycle. 2014;1:117–21.
50. Baker E, Wang B, Bellora N, Peris D, Hulfachor AB, Koshalek JA, et al. The genome sequence of *Saccharomyces eubayanus* and the domestication of lager-brewing yeasts. Mol Biol Evol. 2015;32:2818–31.
51. Andrews S. FastQC: a quality control tool for high throughput sequence data. http://www.bioinformatics.babraham.ac.uk/projects/fastqc/. 2010.
52. Jiang H, Lei R, Ding S-W, Zhu S. Skewer: a fast and accurate adapter trimmer for next-generation sequencing paired-end reads. BMC Bioinform. 2014;15:182.
53. Chiang C, Layer RM, Faust GG, Lindberg MR, Rose DB, Garrison EP, et al. SpeedSeq: ultra-fast personal genome analysis and interpretation. Nat Methods. 2015;12:1–5.
54. Garrison E, Marth G. Haplotype-based variant detection from short-read sequencing. arXiv Prepr. arXiv1207.3907 2012;9. http://arxiv.org/abs/1207.3907.
55. Abyzov A, Urban AE, Snyder M, Gerstein M. CNVnator: an approach to discover, genotype, and characterize typical and atypical CNVs from family and population genome sequencing. Genome Res. 2011;21:974–84.
56. Engel SR, Dietrich FS, Fisk DG, Binkley G, Balakrishnan R, Costanzo MC, et al. The reference genome sequence of *Saccharomyces cerevisiae*: then and now. G3. 2014;4:389–98.
57. Li H, Handsaker B, Wysoker A, Fennell T, Ruan J, Homer N, et al. The sequence alignment/map FORMAT AND SAMtools. Bioinformatics. 2009;25:2078–9.
58. García-Alcalde F, Okonechnikov K, Carbonell J, Cruz LM, Götz S, Tarazona S, et al. Qualimap: evaluating next-generation sequencing alignment data. Bioinformatics. 2012;28:2678–9.
59. Quinlan AR, Hall IM. BEDTools: a flexible suite of utilities for comparing genomic features. Bioinformatics. 2010;26:841–2.
60. Kahm M, Hasenbrink G, Lichtenberg-frate H, Ludwig J, Kschischo M. Grofit: fitting biological growth curves. J Stat Softw. 2010;33:1–21.
61. Pluskal T, Castillo S, Villar-Briones A, Oresic M. MZmine 2: modular framework for processing, visualizing, and analyzing mass spectrometry-based molecular profile data. BMC Bioinform. 2010;11:395.
62. Coghe S, Benoot K, Delvaux F, Vanderhaegen B, Delvaux FR. Ferulic acid release and 4-vinylguaiacol formation during brewing and fermentation: indications for feruloyl esterase activity in *Saccharomyces cerevisiae*. J Agric Food Chem. 2004;52:602–8.
63. McMurrough I, Madigan D, Donnelly D, Hurley J, Doyle A-M, Hennigan G, et al. Control of ferulic acid and 4-vinyl guaiacol in brewing. J Inst Brew. 1996;102:327–32.
64. Thévenot EA, Roux A, Xu Y, Ezan E, Junot C. Analysis of the human adult urinary metabolome variations with age, body mass index, and gender by implementing a comprehensive workflow for univariate and OPLS statistical analyses. J Proteome Res. 2015;14:3322–35.
65. Meilgaard MC. Prediction of flavor differences between beers from their chemical composition. J Agric Food Chem. 1982;30:1009–17.
66. Engan S. Organoleptic threshold values of some alcohols and esters in beer. J Inst Brew. 1972;78:33–6.

8

Dynamic regulation of fatty acid pools for improved production of fatty alcohols in *Saccharomyces cerevisiae*

Paulo Gonçalves Teixeira[1,2], Raphael Ferreira[1,2], Yongjin J. Zhou[1,2], Verena Siewers[1,2] and Jens Nielsen[1,2,3*]

Abstract

Background: In vivo production of fatty acid-derived chemicals in *Saccharomyces cerevisiae* requires strategies to increase the intracellular supply of either acyl-CoA or free fatty acids (FFAs), since their cytosolic concentrations are quite low in a natural state for this organism. Deletion of the fatty acyl-CoA synthetase genes *FAA1* and *FAA4* is an effective and straightforward way to disable re-activation of fatty acids and drastically increase FFA levels. However, this strategy causes FFA over-accumulation and consequential release to the extracellular medium, which results in a significant loss of precursors that compromises the process yield. In the present study, we aimed for dynamic expression of the fatty acyl-CoA synthetase gene *FAA1* to regulate FFA and acyl-CoA pools in order to improve fatty alcohol production yields.

Results: We analyzed the metabolite dynamics of a *faa1Δ faa4Δ* strain constitutively expressing a carboxylic acid reductase from *Mycobacterium marinum* (MmCAR) and an endogenous alcohol dehydrogenase (Adh5) for in vivo production of fatty alcohols from FFAs. We observed production of fatty acids and fatty alcohols with different rates leading to high levels of FFAs not being converted to the final product. To address the issue, we expressed the MmCAR + Adh5 pathway together with a fatty acyl-CoA reductase from *Marinobacter aquaeolei* to enable fatty alcohol production simultaneously from FFA and acyl-CoA, respectively. Then, we expressed *FAA1* under the control of different promoters in order to balance FFA and acyl-CoA interconversion rates and to achieve optimal levels for conversion to fatty alcohols. Expressing *FAA1* under control of the *HXT1* promoter led to an increased accumulation of fatty alcohols per OD_{600} up to 41% while FFA levels were decreased by 63% compared with the control strain.

Conclusions: Fine-tuning and dynamic regulation of key metabolic steps can be used to improve cell factories when the rates of downstream reactions are limiting. This avoids loss of precursors to the extracellular medium or to competing reactions, hereby potentially improving the process yield. The study also provides knowledge of a key point of fatty acid regulation and homeostasis, which can be used for future design of cells factories for fatty acid-derived chemicals.

Keywords: Fatty alcohols, Fatty acid activation, *FAA1*, Dynamic control, Yeast, Metabolic engineering

Background

Society's need for sustainable production of liquid fuels and oleochemicals is indisputable. The use of plants for the extraction of lipid molecules for conversion to biofuels and oleochemicals is often not sustainable in a long term due to requirements for large areas of fertile land together with extensive extraction and chemical conversion processes. Therefore, there is a need for alternative production routes for petrol and plant lipid-derived chemicals that can simultaneously offer a sustainable life cycle and assure a stable supply. The development of microbial cell factories proposes a substitute production path by redirecting cell metabolism towards a product of interest [1, 2]. Many studies have already shown how

*Correspondence: nielsenj@chalmers.se
[1] Department of Biology and Biological Engineering, Chalmers University of Technology, 412 96 Gothenburg, Sweden
Full list of author information is available at the end of the article

Saccharomyces cerevisiae can be successfully engineered for production of different fatty acid-derived chemicals such as fatty alcohols, alkanes, alkenes, fatty acyl ethyl esters and triacylglycerols [3–6].

For an efficient production of a desired chemical through fermentation, there is usually a need to engineer the production organism for reduced formation of byproducts. Byproduct formation diverts nutrients towards unwanted molecules, reducing the yield of the desired product and compromising the process efficiency [7–9]. In metabolic engineering, byproduct formation easily arises when fluxes are not properly balanced and pathway precursors or intermediates accumulate in the cell, becoming abundant substrates for side reactions and processes. For this reason, balancing fluxes through enzyme modulation and controlling metabolite pool levels usually play an important role with regard to the final product yield.

Fatty alcohols in *S. cerevisiae* can be produced either from acyl-CoA or free fatty acids (FFAs) in two steps. In either case, the precursor is first converted to a fatty aldehyde through an acyl-CoA reductase [10] for acyl-CoA reduction, or a carboxylic acid reductase [11] for FFA reduction. The formed fatty aldehydes are then reduced to a primary alcohol by endogenous alcohol dehydrogenases [12]. An alternative route for fatty alcohol production from acyl-CoA is a four-electron reduction catalyzed by a bifunctional fatty acyl-CoA reductase [13], in which an aldehyde is also formed as an intermediary metabolite but the enzyme is capable of catalyzing both reaction steps.

The most successful strategies for production of fatty alcohols so far rely on using the FFA pathway, in which the main factor of success is the possibility to accumulate FFAs in the cytosol at levels several orders of magnitude higher compared with acyl-CoA [14–16]. Here, one of the major strategies used for accumulation of FFAs is the simultaneous deletion of the fatty acyl-CoA synthetase genes *FAA1* and *FAA4*, encoding the main responsible enzymes for the activation of FFAs to fatty acyl-CoA [17, 18]. Simultaneous deletion of these two genes together with *POX1* (encoding fatty acyl-CoA oxidase, the first enzyme of the fatty acid beta-oxidation pathway), allows production of up to 490 mg/L of FFA, whereas the wild-type reference strain only produces around 3 mg/L FFAs [5, 19]. Although there remains some discussion as to the origin of these fatty acids, evidence shows that these are produced mainly from hydrolysis of acyl chains from storage and membrane lipids, and not directly from acyl-CoA due to thioesterase activity or spontaneous hydrolysis [18].

Increased FFA production in *faa1Δ faa4Δ* deletion strains results in high levels of these FFA accumulating in the extracellular medium. Further increasing production levels in these same strains results in an increase of extracellular FFA while intracellular levels are not significantly changed, which strongly suggests a limit to intracellular FFA accumulation in the cell [5, 19]. It is unclear if this release of FFA to the medium is carried out by uncharacterized transporters or if it is a process of transmembrane diffusion due to very high concentration of FFAs in the cytosol. In either case, released FFA are inaccessible to the cell due to lack of re-import and fatty acid activation mechanisms, for which *FAA1* and *FAA4* are responsible [18, 20, 21].

Although an efficient accumulation of FFAs is most beneficial for the production of chemicals derived thereof, enzymes identified so far for production of fatty acid-derived chemicals such as fatty alcohols are often not efficient enough to keep up with the flux of FFA formation. This creates a situation where FFAs are produced faster than they can be converted to the final product [5].

In order to qualify as an industrially feasible process, production of fatty acid-derived chemicals requires achieving the maximum production yield possible. Thus far, these unbalanced fluxes create an intra- and extracellular over-accumulation of a large amounts of FFAs, which are ultimately not converted to the product of interest. This constitutes a serious loss of carbon and reducing cofactors that might hinder the production yields for fatty alcohols and other fatty acid-derived chemicals.

We therefore studied how fine-tuning of a single step in this pathway would allow control of the involved metabolite levels and how this could be a possible approach to reduce the described problem. A successful application of this strategy would allow for an improvement in final product yields and production titers since it provides an optimization of resource usage during cultivation of engineered cell factories.

Methods

Plasmid construction

The *FAA1* gene was amplified by PCR from *Saccharomyces cerevisiae* CEN.PK113-11C and cloned into p413TEF using restriction enzymes BamHI and XhoI, resulting in plasmid pTEF-*FAA1*. For construction of plasmids pHXT1-*FAA1* and pHXT7-*FAA1*, the *HXT7* and the *HXT1* promoters were amplified from *S. cerevisiae* CEN.PK113-11C and fused to the backbone p413 by PCR. The p413 backbone plasmid was amplified from p413TEF. The *FAA1* gene and *HXT1/HXT7* promoters were amplified by PCR to generate the complementary overhangs for insertion into the plasmid by Gibson assembly (New England Biolabs, Ipswich, Massachusetts, United States). Strains and plasmids generated and used in this study are presented in Table 1 and Table 2, respectively. A list of primers used for plasmid construction is shown in Additional file 1: Table S1.

Table 1 *Saccharomyces cerevisiae* strains used in this study

Strain	Genotype	References
YJZ08	MATa MAL2-8c SUC2 *his3Δ1 ura3-52 hfd1Δ pox1Δ faa1Δ faa4Δ*	Zhou et al. [5, 6]
YZFOH1	YJZ08 pAOH3	This study
YZFOH2	YJZ08 pAOH9	This study
p413	YJZ08 pAOH9 p413	This study
HXT1p-FAA1	YJZ08 pAOH9 pHXT1-FAA1	This study
HXT7p-FAA1	YJZ08 pAOH9 pHXT7-FAA1	This study
TEF1p-FAA1	YJZ08 pAOH9 pTEF1-FAA1	This study
CUP1p-FAA1	YJZ08 pAOH9 pCUP1-FAA1	This study

Growth medium

Saccharomyces cerevisiae strains with uracil and histidine auxotrophies were grown on YPD plates containing 20 g/L glucose, 10 g/L yeast extract, 20 g/L peptone from casein and 20 g/L agar. Plasmid carrying strains were grown on selective growth medium containing 6.9 g/L yeast nitrogen base w/o amino acids (Formedium, Hunstanton, UK), 0.77 g/L complete supplement mixture w/o histidine and uracil (Formedium), 20 g/L glucose and 20 g/L agar. Shake flask cultivations were performed in minimal medium containing 20 g/L glucose, 5 g/L $(NH_4)_2SO_4$, 14.4 g/L KH_2PO_4, 0.5 g/L $MgSO_4{\cdot}7H_2O$. After sterilization, 2 mL/L trace element solution and 1 mL/L of vitamin solution were added. The composition of the trace element and vitamin solution has been reported earlier [32].

Shake flask cultivations

All experiments were performed with strains cultivated as biological triplicates. This means that three independent transformants were used to start pre-cultures. For these, 3 mL of minimal medium in a 15 mL tube, or in 5 mL in a 50 mL tube, were inoculated for the first experiment, and cultivated at 200 rpm and 30 °C for 18 h. Subsequently, the pre-culture was used to inoculate 20 mL of minimal medium in a 100 mL shake flask, or 100 mL of minimal medium in a 500 mL shake flask for the first experiment, at an OD_{600} of 0.1. Shake flasks were incubated at 200 rpm and 30 °C for 72 h.

A spectrophotometer (Genesis20, Thermo Fisher Scientific, Waltham, MA, USA) was used to measure cell growth at designated time points and at the end of the shake flask cultivations. Optical density (OD_{600}) was measured by absorbance at 600 nm of a diluted culture sample.

Growth rates on glucose were calculated using OD_{600} data points between 6 and 12 h, from which the slope of the plotted log (OD_{600}) values over time interval was calculated.

Quantification of extracellular metabolites

For quantification of glucose and ethanol, 1 mL samples were taken throughout the culture. The biomass was removed by filtration using a 0.45 μm nylon filter (VWR International AB, Stockholm, Sweden). Sample analysis was performed by HPLC using a Dionex Ultimate 3000 (Dionex, Sunnyvale, CA, USA) together with an Aminex HPX-87H column (300 × 7.8 mm, Bio-Rad Laboratories, Hercules, CA, USA) and a refractive index detector (512 μRIU). The column temperature was kept constant at 45 °C and 15 μL were injected into the mobile phase consisting of 5 mM H_2SO_4. The flow rate was set to 0.6 mL/min.

Quantification of lipids

Samples for lipid analysis were taken as 5 mL of culture at the end of the shake flask cultivations, after 72 h. Subsequently, the samples were centrifuged at 4000 rpm and the supernatant was discarded. The pellets were kept at −20 °C for 5 min and then freeze-dried using a Christ alpha 2–4 LSC (Christ Gefriertrocknungsanlagen, Osterode, Germany). The samples were analyzed as described previously [33] using 10 mg of dry cell biomass.

Quantification of FFAs and fatty-alcohols

FFAs were simultaneously extracted and methylated by dichloromethane containing methyl iodide as methyl

Table 2 Plasmids used in this study

Plasmid	Genotype/features	References
pYX212	pYX212 empty plasmid: 2 μm, ampR, URA3, TPIp, pYX212t	R&D systems
pAOH3	pYX212-(TPIp-*npgA*-FBA1t)-(TDH3p-*MmCAR*-ADH1t)-(TEF1p-*ADH5*-pYX212t)	Zhou et al. [5, 6]
pAOH9	pYX212-(TPIp-*npgA*-FBA1t)-(TDH3p-*MmCAR*-ADH1t)-(tHXT7p-*ADH5*-CYC1t)-(TEF1p-*FaCoAR*-pYX212t)	Zhou et al. [5, 6]
p413	p413TEF empty plasmid: CEN.ARS, ampR, HIS3, TEF1p, CYC1t	ATCC® 87362
pTEF1-FAA1	p413-TEF1p-*FAA1*-CYC1t	This study
pHXT1-FAA1	p413-HXT1p-*FAA1*-CYC1t	This study
pHXT7-FAA1	p413-HXT7p-*FAA1*-CYC1t	This study
pCUP1-FAA1	p413-CUP1p-*FAA1*-CYC1t	This study

donor [34]. Briefly, 200 μL aliquots of whole cell culture (cells + supernatant) were taken into glass vials, then 10 μL 40% tetrabutylammonium hydroxide (base catalyst) was added immediately followed by addition of 200 μL dichloromethane containing 200 mM methyl iodide as methyl donor and 100 mg/L pentadecanoic acid as an internal standard. The mixtures were shaken for 30 min at 1400 rpm. By using a vortex mixer, and then centrifuged at 5000*g* to promote phase separation. A 160 μL dichloromethane layer was transferred into a GC vial with glass insert, and evaporated 4 h to dryness. The extracted methyl esters were resuspended in 160 μL hexane and then analyzed by gas chromatography (Focus GC, ThermoFisher Scientific) equipped with a Zebron ZB-5MS GUARDIAN capillary column (30 m × 0.25 mm × 0.25 μm, Phenomenex) and a Flame Ionization Detector (FID, ThermoFisher Scientific). The GC program was as follows: initial temperature of 50 °C, hold for 2 min; ramp to 140 °C at a rate of 30 °C per minute, then raised to 280 °C at a rate of 10 °C per min and hold for 3 min. The temperature of inlet was kept at 280 °C. The injection volume was 1 μL. The flow rate of the carrier gas (helium) was set to 1.0 mL/min. Final quantification was performed using the Xcalibur software.

For fatty alcohol quantification, cell pellets were collected from 5 mL cell culture and then freeze-dried for 48 h. Metabolites were extracted by 2:1 chloroform:methanol solution [33], which contained pentadecanol as internal standard. The extracted fraction was dried by rotary evaporation and dissolved in ethyl acetate. Quantification of fatty alcohols was performed on the same GC–FID system as used for fatty acid analysis. The GC program for fatty alcohol quantification was as follow: initial temperature of 45 °C hold for 2 min; then ramp to 220 °C at a rate of 20 °C per min and hold for 2 min; ramp to 300 °C at a rate of 20 °C per min and hold for 5 min. The temperature of the inlet was kept at 250 °C. The injection volume was 1 μL. The flow rate of the carrier gas (helium) was set to 1.0 mL/min. Final quantification was performed with Xcalibur software.

Results

FFA accumulation during fatty alcohol production in a *faa1Δ faa4Δ* strain

Before designing and engineering a flux control strategy, the dynamics of metabolite levels throughout the batch culture of the initial fatty alcohol-producing strain was studied. The initial strain YJZ08 (CEN.PK 113-11C *hfd1Δ faa1Δ faa4Δ pox1Δ*) carried deletions in the genes *FAA1* and *FAA4* and as such was unable to reconvert FFAs to acyl-CoAs. Other genes relevant for the process were also deleted such as *POX1*, encoding the first enzyme of the beta-oxidation pathway responsible for the degradation of acyl-CoAs, and *HFD1*, a fatty aldehyde dehydrogenase responsible for conversion of fatty aldehydes into fatty acids. Deletion of *HFD1* has been reported as important for ensuring flux from fatty aldehydes to fatty alcohols in *S. cerevisiae*, since the encoded enzyme efficiently converts produced aldehydes back to their FFA form [12]. YJZ08 was transformed with the 2 μm plasmid pAOH3 for constitutive strong expression of *MmCAR* [11] encoding a carboxylic acid reductase from *Mycobacterium marinum*, which converts long chain FFAs into the respective aldehydes, and *ADH5*, encoding a native alcohol dehydrogenase from *S. cerevisiae*, which efficiently reduces long chain aldehydes to the respective alcohols [5] (Fig. 1a). The resulting strain YZFOH1 was cultivated for 72 h in minimal media with 2% glucose and analyzed by quantification of optical density (OD) and relevant internal and external metabolites, i.e. glucose, ethanol, total FFAs and fatty alcohols. Samples were taken every 3–6 h (Fig. 1b). Due to the ability of fatty acids to form emulsions in the culture media and adsorb to cell membranes, which challenges an accurate distinction between intra- and extracellular FFA, FFA were extracted and quantified from a total volume of culture sample resulting in quantification of total FFA levels.

Measured metabolite profiles showed glucose being entirely consumed at 18 h, while ethanol was gradually formed up to 6 g/L. When glucose was depleted, ethanol started being consumed until its depletion after 60 h. Fatty alcohols and FFAs were constantly produced throughout the culture from either glucose or ethanol as carbon sources. When normalized by the OD values, FFAs showed a constant increase throughout the culture while fatty alcohols reached and oscillated in a plateau from the 21 h time point onwards (Fig. 1c).

The final titers show that from the total formed fatty acids, only 20% were actually converted to fatty alcohols. Furthermore, we observed turbidity in the culture supernatant (not shown) from accumulation of extracellular FFA in form of micelles. This is consistent to what was observed in other studies where similar strains were analyzed [5, 19].

Designing a dynamic substrate pathway for fatty alcohol production

In order to increase the yield of fatty alcohol production, our primary goal was to fine-tune the FFA to acyl-CoA conversion rates. This would allow FFA levels in the cell to be high enough for optimal conversion to fatty alcohols without reaching the excessive level that causes release of these precursors to the extracellular medium.

Furthermore, strategies explored so far focus on production of fatty alcohols either from acyl-CoA or

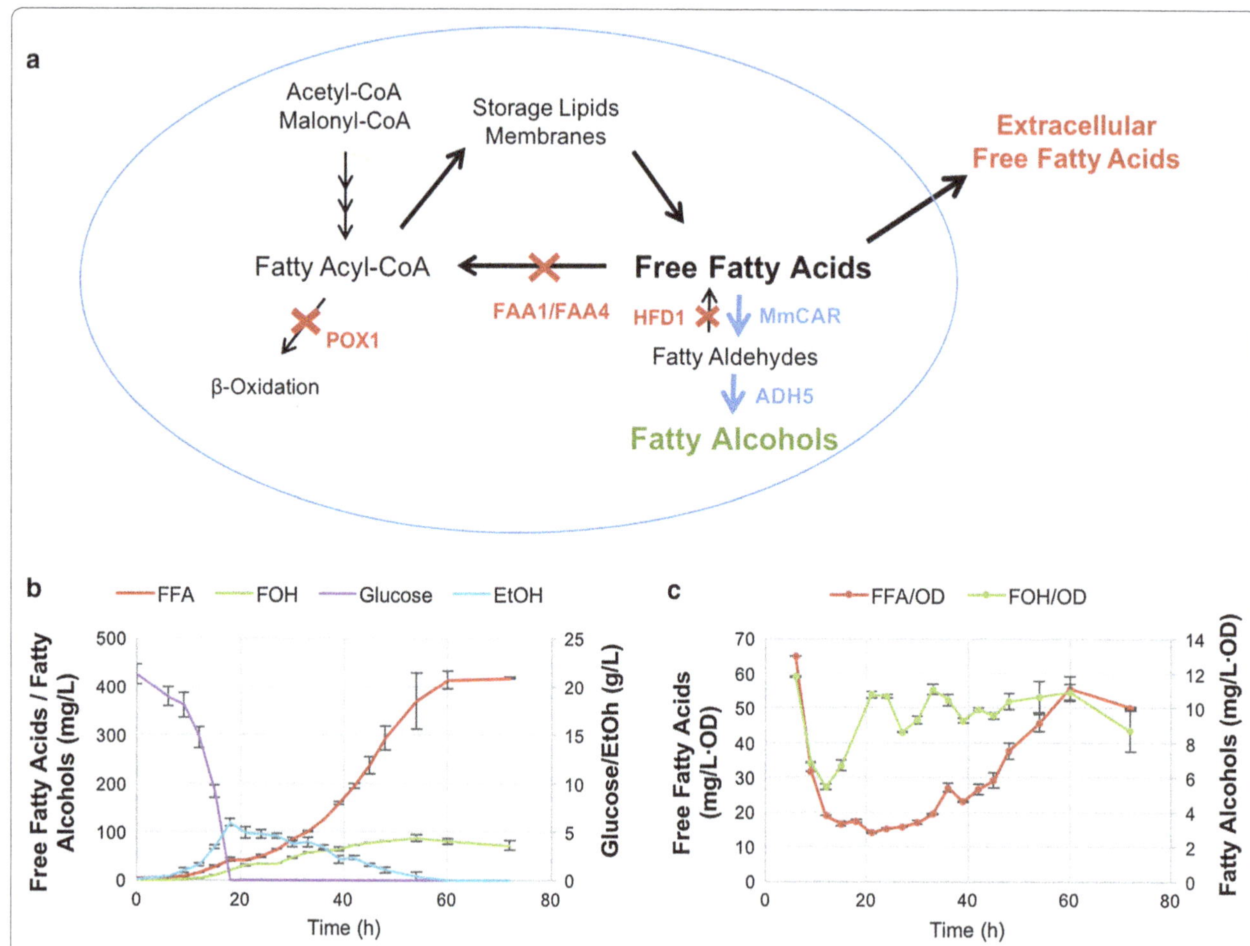

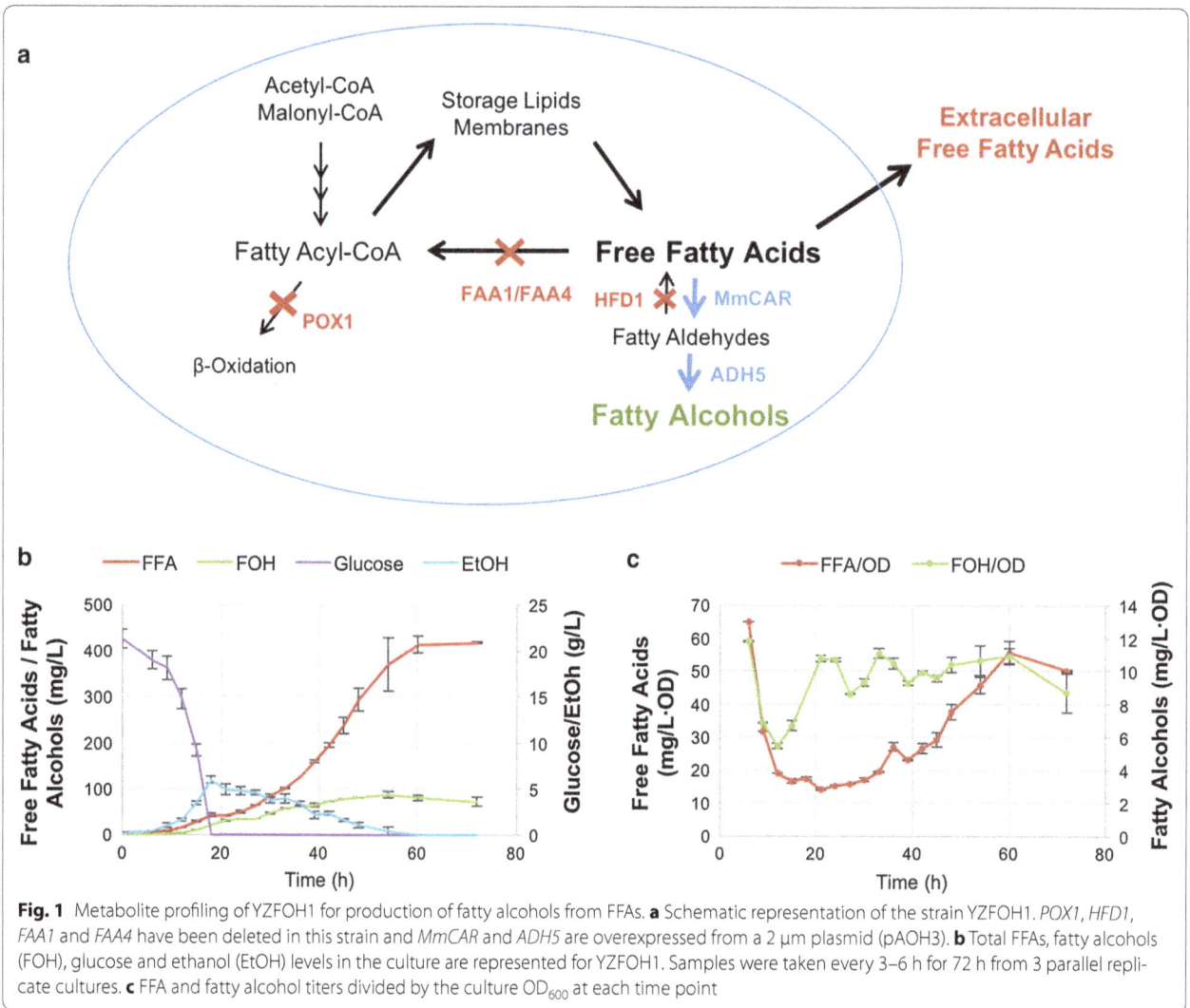

Fig. 1 Metabolite profiling of YZFOH1 for production of fatty alcohols from FFAs. **a** Schematic representation of the strain YZFOH1. *POX1*, *HFD1*, *FAA1* and *FAA4* have been deleted in this strain and *MmCAR* and *ADH5* are overexpressed from a 2 µm plasmid (pAOH3). **b** Total FFAs, fatty alcohols (FOH), glucose and ethanol (EtOH) levels in the culture are represented for YZFOH1. Samples were taken every 3–6 h for 72 h from 3 parallel replicate cultures. **c** FFA and fatty alcohol titers divided by the culture OD_{600} at each time point

from FFAs. Here, we explored the possibility of having a combined expression of both pathways that could benefit from high and balanced levels of both fatty acid pools. Production of fatty alcohols from FFAs through MmCAR + Adh5 has a higher potential to achieve high conversion rates compared to conversion from acyl-CoA, since FFAs can accumulate to much higher levels compared to -CoA metabolites [5, 14–16]. On the other hand, the use of acyl-CoA prevents loss and diffusion of fatty acids to the extracellular medium and other subcellular compartments. Acyl-CoA also has the benefit of being the direct product of the fatty acid biosynthesis pathway, while FFAs are a product of lipid recycling with many intermediary steps and higher energy requirements.

We first used the previously described strain YZFOH1 already overexpressing *MmCAR* and *ADH5* and additionally introduced a fatty acyl-CoA reductase gene (*FaCoAR*) from *Marinobacter aquaeolei VT8* [13] expressed from a 2µ plasmid (pAOH9), resulting in strain YZFOH2. This strain would then be able to convert both acyl-CoA directly to fatty alcohols through FaCoAR, convert FFA to fatty aldehydes through MmCAR, and use both Adh5 and FaCoAR to convert fatty aldehydes to fatty alcohols.

While *faa1Δ faa4Δ* strains lack the ability to recycle FFAs back to acyl-CoA, wild type strains expressing both *FAA1* and *FAA4* have a too high acyl-CoA synthetase activity. Therefore, we needed strains that would have different expression patterns of fatty acyl-CoA synthetase so that FFA and acyl-CoA levels could be simultaneously balanced.

Faergeman et al. [20] showed that Faa1 can fully compensate for the loss of Faa4 activity, which means that expression of *FAA1* is sufficient to restore the acyl-CoA activation and import activity lost by the deletion of the two genes.

In our strategy, we wanted to induce *FAA1* expression at specific time points using different native promoters. This would allow us to study the effects of the regulation of FFA and acyl-CoA pools on the conversion of the two precursors to the final product. For that purpose, *FAA1* was cloned on a CEN.ARS plasmid under the control of a copper-induced promoter (*CUP1* promoter) so that expression could be enabled by the addition of Cu^{2+} to the culture medium at specific time points [22, 23]. In parallel, *FAA1* was also put under control of a promoter induced by high glucose and repressed by low glucose concentrations (*HXT1* promoter) and a promoter repressed by high glucose and induced by low glucose conditions (*HXT7* promoter), respectively [24, 25]. As controls, an empty plasmid not expressing *FAA1* (p413) and a plasmid with strong constitutive expression using the *TEF1* promoter (pTEF1-FAA1) (Fig. 2) were used.

Dynamic control of *FAA1* expression using the copper-inducible promoter *CUP1p*

YZFOH2 strains expressing *FAA1* under control of the *TEF1* promoter (*TEF1p-FAA1*), *CUP1* promoter (*CUP1p-FAA1*), or containing an empty plasmid (p413) were cultured for 72 h in minimal media with 2% glucose. Fatty alcohols, FFAs and OD were analyzed at 24, 48 and 72 h (Fig. 3a, b respectively).

Strong constitutive expression of *FAA1* under control of the *TEF1* promoter yielded the lowest amount of FFA and fatty alcohols levels compared with the p413 control. After 72 h of culture, lack of expression of *FAA1* led to production of tenfold more FFAs and around threefold more fatty alcohols compared to constitutive expression of this gene. Also, there was a significant increase in production of both FFAs and fatty alcohol after 24 h in the p413 control strain. The *TEF1p-FAA1* strain on the other

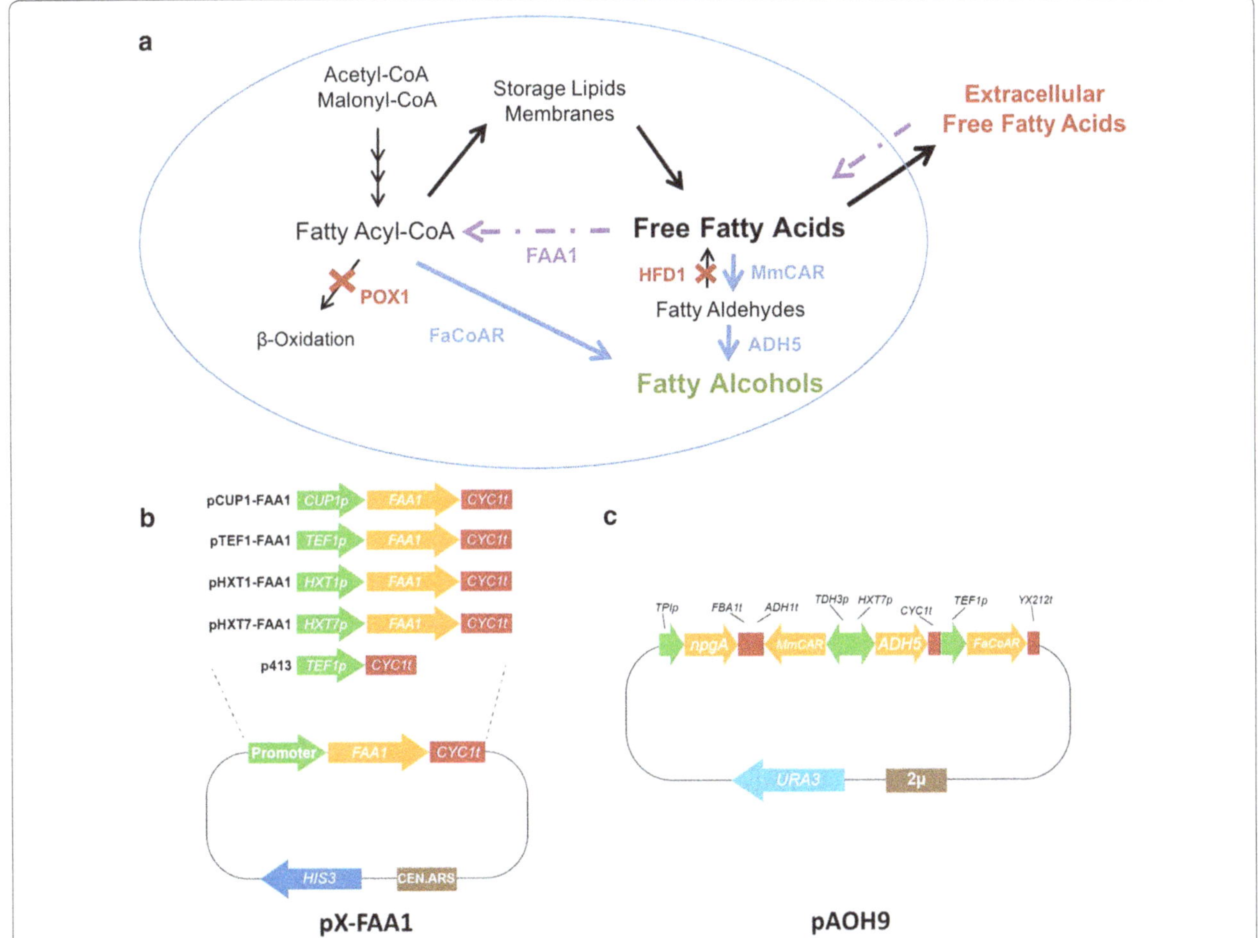

Fig. 2 Design of strain producing fatty alcohols from both fatty acyl-CoA and FFAs using dynamically controlled *FAA1* expression. **a** Schematic representation of strain YZFOH2 (YJZ08 pAOH9) expressing *FAA1* under different promoters. *POX1*, *HFD1*, *FAA1* and *FAA4* have been deleted in this strain and *MmCAR*, *ADH5* and *FaCoAR* are overexpressed. **b** Schematic representation of the plasmids constructed for expression of *FAA1*. *FAA1* is expressed from a CEN.ARS plasmid under the control of promoters *CUP1p*, *TEF1p*, *HXT1p* or *HXT7p*, represented as pX-FAA1 for pCUP1-FAA1, pTEF1-FAA1, pHXT1-FAA1 or pHXT7-FAA1, respectively. **c** Schematic representation of the plasmid expressing the enzymes for production of fatty alcohols. npgA, MmCAR, ADH5 and FaCoAR are expressed under strong constitutive promoters from a 2µ plasmid

hand presented a much less steep increase and from the 48 h time point the FFA levels decreased even though fatty alcohol levels did not increase.

Through the use of a copper-inducible promoter *CUP1p*, we induced *FAA1* expression at specific time points by adding 200 μM Cu^{2+} either at 24 or 48 h and compared it with a non-induced culture (no addition of Cu^{2+}).

All *CUP1p-FAA1* strains produced considerably lower levels of free fatty acids independent of induction time and did not show a decrease of FFA levels after the 48 h time point as it was observed in the *TEF1p-FAA1* strain. The non-induced strain produced only 22.6% FFAs compared to the p413 control strain (empty plasmid). When induced at 24 and at 48 h, FFA production levels are 27.7 and 31.1%, respectively, compared with the p413 control strain.

Inducing expression of *FAA1* at 48 h resulted in the highest fatty alcohol titer of 52 mg/L, which is approximately the same titer achieved by the non-induced strain, and higher than the control strain of 47.4 mg/L. Inducing expression of *FAA1* at 48 h resulted in the highest specific fatty alcohol titers of 8.2 mg/L·OD, and while this is slightly higher than the control strain of 7.6 mg/L·OD, the difference between the two is not statistically significant ($p > 0.05$) (Fig. 3c). Furthermore, no turbidity resulting from FFA micelles was observed in the culture supernatant for *TEF1p-FAA1* or any *CUP1p-FAA1* strains, while presence of this was evident for the p413 control strain, indicating a very significant reduction in excreted FFAs.

These results suggest that presence of low Faa1 levels might be beneficial through most of the culture for balancing the fatty acid pools. However, it is clear that a strong induction of expression early in the process causes a too high flux towards conversion of FFA to acyl-CoA, therefore compromising pathway balance.

Dynamic regulation of FFA and acyl-CoA pools using glucose-regulated promoters

As an alternative strategy to copper-induced promoters, we sought to control *FAA1* expression with glucose-responsive promoters, which allows syncing expression levels with growth phases and carbon source usage. For this purpose, we chose the hexose transporter gene promoters *HXT1p* and *HXT7p*. The promoter *HXT1p* is induced at high glucose concentration and repressed at low glucose concentration [24]. This means that in the *HTX1p-FAA1* system Faa1 is produced during the initial high glucose phase of the batch culture (approximately the first 18 h, Fig. 1b), followed by a tight repression during the ethanol phase, which leads to very low levels of Faa1 present in the cytosol through most of the culture. On the other hand, *HXT7p* is induced under low glucose conditions, while being repressed at high glucose levels [25]. This way, *FAA1* is only expressed in the end of the glucose phase when glucose levels are very low so that FFAs are converted to acyl-CoA mostly during the ethanol phase.

OD, fatty alcohol and FFA levels were analyzed at 24, 48 and 72 h for these strains (Fig. 4a, b respectively). Final FFA levels at 72 h showed a decrease by 63 and 87% for *HXT1p-FAA1* and *HXT7p-FAA1* strains, respectively. *HXT7p-FAA1* showed a small decrease in FFA levels after 48 h, which is similar to *TEF1p-FAA1*, while in the *HXT1p-FAA1* system, FFAs showed a linear increase throughout the culture. Fatty alcohol final titers in the *HXT7p-FAA1* strain were only 61% of the p413 control strain while the *HXT1p-FAA1* strain achieved a fatty alcohol titer of 43.6 mg/L, the same level as the p413 control strain. Furthermore, the specific fatty alcohol titer (production level per OD) of *HXT1p-FAA1* was improved by 41% compared with the control strain (Fig. 4c), which suggests a higher cellular metabolic flux toward fatty alcohol biosynthesis.

Analysis of chain length and saturation level of formed fatty alcohols shows that increased expression levels of *FAA1* decreased the percentage of C16:0 fatty alcohols produced and increased the percentage of unsaturated fatty alcohols C16:1 and C18:1. In addition, there was a 4.5-fold increase in decanol production exclusively in the *HXT1-FAA1* strain (Table 3).

Since we observed a substantial decrease in measured FFAs while fatty alcohol levels were kept stable, we investigated the possibility of fatty acids being accumulated in other forms such as storage lipids or phospholipids. For that, we analyzed the total intracellular lipid content on these strains. Between the control strain p413 and *TEF1p-FAA1*, the lipid profile was radically changed. It is clear that expression of *FAA1* led to a higher accumulation of storage and membrane lipids, since *TEF1p-FAA1* showed a 45% increase in sterol esters, an 86% increase in triacylglycerols and at least a threefold increase in every form of phospholipid measured. The *HXT7p-FAA1* system showed a lipid profile very close to *TEF1p-FAA1*, while the profile of *HXT1p-FAA1* was very close to the control strain p413, with the exception of sterol esters, which in *HXT1p-FAA1* was increased by 37% (Additional file 1: Figure S2).

Growth kinetics for the strains p413, *TEF1p-FAA*, *HXT1p-FAA1* and *CUP1p-FAA1* (not induced by $CuSO_4$) were also studied by measuring OD_{600} values at several timepoints over 72 h. Growth rates on glucose ranged from 0.16 h^{-1} for *HXT1p-FAA1* to 0.19 h^{-1} for *TEF1p-FAA1* but with no statistically significant difference (p value >0.05 for t test analysis) between p413 control and any other strain. Final OD_{600} values at 72 h for

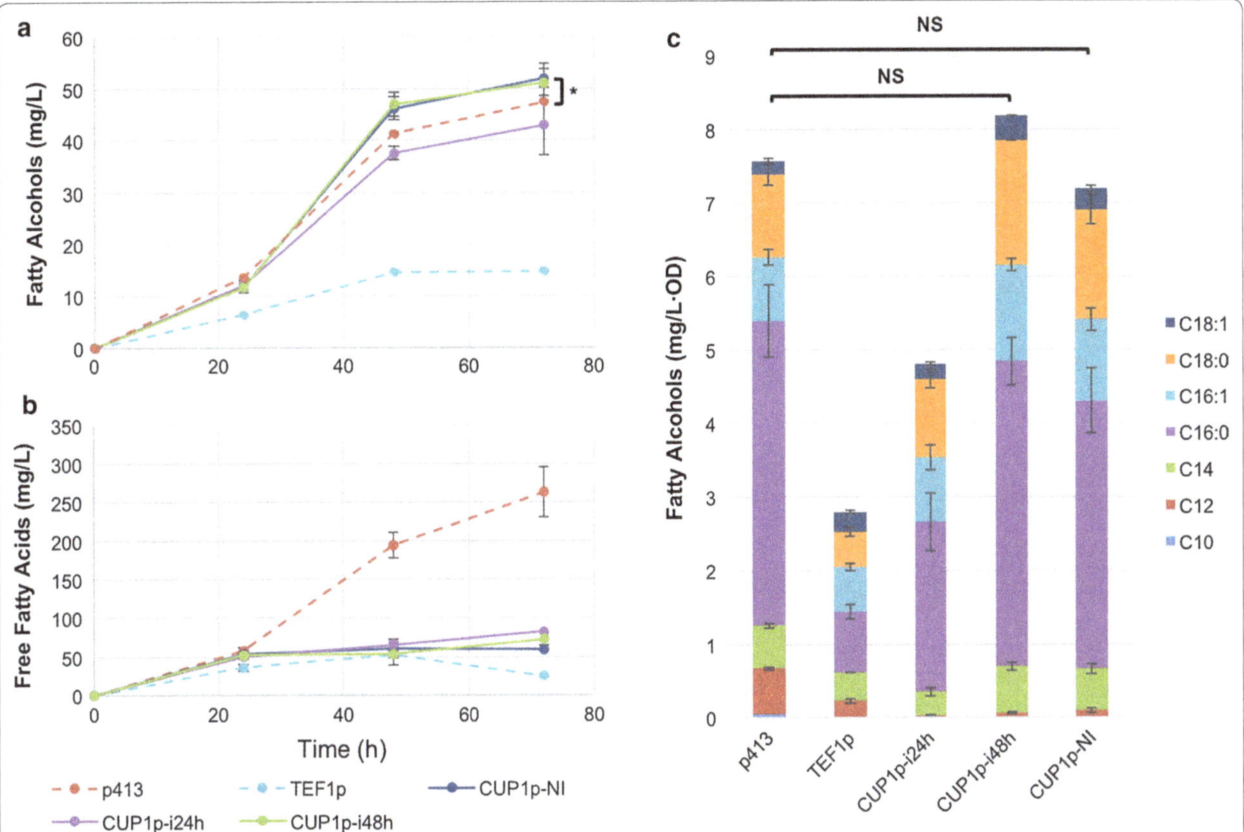

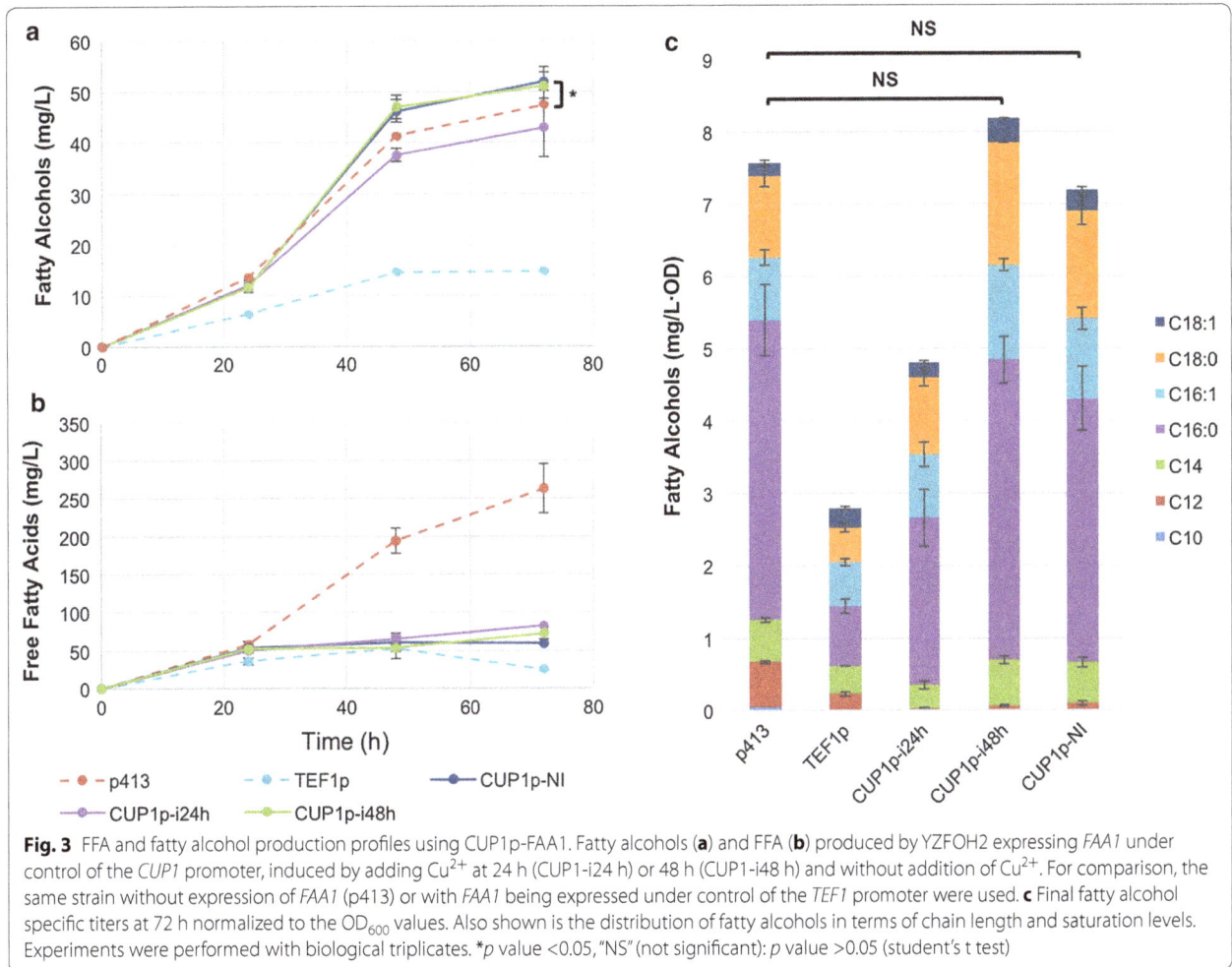

Fig. 3 FFA and fatty alcohol production profiles using CUP1p-FAA1. Fatty alcohols (**a**) and FFA (**b**) produced by YZFOH2 expressing *FAA1* under control of the *CUP1* promoter, induced by adding Cu^{2+} at 24 h (CUP1-i24 h) or 48 h (CUP1-i48 h) and without addition of Cu^{2+}. For comparison, the same strain without expression of *FAA1* (p413) or with *FAA1* being expressed under control of the *TEF1* promoter were used. **c** Final fatty alcohol specific titers at 72 h normalized to the OD_{600} values. Also shown is the distribution of fatty alcohols in terms of chain length and saturation levels. Experiments were performed with biological triplicates. *p value <0.05, "NS" (not significant): p value >0.05 (student's t test)

HXT1p-FAA1 were of 4.9, which is lower than the values reached by the other strains (6.3 for p413 strain, 5.7 for *TEF1p-FAA1, 7.3 for CUP1p-FAA1)* (Additional file 1: Figure S3).

Discussion

Metabolic balancing is essential for production of target molecules with high yield. In this study, metabolic profiling of a fatty alcohol producing strain YZFOH1 (Fig. 1) suggested a limitation in fatty alcohol production and accumulation of fatty acids during the course of the batch culture. The limitation in conversion of FFAs to fatty alcohols results in secretion of FFA to the culture medium, causing loss of precursors needed for fatty alcohol biosynthesis and consequential loss of yield.

We thus engineered fatty alcohol-producing strains using dynamically controlled expression of the fatty acyl-CoA synthetase gene *FAA1* in order to balance the levels of FFAs and acyl-CoA in the cell. This allowed us not only to optimize the intracellular fatty acid levels for conversion to the desired product, but also to explore a potential additive or synergistic effect of using both acyl-CoA and FFA-consuming pathways for fatty alcohol production.

Controlling expression of *FAA1* using the *CUP1* promoter with different induction times resulted in FFA production levels approximately between 20 and 30% of the levels measured for the control strain. On the cases of late induction at 48 h and using only basal expression levels the final fatty alcohol titers were increased by approximately 10% (Fig. 3). On the other hand, early expression of *FAA1* was detrimental to the final fatty alcohol titers and yield. A similar effect could be observed by expressing *FAA1* under control of a *HXT1* promoter, where FFA levels were drastically reduced while fatty alcohol levels per biomass were improved. The *CUP1* promoter has been previously characterized and in a single-copy plasmid it is known to have basal expression levels up to 7% (based on enzyme activity level) compared to induction with 100 μM Cu^{2+} [22, 23]. Furthermore, the minimal culture medium used

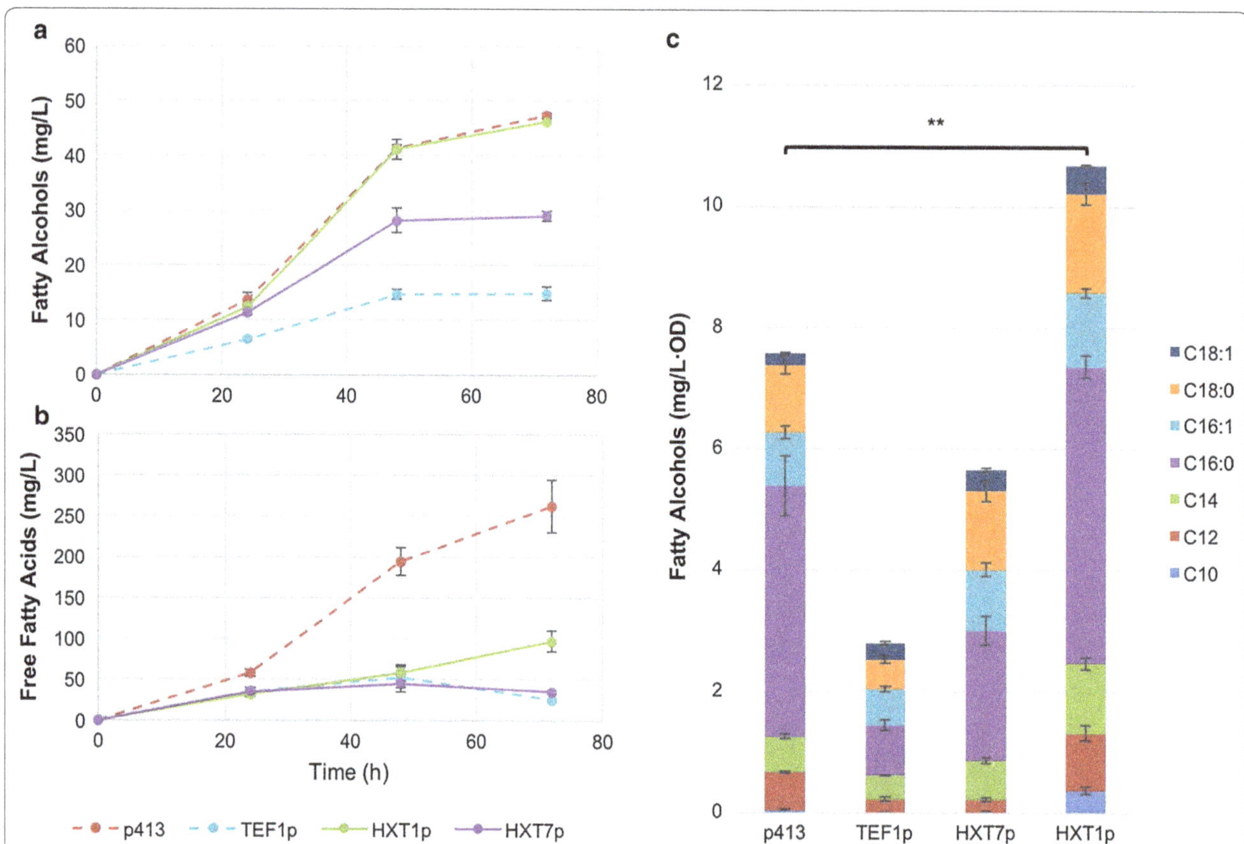

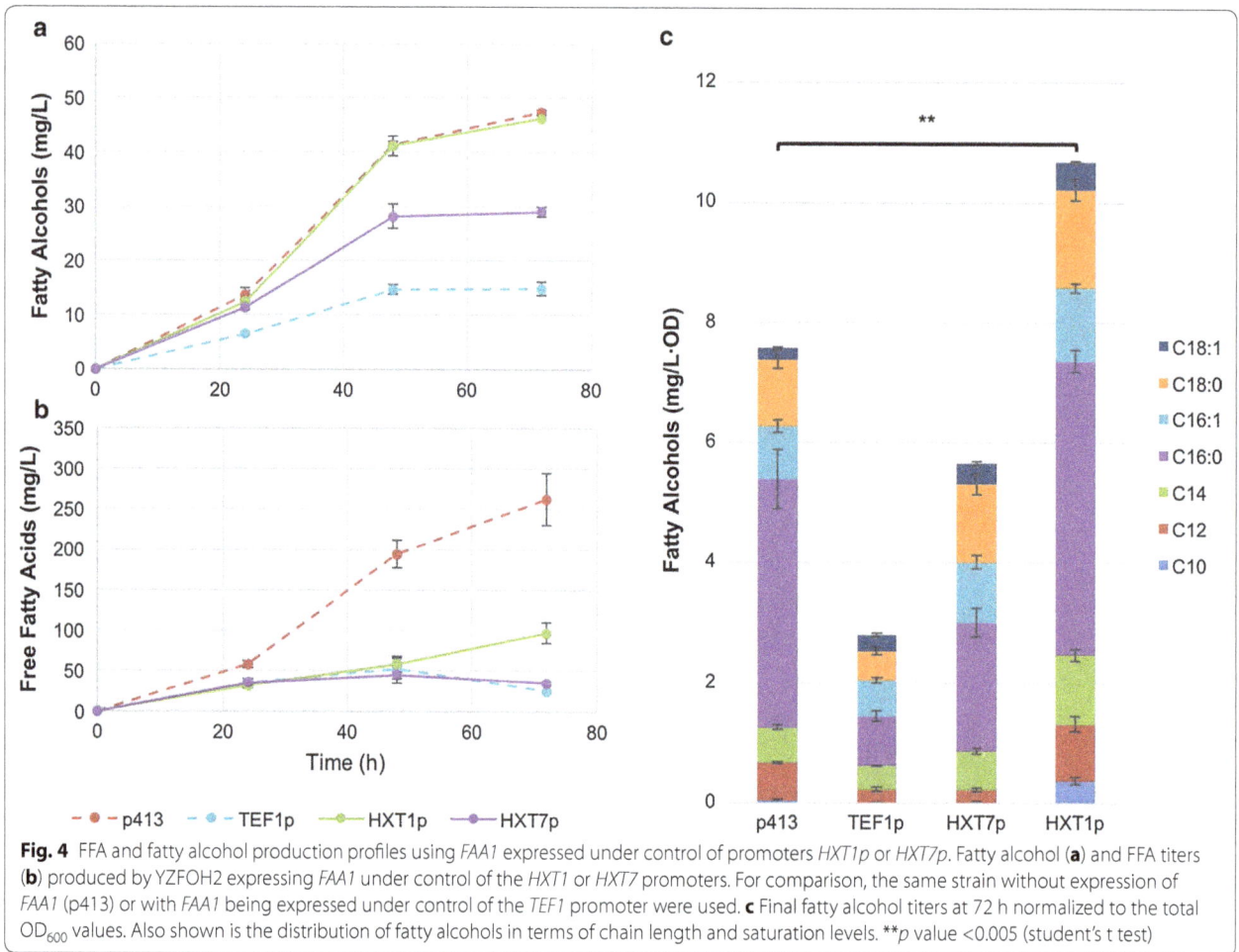

Fig. 4 FFA and fatty alcohol production profiles using *FAA1* expressed under control of promoters *HXT1p* or *HXT7p*. Fatty alcohol (**a**) and FFA titers (**b**) produced by YZFOH2 expressing *FAA1* under control of the *HXT1* or *HXT7* promoters. For comparison, the same strain without expression of *FAA1* (p413) or with *FAA1* being expressed under control of the *TEF1* promoter were used. **c** Final fatty alcohol titers at 72 h normalized to the total OD_{600} values. Also shown is the distribution of fatty alcohols in terms of chain length and saturation levels. **p value <0.005 (student's t test)

Table 3 Fatty alcohol distribution in strains expressing FAA1 under control of different promoters

% Total fatty alcohols							
Strain[a]	C10	C12	C14	C16:0	C16:1	C18:0	C18:1
p413	0.65 ± 0.11	8.25 ± 0.24	7.67 ± 0.47	54.62 ± 6.48	11.52 ± 1.43	14.74 ± 1.91	2.55 ± 0.33
HXT1p-FAA1	3.50 ± 0.52	8.81 ± 1.20	10.81 ± 0.89	45.81 ± 1.73	11.37 ± 0.66	15.29 ± 1.70	4.40 ± 0.11
HXT7p-FAA1	0.38 ± 0.04	3.62 ± 0.52	11.25 ± 0.76	38.08 ± 4.21	17.64 ± 2.05	22.98 ± 2.98	6.06 ± 0.77
TEF1p-FAA1	0.64 ± 0.10	7.48 ± 1.23	14.05 ± 0.14	29.49 ± 3.35	21.48 ± 1.59	17.33 ± 2.24	9.54 ± 0.94

[a] Experiment is a result of biological triplicates, standard deviation is presented for each value

in this study contains approximately 6 μM Cu^{2+}, therefore contributing to a moderate expression level of *FAA1* on the non-induced culture. These results suggest that constitutive basal expression levels of the *CUP1* promoter in our culture conditions produced enough enzyme to regulate the acyl-CoA/FFA levels in a favourable way. The later (48 h) induced expression or not-induced expression of *FAA1* resulted in much higher fatty alcohol production compared to the earlier induction (24 h) (Fig. 3c), which suggested that sustained high expression of *FAA1* created a too high flux of FFA conversion to acyl-CoA, reducing FFA supply for fatty alcohol biosynthesis. This was corroborated when *FAA1* was expressed under control of the *HXT7* promoter. Although main expression of *HXT7* promoter occurs only at the end of the glucose phase, basal expression levels are quite high when compared with *HXT1p* [24, 25], which creates a phenotype much closer to *TEF1p-FAA1*.

Overall interpretation of the results indicates that *TEF1p-FAA1*, *HXT7p-FAA1* and early induced *CUP1p-FAA1* express high levels of *FAA1*. High levels of FAA1 expression created a high flux of FFA towards acyl-CoA formation, which lowered FFA availability for fatty alcohol production through *MmCAR*. Although this could promote higher flux through FaCoAR due to a higher substrate supply, it is generally accepted that acyl-CoA cannot be accumulated in high concentrations due to limited -CoA availability in the cell. Additionally, an increase of acyl-CoA levels in yeast has been previously shown to be related with feedback inhibition of fatty acid biosynthesis at the level of the acyl-CoA carboxylase Acc1 [26], and therefore having a negative effect on fatty acid and fatty acid-derived chemicals production. Furthermore, in contrast to FFAs, acyl-CoA is consumed in many competing reactions, which can be demonstrated by the increased levels of certain storage and phospholipids in those strains with higher *FAA1* expression.

The *HXT1* promoter, on the other hand, allows for expression of *FAA1* in high glucose conditions, which is the case during the very early stage of the batch culture, not longer than 15 h of culture. From that point on, *HXT1p* presumably keeps the expression level of *FAA1* low, which apparently represents a favourable flux balance to maintain the acyl-CoA/FFA pools for efficient conversion of these metabolites into fatty alcohols.

When the distribution of produced fatty alcohols was analyzed in the resulting strains, it was possible to observe a shift in saturation levels and chain length related to the promoter strength expressing FAA1. By respective order, *HXT1p-FAA1*, *HXT7p-FAA1* and *TEF1p-FAA1* strains showed a higher ratio of C16:1/C16:0 and C18:1/C18:0 as well as a higher percentage of C14 fatty alcohols compared with the control strain p413 (Table 3). This is probably a result of Faa1 specificity, since most enzymes have higher affinity towards specific fatty acid chain lengths and saturation levels. Affinity of Faa1 towards different fatty acids would change the balance of corresponding acyl-CoA/FFA for those fatty acids in question and therefore influence the final product distribution. This is a valuable tool if production of a specific fatty alcohol is desired.

Furthermore, variations on *FAA1* expression levels seem to have a minimal impact on growth kinetics. A small decrease in OD_{600} at 72 h was observed for *HXT1-FAA1*, but not for *CUP1-FAA1*. Since *HXT1-FAA1* shows a higher level of accumulated FFA at 72 h compared to *CUP1-FAA1* and nearly identical fatty alcohol titers, the growth defect might suggest a certain toxicity from high FFA levels during stationary phase when *FAA1* is present at a minimal level.

Together, our set of experiments indicates a need for tight regulation of *FAA1* expression in order to achieve an optimal balance of acyl-CoA and FFAs for fatty acid production from a combined use of our two pathways. We see a favourable effect in using very low levels of *FAA1* expression, which can be hard to achieve using native promoters whose activity may be influenced by many factors. Here, the use of technologies that allow for a more predictable and fine-tuned expression such as synthetic promoters [27, 28], CRISPRi [29, 30] and acyl-CoA biosensors [31] for dynamic gene expression could be beneficial for future optimisation on similar systems.

Nevertheless, our final strain represents an improvement over the original one due to the capacity of being able to produce fatty alcohols in similar or slightly improved final titers but with better product-to-biomass yield and without high production of FFA by-products. Further improvement of these strains for increased performance is now possible since precursor accumulation is not an issue that compromises yield.

Conclusions

In this work, we exemplified how one can address the arising problem of unbalanced fluxes in a synthetic pathway. We addressed a situation where kinetics of downstream steps is a limiting factor on the product formation, which leads to accumulation and loss of precursors to the extracellular medium, compromising final yields where these are a fundamental aspect of the process. In many cases accumulated precursors might end up as by-products that comprise the entire production process or represent a major carbon loss. In this aspect, pathway fine-tuning is a fundamental asset for applicability of existing pathway prototypes in microbial cell factories that must find its place in the design process.

On the other hand, the work also provides general knowledge for the development of cell factories for fatty acid-derived products. Fatty acid recycling and regulation of FFA/acyl-CoA interconversion is a central aspect, which is not completely understood and our analysis provide new insight into how fluxes are controlled in this pathway.

Author's contributions

PGT designed the study, performed experiments, evaluated results and wrote the manuscript. RF performed experiments and assisted in designing the study, evaluating results and writing the manuscript. YJZ, VS and JN took part in designing the study, evaluating the results and writing the manuscript. This manuscript has been approved by all the authors listed. All authors read and approved the final manuscript.

Author details
[1] Department of Biology and Biological Engineering, Chalmers University of Technology, 412 96 Gothenburg, Sweden. [2] Novo Nordisk Foundation Center for Biosustainability, Chalmers University of Technology, 412 96 Gothenburg, Sweden. [3] Novo Nordisk Foundation Center for Biosustainability, Technical University of Denmark, 2800 Kgs. Lyngby, Denmark.

Acknowledgements
We would like to acknowledge Jiufu Qin for valuable input and ideas on experimental design and the Chalmers Mass Spectrometry Infrastructure (CMSI) for assisting with the analytic procedures.

Competing interests
The authors declare that they have no competing interests.

Funding
The authors would like to acknowledge the Novo Nordisk Foundation and Vetenskapsrådet for funding support.

References

1. Nielsen J, Keasling JD. Engineering cellular metabolism. Cell. 2016;164:1185–97.
2. Krivoruchko A, Zhang Y, Siewers V, Chen Y, Nielsen J. Microbial acetyl-CoA metabolism and metabolic engineering. Metab Eng. 2015;28:28–42.
3. Runguphan W, Keasling JD. Metabolic engineering of *Saccharomyces cerevisiae* for production of fatty acid-derived biofuels and chemicals. Metab Eng. 2014;21:103–13.
4. Zhou YJ, Buijs NA, Siewers V, Nielsen J. Fatty acid-derived biofuels and chemicals production in *Saccharomyces cerevisiae*. Front Bioeng Biotechnol. 2014;2:32.
5. Zhou YJ, Buijs NA, Zhu Z, Qin J, Siewers V, Nielsen J. Production of fatty acid-derived oleochemicals and biofuels by synthetic yeast cell factories. Nat Commun. 2016;7:11709.
6. Zhou YJ, Buijs NA, Zhu Z, Gómez DO, Boonsombuti A, Siewers V, et al. Harnessing yeast peroxisomes for biosynthesis of fatty-acid-derived biofuels and chemicals with relieved side-pathway competition. Soc: J Am Chem; 2016.
7. Nakayama S, Morita T, Negishi H, Ikegami T, Sakaki K, Kitamoto D. *Candida krusei* produces ethanol without production of succinic acid; a potential advantage for ethanol recovery by pervaporation membrane separation. FEMS Yeast Res. 2008;8:706–14.
8. Balzer GJ, Thakker C, Bennett GN, San K-Y. Metabolic engineering of *Escherichia coli* to minimize byproduct formate and improving succinate productivity through increasing NADH availability by heterologous expression of NAD(+)-dependent formate dehydrogenase. Metab Eng. 2013;20:1–8.
9. Kim TY, Park JM, Kim HU, Cho KM, Lee SY. Design of homo-organic acid producing strains using multi-objective optimization. Metab Eng. 2015;28:63–73.
10. Reiser S, Somerville C. Isolation of mutants of *Acinetobacter calcoaceticus* deficient in wax ester synthesis and complementation of one mutation with a gene encoding a fatty acyl coenzyme A reductase. J Bacteriol. 1997;179:2969–75.
11. Akhtar MK, Turner NJ, Jones PR. Carboxylic acid reductase is a versatile enzyme for the conversion of fatty acids into fuels and chemical commodities. Proc Natl Acad Sci USA. 2013;110:87–92.
12. Buijs NA, Zhou YJ, Siewers V, Nielsen J. Long-chain alkane production by the yeast *Saccharomyces cerevisiae*. Biotechnol Bioeng. 2015;112:1275–9.
13. Willis RM, Wahlen BD, Seefeldt LC, Barney BM. Characterization of a fatty acyl-CoA reductase from *Marinobacter aquaeolei VT8*: a bacterial enzyme catalyzing the reduction of fatty acyl-CoA to fatty alcohol. Biochemistry. 2011;50:10550–8.
14. Feng X, Lian J, Zhao H. Metabolic engineering of *Saccharomyces cerevisiae* to improve 1-hexadecanol production. Metab Eng. 2015;27:10–9.
15. Tang X, Chen WN. Enhanced production of fatty alcohols by engineering the TAGs synthesis pathway in *Saccharomyces cerevisiae*. Biotechnol Bioeng. 2015;112:386–92.
16. Schjerling CK, Hummel R, Hansen JK, Borsting C, Mikkelsen JM, Kristiansen K, et al. Disruption of the gene encoding the acyl-CoA-binding protein (ACB1) perturbs acyl-CoA metabolism in *Saccharomyces cerevisiae*. J Biol Chem. 1996;271:22514–21.
17. Black PN, DiRusso CC. Yeast acyl-CoA synthetases at the crossroads of fatty acid metabolism and regulation. Biochim Biophys Acta. 2007;1771:286–98.
18. Scharnewski M, Pongdontri P, Mora G, Hoppert M, Fulda M. Mutants of *Saccharomyces cerevisiae* deficient in acyl-CoA synthetases secrete fatty acids due to interrupted fatty acid recycling. FEBS J. 2008;275:2765–78.
19. Leber C, Polson B, Fernandez-Moya R, Da Silva NA. Overproduction and secretion of free fatty acids through disrupted neutral lipid recycle in *Saccharomyces cerevisiae*. Metab Eng. 2015;28:54–62.
20. Faergeman NJ, Black PN, Zhao XD, Knudsen J, DiRusso CC. The Acyl-CoA synthetases encoded within FAA1 and FAA4 in *Saccharomyces cerevisiae* function as components of the fatty acid transport system linking import, activation, and intracellular utilization. J Biol Chem. 2001;276:37051-9.
21. Knoll LJ, Johnson DR, Gordon JI. Complementation of *Saccharomyces cerevisiae* strains containing fatty acid activation gene (FAA) deletions with a mammalian acyl-CoA synthetase. J Biol Chem. 1995;270:10861-7.
22. Gorman JA, Clark PE, Lee MC, Debouck C, Rosenberg M. Regulation of the yeast metallothionein gene. Gene. 1986;48:13–22.
23. Butt TR, Sternberg EJ, Gorman JA, Clark P, Hamer D, Rosenberg M, et al. Copper metallothionein of yeast, structure of the gene, and regulation of expression. Proc Natl Acad Sci USA. 1984;81:3332–6.
24. Özcan S, Johnston M. Function and regulation of yeast hexose transporters. Microbiol Mol Biol Rev. 1999;63:554–69.
25. Ye L, Berden JA, van Dam K, Kruckeberg AL. Expression and activity of the Hxt7 high-affinity hexose transporter of *Saccharomyces cerevisiae*. Yeast. 2001;18:1257–67.
26. Faergeman NJ, Knudsen J. Role of long-chain fatty acyl-CoA esters in the regulation of metabolism and in cell signalling. Biochem J. 1997;323(Pt 1):1–12.
27. Rajkumar AS, Liu G, Bergenholm D, Arsovska D, Kristensen M, Nielsen J, et al. Engineering of synthetic, stress-responsive yeast promoters. Nucleic Acids Res. 2016;44:e136.
28. Redden H, Alper HS. The development and characterization of synthetic minimal yeast promoters. Nat Commun. 2015;6:7810.
29. Smith JD, Suresh S, Schlecht U, Wu M, Wagih O, Peltz G, et al. Quantitative CRISPR interference screens in yeast identify chemical-genetic interactions and new rules for guide RNA design. Genome Biol. 2016;17:45.
30. Larson MH, Gilbert LA, Wang X, Lim WA, Weissman JS, Qi LS. CRISPR interference (CRISPRi) for sequence-specific control of gene expression. Nat Protoc. 2013;8:2180–96.
31. Zhang F, Carothers JM, Keasling JD. Design of a dynamic sensor-regulator system for production of chemicals and fuels derived from fatty acids. Nat Biotechnol. 2012;30:354–9.
32. Verduyn C, Postma E, Scheffers WA, Van Dijken JP. Effect of benzoic acid on metabolic fluxes in yeasts: a continuous-culture study on the regulation of respiration and alcoholic fermentation. Yeast. 1992;8:501–17.
33. Khoomrung S, Chumnanpuen P, Jansa-Ard S, Ståhlman M, Nookaew I, Borén J, et al. Rapid quantification of yeast lipid using microwave-assisted total lipid extraction and HPLC-CAD. Anal Chem. 2013;85:4912–9.
34. Haushalter RW, Kim W, Chavkin TA, The L, Garber ME, Nhan M, et al. Production of anteiso-branched fatty acids in *Escherichia coli*; next generation biofuels with improved cold-flow properties. Metab Eng. 2014;26:111–8.

9

Large scale validation of an efficient CRISPR/Cas-based multi gene editing protocol in *Escherichia coli*

Francesca Zerbini[1], Ilaria Zanella[1], Davide Fraccascia[1], Enrico König[1], Carmela Irene[1], Luca F. Frattini[1], Michele Tomasi[1], Laura Fantappiè[1], Luisa Ganfini[1], Elena Caproni[1], Matteo Parri[2], Alberto Grandi[2] and Guido Grandi[1*]

Abstract

Background: The exploitation of the CRISPR/Cas9 machinery coupled to lambda (λ) recombinase-mediated homologous recombination (recombineering) is becoming the method of choice for genome editing in *E. coli*. First proposed by Jiang and co-workers, the strategy has been subsequently fine-tuned by several authors who demonstrated, by using few selected loci, that the *efficiency of mutagenesis* (number of mutant colonies over total number of colonies analyzed) can be extremely high (up to 100%). However, from published data it is difficult to appreciate the *robustness* of the technology, defined as the number of successfully mutated loci over the total number of targeted loci. This information is particularly relevant in high-throughput genome editing, where repetition of experiments to rescue missing mutants would be impractical. This work describes a "brute force" validation activity, which culminated in the definition of a robust, simple and rapid protocol for single or multiple gene deletions.

Results: We first set up our own version of the CRISPR/Cas9 protocol and then we evaluated the mutagenesis efficiency by changing different parameters including sequence of guide RNAs, length and concentration of donor DNAs, and use of single stranded and double stranded donor DNAs. We then validated the optimized conditions targeting 78 "dispensable" genes. This work led to the definition of a protocol, featuring the use of double stranded synthetic donor DNAs, which guarantees mutagenesis efficiencies consistently higher than 10% and a robustness of 100%. The procedure can be applied also for simultaneous gene deletions.

Conclusions: This work defines for the first time the robustness of a CRISPR/Cas9-based protocol based on a large sample size. Since the technical solutions here proposed can be applied to other similar procedures, the data could be of general interest for the scientific community working on bacterial genome editing and, in particular, for those involved in synthetic biology projects requiring high throughput procedures.

Keywords: CRISPR-Cas9, High-throughput genome editing, Synthetic biology

Background

Escherichia coli is one of the most extensively studied living organisms on earth and as such has become an instrumental model system for the understanding of a plethora of gene functions and regulations in both prokaryotes and eukaryotes. Moreover, *E. coli* also plays an invaluable role in modern biological engineering and industrial microbiology: it is a very versatile host for the production of heterologous proteins and their mass-production in industrial fermentation systems. Genetically modified *E. coli* cells are currently used in a wide range of processes, such as vaccine development, bioremediation, production of biofuels and production of immobilized enzymes. Additional biotechnological applications are expected to be developed thanks to the exploitation of modern techniques of metabolic engineering and synthetic biology.

*Correspondence: guido.grandi@unitn.it
[1] Synthetic and Structural Vaccinology Unit, CIBIO, University of Trento, Via Sommarive, 9, Povo, 38123 Trento, Italy
Full list of author information is available at the end of the article

To fully exploit the use of synthetic biology in *E. coli*, the availability of efficient genome editing systems, also applicable in high-throughput modalities is necessary. Currently, there are three main approaches for manipulation of chromosomal DNA in *E. coli*, all utilizing phage recombinase-mediated homologous recombination (recombineering), using either the Rac prophage system [1, 2] or the three bacteriophage λ Red proteins Exo, Beta, and Gam [3–5]. Classically, gene knockout mutants are created by inserting antibiotic resistance markers (or other selection markers) between double-stranded DNA (ds-DNA) PCR products derived from the upstream and downstream regions of the target gene. Mutant colonies are isolated in the appropriate selective medium after transformation with linear or circular constructs and, when necessary, the selection marker is subsequently eliminated by counter-selection, leaving a "scarless" chromosomal mutation. Court and co-workers elegantly demonstrated that chromosomal gene mutations can be achieved without the need of selection markers and using synthetic single stranded DNAs (ss-DNAs) or ds-DNAs, which anneal to their complementary chromosomal regions during replication and mediate recombination and gene modification [6, 7]. While effective, these two approaches are not ideal for high-throughput applications since they are laborious and time consuming (in the case of the first approach) and feature mutagenesis efficiencies often below 1% (in the case of the second approach) [8–10]. More recently, a third approach, proposed for the first time by Jiang and co-workers [9], makes use of the CRISPR/Cas9 technology [11–13]. Briefly, the strain to be modified is first genetically manipulated to express the Cas9 nuclease and the λ Red machinery, and subsequently the strain is co-transformed with (i) a plasmid (pCRISPR) encoding the guide RNA, which anneals with the chromosomal region to be modified and promotes a site-specific DNA cleavage by the Cas9, and (ii) a donor DNA (PCR-derived or chemically synthesized) partially homologous to the cleaved extremities, which promotes the repair of the double stranded break through λ Red-mediated recombination thereby introducing the desired mutation. The presence of the λ Red machinery plays an important role in the process since in its absence the mutagenesis efficiency was shown to drop quite substantially [10]. With this strategy, Jiang and co-workers reported mutation efficiencies as high as 65%. Subsequently, other authors fine-tuned the Jiang et al. protocol by adding innovative solutions and expanding its application for extensive gene deletions and replacements [14–17]. Thanks to the contribution of all these authors, CRISPR/Cas9 coupled to recombineering is becoming the most effective approach for genome editing in bacteria and in particular in *E. coli*.

However, one aspect of the CRISPR/Cas9 technology that still remains to be thoroughly addressed is the definition of its "*robustness*". In fact, while it has been extensively demonstrated that the *efficiency of mutagenesis* (number of mutant colonies/total number of colonies analyzed) can be extremely high (close to 100%), the *robustness* (number of mutated loci/total number targeted loci) has not yet been defined experimentally in a rigorous manner. Knowing the robustness of the specific CRISPR/Cas-based protocol in use is particularly relevant in high-throughput applications, where the repetition of mutagenesis experiments and/or the analysis of large numbers of colonies would be impractical. There is a paucity of papers addressing the "robustness issue" and even in these works the number of targeted loci is limited. For instance, to evaluate the robustness of their protocol, Reisch and Prather [18] using an innovative two plasmid system, one expressing a tightly regulated Cas9 and the other the gRNA, demonstrated that point mutations could be introduced into two dispensable genes with efficiencies close to 100%. On the basis of these data the authors concluded that their system is robust enough to make point mutations (and also larger deletions) in any genomic location carrying an appropriate PAM site. To support their conclusion, the authors also claimed that using the same protocol they successfully inactivated five additional genes. Another recent study describes a CRISPR-based strategy that allowed the integration of entire metabolic pathways in seven distinct loci using ds-DNAs encoding homologous regions to the insertion site of different size with 75–100% efficiencies [19].

In the present paper, we present a "brute force" validation effort, which culminated in the definition of a highly robust, simple and rapid protocol for both single and multiple scarless chromosome manipulations. First, we tested several experimental conditions targeting four dispensable gene loci with the aim of investigating the parameters that mostly influence mutagenesis efficiency. On the basis of this analysis we defined a mutagenesis protocol, which we subsequently validated on a panel of 78 additional genes. The overall approach, which involved the construction of approximately one hundred pCRISPR plasmids, the execution of a few hundred transformation experiments and the analysis of several thousand colony PCR and sequencing, led to the definition of a high fidelity and rapid protocol, which guarantees the generation of single and multiple mutations of dispensable genes with a 100% confidence.

Results

Components of our CRISPR/Cas9 genome editing protocol

Figure 1 schematically represents our CRISPR/Cas9-based *E. coli* genome editing protocol. Briefly, the

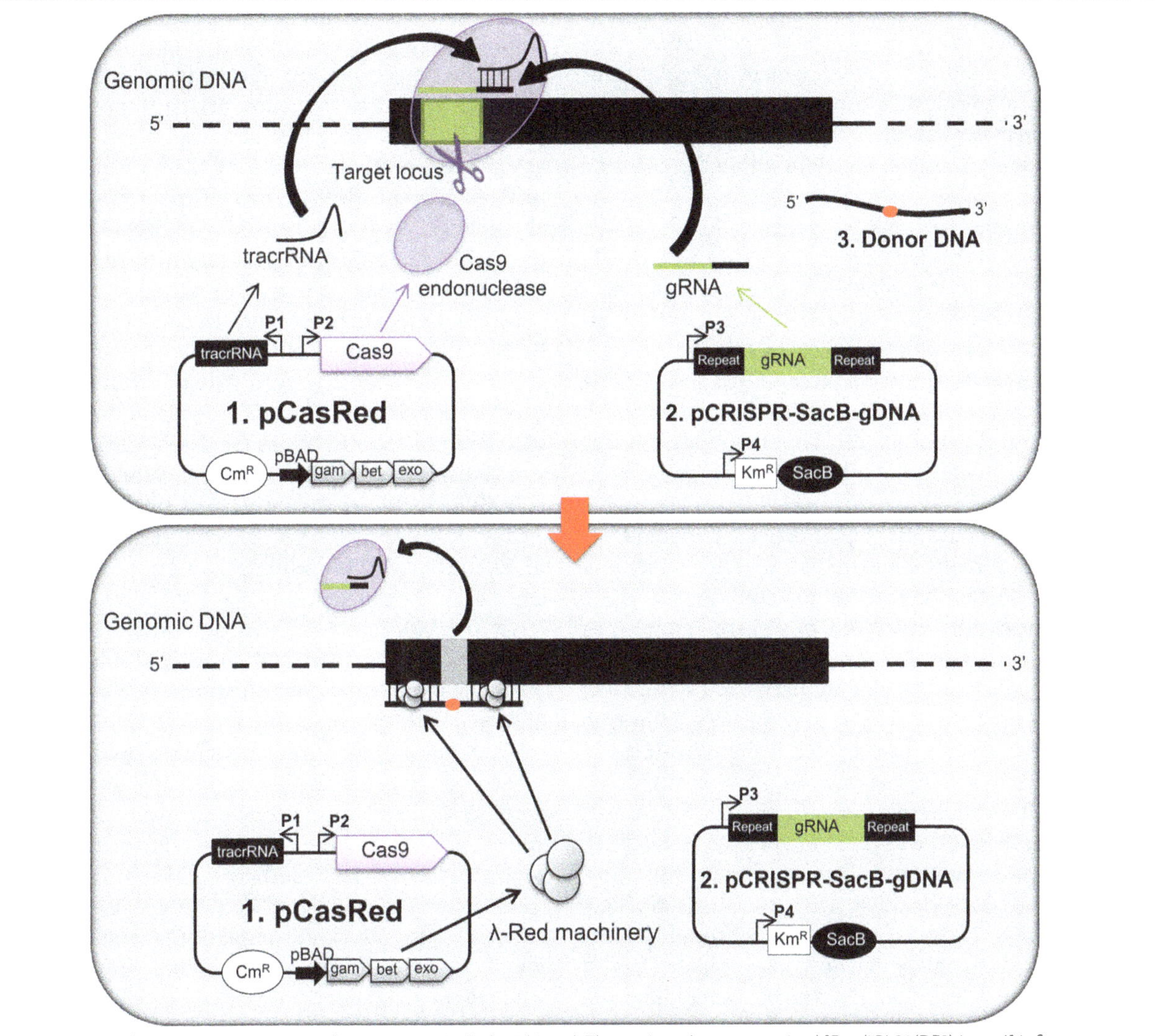

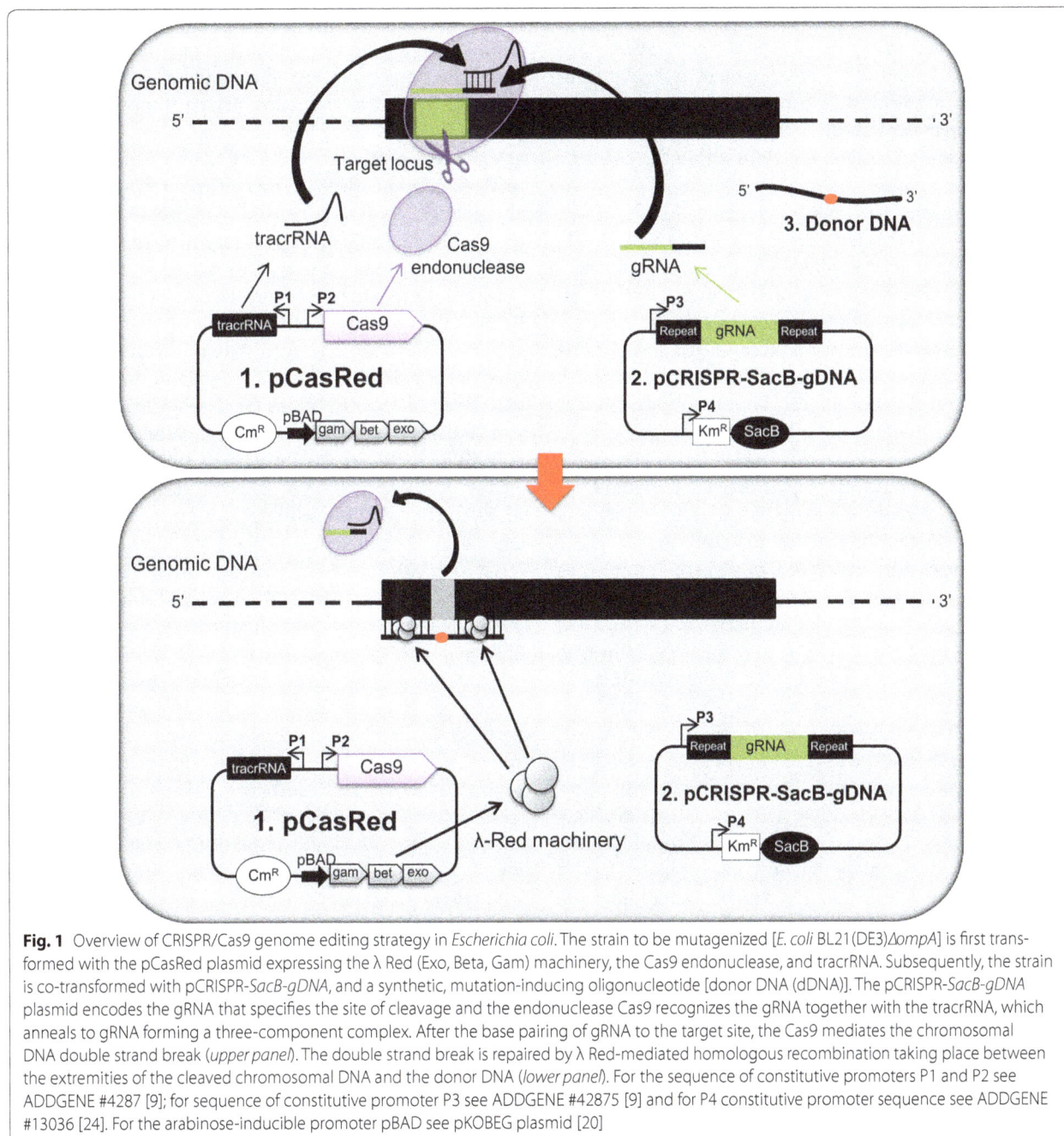

Fig. 1 Overview of CRISPR/Cas9 genome editing strategy in *Escherichia coli*. The strain to be mutagenized [*E. coli* BL21(DE3)*ΔompA*] is first transformed with the pCasRed plasmid expressing the λ Red (Exo, Beta, Gam) machinery, the Cas9 endonuclease, and tracrRNA. Subsequently, the strain is co-transformed with pCRISPR-*SacB-gDNA*, and a synthetic, mutation-inducing oligonucleotide [donor DNA (dDNA)]. The pCRISPR-*SacB-gDNA* plasmid encodes the gRNA that specifies the site of cleavage and the endonuclease Cas9 recognizes the gRNA together with the tracrRNA, which anneals to gRNA forming a three-component complex. After the base pairing of gRNA to the target site, the Cas9 mediates the chromosomal DNA double strand break (*upper panel*). The double strand break is repaired by λ Red-mediated homologous recombination taking place between the extremities of the cleaved chromosomal DNA and the donor DNA (*lower panel*). For the sequence of constitutive promoters P1 and P2 see ADDGENE #4287 [9]; for sequence of constitutive promoter P3 see ADDGENE #42875 [9] and for P4 constitutive promoter sequence see ADDGENE #13036 [24]. For the arabinose-inducible promoter pBAD see pKOBEG plasmid [20]

procedure makes use of three main elements: the pCasRed plasmid, the pCRISPR-*SacB-gDNA* plasmid, and the synthetic, mutation-inducing oligonucleotide [donor DNA (dDNA)]. The pCasRed plasmid carries the chloramphenicol resistance gene (Cm^R) and encodes the Cas9 nuclease, the λ Red (Exo, Beta, Gam) cassette and the tracrRNA. The *cas9* gene and the tracrRNA coding sequence are under the control of constitutive promoters while the λ Red gene cassette is transcribed from the pBAD arabinose-inducible promoter [20]. The pCRISPR-*SacB-gDNA* plasmid derives from a pCRISPR plasmid [9], where the kanamycin resistance gene (Km^R) is fused to the *sacB* gene encoding the *Bacillus subtilis* levansucrase. *SacB* is toxic in *E. coli* if grown in media containing 5% sucrose [21–24] and, as previously demonstrated by Hale and co-workers, plasmid-borne sucrose toxicity can be exploited to cure high copy number plasmids [25]. We used this strategy to remove the pCRISPR-*SacB-gDNA* plasmid after the gene mutation has been introduced. In addition to the *sacB* gene, pCRISPR-*SacB-gDNA* plasmid carries the synthetic DNA fragment (gDNA), transcribed from a constitutive promoter [9], encoding the RNA guide necessary to drive the Cas9-dependent double stranded break to the desired site within the bacterial genome.

In the following sections we provide details of the experimental data that led us to define a protocol, which, under the conditions described here, guarantees the inactivation of any dispensable gene and, after construction of the pCRISPR plasmids and dDNAs (see "Methods"), allows the sequential mutation of genes at a pace of one mutation every other day. Finally, we also describe a highly efficient protocol for the simultaneous introduction of multiple mutations.

Definition of the robustness of gDNA selection method

gDNAs within a target gene are usually selected among those 30 nucleotide sequences followed by an NGG (PAM) trinucleotide, which do not share homologies with other regions in the chromosome (to avoid off-target cleavage). Routine bioinformatics tools (BLAST) are generally used to identify such sequences. However, what has not been fully investigated is the robustness of bioinformatics in gDNA selection. In other words, do all predicted gDNAs promote an efficient CRISPR/Cas9-mediated cleavage at the selected target site? To test this, we selected several gDNAs and we analyzed their capacity to promote Cas9 cleavage. Theoretically, if a gDNA efficiently drives Cas9 cleavage no colonies should be isolated since non-homologous end-joining repair (NHEJ) works poorly in *E. coli* [26, 27]. Practically, colonies are still recoverable and represent "escapers" in which the gDNA from pCRISPR-*SacB-gDNA* is lost, or mutations in the PAM region and/or in the seed region (8–12 nucleotides upstream from the PAM region) [9] have occurred. Alternatively, as suggested by Cui and Bikard [28], if Cas9-mediated cleavage is not efficient, homologous recombination between the cleaved chromosome and the un-cleaved sister chromosome could take place. Finally, gDNA promoter mutations and mutations to the *cas9* gene itself could occur, leading to a basal level of colonies lacking of dsDNA breaks. However, whatever mechanism is involved, the number of "escaper" colonies should be orders of magnitude lower compared to the colonies obtained by transforming the strain with the "empty" pCRISPR-*SacB* plasmid. Therefore, the drop in transformation efficiency between pCRISPR-*SacB* and pCRISPR-*SacB-gDNA* indicates the quality of the gDNA.

To establish the effectiveness of gDNAs in our protocol, we created sets of pCRISPR-*gDNA* plasmids (Additional file 1: Table S1) by cloning synthetic gDNAs (Additional file 1: Table S2) between the two tracrRNA-complementary repeat regions (Fig. 1) and we compared the transformation efficiencies of the recombinant plasmids with the efficiency of the "empty" pCRISPR vector (Fig. 2). In particular, we selected three target genes, *ompF*, *lpp* and *fecA*, and we designed four sets of different gDNAs, two of them targeting different sites at the 5′ and 3′ regions of *ompF* gene, respectively, and the other two sets targeting *lpp* and *fecA* at different positions (Fig. 2). Overall, 20 pCRISPR-*gDNA* plasmids were generated and used to transform *E. coli* BL21(DE3)Δ*ompA*(pCasRed). The BL21(DE3)Δ*ompA* was used since we observed that the strain maintained good transformation efficiencies even in the presence of pCasRed plasmid. Colonies were selected on LB plates supplemented with Cm and Km. While the "empty" pCRISPR vector routinely gave a transformation efficiency of 0.5–2 × 10^6 CFUs/μg of plasmid DNA, most of pCRISPR-*gDNA* plasmids had transformation efficiencies in the range of 0.5–2 × 10^3 CFUs/μg of plasmid DNA (Fig. 2). However, two out of the 20 pCRISPR-*gDNA* plasmids (pCRISPR-*ompF_5′C* and pCRISPR-*lpp_C*) gave a transformation efficiency close to the one observed with the empty vector.

We randomly analyzed ten clones from the transformations with pCRISPR-*ompF_5′G* and pCRISPR-*ompF_5′C* (transformation efficiency of 1 × 10^3 and 1 × 10^5 CFUs/μg, respectively) using PCR and sequence analyses of both the chromosomal regions targeted by the corresponding gDNAs and the gDNA regions of pCRISPR-*ompF_5′G* and pCRISPR-*ompF_5′C*. While the chromosome target sites had wild type sequences in both sets of transformant colonies, the colonies transformed with pCRISPR-*ompF_5′G* carried rearranged plasmids missing the gDNA ompF_5′G region, probably due to homologous recombination occurring between the two repeats flanking the gDNA. By contrast, all the colonies analyzed from the transformation with pCRISPR-*ompF_5′C* carried a wild type gDNA sequence (Additional file 2: Figure S1) [9].

From these experiments we conclude that in silico selection allows the identification of gDNAs, which in most cases drive the Cas9-mediated cleavage with high efficiency. However, approximately 10% of selected gDNAs did not cause bacterial killing. "Good and bad" gDNAs can easily be discriminated by colony counting after transformation with the pCRISPR plasmids carrying the selected gDNAs. Therefore, before modifying any target gene we routinely run the colony testing and if transformation efficiencies higher than 10^3 CFUs/μg of plasmid DNA is obtained we change the gDNA sequence of the plasmid.

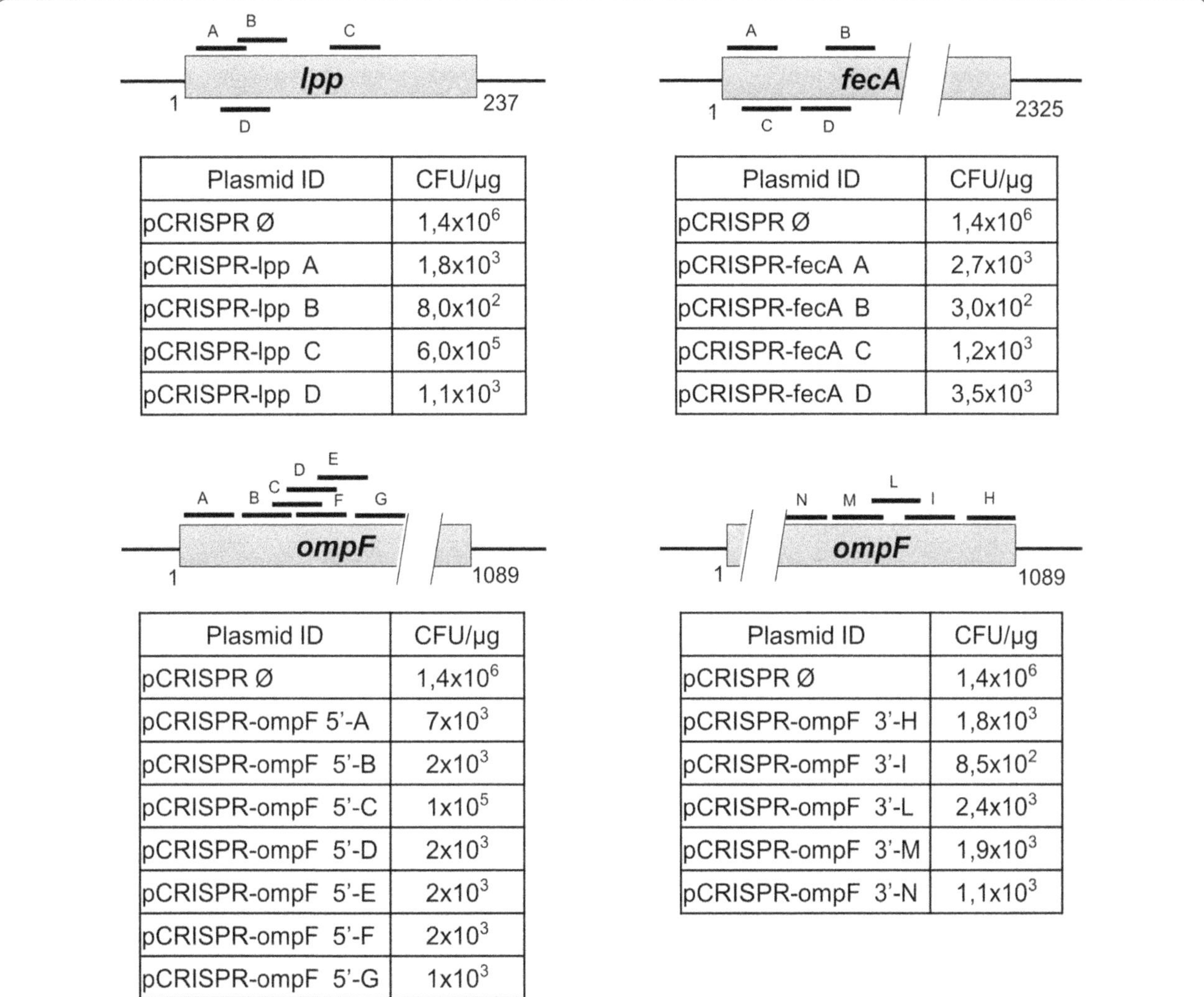

Plasmid ID	CFU/µg
pCRISPR Ø	1,4x10^6
pCRISPR-lpp A	1,8x10^3
pCRISPR-lpp B	8,0x10^2
pCRISPR-lpp C	6,0x10^5
pCRISPR-lpp D	1,1x10^3

Plasmid ID	CFU/µg
pCRISPR Ø	1,4x10^6
pCRISPR-fecA A	2,7x10^3
pCRISPR-fecA B	3,0x10^2
pCRISPR-fecA C	1,2x10^3
pCRISPR-fecA D	3,5x10^3

Plasmid ID	CFU/µg
pCRISPR Ø	1,4x10^6
pCRISPR-ompF 5'-A	7x10^3
pCRISPR-ompF 5'-B	2x10^3
pCRISPR-ompF 5'-C	1x10^5
pCRISPR-ompF 5'-D	2x10^3
pCRISPR-ompF 5'-E	2x10^3
pCRISPR-ompF 5'-F	2x10^3
pCRISPR-ompF 5'-G	1x10^3

Plasmid ID	CFU/µg
pCRISPR Ø	1,4x10^6
pCRISPR-ompF 3'-H	1,8x10^3
pCRISPR-ompF 3'-I	8,5x10^2
pCRISPR-ompF 3'-L	2,4x10^3
pCRISPR-ompF 3'-M	1,9x10^3
pCRISPR-ompF 3'-N	1,1x10^3

Fig. 2 Selection of gDNAs for mutation of *ompF*, *lpp* and *fecA* genes. The *grey bars* are a schematic drawing of the genes *lpp*, *fecA* and *ompF*, and the *black lines* labelled with letters indicate the positions where the gRNAs transcribed from their corresponding gDNAs hybridize within each gene. gDNAs were cloned into pCRISPR, generating the plasmids reported in the Additional file 1: Table S1. The *tables* report the transformation efficiencies (CFU/µg) of each pCRISPR-*gDNA* in BL21(DE3)*ΔompA*(pCasRed)

Definition of the optimal conditions to be used in gene knockout

Next, we investigated the efficiency of gene inactivation using synthetic oligonucleotides (donor DNA, dDNA) under different experimental conditions.

We analyzed three parameters: length and concentration of the mutagenic dDNA and extent of gene deletion. These parameters were first tested targeting the 5′ end of the *ompF* gene using the pCRISPR-*ompF_5′G* plasmid (Fig. 2) and single stranded donor DNAs annealing to the lagging strand (Lg-ss-dDNAs) (Additional file 1: Table S3). In fact, similar to the homologous recombination using synthetic oligonucleotides [5], Lg-ss-dDNAs have been reported to be more efficient in promoting double strand break repair than those targeting the leading strand (Ld-ss-dDNAs) [10, 18]. Donor DNAs were designed with homology arms of equal length upstream and downstream from the deletion. Subsequently, the best conditions were validated targeting three additional gene loci, the 3′ end of the *ompF* gene, and the *fecA* and *lpp* genes, using either single stranded donor DNA (Ld-ss-dDNA) or double stranded donor DNA (ds-dDNA). All parameters were tested in at least three independent experiments and mutation efficiencies were determined by analyzing a total of at least 20 colonies.

Table 1 summarizes the data obtained from all experiments. As far as the use of ss-dDNA is concerned, for short gene deletions (around 30 bp), 10 µg of both 70 and

Table 1 Influence of type of donor DNA (dDNA) (Lg-ss-dDNAs, Ld-ss-dDNAs, ds-dDNA) length of dDNA, concentration of dDNA and size of deletion on mutagenesis efficiency at four chromosomal loci

Target gene-pCRISPR-gDNA	Type of dDNA	dDNA ID	dDNA length (nt)	Mutation (nt)	dDNA quantity (μg)	Efficiency (%)-positive/total
ompF 5′-pCRISPR-ompF_5′G	ss-Lg	ompF_5′G-70-Δ30	70	Δ30	1	86% (19/22)
			70	Δ30	10	95% (40/42)
		ompF_5′G-70-Δ100	70	Δ100	10	5% (2/36)
		ompF_5′G-70-Δ500	70	Δ500	10	0% (0/30)
		ompF_5′G-120-Δ30	120	Δ30	1	64 ± 34% (40/62)
			120	Δ30	10	93% (76/82)
		ompF_5′G-120-Δ100	120	Δ100	10	64 ± 31% (20/31)
		ompF_5′G-120-Δ500	120	Δ500	10	47 ± 19% (14/30)
		ompF_5′G-120-Δ1089	120	Δ1180	10	0% (0/20)
	ss-Ld	ompF_5′G-70-Δ30 R	70	Δ30 nt	10	20% (4/20)
		ompF_5′G-120-Δ30 R	120	Δ30	10	100% (10/10)
		ompF_5′G-120-Δ1089 R	120	Δ1180	10	0% (0/> 100)
	ds	ompF_5′G-120-Δ30 ds	120	Δ30	10	77% (20/26)
		ompF_5′G-120-Δ500 ds	120	Δ500	10	50 ± 14% (15/30)
		ompF_5′G-120-Δ1089 ds	120	Δ1180	10	26 ± 17% (6/23)
ompF 3′-pCRISPR-ompF_3′I	ss-Lg	ompF_3′I-70-Δ30	70	Δ30	10	79% (19/24)
		ompF_3′I-120-Δ30	120	Δ30	10	83% (25/30)
fecA-pCRISPR-fecA_B	ss-Lg	fecA_B-70-Δ30	70	Δ30	10	43 ± 4% (13/30)
		fecA_B-120-Δ30	120	Δ30	10	77% (23/30)
		fecA_B-120-Δ100	120	Δ100	10	20% (6/30)
		fecA_B-120-Δ500	120	Δ500	10	13% (4/30)
		fecA_B-120-Δ2325	120	Δ2325	10	0% (0/30)
	ds	fecA_B-120-Δ2325 ds	120	Δ2325	10	14% (5/35)
lpp-pCRISPR-lpp_B	ss-Lg	lpp_B-70-Δ30 R	70	Δ30	10	20% (4/20)
		lpp_B-120-Δ30 R	120	Δ30	10	55 ± 7% (11/20)
		lpp_B-120-Δ237 R	120	Δ237	10	0% (0/16)
	ss-Ld	lpp_B-70-Δ30	70	Δ30	10	90% (27/30)
		lpp_B-120-Δ30	120	Δ30	10	73% (19/26)
		lpp_B-120-Δ237	120	Δ237	10	0% (0/30)
	ds	lpp_B-120-Δ30 ds	120	Δ30	10	82% (19/23)
		lpp_B-120-Δ237 ds	120	Δ237	10	72% (18/25)

120 bp oligonucleotides were very effective, resulting in efficiencies of mutagenesis >50% for all four genes. Deletions longer than 30 bp and up to approximately 500 bp, could be obtained with relatively good efficiencies (>10%) as long as 10 μg of 120 bp ss-dDNAs were used. However, efficiencies dropped quite substantially (<5%) for deletions longer than 500 bp. No major appreciable difference in deletion efficiencies at all four gene loci was observed when the Ld-ss-dDNA was used in place of the Lg-ss-dDNA. As far as ds-dDNA is concerned, the use of 120 bp ds oligonucleotides was not only as efficient as ss-dDNAs for short deletions but also allowed the generation of the long full gene deletions that failed using ss-dDNAs.

From the experimental data obtained on four gene loci, we conclude that, in order to create relatively short deletions, ss-dDNAs of 70–120 nucleotides targeting either the lagging or the leading strands perform well (100% robustness and >50% efficiency). However, 120 bp-ds-dDNAs showed high robustness and efficiency when extended deletions (of up to 2.200 bp) were required.

Validation of 30 bp gene deletions for mutant library production

To further validate the conclusions of the experiments described above, we attempted the mutation of 78 additional genes that are classified as "dispensable" according to the Keio library [29] (see Additional file 1: Table S4 for the list of genes). First, we constructed the 78 pCRISPR-*SacB-gDNAs* plasmids (Additional file 1: Table S1) encoding the gRNAs targeting the genes next to their 5′ ends to avoid the expression of truncated but nonetheless functional proteins (Additional file 1: Table S2). Subsequently, we verified the effectiveness of the cloned gDNAs in guiding Cas9 cleavage by transformation and colony counting (see above). In line with the statistics previously obtained, approximately 10% of selected gDNAs (7 out of 78, highlighted with an asterisk in Additional file 1: Table S2) did not promote Cas9 cleavage, and therefore other functional gDNAs were selected (Additional file 1: Table S2). After selecting the 78 pCRISPR-*SacB-gDNAs* plasmids with transformation efficiencies three logs lower than pCRISPR-*SacB*, we next designed 78 70 bp-ss-dDNAs to delete 30 bp and introduce a premature stop codon in each gene, downstream the deleted region (Additional file 1: Table S3). Of these 78 dDNAs, 46 targeted the leading strand and 32 the lagging strand. Each pCRISPR-*SacB-gDNA* plasmid, with its corresponding dDNA, was used to transform *E. coli* BL21(DE3)Δ*ompA*(pCasRed) strain, giving transformation efficiencies usually between 2 and 5×10^2 CFUs/μg of plasmid DNA. From each transformation ten colonies were randomly selected and analyzed by PCR to identify those carrying the 30 bp deletion and to establish the "success rate" of mutagenesis. We arbitrarily applied the "ten colony rule" (i.e., mutagenesis efficiencies higher that 10%) because we considered such number still compatible with manual high-throughput procedures. Figure 3a summarizes the results of our screening. Forty-eight of the selected 78 genes were successfully mutated, with efficiencies $\geq$10% (at least one mutant/10 colonies). More specifically, 20 of the 46 Ld-ss-dDNAs created the desired mutations, while Lg-ss-dDNAs mutated 28 out of 32 genes.

The inability to obtain 30 mutations, at least with a sufficiently high frequency to fulfill the "ten colony rule", was partially unexpected, since the missing mutations had been previously obtained in *E. coli* K-12 BW25113 [29]. Therefore, either such mutations are not compatible with the genetic background of our recipient strain [our strain is an *E. coli* BL21(DE3) derivative carrying the deletion of the *ompA* gene and pCasRed plasmid] or the experimental conditions previously defined were not sufficiently robust.

To discriminate between the two possibilities, we first asked the question whether the use of dDNAs targeting the opposite strands could rescue the missing mutants. Figure 3b reports the results of this experiment. The use of Lg-ss-dDNA rescued 14 of the 26 mutants that failed to be isolated with the Ld-ss-dDNA. In contrast, when the mutation of the five genes that were not mutated with the Lg-ss-dDNA was attempted only one mutant was rescued using Ld-ss-dDNA. Overall, the data indicate that 42 out of a total of 58 mutagenesis attempts with the Lg-ss-dDNA were successful, corresponding to a success rate of 72%. By contrast, the use of the Ld-ss-dDNA was successful in 22 of the 51 mutagenesis experiments (success rate 43%).

These data demonstrate that, differently from our preliminary validation experiment but in line with previous studies [10, 18], Lg-ss-dDNA performed better than Ld-ss-dDNA both in terms of success rate (72% rate versus 43%) and efficiency of mutagenesis (average bar heights with Ld-ss-dDNA: 10%; average bar heights with Lg-ss-dDNA: 31%). However, a non-negligible fraction of genes [15 out of 78 (19%)] failed to be mutagenized, regardless the type of the dDNA used.

Considering our experimental data that showed how ds-dDNAs successfully guided long deletions not obtained with ss-dDNAs, we thus re-attempted the deletion of all 78 genes using the same pCRISPR-*SacB-gDNAs*, but making the 70 base dDNAs double stranded. As shown in Fig. 3c, all genes that were successfully mutated using either the Lg-ss-dDNAs or the Ld-ss-dDNAs (63 genes in total) were also inactivated with the ds-dDNA. In addition, seven genes that were not deleted with the ss-dDNAs were mutagenized, bringing the success rate to 90% (70/78). Importantly, when ds-dDNAs were used the mutagenesis efficiency was also improved compared to the efficiencies obtained with ss-dDNAs (average bar heights with ds-dDNA: 48.9%).

As pointed out above, the inability to obtain some mutations classified as "dispensable" according to the Keio library [29] could be due to the fact that such mutations are not compatible with the genetic background of our recipient strain [our strain is an *E. coli* BL21(DE3) derivative carrying the deletion of the *ompA* gene and pCasRed plasmid]. To investigate the possible role of the Δ*ompA* deletion in preventing the inactivation of the eight genes, we tested whether they could be achieved in BL21(DE3) wild type strain. To this aim, pCasRed plasmid was introduced into BL21(DE3) and BL21(DE3)(pCasRed) was subsequently subjected to the gene deletion experiments by co-transforming the strain with the eight pCRISPR/ds-dDNAs couples. As a control, the deletion of the *degP* and *yncD* genes [two genes that were successfully inactivated in BL21(DE3)Δ*ompA* was also attempted]. As shown in Fig. 3 and Additional file 1: Table S6, all eight genes were successfully inactivated, with good efficiencies, thus confirming the incompatibility of these mutations with the Δ*ompA* gene inactivation. It is noteworthy that all the eight genes encode membrane-associated proteins whose inactivation in the absence of one of the major *E. coli* outer membrane proteins might impair the membrane function and/or structural

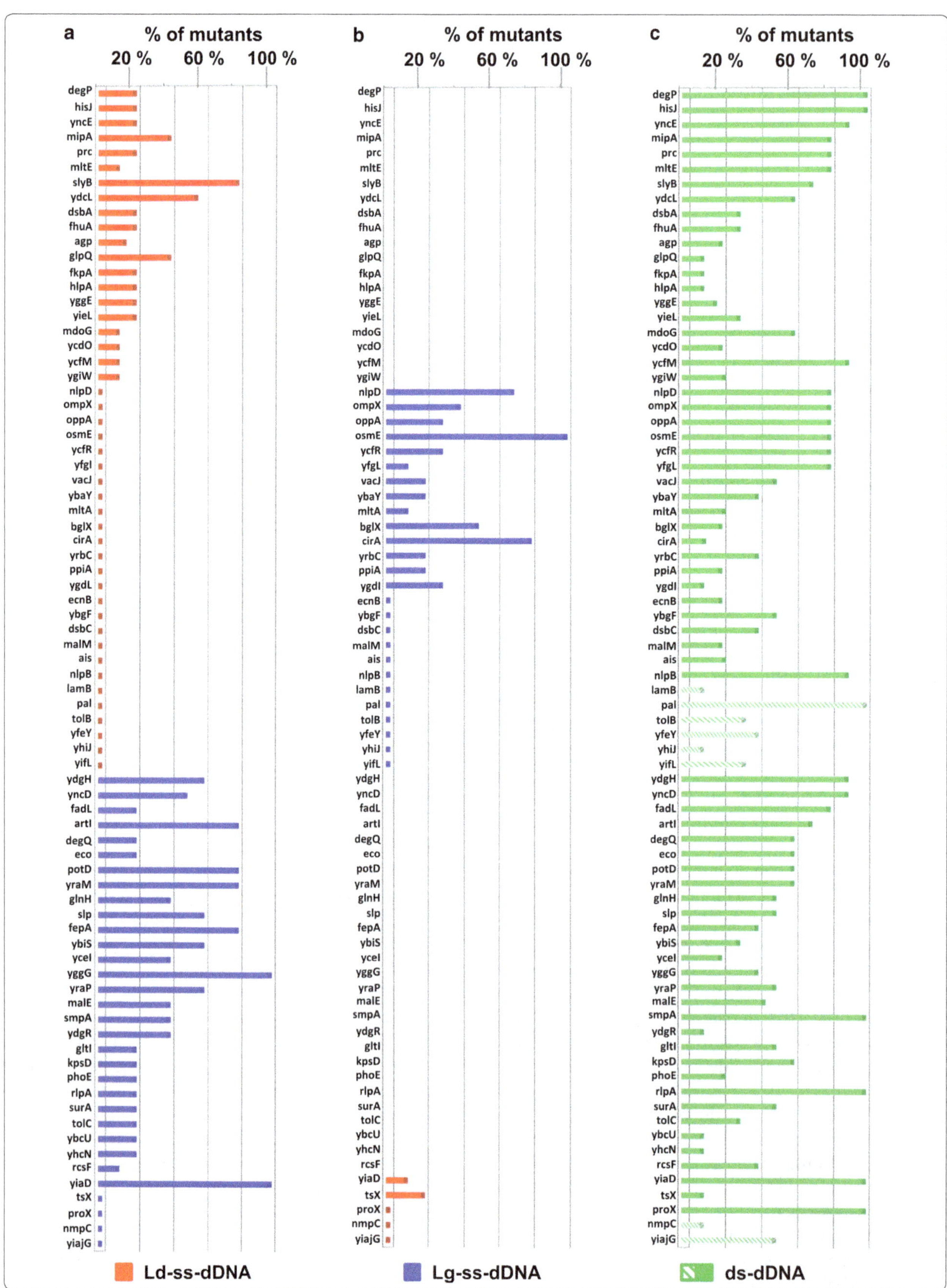
a
b
c
% of mutants
20 %
60 %
100 %
degP
hisJ
yncE
mipA
prc
mltE
slyB
ydcL
dsbA
fhuA
agp
glpQ
fkpA
hlpA
yggE
yieL
mdoG
ycdO
ycfM
ygiW
nlpD
ompX
oppA
osmE
ycfR
yfgL
vacJ
ybaY
mltA
bglX
cirA
yrbC
ppiA
ygdL
ecnB
ybgF
dsbC
malM
ais
nlpB
lamB
pal
tolB
yfeY
yhiJ
yifL
ydgH
yncD
fadL
artI
degQ
eco
potD
yraM
glnH
slp
fepA
ybiS
yceI
yggG
yraP
malE
smpA
ydgR
gltI
kpsD
phoE
rlpA
surA
tolC
ybcU
yhcN
rcsF
yiaD
tsX
proX
nmpC
yiajG
Ld-ss-dDNA
Lg-ss-dDNA
ds-dDNA

(See figure on previous page.)

Fig. 3 Validation of 30 bp deletions on 78 genes. *E. coli* BL21(DE3)Δ*ompA*(pCasRed) or *E. coli* BL21(DE3)(pCasRed) were transformed with different mixtures of pCRISPR-*SacB-gDNAs* and 70-base dDNAs (either ssDNA or dsDNA) to mediate 30 bp deletion at one of the 78 selected gene loci (*y axis*). Ten colonies from each transformation were analyzed by PCR to identify those carrying the deletion. *X axis* indicates the percentage of mutants identified out of the total number of colonies analyzed. *Bar height* indicates the mutation frequency, while the presence of *flat colored squares* above gene names in each graph indicates that no mutants were identified out of ten colonies analyzed. *Absence of bars* or *flat colored squares* above gene names indicate that the gene mutation was not attempted for those specific genes in the experiment indicated in each bar graph. *Red bars/squares* indicate mutation experiments using Ld-ss-dDNAs; *Blue bars/squares* indicate mutation experiments using Lg-ss-dDNAs; *Green bars/squares* indicate mutation experiments using ds-dDNAs; *bars with green downward diagonals* indicate mutation experiments with ds-dDNAs in BL21(DE3)(pCasRed). **a** *Bar graph* representing mutation success rate using 46 Ld-ss-dDNAs (*red bars/squares*) and 32 Lg-ss-dDNAs (*blue bars/squares*). **b** Gene mutations that failed using the Ld-ss-DNAs (26 genes out of 46) and the Lg-ss-DNAs (4 genes out of 32) were re-attempted using ss-dDNAs targeting the opposite strands. The *chart* represents the mutation success rate of this second round of experiments. **c** The *bar graph* represents the mutation success rate in BL21(DE3)Δ*ompA*(pCasRed) (*green bars*) and in BL21(DE3)(pCasRed) (*green downward diagonal bars*) using ds-dDNAs

integrity. An important outcome of the mutagenesis experiments carried out in BL21(DE3) strain is that when ds-dDNA is used, our CRISPR/Cas9 mutagenesis protocol can efficiently inactivate any "dispensable" gene and therefore the protocol has a robustness of 100%.

Creation of multi-gene mutants by stepwise approach

In order to generate multi-gene mutants in a stepwise modality, the pCRISPR-*gDNA* plasmid used to knockout a given gene has to be removed to allow the next round of transformation with a pCRISPR-*gDNA* plasmid targeting another gene locus. The pCRISPR plasmid is a high copy number plasmid (>400 copies/cell in exponential growth phase) and therefore it cannot be easily removed by simply growing bacteria in liquid media deprived of kanamycin, the pCRISPR antibiotic resistance marker. Elegant solutions have been proposed to get rid of high copy number plasmids [10, 18, 30]. We adopted the strategy proposed by Hale and co-workers [25] by introducing the suicide *sacB* gene downstream of the Km cassette into the pCRISPR plasmid, thereby generating the pCRISPR-*SacB* plasmid. *E. coli* strains carrying any pCRISPR-*SacB* plasmid derivative can survive in the presence of sucrose only if they rapidly lose the plasmid [21–24]. The efficiency of pCRISPR-*SacB* plasmid elimination in sucrose containing media can accelerate the entire process of multiple stepwise gene inactivation. Once a mutant colony is selected by colony PCR, it can be inoculated in sucrose-containing LB and the overnight culture can be directly used to make competent cells for the subsequent mutagenesis step. We validated the procedure by creating a BL21(DE3)Δ*ompA*(pCasRed) strain carrying the inactivation of both *fecA* and *lpp* genes. First of all, the gDNA *fecA_B* and gDNA *lpp*_B were inserted into pCRISPR-*SacB* plasmid generating plasmids pCRISPR-*SacB-fecA*_B and pCRISPR-*SacB-lpp*_B plasmids (Additional file 1: Table S1), respectively. Next, BL21(DE3) Δ*ompA*(pCasRed) strain was co-transformed with pCRISPR-*SacB-fecA*_B plasmid and the 120 base oligo fecA_B-120-Δ30 to create the 30 bp *fecA* gene deletion. Seven out of the ten colonies selected on Km/Cm LB plate carried the expected mutation (Fig. 4). One mutant clone was grown overnight in LB supplemented with Cm and 5% sucrose and competent cells, directly prepared from the overnight culture, were co-transformed with pCRISPR-*SacB-lpp*_B plasmid and the double stranded 120 base lpp_B-120-Δ240 oligo designed to eliminate the entire *lpp* gene. BL21(DE3)*ompA*/Δ*fecA*/Δ*lpp* colonies were obtained with an efficiency of 60% (Fig. 4). By the time the two pCRISPR-*SacB-gDNA* plasmids and the two dDNAs were available, four working days were sufficient to generate the BL21(DE3)Δ*ompA* derivative strain carrying the Δ*fecA* and Δ*lpp* double gene deletions.

Simultaneous creation of multi-gene deletions

Finally, we tested the possibility of exploiting the CRISPR/Cas technology to simultaneously create mutations at two separated genome loci. To test this strategy, we created pCRISPR plasmids carrying two gDNAs designed to guide the Cas-mediated genome cleavage at two distant positions within the *E. coli* BL21(DE3)Δ*ompA* genome. The two gDNAs were intercalated by the short repeated regions in the configuration "Repeat-gDNA$_1$-Repeat-gDNA$_2$-Repeat" (Additional file 1: Table S1; Fig. 5). We mimicked the same organization of the CRISPR arrays found in bacteria. The pCRISPR(gDNA)$_2$ plasmid thus obtained was co-transformed with two mutagenic oligonucleotides designed to repair and mutagenize the genome at the two cleaved sites.

First, we tested this approach by creating pCRISPR-*ompF_5′G-ompF_3′I* plasmid to delete 30 bp at both the 5′ and 3′ ends of the *ompF* gene (Fig. 5a). To verify that the plasmid carrying two guide RNAs could drive the simultaneous Cas-mediated cleavage at two chromosomal sites, *E. coli* BL21(DE3)Δ*ompA*(pCasRed) strain was transformed with pCRISPR-*ompF_5′G-ompF_3′I* plasmid in the presence of either ompF_5′G-120-Δ30 or ompF_3′I-120-Δ30 donor DNAs. If pCRISPR-*ompF_5′G-ompF_3′I* mediates the Cas9 cleavage of the *ompF* gene at both sites, co-transformation

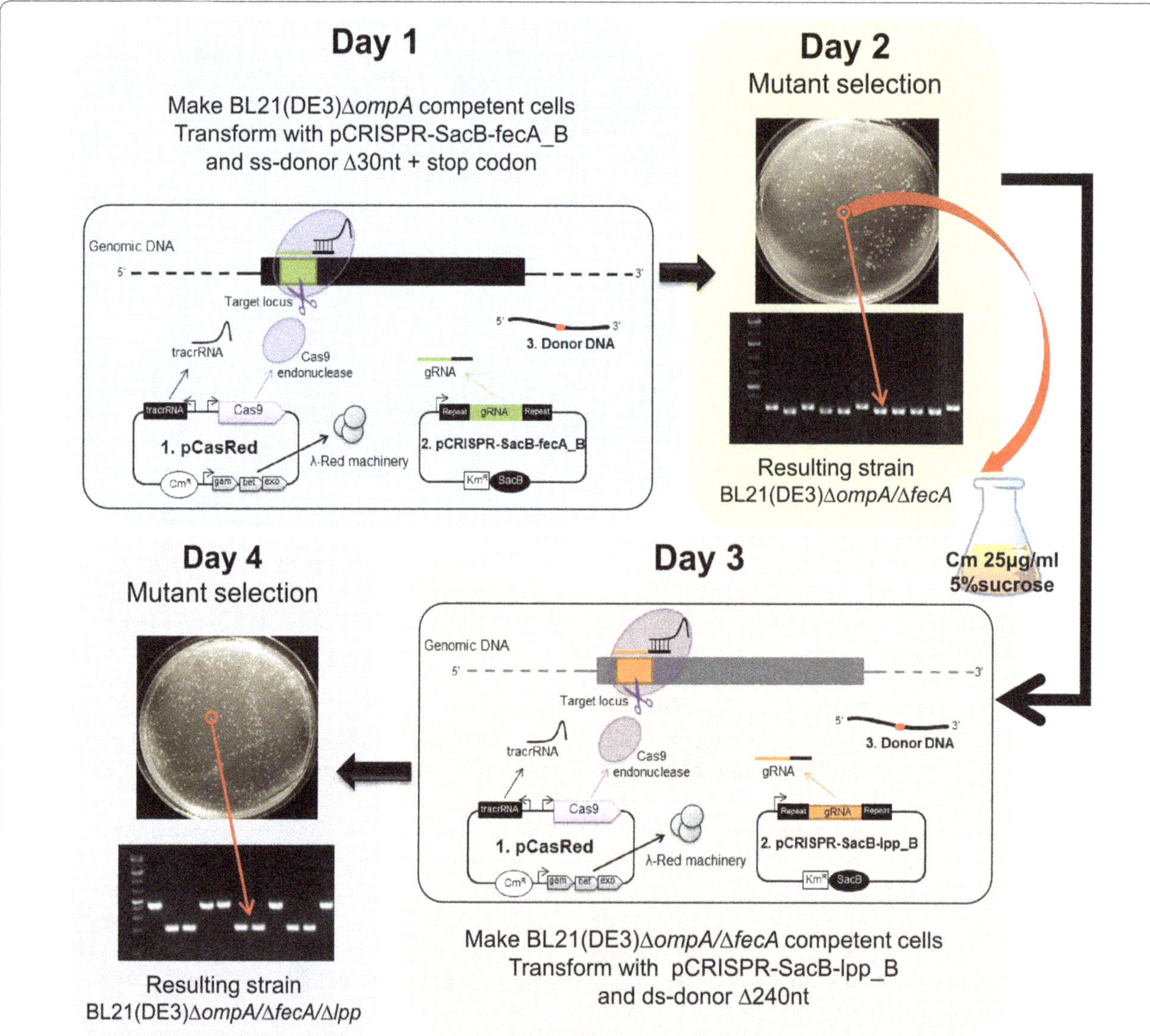

Fig. 4 Representation of the stepwise approach used to isolate strains carrying multiple mutations. Day 1: *E. coli* BL21(DE3)*ΔompA*(pCasRed) was co-transformed with 1 μg/ml of pCRISPR-*SacB-fecA_B* and 10 μg/ml of donor fecA-120-Δ30nt and transformant colonies were selected on LB agar plates supplemented with Cm (25 μg/ml) and Km (50 μg/ml). Day 2: Ten colonies were randomly selected and screened by PCR using primers designed to generate DNA fragments from mutated colonies of 200 bp. PCR products were analyzed on 2% agarose gels. One mutant clone was subsequently inoculated into 5 ml of LB supplemented with 5% sucrose and 25 μg/ml Cm. Day 3: The overnight culture was used to prepare competent cells, which were subsequently co-transformed with 1 μg/ml of pCRISPR-*SacB-lpp_B* and 10 μg of double strand donor DNA lpp-120-Δ237. Day 4: Ten colonies were randomly selected and screened by PCR using primers designed to generate DNA fragments from mutated colonies of 400 bp. PCR products were analyzed on 2% agarose gels

with only one dDNA should not generate mutants. Indeed, co-transformation of pCRISPR-*ompF_5′G-ompF_3′I* with ompF_3′I-120-Δ30 donor DNA gave no mutants (from 20 colonies tested) and the co-transformation of pCRISPR-*ompF_5′G-ompF_3′I* with ompF_5′G-120-Δ30 dDNA gave only two mutants out of 20 screened colonies (Table 2). Next, we co-transformed BL21(DE3)*ΔompA*(pCasRed) with pCRISPR-*ompF_5′G-ompF_3′I* in the presence of both ompF_5′G-120-Δ30 and ompF_3′I-120-Δ30 dDNAs and we analyzed 20 colonies by PCR. Twelve out of 20 colonies carried the deletion at both sites while only one of the 20 colonies analyzed carried a single mutation at the 5′ end of the *ompF* gene (Fig. 5; Table 2).

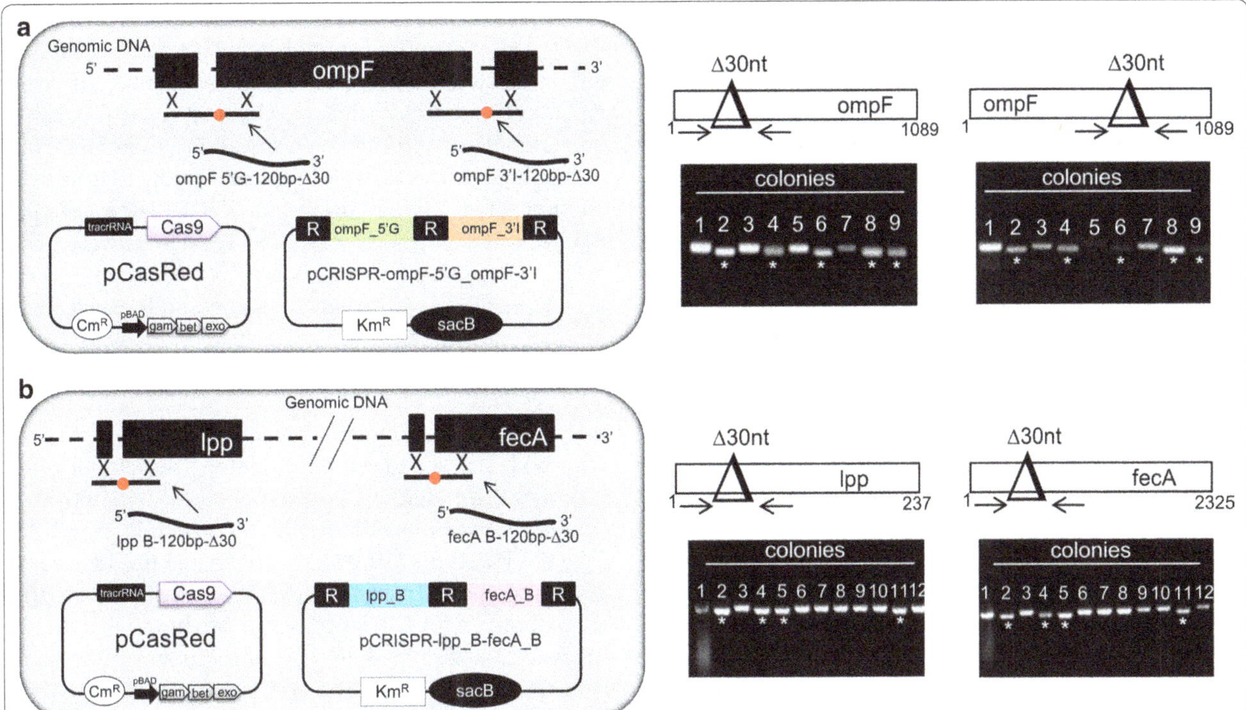

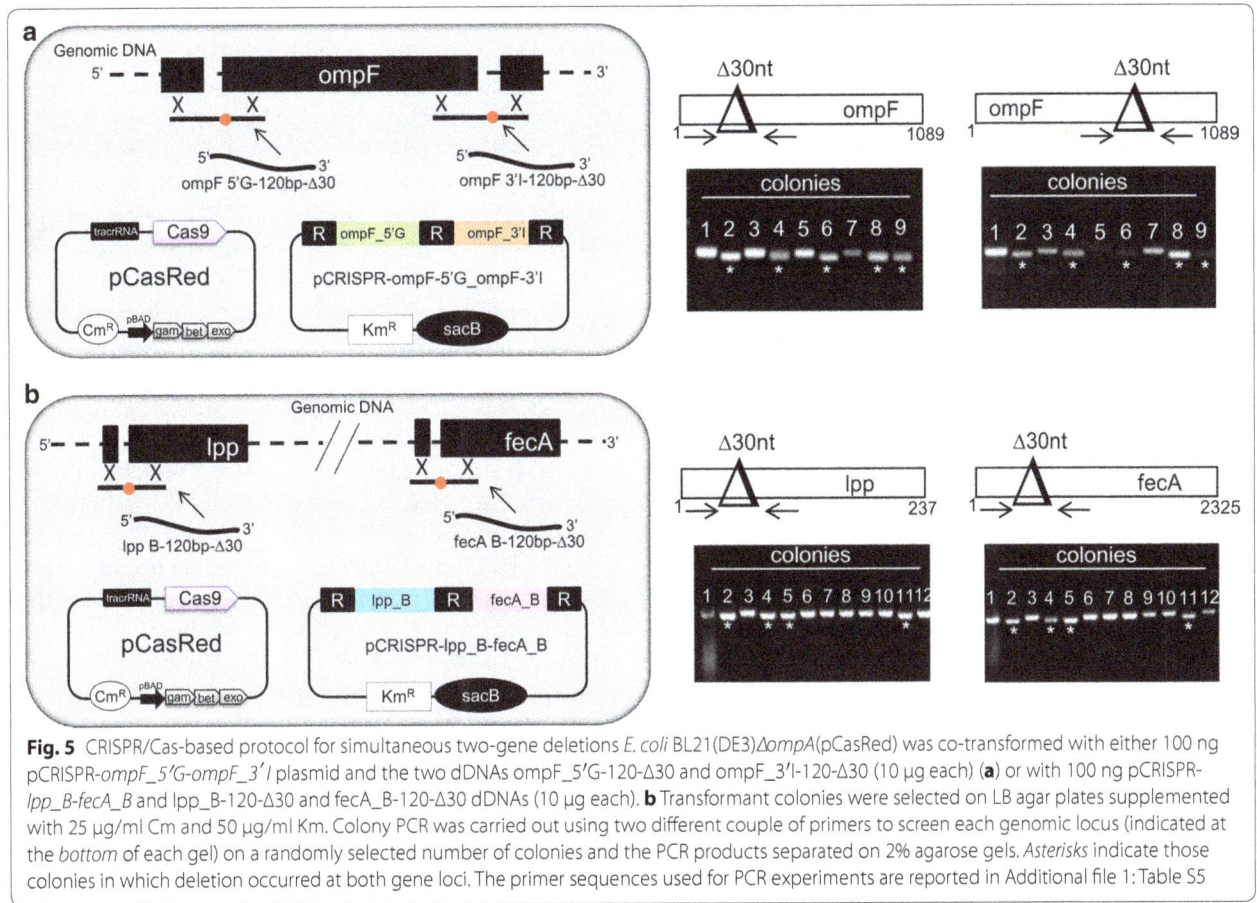

Fig. 5 CRISPR/Cas-based protocol for simultaneous two-gene deletions *E. coli* BL21(DE3)*ΔompA*(pCasRed) was co-transformed with either 100 ng pCRISPR-*ompF_5′G-ompF_3′I* plasmid and the two dDNAs ompF_5′G-120-Δ30 and ompF_3′I-120-Δ30 (10 μg each) (**a**) or with 100 ng pCRISPR-*lpp_B-fecA_B* and lpp_B-120-Δ30 and fecA_B-120-Δ30 dDNAs (10 μg each). **b** Transformant colonies were selected on LB agar plates supplemented with 25 μg/ml Cm and 50 μg/ml Km. Colony PCR was carried out using two different couple of primers to screen each genomic locus (indicated at the *bottom* of each gel) on a randomly selected number of colonies and the PCR products separated on 2% agarose gels. *Asterisks* indicate those colonies in which deletion occurred at both gene loci. The primer sequences used for PCR experiments are reported in Additional file 1: Table S5

Table 2 Efficiency of simultaneous two-loci mutagenesis (ompF 5′/ompF 3′ regions and fecA/lpp genes) using pCRISPR plasmids carrying REPEAT-gDNA$_1$-REPEAT-gDNA$_2$-REPEAT cassette

pCRISPR-gDNA	dDNA ID	Mutation		
pCRISPR-ompF-5′G_ompF-3′I	ompF_5′G-120-Δ30	*Δ30 ompF_5′*	*Δ30 ompF_3′*	*Δ30 ompF_5′/Δ30 ompF_3′*
		10% (2/20)	Not tested	Not tested
pCRISPR-ompF-5′G_ompF-3′I	ompF_3′I-120-Δ30	*Δ30 ompF_5′*	*Δ30 ompF_3′*	*Δ30 ompF_5′/Δ30 ompF_3′*
		Not tested	0% (0/20)	Not tested
pCRISPR-ompF-5′G_ompF-3′I	ompF_5′G-120-Δ30 + ompF_3′I-120-Δ30	*Δ30 ompF_5′*	*Δ30 ompF_3′*	*Δ30 ompF_5′/Δ30 ompF_3′*
		5% (1/20)	0% (0/20)	57% (12/20)
pCRISPR-lpp_B-fecA_B	lpp_B-120-Δ30 + fecA_B-120-Δ30	*Δ30 lpp*	*Δ30 fecA*	*Δ30 lpp/Δ30 fecA*
		3% (1/29)	0% (0/29)	31% (9/29)

To confirm the applicability of the simultaneous double gene knockout, we also tested the concomitant inactivation of *lpp* and *fecA* genes. For this purpose, pCRISPR-*lpp_B-fecA_B* was constructed (Fig. 5b) and used to co-transform *E. coli* BL21(DE3)*ΔompA*(pCasRed) with lpp_B-120-Δ30 and fecA_B-120-Δ30 donor DNAs (Table 2). Here, nine out of 29 colonies were mutated in both genes, while single gene mutations were obtained at frequencies of 0% for *fecA* and 3% for *lpp* (Fig. 5b; Table 2).

Discussion

The cutting edge CRISPR/Cas-technology was elegantly applied to bacteria for the very first time by Jiang and co-workers in 2013 [9]. Subsequently, a number of other groups, using modified versions of the protocol proposed by Jiang et al. demonstrated the broad applicability and flexibility of the technology in several bacterial species. Thanks to these discoveries, the CRISPR/Cas methodology became the most effective strategy to create gene knock-outs and knock-ins in prokaryotes (for a recent review see Choi and Lee [14]). Two are the major breakthroughs of CRISPR/Cas-technology. First, the efficiency of mutagenesis can be as high as 100%. As reported by several authors [9, 10], such efficiency cannot be reached by any other genome editing strategy, including the procedure based on the λ Red/synthetic oligonucleotide-mediated homologous recombination described by Court and co-workers. For instance, when we tested the efficiency of *ompF* 30 bp deletion transforming BL21(DE3)*ΔompA*(pCasRed) with the synthetic dDNA in the absence of the corresponding pCRISPR-*gDNA*, we did not manage to isolate a single mutant colony out of 40 colonies analyzed. Second, even large deletions can be achieved at high frequencies using synthetic oligonucleotides as donor DNAs, thus avoiding the need to generate linear or circular DNA constructs to drive homologous recombination.

However, one missing information from published data was the robustness of the CRISPR/Cas protocols. In other words, how confident could one be that any desired mutation can be consistently achieved at high frequency? This information becomes particularly relevant when CRISPR/Cas genome editing is applied in high-throughput formats where the screening of several colonies and the repetition of mutagenesis experiments to rescue missing mutants could be impractical.

To address this issue, we first set up our own version of CRISPR/Cas9-based genome editing and we defined the best conditions by analyzing the mutagenesis efficiency at four gene loci using different parameters, including (i) sequence of the guide RNA (ii) length and concentration of the dDNA, and (iii) use of ss-dDNA and ds-dDNAs. Subsequently, we challenged the robustness of the protocol by attempting 30 bp deletion in 78 additional genes selected on the basis of their being classified as "dispensable" according to the Keio collection [29]. In this way, and for the first time, we could assign a value to the success rate of a CRISPR/Cas9-based genome editing procedure based on a statistically meaningful number of data (81 target loci in total).

The relevant outcome from our work is that we could establish with a high degree of confidence that our CRISPR/Cas9 protocol features an efficiency of mutagenesis consistently higher than 10% and a robustness of 100%. All 81 genes were successfully mutated with high efficiencies, 73 of them in the *E. coli* BL21(DE3)*ΔompA* background, and the remain eight genes, whose inactivation was incompatible with the concomitant presence of *ompA* mutation, were mutated in *E. coli* BL21(DE3).

Considering that the main components of our CRISPR/Cas9 genome editing protocol (pCRISPR-*SacB-gDNA* and pCasRed) are variations-of-the-theme of what already described by others, we believe that our data can be extrapolated to other procedures and therefore could be of general interest. We confirm, thanks to a massive amount of validation data, that CRISPR/Cas genome editing can be very effective and reliable provided that a few experimental solutions are applied. Among those is the use of double stranded synthetic oligonucleotides as donor DNA, and the pre-testing of gDNAs before attempting gene deletion.

Court and co-workers were the first to demonstrate that "λ Red recombineering" in *E. coli* can be carried out with both single and double stranded synthetic oligonucleotides, and that Lg-ss-dDNAs perform better than Ld-ss-dDNAs in promoting the mutagenesis event [31]. Like other authors [10, 18], we confirmed the superiority of the Lg-ss-dDNA even when "λ Red recombineering" is combined with CRISPR/Cas9, a situation whereby the mechanisms that leads to the introduction of the site-specific mutation must be coupled to the mechanisms of double-stranded DNA repair. However, we also found that the successful mutation of all 81 genes could be achieved, with efficiencies consistently higher than 10%, only if ds-dDNA is used (Fig. 3). Furthermore, we showed that ds-dDNA clearly outperformed ss-Lg-dDNA (and ss-Ld-dDNA) when extended deletions were attempted (Table 1).

We did not investigate on the mechanisms of homologous recombination taking place between ds-dDNA and the two chromosomal extremities generated by the Cas9 cleavage. However, it is reasonable to envisage, in addition to other described mechanisms, the contribution of a single exonuclease activity (possibly the 5′ exonuclease provided by the λ Red machinery) acting on both the ds-dDNA and the extremities of the cleaved chromosomal DNA. The exonuclease activity would release single stranded termini, stabilized by Beta, which would become available for base pairing. Such mechanism would not be effective with ss-dDNAs and therefore this could explain the higher efficiency of mutagenesis of ds-dDNA with respect to ss-dDNA. Alternatively, another possible mechanism is that the double strand break is repaired by dsDNA template (and not by ssDNA template) following a Rec-dependent homologous recombination. Such mechanism can occur even in the absence of the lambda red machinery [28, 32]. To test this possibility, the mutagenesis of *proX* and *yiaD*, two genes successfully mutagenized with ds-dDNA only, and with ssDNAs and dsDNA, respectively (Fig. 3), was re-attempted in BL21(DE3)*ΔompA*(pCas9) in which the λ Red machinery was removed. The strain was transformed with either pCRISPR-*SacB-proX* or pCRISPR-*SacB-yiaD* in the presence of the corresponding ds-dDNAs, and subsequently 100 colonies from each transformation were PCR analyzed. No mutant colonies were identified (data not shown). Therefore, under the experimental conditions we used, the efficiency of dsDNA homologous recombination, if occurring, was at least lower than 1%.

The second factor to consider to achieve high performance is the addition of a rapid experimental step aimed at demonstrating the effectiveness of gDNAs. Theoretically, any sequence in the genome next to the "NGG" (PAM) trinucleotide should represent a potential Cas9 cleavage site. Indeed, most of the gDNAs that we designed and cloned into pCRISPR plasmid produced gRNAs that promoted Cas-mediated cleavage with high specificity. However, 10% of the pCRISPR-*gDNAs* did not work, as judged by their inability to substantially reduce bacterial viability after transformation. We tried to identify the cause root of their failure, such as searching for the presence of sequences forming internal stem-loop structures or partially complementary to other chromosomal regions not carrying the NGG trinucleotides, but we did not find any plausible explanation so far. Whatever reason might be, the use of inefficient gDNAs would affect the overall robustness of the procedure. Therefore, we routinely transform the strain to be mutagenized with all pCRISPR-*gDNAs* before starting mutagenesis. A drop in transformation efficiency of at least three orders of magnitude is indicative of the quality of the gDNAs.

One important aspect to consider in setting up a CRISPR/Cas9-based mutagenesis protocol is the possibility to eliminate the whole machinery once genome editing is terminated. As far as the Cas9/λ Red genes are concerned, since they reside on a low copy number plasmid expressing the Cm^R, they can be easily removed by growing the strain in the absence of the antibiotic. We experimentally confirmed the effectiveness of losing pCasRed by growing strains overnight in the absence of Cm (see "Methods"). To eliminate the high copy number plasmid encoding the gRNA, a few strategies have been devised, the most popular one being the use of a temperature sensitive origin of replication [10]. To get rid of our pCRISPR-*SacB-gDNA* we envisaged to exploit the toxic effect of the *sacB* gene when *E. coli* is grown in media containing high concentrations of sucrose. *SacB* fused to an antibiotic resistance marker (usually chloramphenicol) has been used in chromosomal gene knock-in/knock-out experiments [21–24] and, more recently, for plasmid curing [25]. Indeed, we found that by simply inoculating the mutant colonies in LB supplemented with 5% sucrose, bacterial cells rapidly lose the plasmid and, if necessary, the culture is ready to be used for an additional round of mutagenesis. Multiple mutations can be introduced in the same recipient strain at a pace of one mutation every two working days (Fig. 5).

The pCRISPR-*SacB-gDNA* plasmid can accommodate more than one gRNA coding sequences potentially allowing the simultaneous modification of more than one gene. We demonstrated that by cloning two gDNAs in the configuration repeat-$gDNA_1$-repeat-$gDNA_2$-repeat, double gene inactivation was achieved with very high efficiencies. Therefore, not only bacterial cells could successfully take up all three elements simultaneously [pCRISPR-$(gDNA)_2$-*SacB* and two dDNAs] but also the gRNA was efficiently transcribed and processed to promote the Cas9-mediated cleavage at both chromosomal sites. It will be interesting to investigate the limits of the system and see how many loci can be simultaneously modified.

Methods

Bioinformatics

The "Keio collection" [29], a collection of systematic single-gene knockout mutants of *E. coli*, was used as a basis to evaluate both the biological functions and the non-essentiality of the selected genes listed in Additional file 1: Table S5. To properly select the gDNA and to avoid off-target effects of Cas9, we used the following procedure. First, we identified the PAM sequences in the first half of the target genes. Next, we selected the 30 bp upstream the PAM sequences and we blasted them [33] against *E. coli* BL21(DE3) genome. Those sequences with no complementary nucleotides within the ten nucleotide seed-region and with a homology lower than 15% in the remaining part of the guide were selected.

Bacterial strains and culture conditions

Escherichia coli DH5α strain was routinely grown in LB broth (SIGMA) at 37 °C and used for cloning experiments. *E. coli* BL21(DE3)*ΔompA* strain generated as previously reported [34], was grown in LB broth at 37 or 30 °C when required, and was employed for genome editing experiments. Stock preparations of strains were prepared in LB + 20% glycerol and stored at −80 °C. Each bacterial manipulation was started using an overnight culture from a frozen/glycerol stock. When required, kanamycin and chloramphenicol were added to final concentrations of 50 or 25 μg/ml, respectively.

Construction of plasmids

Information about all primers, plasmids and *E. coli* strains used in this study are provided in Additional file 1: Tables S1, S5.

The pCasRed plasmid carries a chloramphenicol resistance gene (*Cm^R*) and encodes the Cas9 nuclease, the λ Red (Exo, Beta, Gam) cassette and the tracrRNA. The

Cas9 gene and the tracrRNA coding sequence are under the control of constitutive promoters while the λ Red gene cassette is transcribed using an arabinose-inducible promoter.

The pCas9 plasmid (ADDGENE #4287 [9]) was used as template for the construction of pCasRed plasmid as follows. The λ Red cassette was PCR amplified from pKOBEG plasmid [20] and was cloned into pCas9 plasmid using the polymerase incomplete primer extension (PIPE) cloning method [35]. Briefly, the vector pCas9 was linearized by V-PIPE PCR amplification with primers Pipe1 pCAS9-F/ Pipe1 pCAS9-R. The insert, the λ Red cassette, was I-PIPE PCR amplified with primers redF/redR, which contain 5′ sequences complementary to the two distinct ends of the amplified vector. In this manner, annealing occurred directionally by mixing the PCR products, V-PCR and I-PCR, and by transforming *E. coli* HK-100 strain, pCasRed plasmid was generated. The resulting construct was analyzed by DNA sequencing and used to transform *E. coli* BL21(DE3)*ΔompA* electrocompetent cells.

The pCRISPR-*SacB* plasmid, derived from pCRISPR plasmid (ADDGENE #42875, [9]), where a kanamycin resistance gene (Km^R) is fused to the *sacB* gene encoding the *Bacillus subtilis* levansucrase, carries the synthetic DNA fragment (gDNA) coding for the guide RNA necessary to drive the Cas9-dependent double stranded break at the desired site of the bacterial genome. The construction of the pCRISPR-*SacB* plasmid was carried out by the PIPE method in two steps. In a first step, the kanamycin resistance cassette of pCRISPR plasmid was replaced by a "cat-sac cassette" containing the chloramphenicol acetyltransferase gene, along with the *sacB* gene, from pKM154 plasmid (ADDGENE #13035) [24]. The vector pCRISPR was linearized by V-PIPE PCR amplification with primers pipeCRISPR-F and pipeCRISPR-R to exclude the kanamycin gene. The insert, the cat-sac cassette, was I-PIPE PCR amplified with cat/sac-pipeF and cat/sac-pipeR primers, which contain 5′ sequence complementary to the two distinct ends of the amplified vector. In this manner, annealing occurred directionally by mixing the PCR products, V-PCR and I-PCR, and after transformation pCRISPR-CatSacB plasmid was isolated. In a second step, the chloramphenicol resistance cassette of the pCRISPR-CatSacB plasmid was then replaced with the kanamycin gene from the original pCRISPR plasmid. The pCRISPR-*CatSacB* plasmid was linearized by using V-crSac F and V-crSac R primers, excluding the chloramphenicol acetyltransferase gene. The kanamycin gene was amplified from pCRISPR plasmid using primers I-kanaF and I-kanaR. The PCR products were then mixed together and used to transform *E. coli*, generating the pCRISPR-*SacB* plasmid. The resulting construct was analyzed by DNA sequencing.

Plasmids expressing the gRNAs, were constructed by phosphorylation and annealing of oligonucleotides (gDNAs) listed in Additional file 1: Table S2, followed by ligation into pCRISPR (or pCRISPR-*SacB*) digested with *BsaI* (New England BioLabs), generating the plasmids listed in Additional file 1: Table S1. The resulting constructs were used to transform *E. coli* DH5α strain (Invitrogen) and the plasmids prepared by QIAprep Spin Miniprep Kit, Qiagen (QIAGEN kit) were analyzed by DNA sequencing.

Competent cell preparation

For all mutagenesis experiments, *E. coli* BL21(DE3)*ΔompA* carrying pCasRed plasmid was used. To prepare *E. coli* BL21(DE3)*ΔompA*(pCasRed) a 5 ml overnight culture (LB medium) inoculated from a single colony of BL21(DE3)*ΔompA* obtained from an LB-agar plate was grown at 37 °C under vigorous agitation. The overnight culture was diluted 100-fold and grown at 37 °C (200 r.p.m.) until the optical density at 600 nm (OD_{600}) reached 0.6–0.8 (~3 h). Then the cells were harvested at 4000 r.p.m. for 20 min at 4 °C and washed three times with cold MilliQ water. After a final wash in 10% glycerol the cells were aliquoted and stored at −80 °C. 50 μl of competent cells were then electroporated using 1 mm Gene Pulser cuvette (Bio-Rad) at 1.8 kV with 1 ng of pCasRed plasmid. Competent cells of *E. coli* BL21(DE3)*ΔompA*(pCasRed) were prepared by growing a single colony in LB medium with 25 μg/ml chloramphenicol at 37 °C under shaking. The overnight culture was diluted to an OD_{600} of 0.1 and grown at 37 °C under shaking (200 r.p.m.) to an OD_{600nm} of 0.2 and then L-arabinose was added to a final concentration of 0.2% for λ Red induction. After induction, the culture was grown to an OD_{600} of 0.7 and then cells were washed and aliquoted as described above.

The transformation efficiency of BL21(DE3)*ΔompA* (pCasRed) electrocompetent cells was of 0.5–2 × 10^6 CFUs/μg of the "empty" pCRISPR plasmid.

Gene knockout using CRISPR-Cas9

The genes encoding *ompF, lpp* and *fecA* were used as targets to establish the proof of concept for genome editing via CRISPR-Cas9 system in *E. coli*. All mutagenesis oligonucleotides (donor DNA or dDNAs) (Additional file 1: Table S3) (Sigma-Aldrich) were HPLC purified grade. The dDNAs were designed to delete a region ranging from 30 (Δ30) to 2325 (Δ2325) nucleotides from target genes, removing the protospacer and PAM regions, thus disrupting the Cas9 cleavage site and at the same time adding an in-frame stop codon downstream the deleted region.

For the leading and lagging strand design, the oligonucleotides annealing to the 3′ > 5′ strand moving clockwise from *OriC* up to *ter* were Ls-ss-dDNAs, while oligonucleotides annealing to the same strand but moving counterclockwise from *OriC* up to *ter* were Lg-ss-dDNAs. The opposite was true for the oligonucleotides annealing to the 5′ > 3′ strand moving clockwise from *OriC* up to *ter* and for those annealing to the same strand moving counterclockwise from *OriC* to *ter*.

The ds-dDNAs were generated by annealing 10 μg of both forward and reverse oligonucleotides in a total volume of 20 μl at 95 °C for 5 min and allowing the reaction mixture to cool down at room temperature. The annealing reaction was verified by loading 500 ng of each single stranded oligonucleotides and 1 μg of total DNA in the annealing reaction and by visualizing the bands using ATLAS ClearSight DNA Stain (BIOATLAS). The bands corresponding to the single stranded oligonucleotides disappeared in the annealing reaction sample. For CRISPR/Cas9-mediated gene knockouts 50 μl of *E. coli* BL21(DE3)*ΔompA*(pCasRed) or BL21(DE3)(pCasRed) competent cells, corresponding to 10^9 competent cells, were electroporated using 1 mm Gene Pulser cuvette (Bio-rad) at 1.8 kV with 100 ng of pCRISPR-*gDNA* and different quantities of dDNA ranging from 1 to 100 μg. As control, 100 ng of an empty pCRISPR plasmid was used. Cells were then immediately re-suspended in 1 ml of LB medium and allowed to recover at 30 °C for 3 h under agitation before being plated on LB agar with 25 μg/ml chloramphenicol and 50 μg/ml kanamycin and incubated at 37 °C overnight. Mutants were screened by colony PCR using GoTaq master mix (Promega-M7123). Briefly, cells were picked from individual colony using a pipette tip, directly resuspended in PCR reaction mix and DNA amplification was carried out according to the standard cycling GoTaq protocol. The deletion in *ompF* gene was analyzed by using primers: seqs-ompF 1F/ompR (Δ30), seqs-ompF 1F/ompR2 (Δ100), ompF 1F/ompF 4R2 (Δ500, Δ1089) (Additional file 1: Table S5). The deletion of *lpp* was analyzed by using primers: seqs-lpp F/seqs-lpp R (Δ30, Δ237) (Additional file 1: Table S5). The deletion in *fecA* gene was analyzed by using primers: FecA_F1/FecA_R1 (Δ30), dFecA-seqF/dFecA-seqR (Δ500, Δ2325) (Additional file 1: Table S5).

The protocol described above was further validated on 78 genes (Additional file 1: Table S4) using dDNAs (10 μg) designed to delete a region of 30 (Δ30 nt) nucleotides (see above) as described above. For simultaneous gene knockout experiments pCRISPR-*ompF_5′G-ompF_3′I* and pCRISPR-*lpp_B-fecA_B* plasmids were constructed by inserting the synthetic sequences into the *BsaI* site (Additional file 1: Table S2). The plasmids (100 ng) were used to co-transform *E. coli* BL21(DE3)*ΔompA*(pCasRed) with dDNA couples (10 μg each donor) ompF_5′G-120Δ30/ompF_3′I-120-Δ30 and lpp_B-120-Δ30/fecA_B-120-Δ30, respectively (Additional file 1: Table S3). Transformants were analyzed by colony PCR using primers seqs-ompF 1F/ompR (Δ30 at the 5′) and ompF 3′F/ompF 3′R, and seqs-lpp F/seqs-lpp R (Δ30) and FecA_F1/FecA_R1 (Δ30) (Additional file 1: Table S5).

To test mutagenesis efficiency in the absence of the lambda red machinery, E. coli BL21(DE3)*ΔompA*(pCas9) was transformed with either 100 ng pCRISPR-*SacB-proX*/10 μg ds-donor-proX-70-Δ30 or 100 ng pCRISPR-*SacB-yiaD*/10 μg ds-donor-yiaD-70-Δ30 targeting genes *proX* and *yiaD*, respectively. Hundred transformant colonies from each transformation were analyzed by colony PCR using primer s041_proX_F/s041_proX_R and s072_yiaD_F/s072_yiaD_R, respectively.

Plasmid curing and creation of multiple mutations by stepwise approach

To cure mutant strains from pCRISPR plasmid derivatives after each round of mutation, we adopted the strategy proposed by Hale and co-workers [25] by introducing the suicide *sacB* gene downstream of the Km cassette into the pCRISPR plasmid, generating the pCRISPR-*SacB* plasmid. *E. coli* strains carrying any pCRISPR-*SacB* plasmid derivative can survive in the presence of sucrose only if they rapidly lose the plasmid [21–24]. To test the effectiveness of this strategy, we transformed the BL21(DE3)*ΔompA*(pCasRed) strain with pCRISPR-*SacB* and one transformant colony was grown in LB medium supplemented either with 25 μg/ml Cm alone (the antibiotic resistance marker of pCas9-λ Red plasmid) or with 25 μg/ml Cm and 5% sucrose. After overnight growth, 100 μl of each culture were plated on LB agar plates supplemented with either Km and Cm, or Cm only. Not a single colony from the sucrose-containing culture could be isolated on the Km/Cm containing plate (Fig. 6), indicating that all bacteria had lost the pCRISPR-*SacB* plasmid. By contrast, confluent growth was observed on the Km/Cm plate seeded with bacteria grown in sucrose-deprived medium. The loss of pCRISPR plasmid from BL21(DE3)*ΔompA*(pCasRed)(pCRISPR-*SacB*) strain grown in the presence of sucrose was confirmed by plasmid extraction (Fig. 6).

For pCRISPR plasmid curing after each mutation round, the mutant colony, carrying both the pCasRed and pCRISPR-*SacB* plasmids, was inoculated in LB medium containing 5% sucrose and 25 μg/ml chloramphenicol, and grown overnight at 37 °C under shaking conditions. The overnight culture was directly used to prepare competent cells as described above. Briefly, at day 1, 50 μl of *E. coli* BL21(DE3)*ΔompA*(pCasRed) competent cells were co-transformed with 100 ng of

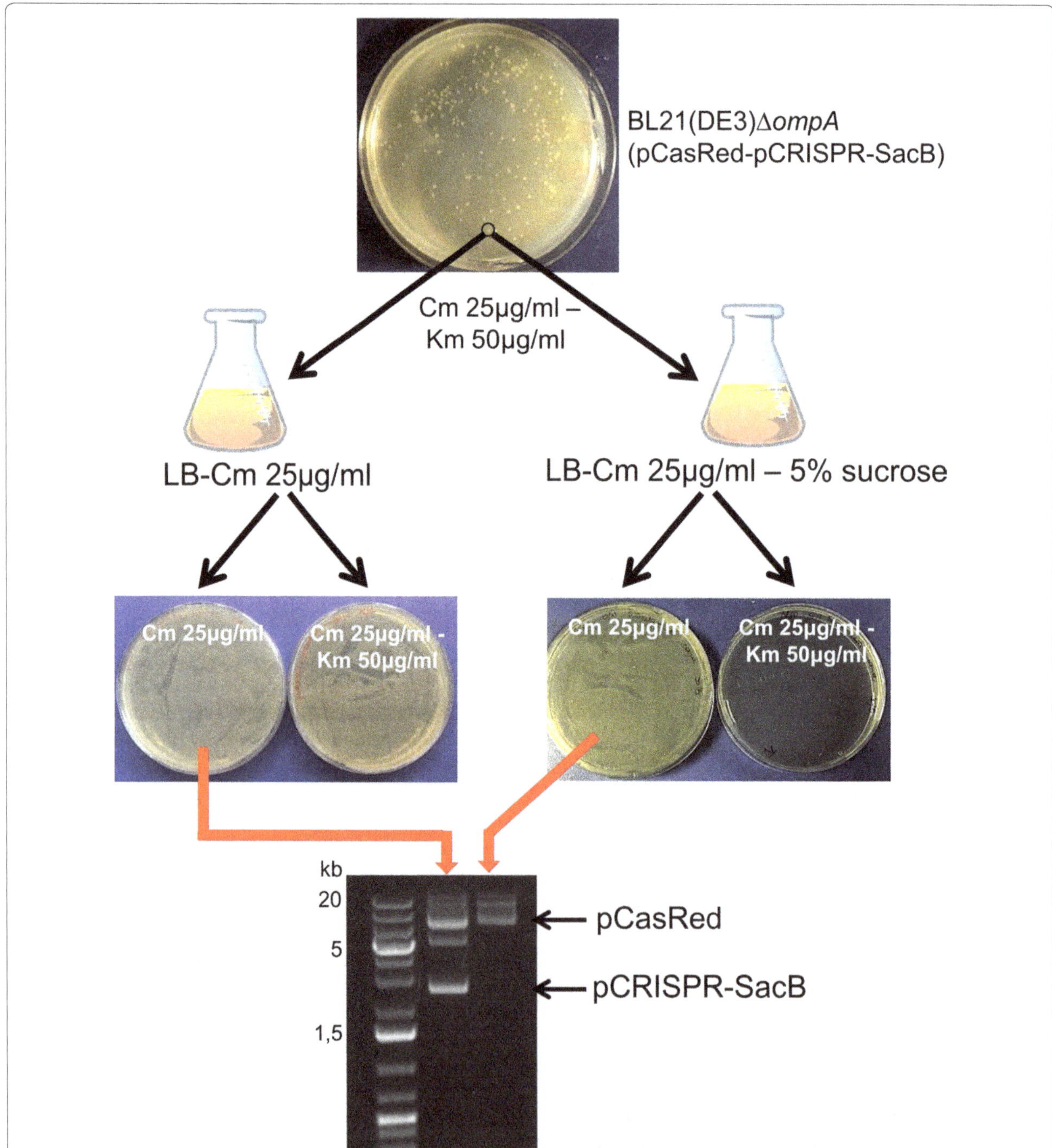

Fig. 6 pCRISPR-*SacB-gDNA* plasmid curing using 5% sucrose containing medium. A single colony from *E. coli* BL21(DE3)Δ*ompA* strain carrying both the pCasRed (Cm resistance) and pCRISPR-*SacB-gDNA* (Km resistance) was grown at 37 °C in LB-medium containing 5% sucrose and 25 µg/ml Cm. After 14 h growth, 100 µl of culture were plated on LB-agar plates containing either Cm (25 µg/ml) + Km (50 µg/ml) or Cm (25 µg/ml) alone. The loss of pCRISPR-*SacB-gDNA* plasmid was verified by 1.5% agarose gel analysis of plasmids extracted from bacteria directly collected from the Cm-containing agar plate. As control, the same colony was grown in the absence of 5% sucrose and plasmid extraction was carried out from bacteria collected from LB-agar plate containing 25 µg/ml Cm

pCRISPR-*SacB-fecA_B* plasmid and 10 µg of fecA_B-120-Δ30 dDNA. At day 2, transformant colonies were PCR screened and one mutant clone was inoculated in 5 ml LB supplemented with 5% sucrose and 25 µg/ml chloramphenicol. The overnight culture was used to prepare competent cells as described above. The new recipient strain BL21(DE3)*ΔompA/ΔfecA*(pCasRed) was co-transformed with 100 ng of pCRISPR-*SacB-lpp_B* plasmid and 10 µg of ds-dDNA lpp_B-120-Δ237 and transformant colonies were selected on LB-Agar 50 µg/ml kanamycin and 25 µg/ml chloramphenicol. The day after transformant colonies were PCR screened to identify the *E. coli* BL21(DE3)*ΔompA/ΔfecA/Δlpp*(pCasRed) mutant.

The curing of the pCasRed plasmid was carried out after overnight growth at 37 °C under shaking in the absence in the culture medium of chloramphenicol. The day after the culture was plated for single colony on LB-agar plate and colonies were analyzed for chloramphenicol resistance. No chloramphenicol resistance colonies were recovered.

Abbreviations

CRISPR: clustered regularly interspaced short palindromic repeats; PAM: protospacer adjacent motif; CM: chloramphenicol; KM: kanamycin; gDNA: guide DNA; NHEJ: non-homologous end-joining; tracrRNA: trans-activating crRNA; dDNA: donor DNA/synthetic oligonucleotides; Lg-ss-dDNA: single strand donor DNA targeting the lagging strand; Ld-ss-dDNA: single strand donor DNA targeting the leading strand; ds-dDNA: double strand donor DNA; PIPE: polymerase incomplete primer extension.

Authors' contributions

FZ designed the experimental set-up and experiments were performed by FZ, IZ, DF, EK. FZ, IZ, DF, EK, CI, LFF and GG analyzed and interpreted the acquired data. The manuscript was primarily prepared by GG and FZ, and includes corrections and intellectual contributions from all other authors. All authors read and approved the final manuscript.

Author details

[1] Synthetic and Structural Vaccinology Unit, CIBIO, University of Trento, Via Sommarive, 9, Povo, 38123 Trento, Italy. [2] Toscana Life Sciences Scientific Park, Via Fiorentina, 1, 53100 Siena, Italy.

Acknowledgements

We want to thank Luciano Marraffini for providing the pCas9 and the pCRISPR plasmids via ADDGENE plasmid repository. In addition, we acknowledge Valeria Cafardi, who designed the lpp guide DNAs for the initial experiments and Dennis Pedri for supporting mutant analysis by PCR screening. Furthermore, we gratefully thank Yaqoub Ashhab for bioinformatic support and Samine Isaac for proof-reading the manuscript.

Competing interests

The authors declare that they have no competing interests.

Funding

The project has been supported by the Advanced ERC Grant OMVac 340915 assigned to Guido Grandi.

References

1. Zhang Y, et al. A new logic for DNA engineering using recombination in *Escherichia coli*. Nat Genet. 1998;20(2):123–8.
2. Datta S, Costantino N, Court DL. A set of recombineering plasmids for gram-negative bacteria. Gene. 2006;379:109–15.
3. Murphy KC. Use of bacteriophage lambda recombination functions to promote gene replacement in *Escherichia coli*. J Bacteriol. 1998;180(8):2063–71.
4. Muyrers JP, et al. Rapid modification of bacterial artificial chromosomes by ET-recombination. Nucleic Acids Res. 1999;27(6):1555–7.
5. Ellis HM, et al. High efficiency mutagenesis, repair, and engineering of chromosomal DNA using single-stranded oligonucleotides. Proc Natl Acad Sci USA. 2001;98(12):6742–6.
6. Yu D, et al. An efficient recombination system for chromosome engineering in *Escherichia coli*. Proc Natl Acad Sci USA. 2000;97(11):5978–83.
7. Yu D, et al. Recombineering with overlapping single-stranded DNA oligonucleotides: testing a recombination intermediate. Proc Natl Acad Sci USA. 2003;100(12):7207–12.
8. Sharan SK, et al. Recombineering: a homologous recombination-based method of genetic engineering. Nat Protoc. 2009;4(2):206–23.
9. Jiang W, et al. RNA-guided editing of bacterial genomes using CRISPR-Cas systems. Nat Biotechnol. 2013;31(3):233–9.
10. Pyne ME, et al. Coupling the CRISPR/Cas9 system with lambda red recombineering enables simplified chromosomal gene replacement in *Escherichia coli*. Appl Environ Microbiol. 2015;81(15):5103–14.
11. Doudna JA, Charpentier E. Genome editing. The new frontier of genome engineering with CRISPR-Cas9. Science. 2014;346(6213):1258096.
12. Sternberg SH, Doudna JA. Expanding the Biologist's Toolkit with CRISPR-Cas9. Mol Cell. 2015;58(4):568–74.
13. Singh V, Braddick D, Dhar PK. Exploring the potential of genome editing CRISPR-Cas9 technology. Gene. 2017;599:1–18.
14. Choi KR, Lee SY. CRISPR technologies for bacterial systems: current achievements and future directions. Biotechnol Adv. 2016;34(7):1180–209.
15. Barrangou R, van Pijkeren JP. Exploiting CRISPR-Cas immune systems for genome editing in bacteria. Curr Opin Biotechnol. 2016;37:61–8.
16. Jakociunas T, Jensen MK, Keasling JD. CRISPR/Cas9 advances engineering of microbial cell factories. Metab Eng. 2016;34:44–59.
17. Mougiakos I, et al. Next generation prokaryotic engineering: the CRISPR-Cas toolkit. Trends Biotechnol. 2016;34(7):575–87.
18. Reisch CR, Prather KL. The no-SCAR (scarless Cas9 assisted recombineering) system for genome editing in *Escherichia coli*. Sci Rep. 2015;5:15096.
19. Bassalo MC, et al. Rapid and efficient one-step metabolic pathway integration in *E. coli*. ACS Synth Biol. 2016;5(7):561–8.
20. Derbise A, et al. A rapid and simple method for inactivating chromosomal genes in Yersinia. FEMS Immunol Med Microbiol. 2003;38(2):113–6.
21. Gay P, et al. Positive selection procedure for entrapment of insertion sequence elements in gram-negative bacteria. J Bacteriol. 1985;164(2):918–21.
22. Gay P, et al. Cloning structural gene sacB, which codes for exoenzyme levansucrase of *Bacillus subtilis*: expression of the gene in *Escherichia coli*. J Bacteriol. 1983;153(3):1424–31.
23. Steinmetz M, et al. Genetic analysis of sacB, the structural gene of a secreted enzyme, levansucrase of *Bacillus subtilis* Marburg. Mol Gen Genet. 1983;191(1):138–44.
24. Murphy KC, Campellone KG, Poteete AR. PCR-mediated gene replacement in *Escherichia coli*. Gene. 2000;246(1–2):321–30.
25. Hale L, et al. An efficient stress-free strategy to displace stable bacterial plasmids. Biotechniques. 2010;48(3):223–8.
26. Shuman S, Glickman MS. Bacterial DNA repair by non-homologous end joining. Nat Rev Microbiol. 2007;5(11):852–61.
27. Wright DG, et al. *Mycobacterium tuberculosis* and *Mycobacterium marinum* non-homologous end-joining proteins can function together to join DNA ends in *Escherichia coli*. Mutagenesis. 2016.
28. Cui L, Bikard D. Consequences of Cas9 cleavage in the chromosome of *Escherichia coli*. Nucleic Acids Res. 2016;44(9):4243–51.
29. Baba T, et al. Construction of *Escherichia coli* K-12 in-frame, single-gene knockout mutants: the Keio collection. Mol Syst Biol. 2006;2:2006 0008.

30. Jiang Y, et al. Multigene editing in the *Escherichia coli* genome via the CRISPR-Cas9 system. Appl Environ Microbiol. 2015;81(7):2506–14.
31. Court DL, Sawitzke JA, Thomason LC. Genetic engineering using homologous recombination. Annu Rev Genet. 2002;36:361–88.
32. Meddows TR, Savory AP, Lloyd RG. RecG helicase promotes DNA double-strand break repair. Mol Microbiol. 2004;52(1):119–32.
33. Altschul SF, et al. Basic local alignment search tool. J Mol Biol. 1990;215(3):403–10.
34. Fantappie L, et al. Antibody-mediated immunity induced by engineered *Escherichia coli* OMVs carrying heterologous antigens in their lumen. J Extracell Vesicles. 2014;3.
35. Klock HE, Lesley SA. The polymerase incomplete primer extension (PIPE) method applied to high-throughput cloning and site-directed mutagenesis. Methods Mol Biol. 2009;498:91–103.

10

Engineering *Halomonas* species TD01 for enhanced polyhydroxyalkanoates synthesis via CRISPRi

Wei Tao[1†], Li Lv[1†] and Guo-Qiang Chen[1,2,3,4,5*]

Abstract

Background: Clustered regularly interspaced short palindromic repeats interference (CRISPRi) has provided an efficient approach for targeted gene inhibition. A non-model microorganism *Halomonas* species TD01 has been developed as a promising industrial producer of polyhydroxyalkanoates (PHA), a family of biodegradable polyesters accumulated by bacteria as a carbon and energy reserve compound. A controllable gene repression system, such as CRISPRi, is needed for *Halomonas* sp. TD01 to regulate its gene expression levels.

Results: For the first time CRISPRi was successfully used in *Halomonas* sp. TD01 to repress expression of *ftsZ* gene encoding bacterial fission ring formation protein, leading to an elongated cell morphology with typical filamentous shape similar to phenomenon observed with *Escherichia coli*. CRISPRi was employed to regulate expressions of *prpC* gene encoding 2-methylcitrate synthase for regulating 3-hydroxyvalerate monomer ratio in PHBV copolymers of 3-hydroxybutyrate (HB) and 3-hydroxyvalerate (HV). Percentages of HV in PHBV copolymers were controllable ranging from less than 1 to 13%. Furthermore, repressions on *gltA* gene encoding citrate synthase channeled more acetyl-CoA from the tricarboxylic acid (TCA) cycle to poly(3-hydroxybutyrate) (PHB) synthesis. The PHB accumulation by *Halomonas* sp. TD01 with its *gltA* gene repressed in various intensities via CRISPRi was increased by approximately 8% compared with the wild type control containing the CRISPRi vector without target.

Conclusions: It has now been confirmed that the CRISPRi system can be applied to *Halomonas* sp. TD01, a promising industrial strain for production of various PHA and chemicals under open and continuous fermentation process conditions. In details, the CRISPRi system was successfully designed in this study to target genes of *ftsZ*, *prpC* and *gltA*, achieving longer cell sizes, channeling more substrates to PHBV and PHB synthesis, respectively. CRISPRi can be expected to use for more metabolic engineering applications in non-model organisms.

Keywords: CRISPRi, PHBV, PHB, Synthetic biology, *ftsZ*, *gltA*, *prpC*

Background

The CRISPRi (clustered regularly interspaced short palindromic repeats interference) system provides an efficient method for targeted gene repression [1]. Deriving from the CRISPR/Cas9 system, the CRISPRi system contains a dCas9 protein co-expressed with a small guide RNA (sgRNA) [1–3]. The Cas9 protein is an RNA-guided DNA endonuclease. In the CRISPR system, the Cas9 protein binds to the sgRNA and form a protein-RNA complex, which will then bind to the targeted DNA sequence. The DNA will be cleaved by the catalytically active Cas9 protein [4]. Mutations in the Cas9 protein result in a catalytically dead dCas9 protein with DNA binding capability [1]. Therefore, the dCas9/sgRNA complex can bind to specific DNA target depending on the designed sequence of sgRNA, block transcriptional elongation, interfere RNA polymerase or transcriptional factor binding [1].

CRISPRi enables convenient and specific gene regulation in microbial metabolic engineering [1]. In our early study, CRISPRi was successfully used in *Escherichia coli*

*Correspondence: chengq@mail.tsinghua.edu.cn
†Wei Tao and Li Lv contributed equally to this work
[1] School of Life Sciences, Tsinghua University, Beijing 100084, China
Full list of author information is available at the end of the article

for regulating polyhydroxyalkanoates (PHA) production via simultaneously repressing multiple genes or multiple targets on one gene [5].

PHA are polyesters synthesized by a wide range of bacteria as carbon and energy source reserves [6]. PHA has been developed into various environmentally friendly plastic products, and it has been formed into an application value chain [7–9]. PHA can be classified into short-chain-length (scl) PHA and medium-chain-length (mcl) PHA [10]. Poly(3-hydroxybutyrate) (PHB) and poly(3-hydroxybutyrate-co-3-hydroxyvalerate) (PHBV) are common PHA already produced in large scale [9]. PHA production cost is still too high compared with petrochemical plastics that are not biodegradable [11].

Halomonas sp. TD01 is a halophile screened from Aydingol Lake in Xinjiang Province, China [12]. It can be grown under conditions of high salt concentrations and high pH, allowing a continuous and open fermentation without contamination. The genome of *Halomonas* sp. TD01 was sequenced and some genetic manipulation technologies have been developed for DNA manipulation [13, 14]. The absence of controllable repression system for gene expression has slowed down more applications for *Halomonas* sp. TD01.

FtsZ is a tubulin-like protein that is of great importance in the cell division process [15, 16]. *FtsZ* assembles to form Z rings in a dynamic state during the cell division process. *FtsZ* inhibition or deletion in *E. coli* leads to cell division repression and results in formation of filamentous cells from bar or spherical shapes [17, 18].

In *Halomonas* sp. TD01, propionic acid is transformed into propionyl-CoA, which can be further catalyzed by 2-methylcitrate synthase to form 2-methylcitrate and then enters the methyl citric acid cycle (MCC cycle) [19]. Propionyl-CoA can also enter the PHBV synthesis pathway to form 3-hydroxyvalerate monomers. 2-Methylcitrate synthase is encoded by *prpC* gene. It was expected that repressions on *prpC* should divert more propionyl-CoA to PHBV synthesis [20, 21].

The tricarboxylic acid (TCA) cycle provides energy and intermediates for synthesis of many important biological compounds [22]. In TCA cycle, acetyl-CoA and oxaloacetate are converted to citrate by citrate synthase encoded by *gltA* gene [22, 23]. Acetyl-CoA is the substrate for PHB synthesis, a high concentration of acetyl-CoA is of great importance to PHB production. Repressions of *gltA* gene should decrease acetyl-CoA consumption by the TCA cycle, therefore, improving substrate conversion to PHB synthesis.

In this study, it was aimed to exploit CRISPRi for enhanced PHA production by engineering *Halomonas* sp. TD01 to achieve regulation of its expression levels of various genes, including *ftsZ*, *prpC* and *gltA*.

Results

Feasibility study of the constructed CRISPRi for *Halomonas* sp. TD01

Gene *ftsZ* encoding bacterial fission ring protein was selected as a reporter gene for feasibility study of CRISPRi system for *Halomonas* sp. TD01. During the bacterial cell division process, FtsZ assembly leads to the formation of Z rings in the middle of a cell. FtsZ inhibitors interact with FtsZ in cytokinesis repressing the cell division progression, resulting in formation of filamentous cells [17, 18].

The sgRNAs were designed in the promoter region or near the ATG sequence in the targeted gene, and were right after a NGG sequence, namely, PAM sequence (protospacer adjacent motif sequence) [4]. All the sgRNAs could bind to the non-template DNA strand with sequence specificity. Thus, two sgRNAs were designed near the ATG sequence in *ftsZ* gene (Fig. 2b). CRISPRi inhibition systems pli-dCas9-ftsZ1 and pli-dCas9-ftsZ2 were constructed. The plasmids were then transferred via *E. coli* conjugation into *Halomonas* sp. TD01, forming the recombinants *Halomonas* sp. TD-ftsZ1 and TD-ftsZ2 strains. *Halomonas* sp. TD01 containing the non-target plasmid pli-dCa9-sgRNA, was named *Halomonas* sp. TD-sgRNA strain. Wild type *Halomonas* sp. TD01 and *Halomonas* sp. TD-sgRNA were used as control groups.

Growth curves of the strains were determined to observe whether cell growth was affected by inhibition of *ftsZ* gene (Additional file 1: Figure S1). Compared with the wild type *Halomonas* sp. TD01 and TD-sgRNA control groups, *Halomonas* sp. TD-ftsZ1 and TD-ftsZ2 exhibited long lag growth and lower cell density, implying that *ftsZ* gene inhibition decreased cell growth rate.

All the strains were cultured in the MM medium with 30 g/L glucose at 37 °C and 200 rpm. After 8 h cultivation, 1 mM IPTG was added for inducing the CRISPRi system. Cells were harvested after an overall cultivation of 48 h. Under an environmental scanning electron microscope (ESEM), *Halomonas* sp. TD-ftsZ1 and TD-ftsZ2 harboring the CRISPRi system showed elongated shapes compared with their controls *Halomonas* sp. TD01 and TD-sgRNA (Fig. 3a), demonstrating that the cell division process was effectively repressed via the CRISPRi. Lengths of bacteria in *Halomonas* sp. TD01 and TD-sgRNA control groups were approximately 1 μm, while lengths of bacteria in *Halomonas* sp. TD-ftsZ1 and TD-ftsZ2 groups varied from 20 to 70 μm, showing 20–70 folds increase in their cell lengths (Fig. 3b). The phenotype changes clearly indicated that the CRISPRi system had been successfully developed in *Halomonas* sp. TD01.

The uses of *Halomonas* CRISPRi system for controlling PHBV monomer ratios

Halomonas sp. TD01 is able to produce PHBV by adding the substrate propionic acid in the presence of glucose [14]. The PHBV synthesis involves the conversion of propionic acid to propionyl-CoA and subsequent transformation into 3-hydroxyvaleryl-CoA by β-ketothiolase (PhaA) and NADPH-dependent acetoacetyl-CoA reductase (PhaB). PHA synthase (PhaC) polymerizes 3-hydroxyvaleryl-CoA with 3-hydroxybutyryl-CoA to form PHBV (Fig. 1). Propionyl-CoA directly leads to the 3HV monomers in PHBV. In *Halomonas* sp. TD01, 3HV monomer amounts in PHBV was very low due to the rapid conversion of propionyl-CoA to 2-methylcitrate, which is converted to the methylcitric acid cycle (MCC cycle) [14]. In *Halomonas* sp. TD01, 2-methylcitrate synthase is encoded by *prpC* gene. *PrpC* knockout *Halomonas* sp. TD01 showed an accumulation of PHBV with a higher 3HV ratio when in presence of 1 g/L propionic acid. However, in this condition, the cells grew poorly and the cell dry weight (CDW) was very low [14]. Therefore, it is expected that the use of a CRISPRi repression system targeting *prpC* gene in *Halomonas* sp. TD01 could channel more propionic acid, or propionyl-CoA, to 3HV monomer in PHBV synthesis without impairing cell growth (Fig. 1).

As shown in Fig. 4a, seven sgRNAs were designed along the *prpC* gene, mostly near the start site of the coding sequence or within 260 bp from the ATG sequence. CRISPRi plasmids are constructed and named as pli-dCas9-prpC1, pli-dCas9-prpC2, pli-dCas9-prpC3, pli-dCas9-prpC4, pli-dCas9-prpC5, pli-dCas9-prpC6 and pli-dCas9-prpC7, respectively (Additional file 1: Table S1; Fig. 4a). Various sgRNA binding sites led to different repressive effects. Two effective inhibition sites, namely prpC6 and prpC7, were combined together to form pli-dCas9-prpC6prpC7 to verify possible enhanced combinatory inhibition effect. All the plasmids were transferred via *E. coli* conjugation into *Halomonas* sp. TD01, forming strains TD-prpC1, TD-prpC2, TD-prpC3, TD-prpC4, TD-prpC5, TD-prpC6, TD-prpC7 and TD-prpC6prpC7, respectively.

To study the impact of *prpC* gene repression on cell growth, growth curves of the strains were established (Additional file 1: Figure S2). Wild type *Halomonas* sp. TD01 and TD-sgRNA were used as the control groups. The growth trends of the bacteria in all study groups were nearly the same, indicating that the CRISPRi system functioned without affecting cell growth.

All the strains were cultured in the MM medium in the presence of 1 g/L propionic acid and 30 g/L glucose at 37 °C and 200 rpm. After 12 h cultivation, 1 mM IPTG was added to induce the CRISPRi system. All bacteria were harvested after an overall cultivation time of 48 h (Table 1). All the strains grew well with CDW reaching approximately 14 g/L and accumulating approximately 75% PHBV. Meanwhile, *Halomonas* sp. TD01 and TD-sgRNA strains were used as controls, accumulating around 1% 3HV in PHBV copolymer. The percentage of 3HV monomer in PHBV copolymer varied from less than 1% to nearly 13% depending on the *prpC* repression intensity. In the single sgRNA inhibition system, *Halomonas* strains TD-prpC2, TD-prpC3, TD-prpC4 and TD-prpC5 all showed an obvious improvement in 3HV ratio compared to the controls. *Halomonas* strain TD-prpC2 accumulated nearly 5% 3HV in the PHBV, while 3HV ratios in PHBV copolymer accumulated by *Halomonas* strains TD-prpC3 and TD-prpC4 were around 6%. 3HV ratio in the PHBV produced by *Halomonas* strain TD-prpC5 was five times higher than that in the

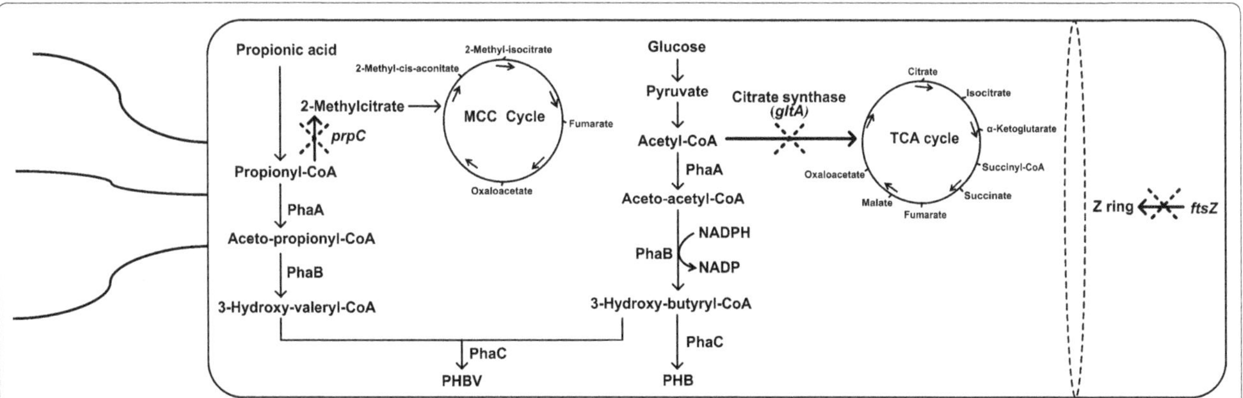

Fig. 1 PHBV and PHB metabolic pathways. *Dotted crosses* show the repressed pathways in *Halomonas* sp. TD01 when targeting the indicated genes using the CRISPRi repression system. MMC, methylcitric acid cycle; *prpC*, 2-methylcitrate synthase; *phaA*, β-ketothiolase; *phaB*, NADPH-dependent acetoacetyl-CoA reductase; *phaC*, PHA synthase gene; TCA, tricarboxylic acid cycle; *ftsZ*, filamenting temperature-sensitive mutant Z

Table 1 Shake flask PHBV production by recombinants *Halomonas* sp. TD01 controlling *prpC* gene expression via CRISPRi

Recombinant TD01 strains	CDW (g/L)	PHBV (wt%)	3HV (mol%)
TD01	13.22 ± 0.20	76.08 ± 2.81	0.83 ± 0.08
TD-sgRNA	13.95 ± 0.71	74.73 ± 2.25	1.29 ± 0.47
TD-prpC1	14.99 ± 0.30	72.14 ± 2.37	1.79 ± 0.02
TD-prpC2	14.43 ± 0.45	75.06 ± 3.20	4.78 ± 0.83
TD-prpC3	14.08 ± 1.26	73.96 ± 7.18	5.72 ± 0.30
TD-prpC4	13.58 ± 0.57	80.14 ± 8.62	6.44 ± 0.43
TD-prpC5	13.59 ± 0.51	80.12 ± 3.27	8.16 ± 0.31
TD-prpC6	14.17 ± 0.23	82.20 ± 3.82	11.94 ± 0.80
TD-prpC7	13.54 ± 0.31	76.74 ± 0.80	12.15 ± 0.31
TD-prpC6prpC7	14.67 ± 0.78	73.78 ± 4.82	12.70 ± 0.27

The recombinants harboring CRISPRi system were cultivated in MM medium containing 30 g/L glucose and 1 g/L propionic acid at 37 °C for 48 h as described in "Methods". CDW, cell dry weight; PHBV (wt%), the weight percent of PHBV in CDW; TD01, *Halomonas* sp. TD wild type strain; TD-sgRNA, TD01 strain harboring the pli-dCas9-sgRNA plasmid without any DNA target site; TD-prpC1, TD-prpC2, TD-prpC3, TD-prpC4, TD-prpC5, TD-prpC6, TD-prpC7, TD-prpC6prpC7, TD01 strains harboring the pli-dCas9-sgRNA plasmid with different sgRNA targets of gene *prpC*, respectively. All data were the average of three independent studies with standard deviations. Mean ± SE (n = 3)

control groups, reaching 8% (Table 1). *Halomonas* strains TD-prpC6 and TD-prpC7 exhibited much higher repression efficiency, with around 12% 3HV in PHBV copolymer. In the two targets inhibition system, *Halomonas* strain TD-prpC6prpC7 produced PHBV copolymer consisting of 12.7% 3HV monomer (Table 1). The combination of prpC6 and prpC7 targets showed a slightly enhanced repression effect compared with that of the individual single sgRNA repression systems.

The mRNA expression levels RT-PCR agreed with the shake flask results (Fig. 4b). Wild type *Halomonas* sp. TD01 and the control TD-sgRNA showed a similar *prpC* mRNA expression level, indicating that the non-target system did not influence the gene expression in the strain *Halomonas* sp. TD-sgRNA. Strains *Halomonas* sp. TD-prpC1 and TD-prpC2 showed a higher mRNA expression level compared with strains TD-prpC3, TD-prpC4 and TD-prpC5, while strains TD-prpC6, TD-prpC7 and TD-prpC6prpC7 revealed a much lower mRNA expression level. This phenomenon again demonstrated the clear feasibility of the CRISPRi system for *Halomonas* sp. TD01. In addition, considering that *prpC* mRNA expression level in *Halomonas* strains TD-prpC6 and TD-prpC7 were already repressed to a very low level, it was hypothesized that further repression could hardly be expected when combining prpC6 and prpC7 inhibition targets. Thus the 3HV ratio in PHBV copolymer produced by *Halomonas* strain TD-prpC6prpC7 was almost the same as that in *Halomonas* strain TD-prpC6 or TD-prpC7.

The uses of *Halomonas* CRISPRi system for enhanced PHB synthesis

Halomonas sp. TD01 is able to produce PHB using glucose as substrate [13]. Acetyl-CoA, as a PHB substrate, is generated from pyruvate after glycolysis and oxidized in the TCA cycle. The citrate synthase catalyzes conversion of acetyl-CoA and oxaloacetate to citrate (Fig. 1). Gene *gltA* encoding citrate synthase, can not be completely repressed in *Halomonas* sp. TD01. However, partial repression on *gltA* should decrease the consumption of acetyl-CoA for TCA cycle and thus save some acetyl-CoA substrate for PHB synthesis (Fig. 1).

Four sgRNAs targeting *gltA* were designed. GltA1 was located in the promoter region, gltA2 sequence started with ATG, while gltA3 was just three base pairs away from gltA2. Finally, gltA4 was designed within 200 bp from the ATG sequence. The respective plasmids were constructed and named as pli-dCas9-gltA1, pli-dCas9-gltA2, pli-dCas9-gltA3 and pli-dCas9-gltA4. They were transferred into *Halomonas* sp. TD01 via *E. coli* conjugation, forming *Halomonas* strains TD-gltA1, TD-gltA2, TD-gltA3 and TD-gltA4.

Growth curves of the strains were established to investigate the effect of *gltA* inhibition on cell growth (Additional file 1: Figure S3). The bacterial growth curves showed that *Halomonas* strains TD-gltA1, TD-gltA2, TD-gltA3 and TD-gltA4 exhibited long lag growth phase in the beginning of the culture compared with the growth curves of *Halomonas* sp. TD01 and TD-sgRNA strains. Also, a similar cell density was reached after 15 h of cultivation, indicating that *gltA* repression prolonged the lag phase in cell growth, yet it did not affect cell density after the overnight cultivation.

All strains were cultured in the MM medium with 30 g/L glucose at 37 °C and 200 rpm. After 12 h cultivation, 1 mM IPTG was added to induce the CRISPRi system. Compared with the control *Halomonas* sp. TD-sgRNA, PHB content in *Halomonas* sp. TD-gltA2 showed a nearly 8% improvement, while strain TD-gltA3 had a 5% increase (Table 2). The mRNA expression levels from RT-PCR demonstrated that the *gltA* was repressed in the recombinant strains harboring the CRISPRi inhibition system targeting *gltA* gene (Additional file 1: Figure S4). Wild type *Halomonas* sp. TD01 and TD-sgRNA showed a similar *gltA* mRNA expression level, yet *Halomonas* strains TD-gltA1, TD-gltA2, TD-gltA3 and TD-gltA4 all showed a decreased *gltA* mRNA expression level. Therefore, partial CRISPRi repression on *gltA* had indeed reduced the consumption of acetyl-CoA thus improved PHB production in *Halomonas* sp. TD-sgRNA. Once again, *gltA* repressions demonstrated that the CRISPRi system was useful for metabolic engineering of *Halomonas* sp. TD01.

Table 2 Shake flask PHB production by recombinants *Halomonas* sp. TD01 with controllable *gltA* gene expression via CRISPRi

Recombinant TD01 strains	CDW (g/L)	PHB (wt%)
TD01	10.22 ± 0.25	77.68 ± 3.75
TD-sgRNA	13.28 ± 0.57	63.80 ± 3.08
TD-gltA1	13.68 ± 0.21	66.79 ± 1.59
TD-gltA2	13.53 ± 0.41	71.77 ± 7.16
TD-gltA3	13.18 ± 0.33	69.22 ± 3.43
TD-gltA4	13.10 ± 0.42	66.16 ± 8.11

All strains were cultivated in MM medium containing 30 g/L glucose at 37 °C for 48 h as described in "Methods". CDW, cell dry weight; PHBV (wt%), the weight percent of PHBV in CDW; TD01, *Halomonas* sp. TD wild type strain; TD-sgRNA, TD01 strain harboring the pli-dCas9-sgRNA plasmid without any DNA target site; TD-gltA1, TD-gltA2, TD-gltA3, TD-gltA4, TD01 strains harboring the pli-dCas9-sgRNA plasmid with different sgRNA targets of gene *gltA*, respectively. All data were the average of three independent studies with standard deviations. Mean ± SE (n = 3)

Discussion

Halomonas sp. TD01 has been demonstrated to be a promising strain for PHA production due to its tolerance to high pH and high salt concentration. Therefore, allowing an open and continuous fermentation process without contamination [13]. This improves competitiveness of *Halomonas* sp. TD01 based PHA production process [24, 25]. However, as a non-model microorganism, *Halomonas* sp. TD01 still requires an inducible gene repression system for better performances.

CRISPRi has been used to regulate expression of desired genes without affecting the normal growth of engineered cells [1, 26–28]. In this study, an effective CRISPRi platform for genome editing in *Halomonas* sp. TD01 was developed. Pli-dCas9-sgRNA suitable for *Halomonas* sp. TD01 was designed to insert various sgRNAs, which form numerous CRISPRi repression plasmids. Multiple sgRNAs could also be inserted into pli-dCas9-sgRNA (Fig. 2a). The IPTG inducible CRISPRi system functioned well in *Halomonas* sp. TD01 for manipulating gene expression levels, achieving elongation of cell sizes, controllable PHBV copolymer monomer ratios, and enhanced PHB synthesis (Figs. 3, 4; Tables 1, 2).

FtsZ plays a crucial role in cytokinesis and assembles to form the Z rings during the cell division process [17]. Inhibition of *ftsZ* gene in *Halomonas* sp. TD01 resulted in formation of filamentous cells compared with short bar wild type of control groups (Fig. 3a). This phenomenon demonstrated the effectiveness of the established CRISPRi system for *Halomonas* sp. TD01, even though

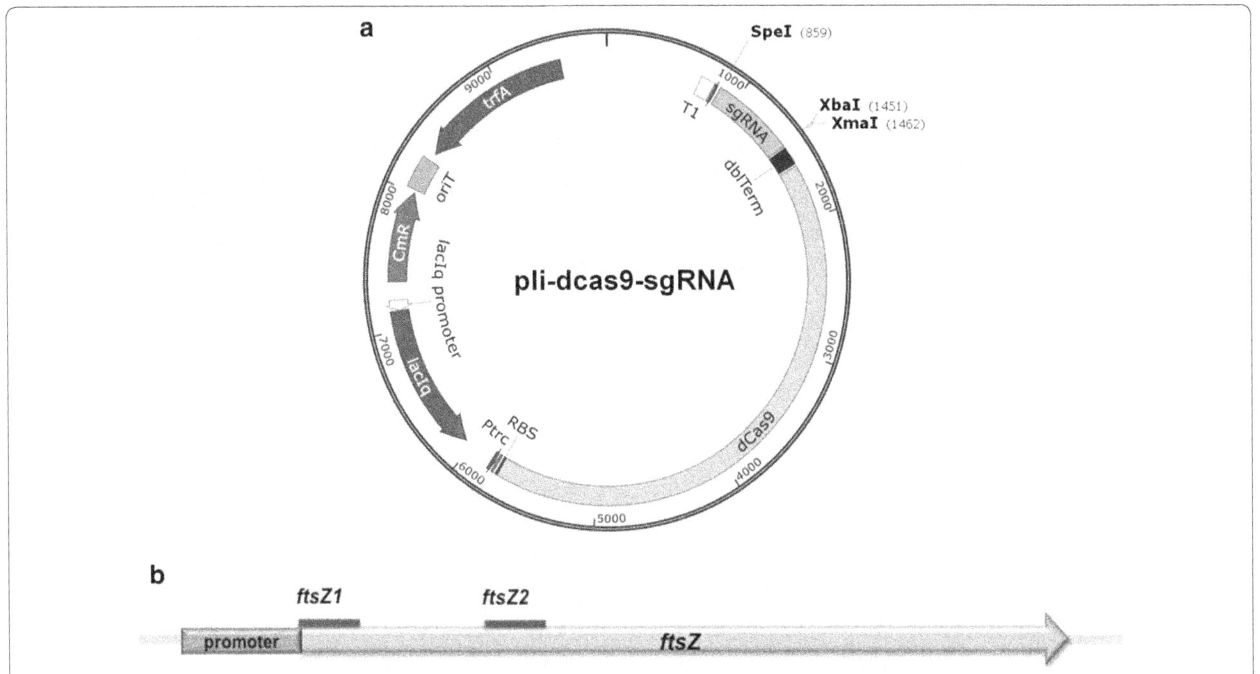

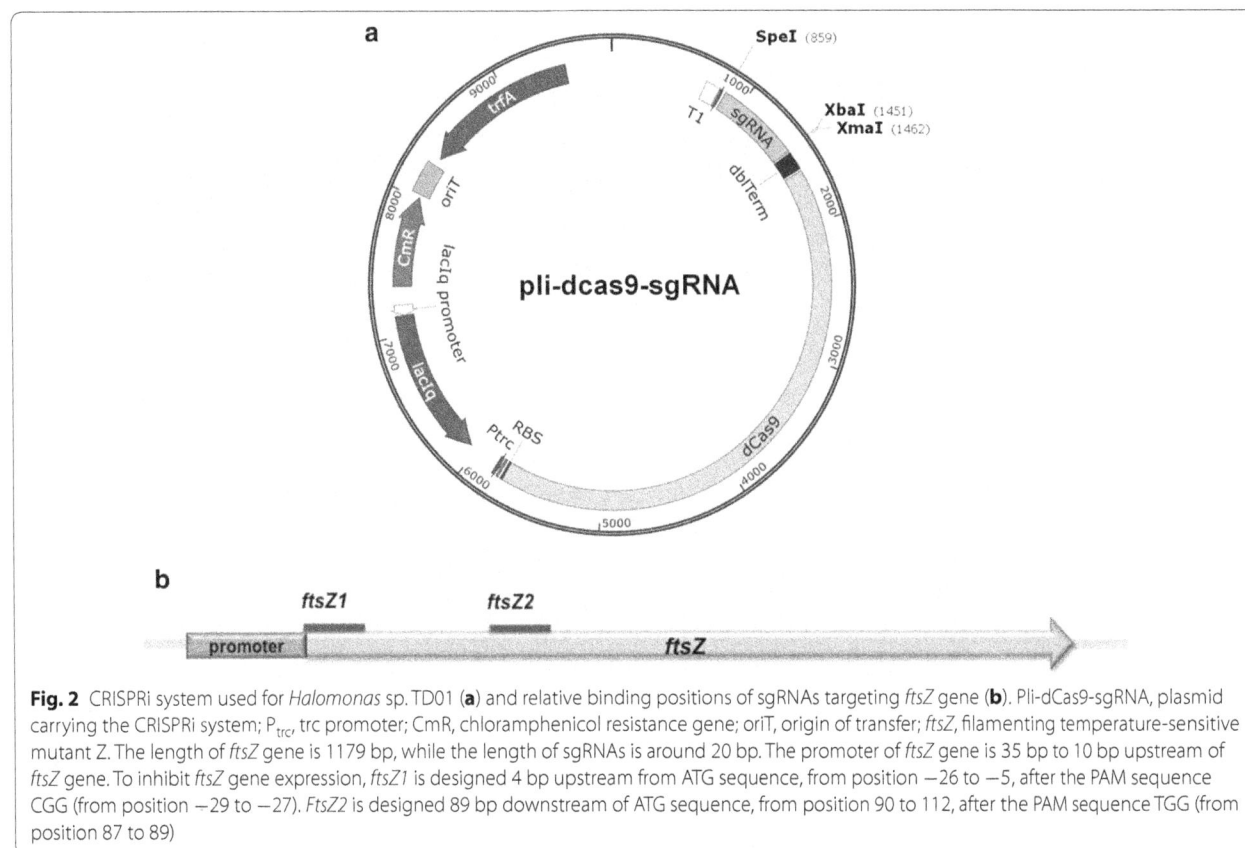

Fig. 2 CRISPRi system used for *Halomonas* sp. TD01 (**a**) and relative binding positions of sgRNAs targeting *ftsZ* gene (**b**). Pli-dCas9-sgRNA, plasmid carrying the CRISPRi system; P_{trc}, trc promoter; CmR, chloramphenicol resistance gene; oriT, origin of transfer; *ftsZ*, filamenting temperature-sensitive mutant Z. The length of *ftsZ* gene is 1179 bp, while the length of sgRNAs is around 20 bp. The promoter of *ftsZ* gene is 35 bp to 10 bp upstream of *ftsZ* gene. To inhibit *ftsZ* gene expression, *ftsZ1* is designed 4 bp upstream from ATG sequence, from position −26 to −5, after the PAM sequence CGG (from position −29 to −27). *FtsZ2* is designed 89 bp downstream of ATG sequence, from position 90 to 112, after the PAM sequence TGG (from position 87 to 89)

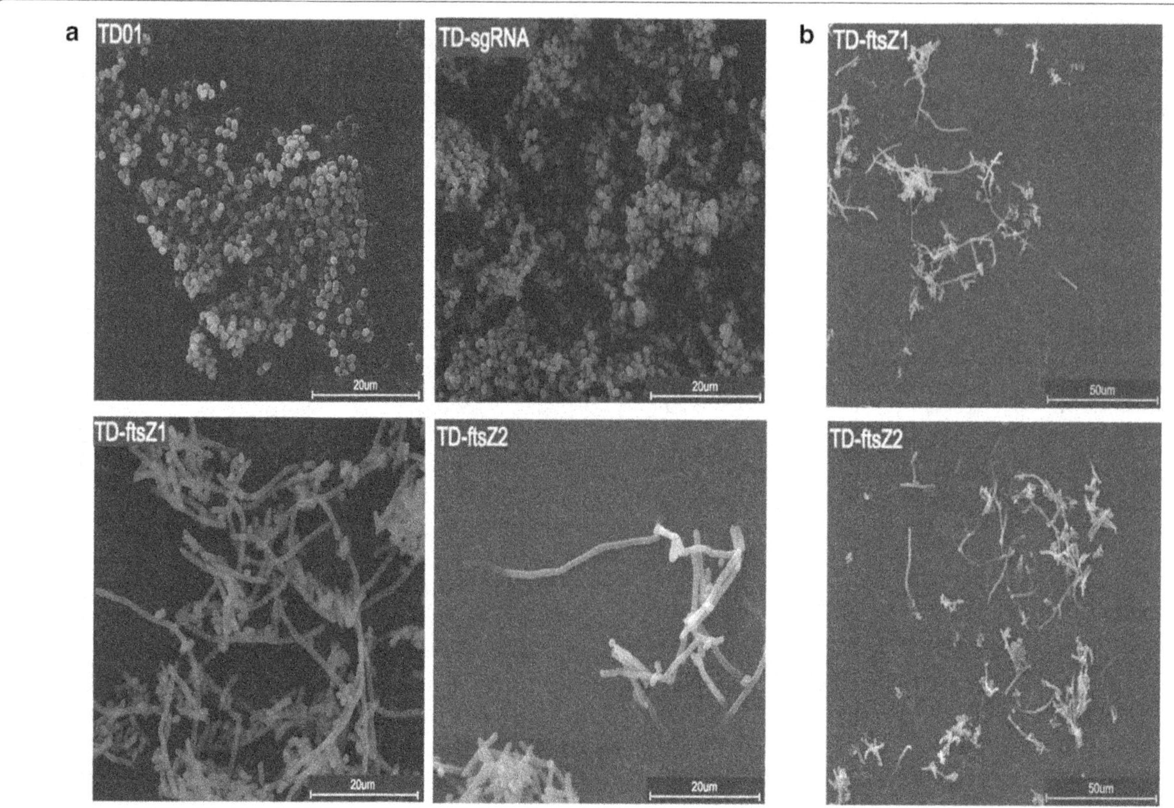

Fig. 3 Scanning electron microscopy study on *Halomonas* sp. TD01 with its *ftsZ* gene repressed via CRISPRi under 5000 (**a**) and 2000 (**b**) times magnification. TD01: wild type *Halomonas* sp. TD used as a control group; TD-sgRNA, TD01 strain harboring the pli-dCas9-sgRNA plasmid without any DNA target site; TD-ftsZ1, TD-ftsZ2, TD01 strain harboring the CRISPRi plasmids pli-dCas9-ftsZ1 and pli-dCas9-ftsZ2 that regulated the expression level of fission ring protein *ftsZ* gene, respectively. *Scale bars* are 50 or 20 μm, as indicated

some *Halomonas* cells in the experimental groups still maintained their short bar shape possibly due to plasmid instability (Fig. 3a). Eventually, it is important to integrate the CRISPRi elements into the genome of *Halomonas* sp. TD01 to lower metabolic pressure on the bacteria, and to reduce the cost of antibiotics, as well as to improve the plasmid stability.

Halomonas sp. TD01 is able to produce PHBV using propionic acid as the precursor [14]. By increasing propionic acid concentration in the culture medium, *Halomonas* sp. TD01 can accumulate PHBV with a slightly improved ratio of 3HV (3% 3HV per 1 g/L propionic acid) [14]. The *prpC* gene knockout *Halomonas* sp. TD01 was very sensitive to 1 g/L propionic acid, growing slowly yet accumulated PHBV with higher 3HV content [14]. In this study, we achieved to regulate *prpC* gene at different expression levels without affecting cell growth (Table 1): all the recombinants and the controls grew well reaching approximately 14 g/L CDW that contains around 75% PHBV in the presence of 1 g/L propionic acid. Different CRISPRi inhibition targets resulted in variable repression effects (Table 1). Among the *prpC* gene inhibition targets, prpC6, prpC7 and their combined inhibition sites led to the highest improvement on 3HV ratio in PHBV copolymers (Table 1), reaching around 12–13% 3HV in the PHBV. Compared with *prpC* gene knockout approach which generates a fixed 3HV ratio in the PHBV, the CRISPRi platform provided a flexible regulation on 3HV contents in the PHBV (Table 1).

TCA cycle plays a crucial role in cell growth [22]. Genes involved in the TCA cycle are essential, as is the case of *gltA* gene, and therefore, they cannot be deleted from the genome of *Halomonas* sp. TD01. Thus, CRISPRi was employed to partially repress *gltA* expression under controlled intensities, allowing reduced consumption of acetyl-CoA for TCA cycle and diverging more acetyl-CoA to improve PHB production. Therefore, the constructed CRISPRi system can be used to regulate essential gene expression in *Halomonas* sp. TD01 for achieving multiple metabolic engineering goals.

Conclusions

A CRISPRi system dedicated to the non-model organism *Halomonas* sp. TD01 was successfully constructed and proven feasible, as evidenced by changing cell morphology, copolymer PHBV structures and homopolymer

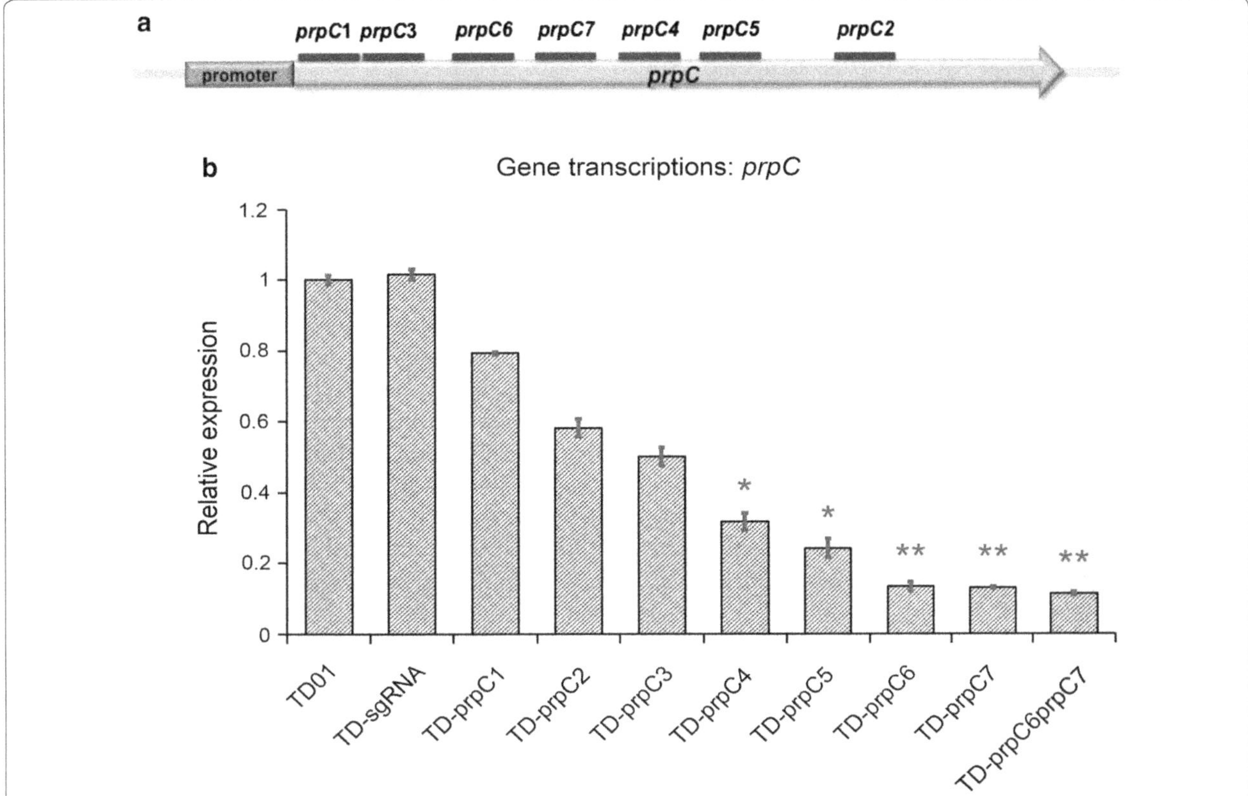

Fig. 4 Controllable repression of *prpC* gene transcription in recombinant *Halomonas* sp. TD01. The relative binding positions of sgRNAs targeting *prpC* gene (**a**) and RT-PCR study of *prpC* transcription levels (**b**). All data were the average of three independent studies with standard deviations. Mean ± SE (n = 3). *p < 0.05 and **p < 0.01. TD01, *Halomonas* sp. TD wild type; TD-sgRNA, TD01 strain harboring the pli-dCas9-sgRNA plasmid without any target site; TD-prpC1, TD-prpC2, TD-prpC3, TD-prpC4, TD-prpC5, TD-prpC6, TD-prpC7, TD-prpC6prpC7, TD01 strains harboring the pli-dCas9-sgRNA plasmid with different DNA targets on gene *prpC*

PHB synthesis when the genes *ftsZ*, *prpC* or *gltA* were repressed under different intensities. Considering the promising application prospect of *Halomonas* sp. TD01 for PHA and the chemical industry, the established CRISPRi system is expected to be useful for more metabolic engineering applications.

Methods

Strains, plasmids and culture conditions

Halomonas sp. TD01 was isolated from Aydingol Lake of Xinjiang Province, China, and stored in CGMCC (China General Micro-biological Culture Collection Center, Beijing). The collection number is 4353. *E. coli* S17-1 was used as a vector donor strain in conjugation. *E. coli* S17-1 was cultured in LB-20 medium. The ingredients of LB-20 medium are (g/L): 20 NaCl, 10 tryptone, 5 yeast extract. *Halomonas* sp. TD01 and its derivative strains were all cultivated in LB-60 medium. The ingredients of LB-60 medium are (g/L): 60 NaCl, 10 tryptone, 5 yeast extract. Chloramphenicol concentration used in this study was 25 μg/mL. All the strains and plasmids used in this study are listed (Additional file 1: Table S1).

Construction of recombinant strains

Plasmid construction

The plasmids used in this study are listed in Additional file 1: Table S1. Molecular cloning experiments were carried out according to manufacturers' instructions or standard procedures. Kits for DNA purification and isolation of high quality plasmids were purchased from Qiagen (Shanghai, China). Restriction enzymes and DNA modification enzymes were provided by New England Biolabs (USA).

Based on plv-dCas9-sgRNA constructed in our previous study and pSEVA321 (kindly donated by Dr. Victor de Lorenzo of CSIC, Spain) [5, 29], a new plasmid termed pli-dCas9-sgRNA was successfully constructed, containing the dCas9 protein, restriction enzyme sites for sgRNA sequence insertion, P_{trc} promoter, RK2 origin, chloramphenicol resistance selection marker, and the origin of transfer (oriT) for conjugation (Fig. 2a). Compared with the P_{tet} promoter induction system in our previous study [5], P_{trc} promoter in plv-dCas9-sgRNA plasmid was found to be more effective as P_{tet} promoter could not function well in *Halomonas* sp. TD01. The IPTG inducible P_{trc} promoter was more sensitive for induction.

To construct pli-dCas9-sgRNA, DNA fragments containing the dCas9 and sgRNA domain were amplified from PCR using plv-dCas9-sgRNA as the template, and then inserted into pSEVA321, forming pSEVA-dCas9-sgRNA. Multiple cloning sites (MCS) including *XmaI*, *XbaI* and *SpeI* were introduced into pSEVA-dCas9-sgRNA to form pli-dCas9-sgRNA. *XmaI* and *XbaI* were introduced upstream of the sgRNA expression cassette, while *SpeI* was inserted downstream of it. *XbaI* and *SpeI* are isocaudomers that create the same cohesive end after digestion and are required for sgRNA biobrick assembly (Fig. 2). By digesting pli-dCas9-sgRNA1 vector and PCR fragment containing sgRNA2 from pli-dCas9-sgRNA2 with *XmaI/XbaI* and *XmaI/SpeI*, respectively, pli-dCas9-sgRNA1sgRNA2 plasmid was formed after ligation. In addition, *XmaI* and *XbaI* restriction sites were reconstructed for the next round of sgRNA biobrick insertion (Fig. 2). In this way, multiple sgRNA biobricks could be inserted into one vector backbone for manipulating multiple genes simultaneously.

The 20–23 bp sgRNA complementary sequence was designed via primers (Additional file 1: Table S2). The forward and reverse primers were annealed to be a double-stranded DNA fragment precisely fitting the pli-dCas9-sgRNA vector, which was cleaved by *BspQI* enzyme. The reconstructed plasmid harboring the designed sgRNA sequence was formed after ligation. This technique allowed convenient changes in the complementary region to suit any interested gene.

To construct pli-dCas9-ftsZ(1-2), pli-dCas9-prpC(1-7) and pli-dCas9-gltA(1-4), sgRNA primers were annealed by temperature gradient PCR, the PCR product was then cut by BspQI at 50 °C for 2 h and purified. The pli-dCas9-sgRNA vector was digested by BspQI at 50 °C for 4 h and purified via electrophoresis. The inhibition plasmid was formed after ligation of the vector and PCR product. To prepare electro-competent *E. coli* S17-1, 3 mL volume of overnight cell culture was collected after centrifugation, the cells were then washed twice with ice-cold 10% glycerol. The ligation product was transformed into *E. coli* S17-1 via electroporation. Cells after transformation were placed on a LB-20 plate in the presence of 25 μg/mL chloramphenicol and cultivated overnight. Positive colonies were verified by PCR. Subsequently, the constructed plasmid was transformed into *E. coli* S17-1 and then conjugated into *Halomonas* sp. TD01.

To construct pli-dCas9-prpC6prpC7, DNA fragment containing prpC6 sgRNA, *XmaI* and *SpeI* restriction sites were amplified from PCR as the insert DNA, using pli-dCas9-prpC6 as the template. The insert DNA fragment was then cut by *XmaI* and *SpeI*. Vector pli-dCas9-prpC7 was restricted using *XmaI* and *XbaI*. After ligation of the insert DNA and the vector, pli-dCas9-prpC6prpC7 was formed, containing the reconstructed *XmaI* and *XbaI* restriction sites. The plasmid was then conjugated from *E. coli* S17-1 to *Halomonas* sp. TD01.

Designing sgRNA for repressing *prpC* gene

The sgRNAs were designed in the promoter region or near the ATG sequence in the targeted gene, and were right after a NGG sequence, namely, PAM sequence (protospacer adjacent motif sequence) [4]. All the sgRNAs could bind to the non-template DNA strand with sequence specificity. As shown in Fig. 4a, seven sgRNAs were designed along the *prpC* gene, mostly near the start site of the coding sequence or within 260 bp from the ATG sequence. In details: sgRNAs *prpC1, prpC3, prpC6, prpC7, prpC4, prpC5* and *PrpC2* were designed 15 bp downstream from ATG sequence, from position 16 to 38, after the PAM sequence CGG (from position 13 to 15), 35 bp downstream from ATG sequence, from position 36 to 58, after the PAM sequence CGG (from position 33 to 35), 88 bp downstream from ATG sequence, from position 89 to 111, after the PAM sequence CGG (from position 86 to 88), 127 bp downstream from ATG sequence, from position 128 to 150, after the PAM sequence CGG (from position 125 to 127), 190 bp downstream from ATG sequence, from position 191 to 213, after the PAM sequence CGG (from position 188 to 190), 237 bp downstream from ATG sequence, from position 238 to 260, after the PAM sequence GGG (from position 235 to 237) and 656 bp downstream from ATG sequence, from position 657 to 679, after the PAM sequence GGG (from position 654 to 656), respectively.

Conjugation into *Halomonas* sp. TD01

Plasmids were transferred from *E. coli* S17-1 to *Halomonas* sp. TD01 through conjugation. Both the donor and recipient cells were cultured overnight. Then 10 μL *E. coli* S17-1 and 10 μL *Halomonas* sp. TD01 were mixed and placed on a LB-20 plate without antibiotics and cultured in 37 °C. After four hours, the colonies from the LB-20 plate were picked up and cultured in LB-60 plates with chloramphenicol. The ingredients of LB-20 medium are (g/L): NaCl 20, tryptone 10, yeast extract 5.

Growth curve establishments

The cells were cultivated in LB-60 for 3 h, then 1 mM IPTG was added to induce the CRISPRi system. The total time of cell cultivation was 12 h. Then 200 μl bacterial fluid was inoculated into each well of a 96-well plate, with three parallel samples for each strain. Each sample in the well was diluted with LB-60 medium until OD600 reached 0.001. LB-60 medium was used as the blank control. All the samples was then cultured for 24 h under continuous rotary shaking (Thermo Scientific Varioskan Flash, Thermo Scientific, USA). OD600 was examined every half hour. After a deduction of the blank control, the average OD600 of each sample at each time point was calculated and used for growth curves.

Environmental scanning electron microscope (ESEM) analysis

The cells were harvested by centrifugation at 10,000*g* for 1 min, and were first fixed with 2.5% (v/v) glutaraldehyde for more than 4 h, followed by washing with 0.1 M phosphate-buffered saline (PBS) (pH 7.3) (3 times, 10 min each). Afterwards, the fixed cells were washed by ethanol in a concentration gradient (v/v) of 50, 70, 80, 90 and 100% successively, and then dehydrated by tertiary butyl alcohol mixed with ethanol in a ratio of 1:1. The cells were treated with pure tertiary butyl alcohol and used for imaging after lyophilization. The bacteria were imaged using a environmental scanning electron microscope (FEI Quanta 200, America) and analyzed utilizing XT Microscope Server imaging software.

Shake flask experiments in PHA production

After the plasmid was conjugated into *Halomonas* sp. TD01, the positive colonies were PCR verified. Then we obtained experimental TD strains, TD-ftsZ(1-2), TD-prpC(1-7), TD-prpC6prpC7 and TD-gltA(1-4), with each harboring a CRISPRi inhibition plasmid. Wild type *Halomonas* sp. TD01 and TD-sgRNA were used as the controls.

For production of PHA in *Halomonas* sp. TD01 and its derivate strains, the bacteria were cultivated in mineral medium (MM) containing (g/L): 60 NaCl, 30 glucose, 1 yeast extract, 1 NH_4Cl, 0.2 $MgSO_4$, 9.65 Na_2HPO_4-$12H_2O$, 1.5 KH_2PO_4, 10 mL/L trace element solution I and 1 mL/L trace element solution II. Glucose concentration of the MM medium culturing *Halomonas* sp. TD01, TD-sgRNA, TD-ftsZ1 and TD-ftsZ2 in feasibility study of the constructed CRISPRi for *Halomonas* sp. TD01 was 20 g/L. The composition of trace element solution I was (g/L): 5 Fe(III)-NH_4-citrate, 2 $CaCl_2$, 1 M HCl. The trace element solution II contains (mg/L): 100 $ZnSO_4$-$7H_2O$, 30 $MnCl_2$-$4H_2O$, 300 H_3BO_3, 200 $CoCl_2$-$6H_2O$, 10 $CuSO_4$-$5H_2O$, 20 $NiCl_2$-$6H_2O$, 30 $NaMoO_4$-$2H_2O$, 1 M HCl at pH 9.0 [13].

For shake flask experiments, seed cultures were grown in 37 °C in LB-60 medium for 12 h at 200 rpm on a rotary shaker (HZQ-F160, HDL, Harbin, China). For each experimental group, three parallel samples were set. For each shake flask, 3 mL volume of the seed culture was inoculated in MM medium with 25 μg/mL chloramphenicol. The total volume of the shake flask was 50 mL. After 12 h of cultivation, IPTG was added to a final concentration at 1 mM to induce the CRISPRi inhibition system. Exceptionally, for the feasibility study of CRISPRi in *Halomonas* sp. TD01, TD-sgRNA, TD-ftsZ1 and TD-ftsZ2, the bacteria were cultured for 8 h before induction. The total time of cell culture was 48 h.

The bacteria were harvested, centrifuged at 10,000×*g* and washed once with distilled water. The cells were lyophilized and CDWs were measured. After methanolysis in chloroform at 100 °C for 4 h and cooling to room temperature, 1 mL deionized water was added. The components of the samples were then mixed by vortexing. Stratification then appeared in the sample solution, with the organic phase containing PHA. The samples were stood still for 1.5 h, and 1 mL chloroform containing PHA from the bottom layer was taken by syringe for gas chromatograph analysis. The samples were then analyzed by a gas chromatograph (GC-2014, SHIMADZU, Japan) and GCsolution software was employed to determine the PHA content [5]. Analytically pure PHB and PHBV copolymer (Sigma-Aldrich) were used as the standard samples to investigate 3HB and 3HV monomer quantities, respectively.

Real-time PCR

All the TD strains were cultivated in LB-60 medium for 3 h, then 1 mM IPTG was added to induce the CRISPRi system. The total time of cell cultivation was 12 h. The total RNA was isolated from *Halomonas* sp. TD01 and recombinant *Halomonas* sp. TD01 strains by using the RNA prep pure Cell/Bacteria Kit (Tiangen, Beijing, China). The Fastquant RT Kit (Tiangen, Beijing, China) was used to synthesize the cDNA for mRNA analysis. 16S rRNA was used as the inner standard, real-time PCR (RT-PCR) was carried out for mRNA analysis with SuperReal PreMix (SYBR Green) (Tiangen, Beijing, China).

The total extracted RNA concentration was measured by Nanodrop 2000 spectrophotometer (Thermo Fisher Scientific, USA) to design a concentration gradient for cDNA synthesis (using random primers following standard procedures described in the manufacturer's product specification). The cDNA was used immediately in the RT-PCR analysis. The linear interval of total RNA was analyzed as a standard for the following experiments to adjust the quantity of the template within its linear range, so that the fluorescence quantitative results could be designed within a rational range. All samples were prepared with three parallel groups to obtain results of ΔCt values from the outputs of RT-PCR.

Additional file

Additional file 1: Table S1. Strains and plasmids used in this study. **Table S2.** Primers used for genetic manipulations. **Fig. S1.** Growth of recombinant *Halomonas* sp. TD01 harboring CRISPRi system targeting *ftsZ* gene in LB-60 medium. **Fig. S2.** Growth of recombinant *Halomonas* sp. TD01 with controllable *prpC* gene transcription via CRISPRi in LB-60 medium. **Fig. S3.** Growth of recombinant *Halomonas* sp. TD01 harboring CRISPRi system targeting *gltA* gene in LB-60 medium. **Fig. S4.** RT-PCR tests of *gltA* transcription levels in recombinant *Halomonas* sp. TD01.

Authors' contributions

TW and LL designed and carried out the experiments, analyzed the data and drafted the manuscript. GQC draft the basic idea and supervised the study. All authors read and approved the final manuscript.

Author details

[1] School of Life Sciences, Tsinghua University, Beijing 100084, China. [2] Center for Synthetic and Systems Biology, Tsinghua University, Beijing 100084, China. [3] Tsinghua-Peking Center for Life Sciences, Tsinghua University, Beijing 100084, China. [4] Center for Nano- and Micro-Mechanics, Tsinghua University, Beijing 100084, China. [5] MOE Key Lab of Industrial Biocatalysis, Department Chemical Engineering, Tsinghua University, Beijing 100084, China.

Acknowledgements

We are grateful to the Center of Biomedical Analysis, Tsinghua University for the ESEM analysis. Plasmid pSEVA-321 was kindly donated by Professor Víctor de Lorenzo of Centro Nacional de Biotecnología in Spain.

Competing interests

The authors declare that they have no competing interests.

Funding

Financial supports for this research were provided by the National Natural Science Foundation of China (Grant Nos. 31430003 and 31270146) and a special Tsinghua President Grant dedicated to this project (Grant No. 2015THZ10).

References

1. Barrangou R, Fremaux C, Deveau H, Richards M, Boyaval P, Moineau S, Romero DA, Horvath P. CRISPR provides acquired resistance against viruses in prokaryotes. Science. 2007;315:1709–12.
2. Wiedenheft B, Sternberg SH, Doudna JA. RNA-guided genetic silencing systems in bacteria and archaea. Nature. 2012;482:331–8.
3. Jinek M, Chylinski K, Fonfara I, Hauer M, Doudna JA, Charpentier E. A programmable dual-RNA-guided DNA endonuclease in adaptive bacterial immunity. Science. 2012;337:816–21.
4. Cong L, Ran FA, Cox D, Lin S, Barretto R, Habib N, Hsu PD, Wu X, Jiang W, Marraffini LA, Zhang F. Multiplex genome engineering using CRISPR/Cas systems. Science. 2013;339:819–23.
5. Lv L, Ren YL, Chen JC, Wu Q, Chen GQ. Application of CRISPRi for prokaryotic metabolic engineering involving multiple genes, a case study: controllable P(3HB-co-4HB) biosynthesis. Metab Eng. 2015;29:160–8.
6. Anderson AJ, Dawes EA. Occurrence, metabolism, metabolic role, and industrial uses of bacterial polyhydroxyalkanoates. Microbiol Rev. 1990;54(4):450–72.
7. Martin DP, Williams SF. Medical applications of poly-4-hydroxybutyrate: a strong flexible absorbable biomaterial. Biochem Eng J. 2003;16:97–105.
8. Park SJ, Choi JI, Lee SY. Short-chain-length polyhydroxyalkanoates: synthesis in metabolically engineered *Escherichia coli* and medical applications. J Microbiol Biotechnol. 2005;15(1):206–15.
9. Chen GQ. A microbial polyhydroxyalkanoates (PHA) based bio- and materials industry. Chem Soc Rev. 2009;38:2434–46.
10. Witholt B, Kessler B. Perspectives of medium chain length poly(hydroxyalkanoates), a versatile set of bacterial bioplastics. Curr Opin Biotechnol. 1999;10:279–85.
11. Lee SY. Bacterial polyhydroxyalkanoates. Biotechnol Bioeng. 1996;49(1):1–14.
12. Cai L, Tan D, Aibaidula G, Dong XR, Chen JC, Tian WX, Chen GQ. Comparative genomics study of polyhydroxyalkanoates (PHA) and ectoine relevant genes from *Halomonas* sp. TD01 revealed extensive horizontal gene transfer events and co-evolutionary relationships. Microb Cell Fact. 2011;10:88.
13. Tan D, Xue YS, Aibaidula G, Chen GQ. Unsterile and continuous production of polyhydroxybutyrate by *Halomonas* TD01. Bioresour Technol. 2011;102:8130–6.
14. Fu XZ, Tan D, Aibaidula G, Wu Q, Chen JC, Chen GQ. Development of *Halomonas* TD01 as a host for open production of chemicals. Metab Eng. 2014;23:78–91.
15. Bi E, Lutkenhaus J. Cell division inhibitors SulA and MinCD prevent formation of the *FtsZ* ring. J Bacteriol. 1993;175(4):1118–25.
16. Loose M, Mitchison TJ. The bacterial cell division proteins FtsA and *FtsZ* self-organize into dynamic cytoskeletal patterns. Nat Cell Biol. 2014;16:38–46.
17. Margolin W. *FtsZ* and the division of prokaryotic cells and organelles. Nat Rev Mol Cell Biol. 2005;6:862–71.
18. Jiang XR, Wang H, Shen R, Chen GQ. Engineering the bacterial shapes for enhanced inclusion bodies accumulation. Metab Eng. 2015;29:227–37.
19. Ewering C, Heuser F, Benölken JK, Brämer CO, Steinbüchel A. Metabolic engineering of strains of *Ralstonia eutropha* and *Pseudomonas putida* for biotechnological production of 2-methylcitric acid. Metab Eng. 2006;8:587–602.
20. Silva LF, Gomez JGC, Oliveira MS, Torres BB. Propionic acid metabolism and poly-3-hydroxybutyrate-co-3-hydroxyvalerate (P3HB-co-3HV) production by *Burkholderia* sp. J Biotechnol. 2000;76:165–74.
21. Tan D, Wu Q, Chen JC, Chen GQ. Engineering *Halomonas* TD01 for low cost production of polyhydroxyalkanoates. Metab Eng. 2014;26:34–47.
22. Gest H. Evolution of the citric acid cycle and respiratory energy conversion in prokaryotes. FEMS Microbiol Lett. 1981;12:209–15.
23. Zhang FY, Li JJ, Liu HW, Liang QF, Qi QS. ATP-based ratio regulation of glucose and xylose improved succinate production. PLoS ONE. 2016;11(6):e0157775.
24. Xiao XW, Zheng WT, Hou X, Liang J, Li ZJ. Metabolic engineering of *Escherichia coli* for poly(3-hydroxybutyrate) production under microaerobic condition. Biomed Res Int. 2015;2015:789315.
25. Bäckström BT, Brockelbank JA, Rehm BHA. Recombinant *Escherichia coli* produces tailor-made biopolyester granules for applications in fluorescence activated cell sorting: functional display of the mouse interleukin-2 and myelin oligodendrocyte glycoprotein. BMC Biotechnol. 2007;7:3.
26. Cress BF, Linhardt RJ, Koffas MAG, et al. CRISPathBrick: modular combinatorial assembly of type II-A CRISPR arrays for dCas9-mediated multiplex transcriptional repression in *E. coli*. ACS Synth Biol. 2015;4:987–1000.
27. Wendt KE, Ungerer J, Cobb RE, Zhao H, Pakrasi HB. CRISPR/Cas9 mediated targeted mutagenesis of the fast growing cyanobacterium Synechococcus elongatus UTEX 2973. Microb Cell Fact. 2016;15:115.
28. Cress BF, Linhardt RJ, Koffas MAG, et al. Rapid generation of CRISPR/dCas9-regulated, orthogonally repressible hybrid T7-lac promoters for modular, tuneable control of metabolic pathway fluxes in *Escherichia coli*. Nucleic Acids Res. 2016;44:4472–85.
29. Martínezgarcía E, Aparicio T, et al. SEVA 2.0: an update of the Standard European Vector Architecture for de-/re-construction of bacterial functionalities. Nucleic Acids Res. 2015;43:D1183–9.

11

Dynamic control of *ERG20* expression combined with minimized endogenous downstream metabolism contributes to the improvement of geraniol production in *Saccharomyces cerevisiae*

Jianzhi Zhao[1], Chen Li[1], Yan Zhang[1], Yu Shen[1], Jin Hou[1*] and Xiaoming Bao[1,2*]

Abstract

Background: Microbial production of monoterpenes provides a promising substitute for traditional chemical-based methods, but their production is lagging compared with sesquiterpenes. Geraniol, a valuable monoterpene alcohol, is widely used in cosmetic, perfume, pharmaceutical and it is also a potential gasoline alternative. Previously, we constructed a geraniol production strain by engineering the mevalonate pathway together with the expression of a high-activity geraniol synthase.

Results: In this study, we further improved the geraniol production through reducing the endogenous metabolism of geraniol and controlling the precursor geranyl diphosphate flux distribution. The deletion of *OYE2* (encoding an NADPH oxidoreductase) or *ATF1* (encoding an alcohol acetyltransferase) both involving endogenous conversion of geraniol to other terpenoids, improved geraniol production by 1.7-fold or 1.6-fold in batch fermentation, respectively. In addition, we found that direct down-regulation of *ERG20* expression, the branch point regulating geranyl diphosphate flux, does not improve geraniol production. Therefore, we explored dynamic control of *ERG20* expression to redistribute the precursor geranyl diphosphate flux and achieved a 3.4-fold increase in geraniol production after optimizing carbon source feeding. Furthermore, the combination of dynamic control of *ERG20* expression and *OYE2* deletion in *LEU2* prototrophic strain increased geraniol production up to 1.69 g/L with pure ethanol feeding in fed-batch fermentation, which is the highest reported production in engineered yeast.

Conclusion: An efficient geraniol production platform was established by reducing the endogenous metabolism of geraniol and by controlling the flux distribution of the precursor geranyl diphosphate. The present work also provides a production basis to synthesis geraniol-derived chemicals, such as monoterpene indole alkaloids.

Keywords: Monoterpene, Geraniol, Geranyl diphosphate, *ERG20*, *Saccharomyces cerevisiae*

Background

In nature, isoprenoids are a large group of diverse compounds, that are biosynthesized via both the cytosolic mevalonate (MVA) pathway and the plastidial 2C-methyl-D-erythrtiol 4-phosphate (MEP) pathway in plants and most bacteria, or the mevalonate (MVA) pathway in animals and eukaryotes [1]. The universal precursors of isoprenoids, isopentenyl diphosphate (IPP) and dimethylallyl pyrophosphate (DMAPP), are derived from both pathways. Among them, monoterpenes are a particularly interesting subset of this family, which has been widely used as flavors, fragrances and pharmaceuticals as well as potential fuels [2, 3]. Traditionally, monoterpenes

*Correspondence: houjin@sdu.edu.cn; bxm@sdu.edu.cn
[1] State Key Laboratory of Microbial Technology, School of Life Science, Shandong University, Jinan 250100, China
Full list of author information is available at the end of the article

and their derivatives are produced from plants in low amounts, and the extraction process from plants is costly and highly dependent on the availability of raw materials. Engineering microbial organisms for monoterpene synthesis provides a potential effective route for their production.

Monoterpene geraniol (*trans*-3,7-dimethyl-2,6-octadien-1-ol; $C_{10}H_{18}O$), an acyclic monoterpene, is typically generated from aromatic plants and has many applications in perfume, pharmaceutical, and other chemical industries [4]. Geraniol is also considered as a promising biofuel due to its high energy content, low hygroscopicity and low volatility in comparison with ethanol [5]. In addition, a recent study reported that geraniol can undergo an 11-step heterologous biosynthetic pathway to form strictosidine, which is the common intermediate for production of the anticancer agent monoterpene indole alkaloids (MIAs) [6]. Given its low yield in aromatic plants, geraniol has been successfully synthesized in *Escherichia coli* and *Saccharomyces cerevisiae* through metabolic engineering strategies [7–11]. *S. cerevisiae* has been used widely as a cell factory to produce a diversity of terpenes, and many of them have achieved significant yields [12–15]. However, monoterpene production has been achieved only at low levels thus far [8, 16, 17]. There are two challenges for high level monoterpenes production. One is the availability of intracellular geranyl diphosphate (GPP) precursor, and the other is the toxicity of monoterpenes to cells [18, 19]. Most of monoterpenes are generated from the precursor GPP which is formed by condensation of one molecule of IPP with one molecule of its isomer DMAPP (Fig. 1). Unlike plants, *S. cerevisiae* does not supply enough GPP for production due to the lack of a specific GPP synthase (GPPS). In *S. cerevisiae*, the *ERG20* gene of the endogenous MVA pathway encodes a farnesyl pyrophosphate synthase (FPPS) that, despite having both GPP and FPP synthase activity, only releases a very low amount of GPP from its catalytic site [20]. To increase the flux to GPP, a set of *ERG20* mutants was constructed to screen for GPPS preference mutants. $ERG20^{K197G}$, the most effective mutant, improved geraniol production to 5 mg/L while expressing geraniol synthase (GES) from *Ocimum basilicum* in *S. cerevisiae* [8]. Liu et al. obtained 36 mg/L geraniol by overexpressing two key rate-limiting genes, *tHMG1* encoding a truncated version of 3-hydroxy-3-methylglutaryl coenzyme A and *IDI1* encoding IPP isomerase, to enhance MVA pathway flux and by overexpressing *MAF1*, a negative regulator of *MOD5* encoding tRNA isopentenyltransferase, to decrease the flux to tRNA biosynthesis [10]. In addition, Ignea et al. demonstrated that the double mutant Erg20p (F96W-N127W) had a strong dominant negative ability to decrease the FPPS function of Erg20p, increasing

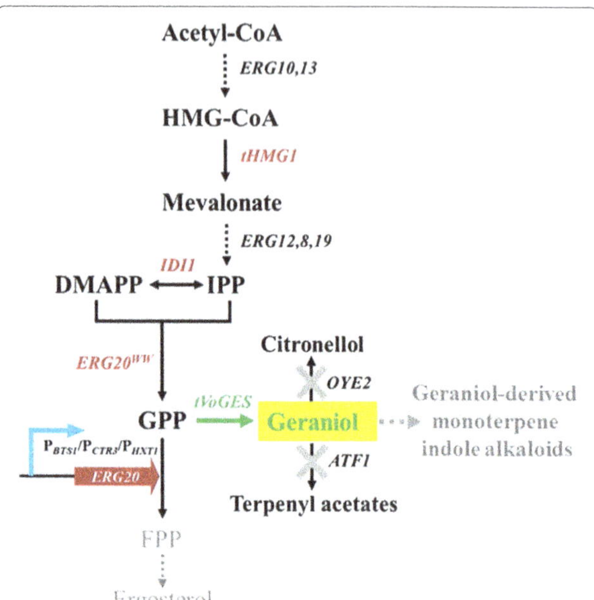

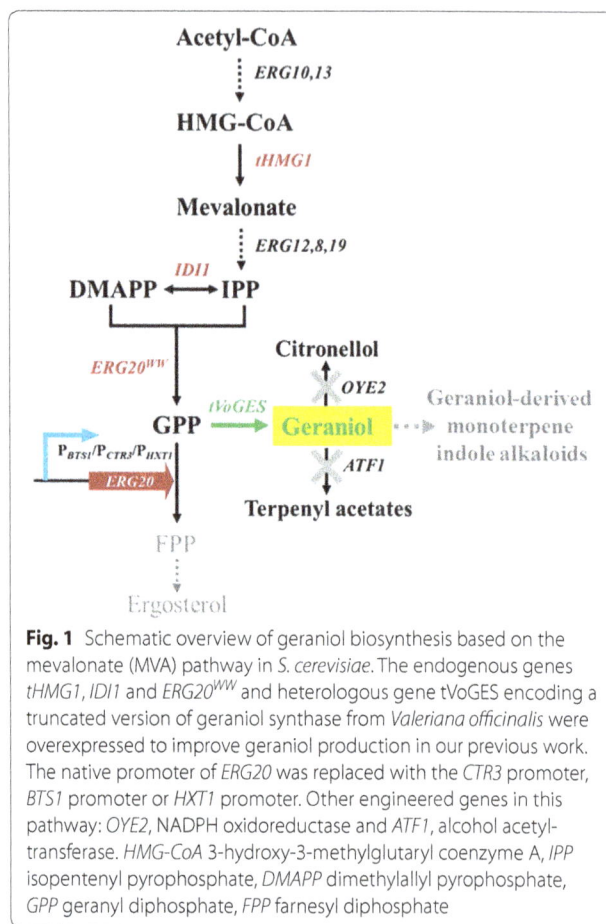

Fig. 1 Schematic overview of geraniol biosynthesis based on the mevalonate (MVA) pathway in *S. cerevisiae*. The endogenous genes *tHMG1*, *IDI1* and $ERG20^{WW}$ and heterologous gene tVoGES encoding a truncated version of geraniol synthase from *Valeriana officinalis* were overexpressed to improve geraniol production in our previous work. The native promoter of *ERG20* was replaced with the *CTR3* promoter, *BTS1* promoter or *HXT1* promoter. Other engineered genes in this pathway: *OYE2*, NADPH oxidoreductase and *ATF1*, alcohol acetyltransferase. *HMG-CoA* 3-hydroxy-3-methylglutaryl coenzyme A, *IPP* isopentenyl pyrophosphate, *DMAPP* dimethylallyl pyrophosphate, *GPP* geranyl diphosphate, *FPP* farnesyl diphosphate

production of the monoterpene sabinene by 10.4-fold to 0.53 mg/L in an industrial diploid strain [16]. In our previous work, we further improved geraniol production through enhancing MVA pathway flux, screening different sources of GESs and GPPSs, and designing fusion proteins of GES and GPPS. Geraniol of 293 mg/L production was achieved in fed-batch cultivation, which is the highest monoterpene production in *S. cerevisiae* ever reported [21]. However, it is still lower than the production of many other terpenes in *S. cerevisiae* [12, 22, 23].

In addition to the low GPP pool, the toxicity to microorganisms of monoterpenes also largely limits their microbial production [2, 24]. To alleviate the toxicity of monoterpenes, a two-phase fermentation system has been extensively applied by adding a non-toxic extractive solvent (e.g., dodecane) [2]. In addition, monoterpenols, usually undergo biotransformation to other terpenoids in aromatic plants, as well as in some wine yeasts [25, 26]. A recent study demonstrated that geraniol is converted into citronellol under the catalysis of the enzyme *OYE2*, and it is acetylated by *ATF1* encoding alcohol acetyltransferase during wine fermentation by *S. cerevisiae* [27].

Intracellular bioconversion of geraniol may be a self-defense response to avoid the toxicity of monoterpenes. In addition, introduction of a heterologous biosynthetic pathway and the overexpression of an endogenous pathway often lead to the accumulation of some intermediate metabolites, that may interfere with the native regulation of metabolic flux and increase metabolic burden, resulting in overproduction of some biosynthetic enzymes and accumulation of some toxic intermediate metabolites. Therefore, a dynamic control system that can sense environmental changes such as metabolic intermediates or nutrient concentrations can be used to avoid the toxicity of intermediate metabolites [28–30].

Previously, we achieved a significant increase in geraniol production through expressing geraniol synthase from *Valeriana officinalis* and regulating GPP synthesis [21]. In this study, we further engineered strains through dynamic control of *ERG20* to fine-tune the GPP flux combined with minimized geraniol endogenous conversion. Combining these strategies together with *LEU2* auxotrophic complementation, the final geraniol production reached 1.69 g/L in fed-batch fermentation, which is the highest reported production in yeast.

Results

Improving geraniol production by minimizing endogenous bioconversion

Geraniol is the main precursor of monoterpenoids in aromatic plants [4]. It also has many derivatives including nerol, neral, geranylgeraniol, geranial, citronellol and terpenyl acetates [31]. Previous studies have demonstrated that wine yeasts are able to convert geraniol into other monoterpenoids, which influence the sensory properties of wine [26, 32]. During *S. cerevisiae* fermentation, Oye2p is the main enzyme involved in the conversion of geraniol to citronellol, and Atf1p is the main contributor to synthesis of terpenyl acetates from geraniol [27]. In our early study, an efficient *S. cerevisiae* strain YZG13-GE1 was constructed to produce geraniol from glucose through engineering of geraniol synthesis and optimizing GPP synthesis. To further improve geraniol production, we attempted to minimize endogenous metabolism of geraniol in *S. cerevisiae*. Thus, we deleted *OYE2* or *ATF1* in YZG13-GE1 to create YZG14 or YZG15, respectively. Batch fermentation was carried out in a two-phase fermentation system using dodecane as extractive solvent. Geraniol was produced at 285.9 mg/L by YZG14 with *OYE2* deletion, which was improved by 1.7-fold compared with the control strain (YZG13-GE1), and 259.8 mg/L was produced by YZG15 with *ATF1* deletion, which was improved by 1.6-fold compared with the control strain (YZG13-GE1) (Fig. 2a). The conversion of geraniol to citronellol was decreased by 55% in YZG14 (Δ*oye2*) (Additional file 1: Figure S1). In addition, we further investigated the effect of double deletion of *OYE2* and *ATF1* in YZG13-GE1 on geraniol production. The results showed that geraniol production dramatically decreased to 32.2 mg/L in YZG16 with *OYE2-ATF1* deletions (Fig. 2a). Cell growth was not affected in either single gene deletion strain (Δ*oye2* or Δ*atf1*) (Additional file 1: Figure S2), and geraniol yields were 57.8 mg/g DCW and 58.2 mg/g DCW, a 1.8-fold and 1.7-fold increase, respectively (Fig. 2b; Table 1). However, double deletion of *OYE2* and *ATF1* decreased the final biomass by 35%, and geraniol yield significantly dropped to 9.6 mg/g DCW (Fig. 2; Table 1).

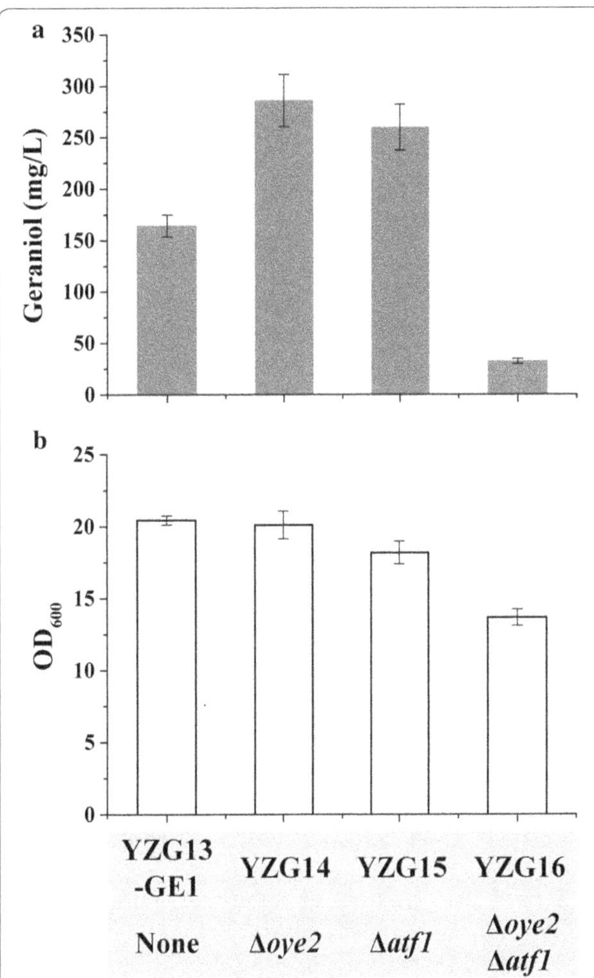

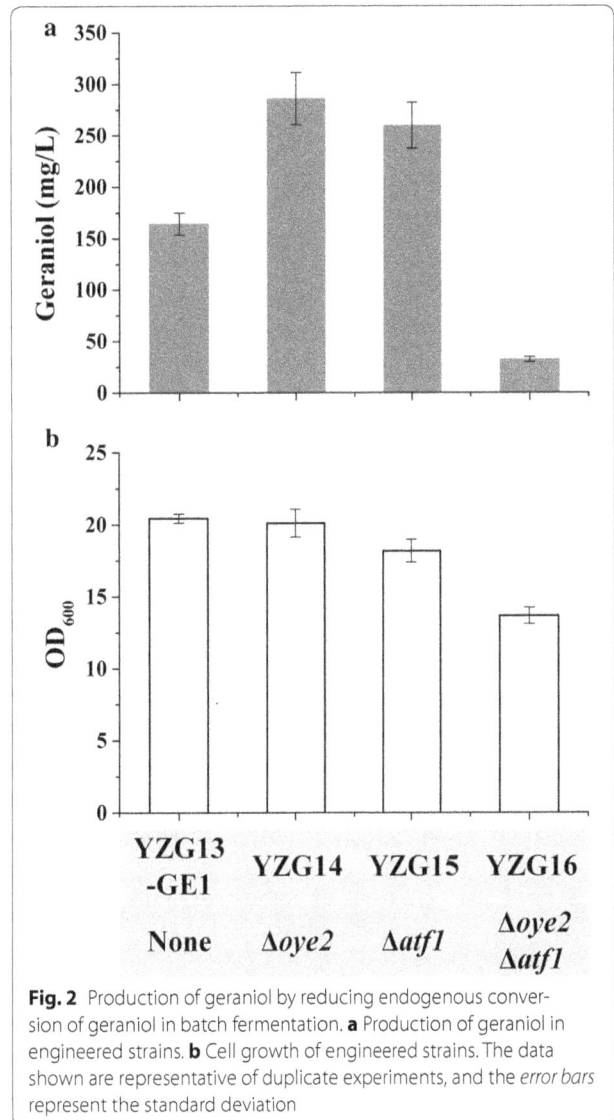

Fig. 2 Production of geraniol by reducing endogenous conversion of geraniol in batch fermentation. **a** Production of geraniol in engineered strains. **b** Cell growth of engineered strains. The data shown are representative of duplicate experiments, and the *error bars* represent the standard deviation

We also compared geraniol production in YZG14 (Δ*oye2*), YZG15 (Δ*atf1*) and YZG16 (Δ*oye2*Δ*atf1*) in fed-batch cultivation using glucose as sole carbon source. The

Table 1 Geraniol yield of the engineered strains in batch fermentation

Strains[a]	Dry cell weight (g/L)	Geraniol yield (mg/g DCW)
YZG13-GE1	5.02 ± 0.31	32.69 ± 2.01
YZG14 (Δ*oye2*)	4.94 ± 0.28	57.83 ± 3.24
YZG15 (Δ*atf1*)	4.47 ± 0.22	58.16 ± 2.59
YZG16 (Δ*oye2*Δ*atf1*)	3.36 ± 0.34	9.58 ± 0.96
YZG17 (P_{BTS1}-*ERG20*)	1.96 ± 0.10	20.58 ± 0.91
YZG18 (P_{CTR3}-*ERG20*)	2.94 ± 0.23	16.61 ± 1.21
YZG19 (P_{HXT1}-*ERG20*)	3.92 ± 0.26	37.87 ± 2.55

[a] Duplicate experiments were performed for each strain, and the error bars the represented the standard deviation

cell growth of YZG14 (Δ*oye2*) and YZG15 (Δ*atf1*) exhibited a similar growth profile compared with the reference strain YZG13-GE1. YZG14 (Δ*oye2*) and YZG15 (Δ*atf1*) produced 421.7 and 371.1 mg/L of geraniol, which represented approximately 2.5-fold and 2.2-fold increases relative to YZG13-GE1 (256.4 mg/L), respectively (Fig. 3). However, geraniol production was dramatically decreased in double deletion of *OYE2* and *ATF1* strain, only producing 46.1 mg/L geraniol (Fig. 3d).

The effect of *ERG20* expression control on geraniol production

In *S. cerevisiae*, synthesis of both the monoterpene precursor GPP and sesquiterpene precursor FPP is catalyzed by a single enzyme, Erg20p, with FPP synthase preference. In the absence of a specific GPP synthase, many efforts have been made to increase the intercellular GPP pool. Expression of heterologous GPP synthase does not increase the GPP-derived products significantly [16, 21]. Although engineering Erg20p into a GPP synthase increased sabinene and geraniol production by 10.4-fold and 2.8-fold, respectively, the titer is still much lower than that of FPP-derived sesquiterpenes, demonstrating the importance of down-regulating FPP synthesis [10, 16]. In this study, the native *ERG20* promoter was first replaced with either the weak *BTS1* promoter or the copper repressible *CTR3* promoter. YZG13-GE1 containing the native *ERG20* promoter was used as a reference strain. The transcription level of *ERG20* using the *BTS1* promoter or *CTR3* promoter was reduced by 90 and 75% (Fig. 4a), respectively, in SD-URA-HIS medium with 2% glucose. Unexpectedly, the replacement of the *ERG20* promoter with that of *BTS1* or *CTR3* decreased geraniol production by 75.6 and 70.3%, resulting in 40.37 and 48.85 mg/L, respectively (Fig. 4b). The *BTS1* and *CTR3* promoters also decreased the final biomass by 60 and 40% (Fig. 4c), repectively. Previously, we also found that overexpression of *ERG20* reduced geraniol production [21]. These results demonstrated that both down-regulation and up-regulation of *ERG20* have a negative impact on geraniol production. The dynamic control of *ERG20* was then attempted by replacing the *ERG20* promoter with the glucose-sensing *HXT1* promoter. This promoter strength is higher than that of *ERG20* promoter when glucose is present in the medium (Fig. 4a), but it is repressed when the glucose concentration is low or absent (Additional file 1: Figure S3). We found that the geraniol titer and final biomass were not markedly affected by *HXT1* promoter replacement in batch cultivation, and the yield of geraniol increased slightly (37.9 mg/g DCW vs. 32.7 mg/g DCW) (Table 1). Considering that the *HXT1* promoter provides both high-glucose induction and low-glucose repression, we attempted both glucose and glucose/ethanol (1:7) mixture feeding strategies to control *ERG20* expression in fed-batch fermentation. Compared with the control strain, *HXT1* promoter replacement did not increase geraniol production when pure glucose was fed; geraniol production only increased slightly (data not shown). In contrast, when a glucose/ethanol (1:7) mixture was fed, the geraniol production achieved a 176% increase in the *HXT1* promoter replacement strain compared with the control strain, leading to 650.8 mg/L (Fig. 4d).

Previously, ethanol has been used as carbon for the production of hydrocortisone, amorpha-4,11-diene and resveratrol [33–35]. In this study, we further compared the impact of feeding different ratios of glucose and ethanol on geraniol production using YZG19 (P_{HXT1}-*ERG20*). As shown in Fig. 5b, the cell growth of YZG19 (P_{HXT1}-*ERG20*) gradually improved as the ethanol percentage in glucose/ethanol mixture increased, and the final production of geraniol also gradually increased. The highest production of geraniol was obtained when pure ethanol was fed, producing 867.7 mg/L of geraniol (Fig. 5a). Based on the results described above, the dynamic control of *ERG20* expression under the *HXT1* promoter together with optimizing the glucose and ethanol ratio improved the production of geraniol by 3.8-fold.

Combining the dynamic control of *ERG20* and reduction of endogenous conversion together with *LEU2* complementation enabled a further increase in geraniol production

To further improve geraniol production, we deleted the *OYE2* gene in YZG19 (P_{HXT1}-*ERG20*) to generate YZG20 (Δ*oye2*, P_{HXT1}-*ERG20*), which produced 984.6 mg/L of geraniol in fed-batch fermentation by feeding ethanol, a 14% increase over YZG19 (P_{HXT1}-*ERG20*) (Fig. 6a). A recent study showed that leucine metabolism may be networked with isoprenoid biosynthesis, and leucine prototrophy was found to improve diterpenoid miltiradiene

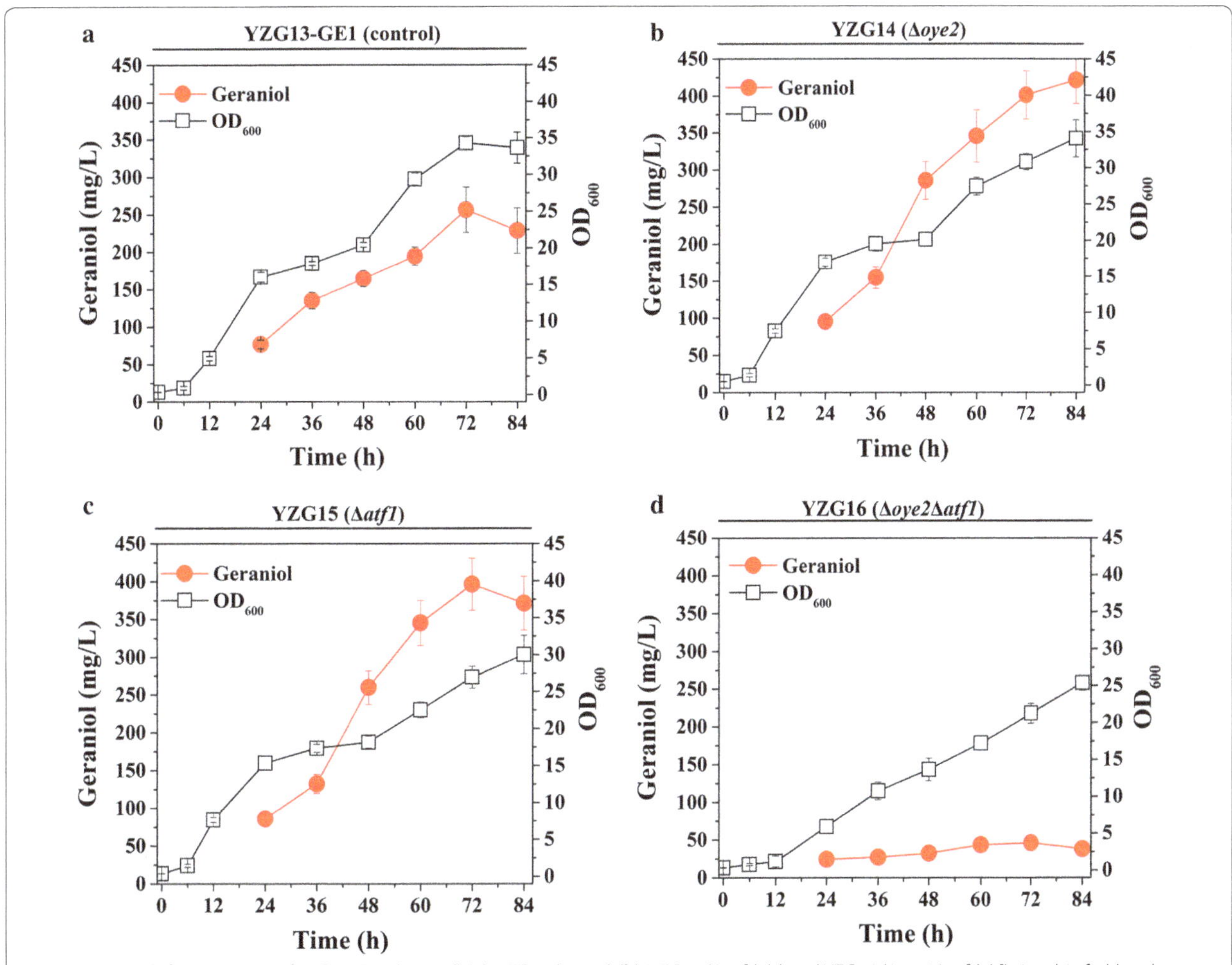

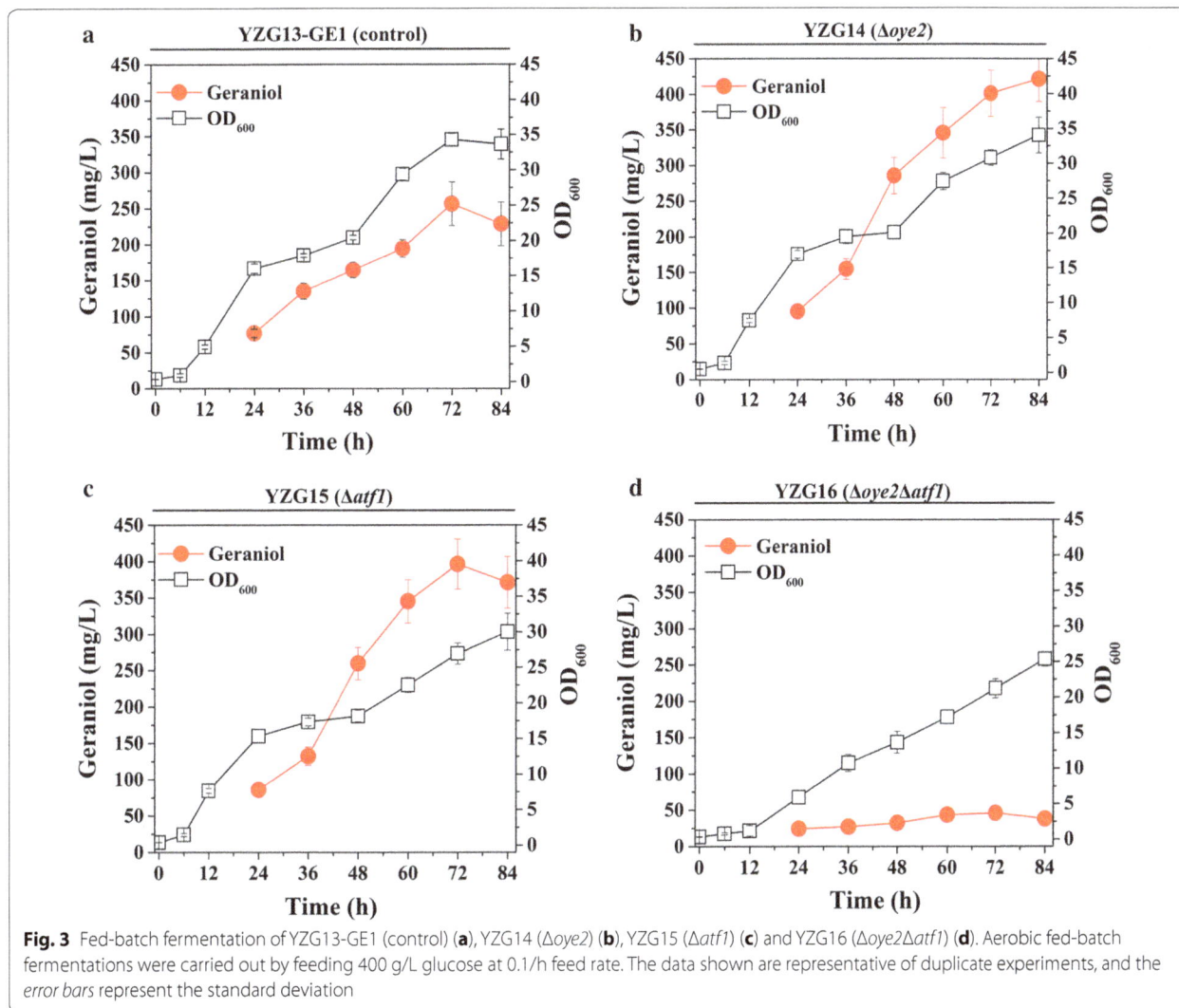

Fig. 3 Fed-batch fermentation of YZG13-GE1 (control) (**a**), YZG14 (*Δoye2*) (**b**), YZG15 (*Δatf1*) (**c**) and YZG16 (*Δoye2Δatf1*) (**d**). Aerobic fed-batch fermentations were carried out by feeding 400 g/L glucose at 0.1/h feed rate. The data shown are representative of duplicate experiments, and the *error bars* represent the standard deviation

production [36]. We further engineered a strain with complemented *LEU2* auxotrophy, and final geraniol production was enhanced up to 1.69 g/L, a further 71% increase over YZG20 (Δ*oye2*, P_{HXT1}-*ERG20*). Meanwhile, we noticed that *LEU2* complementation also improved cell growth and geraniol yield (Fig. 6b). The final geraniol production in this study was improved by 5.8-fold compared with our previous study.

Discussion

Engineering *S. cerevisiae* as a cell factory is a promising and attractive route for rapid and inexpensive biosynthesis of terpenoids [12, 13, 22, 23, 34]. Over the last few years, most work has focused on the production of sesquiterpenes in microorganisms including *S. cerevisiae*. For instance, production of the sesquiterpene amorpha-4,11-diene, the precursor for the antimalarial agent artemisinin, has reached 40 g/L using engineered *S. cerevisiae* in fed-batch fermentation [12]. Both monoterpenes and sesquiterpenes are derived from the MVA pathway. However, monoterpene production in *S. cerevisiae* is much lower than that of sesquiterpenes, mainly due to the poor GPP precursor supply and the toxicity of monoterpenes in yeast. In this work, we further engineered production of the monoterpene geraniol through reduction of endogenous geraniol conversion, dynamic control of *ERG20* expression and leucine biosynthesis complementation based on YZG13-GE1, a geraniol-producing strain from our previous work [21]. Final geraniol production of 1.69 g/L was obtained, which to our knowledge is the highest level reported for engineered yeast.

Monoterpenes are highly toxic to many microorganisms. Cells may reduce their toxicity through the conversion of monoterpenes to other lower toxicity compounds

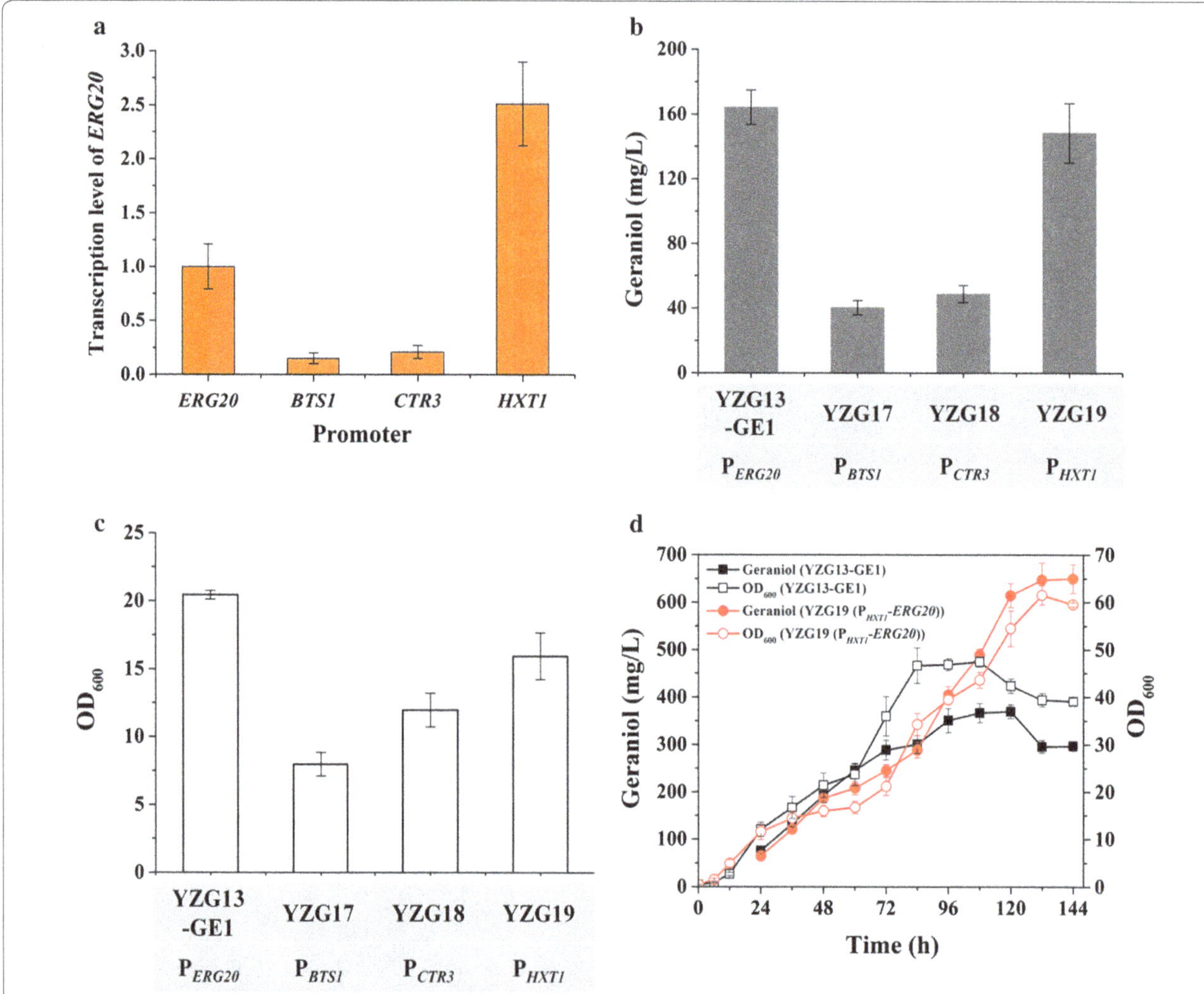

Fig. 4 Production of geraniol in *ERG20* promoter replacement strains. **a** Transcription level of *ERG20* controlled by different promoters. **b** Production of geraniol in engineered strains in batch fermentation. **c** Cell growth of engineered strains in batch fermentation. **d** Geraniol production in YZG13-GE1 and *HXT1* promoter replacement strain (YZG19) in fed-batch fermentation. All strains were grown in SD medium with 2% glucose, and the cells were collected when OD_{600} reached 0.6 to extract mRNA for *ERG20* mRNA and transcription determination. Aerobic fed-batch fermentations were carried out by feeding a 50 g/L glucose and 350 g/L ethanol (1:7) mixture at 0.1/h feed rate. The data shown are representative of duplicate experiments, and the *error bars* represent the standard deviation

by various specific or non-specific enzymes. In our previous study, we confirmed that 180 mg/L of geraniol in culture could strongly inhibit yeast cell growth [21]. Although two-phase extractive fermentation could effectively alleviate monoterpene toxicity in situ, the conversion of monoterpenes to other terpenoids is not avoided completely. In *S. cerevisiae*, a mechanism was proposed in which geraniol was first oxidized to geranial by *OYE2* encoding an NADPH oxidoreductase and its homologous gene *OYE3*, belonging to the old yellow enzyme family, and then was reduced to citronellol [25, 27, 37]. However, the oxido-reductases involved in the reaction of geranial to citronellol are still unknown. Of the two old yellow enzymes, *OYE2* was the main contributor to the reduction reaction. Except for citronellol, terpenyl acetates including citronellyl acetate and geranyl acetate are another group of geraniol-derived metabolites catalyzed by *ATF1* encoding alcohol acetyltransferase. In our study, deletion of *OYE2* or *ATF1* led to a 1.8-fold increase of geraniol production. Interestingly, the *ATF1* deletion strain increased citronellol production compared with the control strain (Additional file 1: Figure S1). A similar phenomenon was observed in a previous study, which indicated that there may be a relationship between the physiological functions of *OYE2* and *ATF1*. Although the old yellow enzymes in yeast may have a physiological role

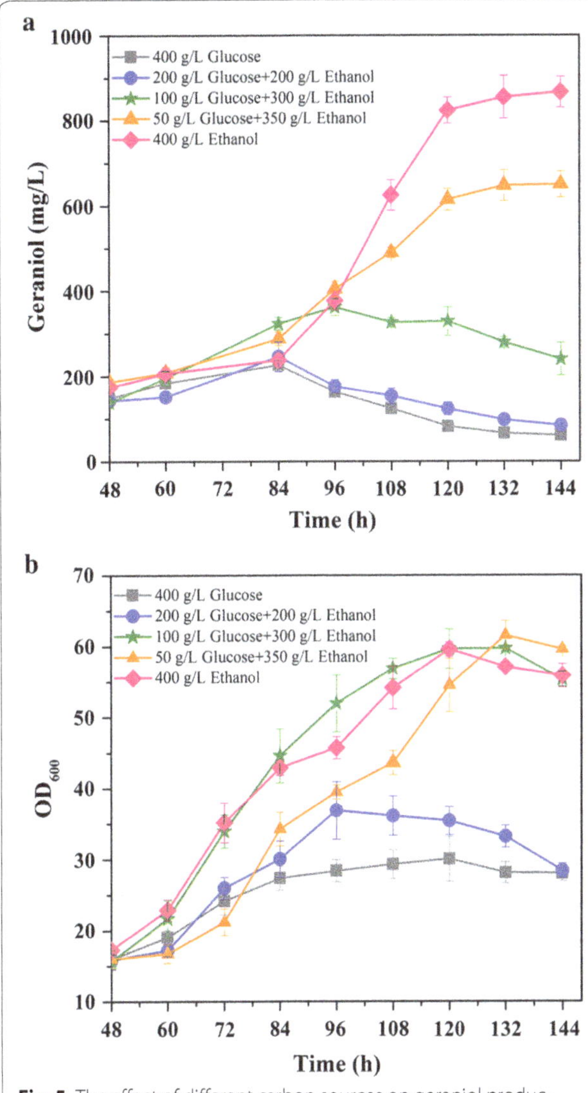

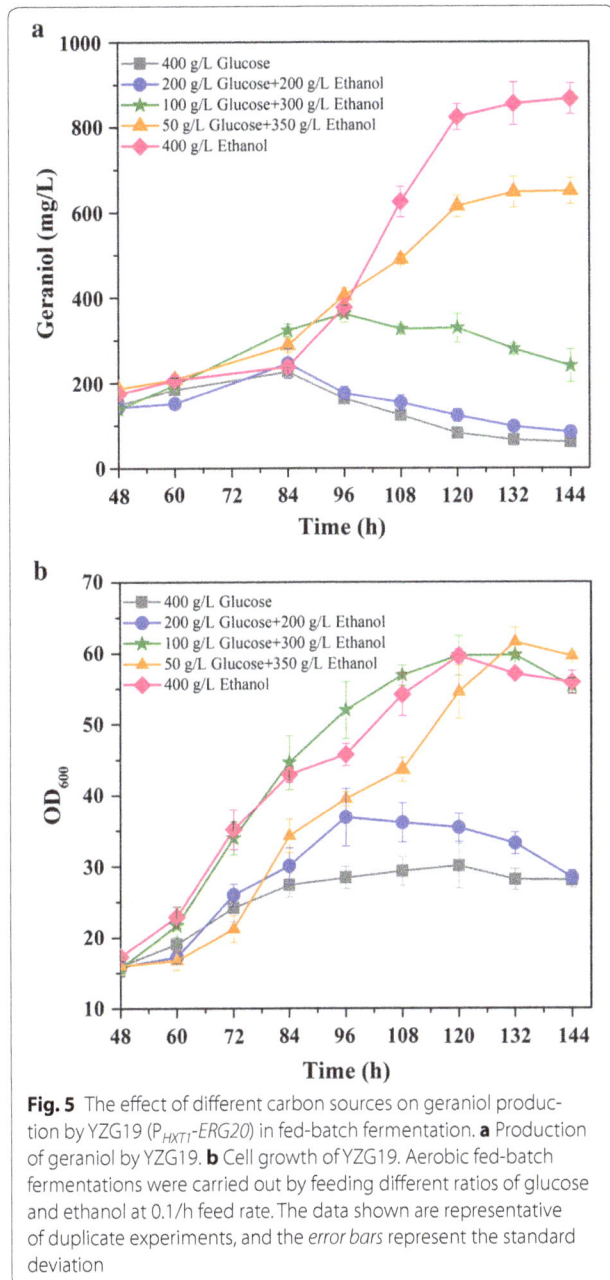

Fig. 5 The effect of different carbon sources on geraniol production by YZG19 (P_{HXT1}-*ERG20*) in fed-batch fermentation. **a** Production of geraniol by YZG19. **b** Cell growth of YZG19. Aerobic fed-batch fermentations were carried out by feeding different ratios of glucose and ethanol at 0.1/h feed rate. The data shown are representative of duplicate experiments, and the *error bars* represent the standard deviation

in detoxification of unsaturated metabolites and reactive oxygen species, both *OYE2* and *ATF1* are thought to be involved in sterol metabolism [38, 39]. Therefore, functional complementation in the regulation of sterol metabolism between *OYE2* and *ATF1* may be a reason for the lower cell growth and geraniol production in the double deletion strain.

In the geraniol-producing engineered *S. cerevisiae*, the reaction catalyzed by *ERG20* is a key branch point towards either geraniol synthesis or downstream ergosterol synthesis for cell growth. Although *ERG20* mutations could improve monoterpene production, the reduction of intercellular FPP synthesis damaged cell growth [8, 16]. In this study, the weak promoter P_{BTS1} and the copper-repressible promoter P_{CTR3} were first tested for *ERG20* down-regulation to redirect GPP flux. The transcriptional level of *ERG20* in the promoter replacement strains was decreased to a one-fifth to one-sixth, and reduced cell growth was also observed when geraniol synthase and the MVA pathway were overexpressed in promoter replacement strains (Fig. 4; Additional file 1: Figure S4). The enhancement of the MVA pathway together with *ERG20* down-regulation possibly led to the accumulation of toxic intermediates such as DMAPP or GPP. The toxicity of isoprenoid precursors has also been reported in *E. coli* and *Bacillus subtilis* [40–42]. Previously, we found that although overexpression of *ERG20* did not affect growth, it decreased geraniol production. Therefore, neither up-regulation or down-regulation of *ERG20* was able to increase geraniol production. A suitable *ERG20* expression strategy that can weaken the downstream FPP flux but not impair cell growth to further improve geraniol production in engineered yeast is needed. To achieve this goal, dynamic control of *ERG20* is proposed, which is likely to not only balance metabolism between product formation and cell growth, but also prevent the accumulation of toxic metabolites. Recently, several dynamic control systems have been developed and applied in metabolic engineering, i.e., a biosensor (sensor-regulator) system [43], a sequential control system based on the medication *GAL* regulation system combined with the *HXT1* promoter [23] and a novel CRISPR-Cas9 interference system [44]. To avoid early accumulation of toxic intermediates and to better balance the utilization of GPP, P_{HXT1}, a glucose-sensing promoter, was employed to dynamically control *ERG20* expression in our study. An *HXT1* promoter replacement strain (YZG19) did not contribute to improve geraniol production in batch or fed-batch fermentation using glucose as sole carbon, but its advantage appeared when the glucose concentration was low (feeding glucose/ethanol mixture carbon). The geraniol production improved stepwise when the glucose percentage decreased in the glucose/ethanol mixture (Figs. 4, 5). These results indicate that the dynamic control of *ERG20* expression by the *HXT1* promoter can balance the flux distribution between cell growth and monoterpene synthesis, thus improving geraniol production with low glucose feeding (Fig. 5). In addition, ethanol feeding possibly improved the geraniol production to some extent due to supplying acetyl-CoA precursors in a more direct metabolic pathway. An ethanol feeding strategy has also been successfully applied in amorphadiene and resveratrol production [34, 35].

Finally, strategies combining the *OYE2* deletion and P_{HXT1}-controlled *ERG20* expression further improved

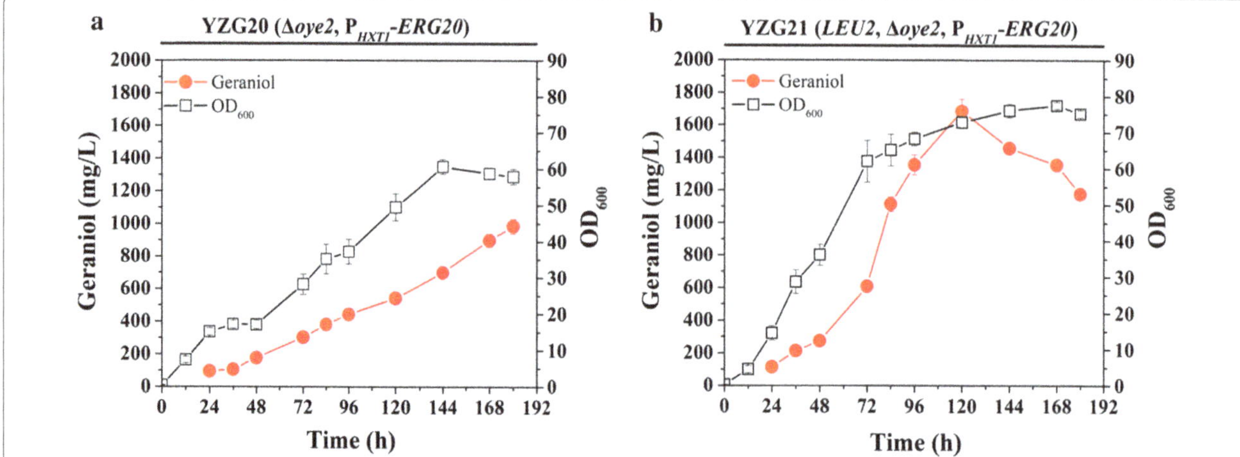

Fig. 6 Fed-batch fermentation of YZG20 (*Δoye2*, P_{HXT1}-*ERG20*) (**a**) and YZG21 (*LEU2*, *Δoye2*, P_{HXT1}-*ERG20*) (**b**) using pure ethanol feeding. Aerobic fed-batch fermentations were carried out by feeding 400 g/L ethanol with 0.1/h feed rate. The data shown are representative of duplicate experiments, and the *error bars* represent standard deviations

geraniol production to 984.6 mg/L (Fig. 6). A previous report showed that leucine metabolism may be networked with sterol biosynthesis, and that leucine was dis-assimilated to form HMG-CoA in *Leishmania mexicana* [45]. In *S. cerevisiae*, the leucine biosynthetic pathway may act as a bypass pathway to supplement additional HMG-CoA. Therefore, we further constructed the prototrophic strain YZG21 by complementing the auxotrophic marker *LEU2*. The *LEU2* complementation ultimately improved the geraniol production to 1.69 g/L, meanwhile also improving cell growth and yield. A similar result was observed in a previous study on miltiradiene production [36].

Although a relatively high level of geraniol was obtained in our work, the cytotoxicity of monoterpenes is still considered as a problem that hinders further improvement of monoterpene production. Improving the resistance of microorganisms to monoterpenes still needs to be addressed to further enhance the monoterpene production. In addition, diploidization is another potential strategy for improving biomass and productivity given that diploid strains generally grow faster and tolerate higher stresses compared with haploid strains [36, 46]. More importantly, constructing a plasmid-free geraniol overproducing strain by integration of all pathway components into the genome is necessary for application in industrial processes.

Conclusions

In summary, we engineered geraniol biosynthesis through the reduction of endogenous geraniol conversion and dynamic control of *ERG20* expression. In addition, we further identified the importance of *LEU2* complementation to geraniol synthesis. Ultimately, 1.69 g/L of geraniol was achieved when these approaches were combined in fed-batch fermentation with pure ethanol feeding. The present work provides a good platform for the production of geraniol and its derived chemicals.

Methods

Medium

The *E. coli* strain Tans5α was used for gene cloning and grown in Luria–Bertani medium (5 g/L yeast extract, 10 g/L tryptone, and 10 g/L NaCl) supplemented with 100 mg/L ampicillin at 37 °C. The engineered yeast strain YZG13-GE1 constructed in our previous work was used as the parent strain for further engineering [21]. Yeast cells were cultivated in yeast extract-peptone-dextrose (YPD) medium (20 g/L glucose, 20 g/L tryptone, and 10 g/L yeast extract), SD-URA or SD-URA-HIS medium (20 g/L glucose, 1.7 g/L yeast nitrogen base, and 5 g/L $(NH_4)_2SO_4$; synthetic complete drop-out medium without uracil and/or histidine). Geneticin (G418) at 400 mg/L was added to the culture medium for gene deletion and promoter replacement.

DNA manipulation and strain construction

All yeast strains constructed in this study are listed in Table 2. The reference strain is YZG13-GE1, which is derived from CEN.PK102-5B and harbored pZGV6-GE1 (2μ ori, *URA3*, P_{TEF1}-*tVoGES*-(GGGS)-*ERG20*^{WW}) and pZMVA4 (2μ ori, *HIS3*, P_{TEF1}-*tHMG1*, P_{PGK1}-*IDI1*, P_{TEF1}-*UPC2.1*) [21]. The primers for all DNA fragment amplifications are listed in Additional file 1: Table S1.

Table 2 Strains used in this study

Strain name	Parent strain	Plasmids/genotype	Source
CEN.PK102-5B		*MATa ura3-52 his3Δ1 leu2-3,112*	Dr. P. Kötter, Frankfurt, Germany
5B-Δoye2	CEN.PK102-5B	*Δoye2::loxP*	This study
5B-Δatf1	CEN.PK102-5B	*Δatf1::loxP*	This study
5B-Δoye2Δatf1	CEN.PK102-5B	*Δoye2::loxP, Δatf1::loxP*	This study
5B-PBTS1	CEN.PK102-5B	ΔP_{ERG20}*::loxP-*P_{BTS1}	This study
5B-PCTR3	CEN.PK102-5B	ΔP_{ERG20}*::loxP-*P_{CTR3}	This study
5B-PHXT1	CEN.PK102-5B	ΔP_{ERG20}*::loxP-*P_{HXT1}	This study
5B-Δoye2-PHXT1	5B-PHXT1	ΔP_{ERG20}*::loxP-*P_{HXT1}, *Δoye2::loxP*	This study
5B-LEU2-Δoye2-PHXT1	5B-Δoye2-PHXT1	ΔP_{ERG20}*::loxP-*P_{HXT1}, *ΔYPRCtau3::*P_{LEU2}*-LEU2-*T_{LEU2}*-loxP*	This study
YZG13-GE1	CEN.PK102-5B	pZGV6-GE1, pZMVA4	[21]
YZG14	5B-Δoye2	pZGV6-GE1, pZMVA4	This study
YZG15	5B-Δatf1	pZGV6-GE1, pZMVA4	This study
YZG16	5B-Δoye2Δatf1	pZGV6-GE1, pZMVA4	This study
YZG17	5B-PBTS1	pZGV6-GE1, pZMVA4	This study
YZG18	5B-PCTR3	pZGV6-GE1, pZMVA4	This study
YZG19	5B-PHXT1	pZGV6-GE1, pZMVA4	This study
YZG20	5B-Δoye2-PHXT1	pZGV6-GE1, pZMVA4	This study
YZG21	5B-LEU2-Δoye2-PHXT1	pZGV6-GE1, pZMVA4	This study

To replace the *ERG20* promoter with different promoters, integration cassettes were constructed via fusion PCR using loxp-*kanMX*-loxp as the selection marker. The *BTS1* promoter (from −1 to −333), *CTR3* promoter (from +30 to −1116), and *HXT1* promoter (from −1 to 1190) replaced the *ERG20* promoter in the chromosome [23, 47, 48]. Similarly, the deletions of genes *ATF1* and *OYE2* were also performed through homologous recombination using loxp-*kanMX*-loxp as the selection marker. The *LEU2* cassette was integrated into the YPRCτ3 site of the chromosome. The DNA fragments were transformed into *S. cerevisiae* CEN.PK102-5B by the standard lithium acetate method [49], and the transformants were selected on YPD agar plates supplemented with 400 mg/L geneticin (G418). The plasmid pSH47 (purchased from Euroscarf) was transformed into *S. cerevisiae* to remove the selection marker. The engineered strains were then transformed with plasmids pZGV6-GE1 and pZMVA4, resulting in a series of geraniol-producing strains (Table 2).

Batch and fed-batch fermentation for geraniol production

To determine the performance of the recombinant yeast strains, batch fermentation and fed-batch fermentation were performed in a 1-L Infors-HT fermenter (Infors AG, Bottmingen, Switzerland). To prepare seed cultures, the strains were grown for 24 h in 5 mL of SD-URA-HIS medium, and then inoculated into fresh SD-URA-HIS medium at an OD_{600} of 0.2 and cultivated for 12 h. For batch fermentation, a seed culture was inoculated into a 1-L Infors-HT fermenter (Infors AG, Bottmingen, Switzerland) containing 0.6 L of SD-URA-HIS medium at an initial OD_{600} of 0.2. Batch cultures were conducted at 30 °C, with the agitation rate at 600 rpm and an airflow rate of 1 vvm. The pH was maintained at 5.0 by automatic addition of 2.5 M NaOH. Dissolved oxygen was maintained above 30% saturation throughout cultivation by setting the stirring speed rate. For fed-batch cultivation, strains were first grown in a 400-mL batch culture at 30 °C with shaking at 600 rpm, an airflow rate of 1 vvm and pH 5.0. After the glucose and ethanol produced in the batch phase were depleted, feeding solution containing 400 g/L glucose, 100 g/L $(NH_4)_2SO_4$, 34 g/L yeast nitrogen base, 1.3 g/L CSM-Ura-His, and 2 g/L leucine was fed to the fermenter at a controlled specific feed rate of 0.1/h. Feeding of a glucose/ethanol mixture or pure ethanol was also used for strains in which the *ERG20* promoter was replaced with the *HXT1* promoter to control *ERG20* expression. In both batch and fed-batch cultures, 20% dodecane was added to the medium after cultivating for 12 h for geraniol extraction. Independent duplicate cultures were conducted for each strain.

RNA extraction and quantitative real-time PCR

Total RNA was prepared from 40 mL exponentially growing cell cultures using the UNIQ-10 Trizol RNA extraction kit (Sangon Biological Engineering, Shanghai,

China). The samples were digested using DNase I treatment (TakaRa, Dalian, China) to avoid DNA contamination. The treated total RNA was used for cDNA synthesis using the PrimeScript RT-PCR Kit (TakaRa, Dalian, China). qPCR was performed using the SYBR Green Master Mix Kit (Roche Molecular Biochemicals, Germany). The *ACT1* gene was chosen as the internal control gene. The relative transcription levels of genes were analyzed using the $2^{-\Delta\Delta CT}$ method.

Analysis of metabolites by HPLC

At each 12-h interval, the cultures were sampled and centrifuged at 12,000 rpm for 5 min. The filtered samples were analyzed using an HPLC equipped with an Aminex HPX-87H ion-exchange column (Bio-Rad, Hercules, CA, USA) at 45 °C with a mobile phase of 5 mM H_2SO_4 at a flow rate of 0.6 mL/min. The peaks of metabolites including glucose, ethanol, acetic acid and glycerol were detected by refractive index (RI) and ultraviolet (UV) detectors.

Characterization of geraniol and citronellol by GC–MS

To quantify titers of geraniol and citronellol in different cultures, 1 mL of the upper layer (dodecane/culture mixture) was sampled and concentrated at 13,000 rpm for 5 min to separate the dodecane phase, and then the dodecane layer was transferred into a GC vial and stored at −20 °C for analysis. The residual geraniol was extracted again by adding 10% (v/v) dodecane into the medium. Geraniol and citronellol were identified using a GC–MS system (Shimadzu Co., Kyoto, Japan) equipped with an HP-5 ms capillary column (30 m × 0.25 mm × 0.25 μm), an AOC-20i auto-injector, and a QP-2010 mass detector, and the operational conditions were as follows. One microliter of each dodecane sample was injected into the system with a split ratio of 10 and the carrier gas helium was set at a constant flow rate of 0.78 mL/min. The oven temperature was first maintained at 60 °C for 2 min, and then gradually increased to 150 °C at a rate of 10 °C/min, held for 10 min, and finally increased to 230 °C at a rate of 20 °C/min and held for 5 min. The mass spectrometer was set to SIM acquisition mode, scanning m/z ions within the range 40–500 for identification of geraniol and citronellol. The total run time was 30 min. Standard compounds of geraniol and citronellol (Sigma-Aldrich) were dissolved in dodecane and used to plot standard curves for quantification.

Biomass determination

Optical density at 600 nm (OD_{600}) was measured using a spectrophotometer (Eppendorf AG, 22331 Hamburg, Germany). Dry cell weight (DCW) was obtained from OD_{600} measurement after calibration as indicated before (1 g/L biomass $= 0.246 \times (OD_{600}) - 0.0012$) [50].

Additional file

Additional file 1: Table S1. Primers used in this study. **Figure S1.** The conversion of geraniol to citronellol in control strain and deletion strains. The data shown are representative of duplicate experiments, and the error bars represent the standard deviation. **Figure S2.** The effect of *OYE2* or/ and *ATF1* deletion on cell growth in batch fermentation. All strains harbored pZGV6-GE1and pZMVA4 plasmids for geraniol production. The data shown are representative of duplicate experiments, and the error bars represent the standard deviation. **Figure S3.** Transcription level of *ERG20* controlled by different promoters in SD-URA-HIS medium with different concentrations glucose. The cells were collected when OD600 reached 0.6 to extract mRNA for *ERG20* mRNA and transcription determination. The data shown are representative of triplicate experiments, and the error bars represent the standard deviation. **Figure S4.** The effect of *ERG20* expression controlled by different promoters on cell growth in batch fermentation. All strains harbored pZGV6-GE1and pZMVA4 plasmids for geraniol production. The data shown are representative of duplicate experiments, and the error bars represent the standard deviation.

Authors' contributions

JZ and JH designed experiments. JZ, CL and YZ carried out the experiments. JZ, XB, JH and YS analyzed data. JZ and JH wrote the manuscript. All authors read and approved the final manuscript.

Author details

[1] State Key Laboratory of Microbial Technology, School of Life Science, Shandong University, Jinan 250100, China. [2] Shandong Provincial Key Laboratory of Microbial Engineering, School of Bioengineering, QiLu University of Technology, Jinan 250353, China.

Acknowledgements

Not applicable.

Competing interests

The authors declare that they have no competing interests.

Funding

This work was supported by the National Natural Science Foundation of China (31470163), the National Key Technology R&D Program of China (2014BAD02B07), the Key R&D Program of Shandong Province (2015GSF121015) and State Key Laboratory of Microbial Metabolism, Shanghai Jiao Tong University (MMLKF16-06).

References

1. Kuzuyama T. Mevalonate and nonmevalonate pathways for the biosynthesis of isoprene units. Biosci Biotechnol Biochem. 2002;66:1619–27.
2. Brennan TC, Turner CD, Kromer JO, Nielsen LK. Alleviating monoterpene toxicity using a two-phase extractive fermentation for the bioproduction of jet fuel mixtures in *Saccharomyces cerevisiae*. Biotechnol Bioeng. 2012;109:2513–22.
3. Ajikumar PK, Tyo K, Carlsen S, Mucha O, Phon TH, Stephanopoulos G. Terpenoids: opportunities for biosynthesis of natural product drugs using engineered microorganisms. Mol Pharm. 2008;5:167–90.

4. Chen W, Viljoen AM. Geraniol—a review of a commercially important fragrance material. S Afr J Bot. 2010;76:643–51.
5. Peralta-Yahya PP, Keasling JD. Advanced biofuel production in microbes. Biotechnol J. 2010;5:147–62.
6. Brown S, Clastre M, Courdavault V, O'Connor SE. De novo production of the plant-derived alkaloid strictosidine in yeast. Proc Natl Acad Sci USA. 2015;112:3205–10.
7. Zhou J, Wang C, Yoon SH, Jang HJ, Choi ES, Kim SW. Engineering *Escherichia coli* for selective geraniol production with minimized endogenous dehydrogenation. J Biotechnol. 2014;169:42–50.
8. Fischer MJ, Meyer S, Claudel P, Bergdoll M, Karst F. Metabolic engineering of monoterpene synthesis in yeast. Biotechnol Bioeng. 2011;108:1883–92.
9. Liu W, Xu X, Zhang R, Cheng T, Cao Y, Li X, Guo J, Liu H, Xian M. Engineering *Escherichia coli* for high-yield geraniol production with biotransformation of geranyl acetate to geraniol under fed-batch culture. Biotechnol Biofuels. 2016;9:58.
10. Liu J, Zhang W, Du G, Chen J, Zhou J. Overproduction of geraniol by enhanced precursor supply in *Saccharomyces cerevisiae*. J Biotechnol. 2013;168:446–51.
11. Javelot C, Girard P, Colonna-Ceccaldi B, Vladescu B. Introduction of terpene-producing ability in a wine strain of *Saccharomyces cerevisiae*. J Biotechnol. 1991;21:239–51.
12. Paddon CJ, Westfall PJ, Pitera DJ, Benjamin K, Fisher K, McPhee D, Leavell MD, Tai A, Main A, Eng D, et al. High-level semi-synthetic production of the potent antimalarial artemisinin. Nature. 2013;496:528–32.
13. Tippmann S, Scalcinati G, Siewers V, Nielsen J. Production of farnesene and santalene by *Saccharomyces cerevisiae* using fed-batch cultivations with RQ-controlled feed. Biotechnol Bioeng. 2016;113:72–81.
14. Zhuang X, Chappell J. Building terpene production platforms in yeast. Biotechnol Bioeng. 2015;112:1854–64.
15. Ignea C, Cvetkovic I, Loupassaki S, Kefalas P, Johnson CB, Kampranis SC, Makris AM. Improving yeast strains using recyclable integration cassettes, for the production of plant terpenoids. Microb Cell Fact. 2011;10:4.
16. Ignea C, Pontini M, Maffei ME, Makris AM, Kampranis SC. Engineering monoterpene production in yeast using a synthetic dominant negative geranyl diphosphate synthase. ACS Synth Biol. 2014;3:298–306.
17. Jongedijk E, Cankar K, Ranzijn J, van der Krol S, Bouwmeester H, Beekwilder J. Capturing of the monoterpene olefin limonene produced in *Saccharomyces cerevisiae*. Yeast. 2015;32:159–71.
18. Khor GK, Uzir MH. *Saccharomyces cerevisiae*: a potential stereospecific reduction tool for biotransformation of mono- and sesquiterpenoids. Yeast. 2011;28:93–107.
19. Brennan TC, Williams TC, Schulz BL, Palfreyman RW, Kromer JO, Nielsen LK. Evolutionary engineering improves tolerance for replacement jet fuels in *Saccharomyces cerevisiae*. Appl Environ Microbiol. 2015;81:3316–25.
20. Oswald M, Fischer M, Dirninger N, Karst F. Monoterpenoid biosynthesis in *Saccharomyces cerevisiae*. FEMS Yeast Res. 2007;7:413–21.
21. Zhao J, Bao X, Li C, Shen Y, Hou J. Improving monoterpene geraniol production through geranyl diphosphate synthesis regulation in *Saccharomyces cerevisiae*. Appl Microbiol Biotechnol. 2016;100:4561–71.
22. Dai Z, Liu Y, Zhang X, Shi M, Wang B, Wang D, Huang L, Zhang X. Metabolic engineering of *Saccharomyces cerevisiae* for production of ginsenosides. Metab Eng. 2013;20:146–56.
23. Xie W, Ye L, Lv X, Xu H, Yu H. Sequential control of biosynthetic pathways for balanced utilization of metabolic intermediates in *Saccharomyces cerevisiae*. Metab Eng. 2015;28:8–18.
24. Brennan TC, Kromer JO, Nielsen LK. Physiological and transcriptional responses of *Saccharomyces cerevisiae* to D-limonene show changes to the cell wall but not to the plasma membrane. Appl Environ Microbiol. 2013;79:3590–600.
25. Marmulla R, Harder J. Microbial monoterpene transformations—a review. Front Microbiol. 2014;5:346.
26. Gamero A, Manzanares P, Querol A, Belloch C. Monoterpene alcohols release and bioconversion by *Saccharomyces* species and hybrids. Int J Food Microbiol. 2011;145:92–7.
27. Steyer D, Erny C, Claudel P, Riveill G, Karst F, Legras JL. Genetic analysis of geraniol metabolism during fermentation. Food Microbiol. 2013;33:228–34.
28. Liu D, Evans T, Zhang F. Applications and advances of metabolite biosensors for metabolic engineering. Metab Eng. 2015;31:35–43.
29. Yuan J, Ching CB. Dynamic control of ERG9 expression for improved amorpha-4,11-diene production in *Saccharomyces cerevisiae*. Microb Cell Fact. 2015;14:38.
30. Williams TC, Espinosa MI, Nielsen LK, Vickers CE. Dynamic regulation of gene expression using sucrose responsive promoters and RNA interference in *Saccharomyces cerevisiae*. Microb Cell Fact. 2015;14:43.
31. King A, Richard Dickinson J. Biotransformation of monoterpene alcohols by *Saccharomyces cerevisiae*, Torulaspora delbrueckii and Kluyveromyces lactis. Yeast. 2000;16:499–506.
32. Pardo E, Rico J, Gil JV, Orejas M. De novo production of six key grape aroma monoterpenes by a geraniol synthase-engineered *S. cerevisiae* wine strain. Microb Cell Fact. 2015;14:136.
33. Szczebara FM, Chandelier C, Villeret C, Masurel A, Bourot S, Duport C, Blanchard S, Groisillier A, Testet E, Costaglioli P, et al. Total biosynthesis of hydrocortisone from a simple carbon source in yeast. Nat Biotechnol. 2003;21:143–9.
34. Westfall PJ, Pitera DJ, Lenihan JR, Eng D, Woolard FX, Regentin R, Horning T, Tsuruta H, Melis DJ, Owens A, et al. Production of amorphadiene in yeast, and its conversion to dihydroartemisinic acid, precursor to the antimalarial agent artemisinin. Proc Natl Acad Sci USA. 2012;109:E111–8.
35. Li M, Kildegaard KR, Chen Y, Rodriguez A, Borodina I, Nielsen J. De novo production of resveratrol from glucose or ethanol by engineered *Saccharomyces cerevisiae*. Metab Eng. 2015;32:1–11.
36. Zhou YJ, Gao W, Rong Q, Jin G, Chu H, Liu W, Yang W, Zhu Z, Li G, Zhu G, et al. Modular pathway engineering of diterpenoid synthases and the mevalonic acid pathway for miltiradiene production. J Am Chem Soc. 2012;134:3234–41.
37. Odat O, Matta S, Khalil H, Kampranis SC, Pfau R, Tsichlis PN, Makris AM. Old yellow enzymes, highly homologous FMN oxidoreductases with modulating roles in oxidative stress and programmed cell death in yeast. J Biol Chem. 2007;282:36010–23.
38. Trotter EW, Collinson EJ, Dawes IW, Grant CM. Old yellow enzymes protect against acrolein toxicity in the yeast *Saccharomyces cerevisiae*. Appl Environ Microbiol. 2006;72:4885–92.
39. Mason AB, Dufour JP. Alcohol acetyltransferases and the significance of ester synthesis in yeast. Yeast. 2000;16:1287–98.
40. Martin VJ, Pitera DJ, Withers ST, Newman JD, Keasling JD. Engineering a mevalonate pathway in *Escherichia coli* for production of terpenoids. Nat Biotechnol. 2003;21:796–802.
41. Sivy TL, Fall R, Rosenstiel TN. Evidence of isoprenoid precursor toxicity in *Bacillus subtilis*. Biosci Biotechnol Biochem. 2011;75:2376–83.
42. Sarria S, Wong B, Garcia Martin H, Keasling JD, Peralta-Yahya P. Microbial synthesis of pinene. ACS Synth Biol. 2014;3:466–75.
43. Xu P, Li L, Zhang F, Stephanopoulos G, Koffas M. Improving fatty acids production by engineering dynamic pathway regulation and metabolic control. Proc Natl Acad Sci USA. 2014;111:11299–304.
44. Kim SK, Han GH, Seong W, Kim H, Kim SW, Lee DH, Lee SG. CRISPR interference-guided balancing of a biosynthetic mevalonate pathway increases terpenoid production. Metab Eng. 2016;38:228–40.
45. Ginger ML, Chance ML, Sadler IH, Goad LJ. The biosynthetic incorporation of the intact leucine skeleton into sterol by the trypanosomatid *Leishmania mexicana*. J Biol Chem. 2001;276:11674–82.
46. Lv X, Wang F, Zhou P, Ye L, Xie W, Xu H, Yu H. Dual regulation of cytoplasmic and mitochondrial acetyl-CoA utilization for improved isoprene production in *Saccharomyces cerevisiae*. Nat Commun. 2016;7:12851.
47. Knight SA, Labbe S, Kwon LF, Kosman DJ, Thiele DJ. A widespread transposable element masks expression of a yeast copper transport gene. Genes Dev. 1996;10:1917–29.
48. Chen Y, Partow S, Scalcinati G, Siewers V, Nielsen J. Enhancing the copy number of episomal plasmids in *Saccharomyces cerevisiae* for improved protein production. FEMS Yeast Res. 2012;12:598–607.
49. Gietz RD, Woods RA. Transformation of yeast by lithium acetate/single-stranded carrier DNA/polyethylene glycol method. Methods Enzymol. 2002;350:87–96.
50. Hou J, Suo F, Wang C, Li X, Shen Y, Bao X. Fine-tuning of NADH oxidase decreases byproduct accumulation in respiration deficient xylose metabolic *Saccharomyces cerevisiae*. BMC Biotechnol. 2014;14:13.

12

Functional screening of aldehyde decarbonylases for long-chain alkane production by *Saccharomyces cerevisiae*

Min-Kyoung Kang[1†], Yongjin J. Zhou[1,2,5†], Nicolaas A. Buijs[1,6] and Jens Nielsen[1,2,3,4*]

Abstract

Background: Low catalytic activities of pathway enzymes are often a limitation when using microbial based chemical production. Recent studies indicated that the enzyme activity of aldehyde decarbonylase (AD) is a critical bottleneck for alkane biosynthesis in *Saccharomyces cerevisiae*. We therefore performed functional screening to identify efficient ADs that can improve alkane production by *S. cerevisiae*.

Results: A comparative study of ADs originated from a plant, insects, and cyanobacteria were conducted in *S. cerevisiae*. As a result, expression of aldehyde deformylating oxygenases (ADOs), which are cyanobacterial ADs, from *Synechococcus elongatus* and *Crocosphaera watsonii* converted fatty aldehydes to corresponding C_{n-1} alkanes and alkenes. The CwADO showed the highest alkane titer (0.13 mg/L/OD_{600}) and the lowest fatty alcohol production (0.55 mg/L/OD_{600}). However, no measurable alkanes and alkenes were detected in other AD expressed yeast strains. Dynamic expression of SeADO and CwADO under GAL promoters increased alkane production to 0.20 mg/L/OD_{600} and no fatty alcohols, with even number chain lengths from C8 to C14, were detected in the cells.

Conclusions: We demonstrated in vivo enzyme activities of ADs by displaying profiles of alkanes and fatty alcohols in *S. cerevisiae*. Among the AD enzymes evaluated, cyanobacteria ADOs were found to be suitable for alkane biosynthesis in *S. cerevisiae*. This work will be helpful to decide an AD candidate for alkane biosynthesis in *S. cerevisiae* and it will provide useful information for further investigation of AD enzymes with improved activities.

Keywords: Metabolic engineering, *Saccharomyces cerevisiae*, Alkane biosynthesis, Aldehyde decarbonylase, Biofuels

Background

Global warming and depletion of fossil fuels are two urgent matters. Fossil fuels are finite energy resources, but the world energy demand has been increasing along with economic development and population growth. Moreover, increase in carbon dioxide emissions have caused the global temperature to rise resulting in dramatic environmental changes. Therefore, there has been growing interest in sustainable production of biofuels and bio-based chemicals using microorganisms, so called cell factories. Advances in metabolic engineering and synthetic biology enables the production of bio-based chemicals using microbial cell factories [1–5].

One of the most important microbial cell factories, *Saccharomyces cerevisiae* is generally recognized as safe (GRAS) and, it is an extremely well-characterized and tractable organism. Because of its robustness and tolerance towards various stress conditions, it has been intensively used to produce several advanced biofuels and chemicals [6–9].

Alkanes are indispensable chemicals in our daily lives. As major components of current petroleum fuels, the chain lengths of alkanes determine their applications, such as gas (C1–C4), gasoline (C4–C9), jet fuel (C8–C16), diesel (C10–18), and lubricants (C16–C30) [10]. In nature, a variety of organisms synthesize alkanes to protect them against threatening environmental conditions,

*Correspondence: nielsenj@chalmers.se
†Min-Kyoung Kang and Yongjin J. Zhou equally contributed to this work
1 Department of Biology and Biological Engineering, Chalmers University of Technology, Kemivägen 10, 412 96 Gothenburg, Sweden
Full list of author information is available at the end of the article

or to sustain growth [11–13]. However, the alkane production level from natural producers is very low and the alkane formulas are not suitable to replace current petroleum-based alkanes [1, 2]. In addition, current alkane needs are only fulfilled after the challenging and costly cracking processes of crude petroleum. Therefore, many efforts have been made to engineer microorganisms to produce desirable types of alkanes. Several alkane biosynthetic routes have been discovered and various enzymes are available to synthesize alkanes in heterologous hosts [14, 15]. To date, three major precursors, fatty acyl-ACP (or CoA), fatty acids, and fatty aldehydes have been utilized to demonstrate alkane production in engineered microorganisms [14, 15]. Aldehyde decarbonylases (ADs), which were discovered in plants, insects, and cyanobacteria, can convert fatty aldehydes to the corresponding C_{n-1} alkanes by co-producing carbon monoxide (CO), carbon dioxide (CO_2), or formate, respectively. In engineered microbial strains, expression of ADs from a plant (Arabidopsis CER1), an insect (*Drosophila melanogaster* CYP4G1), and various species of cyanobacteria (ADOs) displayed long-chain alkane products [12, 13, 16, 17]. However, the low enzyme activities of cyanobacteria ADs have been noticed and only allow for low alkane titers in *S. cerevisiae* [17–20]. To date, no direct comparative study of ADs from different origins for alkane biosynthesis has been carried out, so we performed a functional screening of different ADs to identify applicable enzyme candidates that can increase alkane production in *S. cerevisiae*. We constructed AD expressing yeast strains and presented the cell metabolite profiles of alkanes and fatty alcohols from each construct. In light of these results, we suggested the most efficient AD enzyme and proposed a strategy to enhance alkane production. As the mechanisms of AD enzymes are not clearly elucidated, our study explored to develop ideal AD enzymes for alkane biosynthesis in yeast cell factories. We anticipate the strategy described here will provide a feasible strategy to functional screening of other AD enzymes for various microbial cell factories.

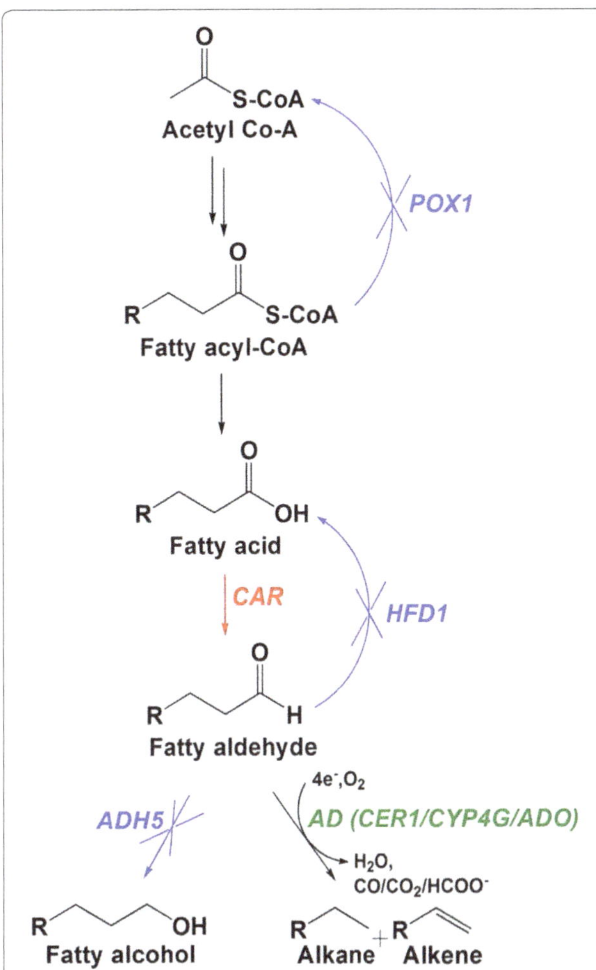

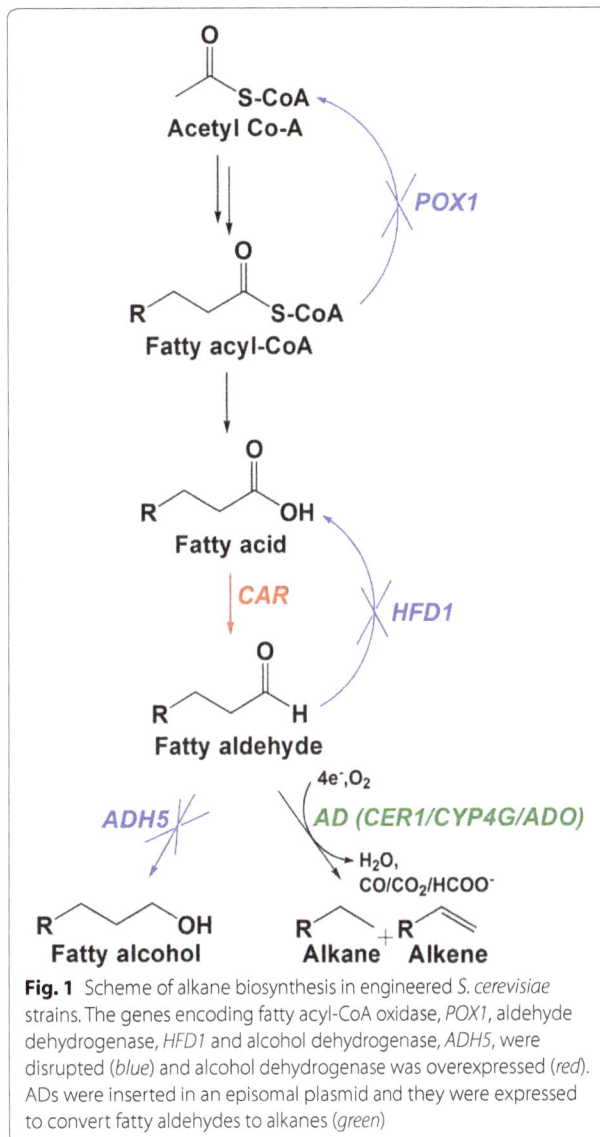

Fig. 1 Scheme of alkane biosynthesis in engineered *S. cerevisiae* strains. The genes encoding fatty acyl-CoA oxidase, *POX1*, aldehyde dehydrogenase, *HFD1* and alcohol dehydrogenase, *ADH5*, were disrupted (*blue*) and alcohol dehydrogenase was overexpressed (*red*). ADs were inserted in an episomal plasmid and they were expressed to convert fatty aldehydes to alkanes (*green*)

Results

Construction of alkane biosynthetic pathways

In our previous study, the fatty acid biosynthetic pathway was engineered to supply sufficient fatty aldehydes in *S. cerevisiae* [18]. Here we used the engineered strain YJZ60 from this study as the background strain. The strain was optimized to accumulate fatty aldehydes in cells by deleting reversible reactions (*POX1* and *HFD1*) and expressing carboxylic acid reductase (*CAR*). One of the competing enzymes, alcohol dehydrogenase, Adh5, was deleted to reduce fatty alcohol accumulation (Fig. 1). In addition, the FNR/Fd reducing systems were expressed to supply sufficient electrons. Figure 1 and Table 1 summarizes information of YJZ60. To enable *S. cerevisiae* to convert the synthesized aldehydes to alkanes, we expressed various ADs by using the episomal plasmid pYX212 in the background strain YJZ60. We introduced three different types of ADs, the *ECERIFERUM1* (CER1) from Arabidopsis plant [16, 21], insect cytochrome p450s (CYP4G1 and CYP4G2) from *D. melanogaster* and house fly [12], and cyanobacteria aldehyde deformylating oxidases (ADOs) from *S. elongatus* [17, 18], *Crocosphaera watsonii*, *Thermosynechococcus elongatus*, and *Cyanothece* sp. PCC 7425 [22] (Table 1; Additional file 1: Figure S1). All AD candidates were selected by literature reviews and preliminary data. Codon-optimized ADO and CER1 genes were expressed under the control of

Table 1 Strains and plasmids used in this study

Name	Description	Reference
Plasmids		
pYX212	2 μm, AmpR, URA3, TPIp, pYX212t	R&D systems
pAlkane78	pYX212-(TPIp-Mdb5-FBA1t-CYC1t-MdCPR-TDH3p-tHXT7P-CYP4G1-pYX212t)	This study
pAlkane8	pYX212-(TPIp-Mdb5-FBA1t-CYC1t-MdCPR-TDH3p-tHXT7P-CYP4G2-pYX212t)	This study
pAlkane71	pYX212-(eTDH3p-CER1-Syn27t-pYX212t)	This study
pAlkane67	pYX212-(eTDH3p-SeADO-pYX212t)	[17]
pAlkane83	pYX212-(eTDH3p-CwADO-pYX212t)	This study
pAlkane84	pYX212-(eTDH3p-TeADO-pYX212t)	This study
pAlkane85	pYX212-(eTDH3p-CyADO-pYX212t)	This study
pAlkane86	pYX212-(CYC1t-CwADO-Gal10p-Gal1p-SeADO-pYX212t)	This study
Strains		
DH5α	F^- (80d *lacZ* M15) (*lacZYA-argF*) U169 *hsd*R17(r^- m^+) *recA*1 *endA*1 *relA*1 *deoR*96	
YJZ60	MATa MAL2-8c SUC2 his3Δ1ura3-52 hfd1Δpox1Δ Gal80Δ:: SeFNR + SeFd adh5Δ::(TPIp-MmCAR-FBA1t) + (PGK1p-EcFNR-CYC1t) + (TEF1p-EcFD-TDH2t) + (tHXT7p-npgA-ADH5t)	[17]
Con	YJZ60 strain harboring pYX212	This study
CYP4G1	YJZ60 strain harboring pAlkane78	This study
CYP4G2	YJZ60 strain harboring pAlkane8	This study
CER1	YJZ60 strain harboring pAlkane71	This study
SeADO	YJZ60 strain harboring pAlkane67	This study
CwADO	YJZ60 strain harboring pAlkane83	This study
TeADO	YJZ60 strain harboring pAlkane84	This study
CyADO	YJZ60 strain harboring pAlkane85	This study
CSADO	YJZ60 strain harboring pAlkane86	This study

the enhanced TDH3 promoter [23], while CYP4G1 and CYP4G2 were expressed under the control of the truncated HXT7 promoter, tHXT7p [24], yeast *S. cerevisiae* allowing constitutive expression independent of extracellular glucose levels. Additional file 1: Figure S1 provides brief features of the used gene expression modules. In the CSADO strain, *C. watsoni* and *S. elongatus* ADOs were co-expressed under the control of the *GAL1* and *GAL10* promoters, respectively (Table 1; Additional file 1: Figure S1) to alleviate the growth inhibition by separating cell growth and gene expression.

Evaluation of ADs for alkane biosynthesis in *S. cerevisiae*

After the introduction of ADs in YJZ60, we carried out functional evaluation of three different types of ADs (CER1, CYP4G, and ADO). Among all the AD constructs tested, only two cyanobacterial ADOs from *S. elongatus* (SeADO) and *C. watsonii* (CwADO) produced long-chain alkanes and alkenes. Expression of ADOs from *S. elongatus* and *C. watsonii*, reached 0.11 and 0.13 mg/L/OD_{600} of total alkanes and alkenes, respectively, with different odd chain lengths from C11 to C17 (Fig. 2a; Additional file 1: Figure S2a). The major compounds in both strains were pentadecane (C15) and 7-pentadecene (C15:1) (Additional file 1: Figure S2a).

We found accumulation of fatty alcohols in all the engineered strains (Fig. 2b). This is consistent with previous observations that fatty alcohols are produced as significant by-products in engineered *S. cerevisiae* expressing alkane biosynthesis, and might be caused by endogenous aldehyde reductases (ALRs) and alcohol dehydrogenases (ADHs) [17, 18].The control strain Con without AD had the highest fatty alcohol accumulation (1.71 mg/L/OD_{600}, Fig. 2) with even number chain lengths from C8 to C18 and the CwADO strain produced the lowest amount of fatty alcohols (0.55 mg/L/OD_{600}) in the cells (Additional file 1: Figure S2). Other AD expressing strains produced fatty alcohol levels in between these strains, i.e. TeADO: 1.29 mg/L/OD_{600}, CER1: 1.44 mg/L/OD_{600}, CYP4G1: 1.33 mg/L/OD_{600}, and CYP4G2: 0.97 mg/L/OD_{600} (Fig. 2b). The alkane production is much lower than the decrease in fatty alcohol accumulation when the CwADO and SeADO strains are compared with the control strain (Fig. 2a, b), and suggests that the functional ADs have a high binding affinity for fatty aldehydes, but low catalytic efficiency for alkane biosynthesis.

Though the CwADO strain had the highest alkane production and the lowest fatty alcohol production, this strain showed very poor growth (OD_{600} of 3.5 at 72 h) compared with the SeADO (OD_{600} of 6.1 at 72 h) and

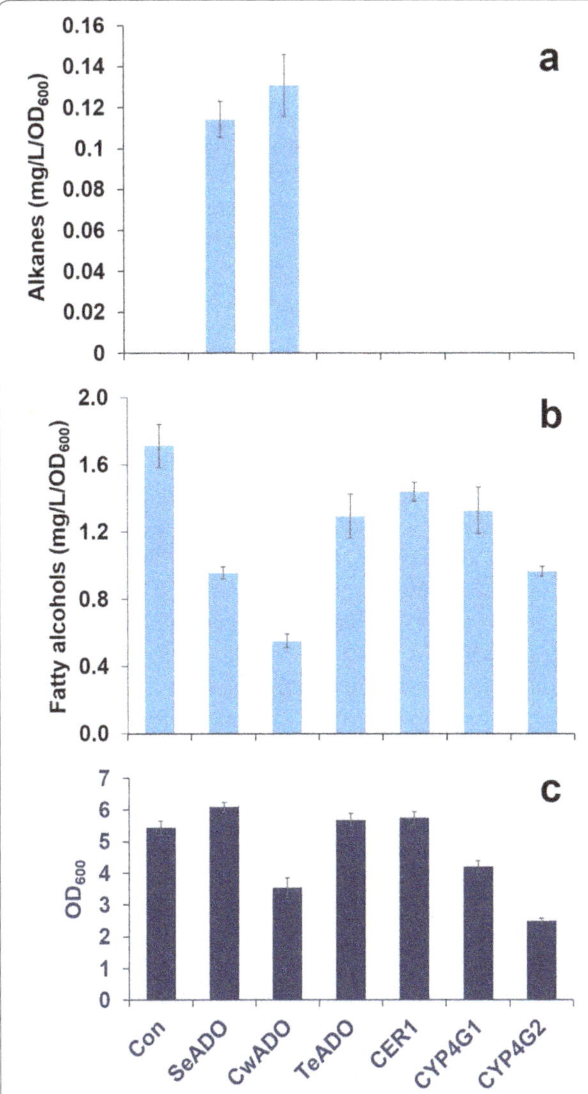

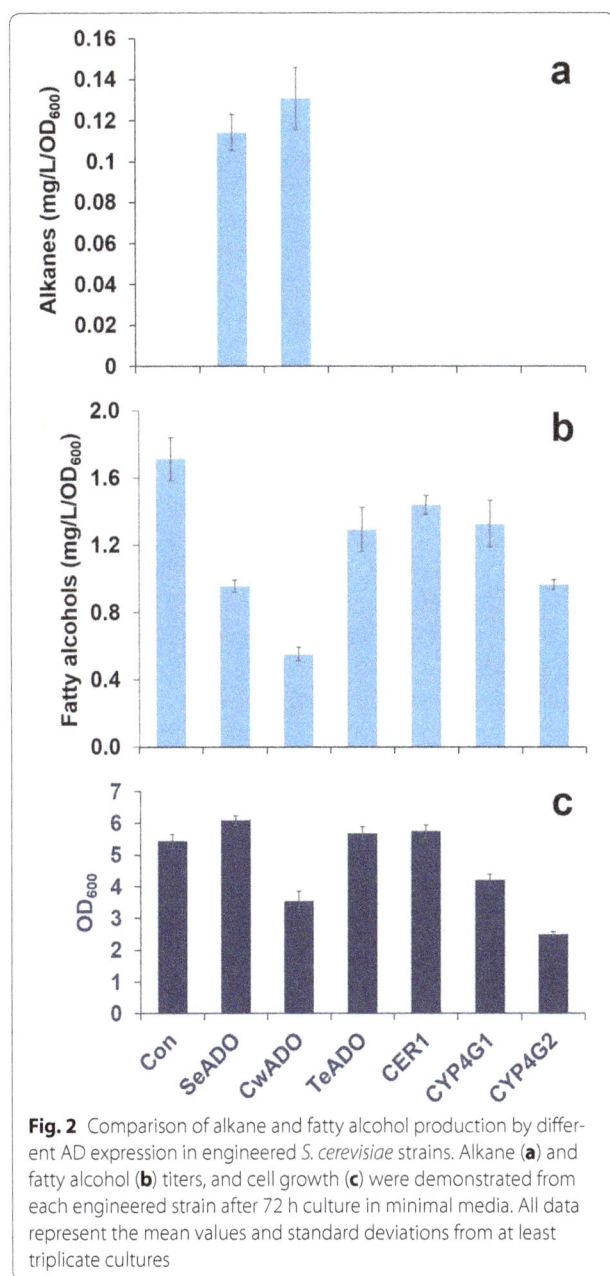

Fig. 2 Comparison of alkane and fatty alcohol production by different AD expression in engineered *S. cerevisiae* strains. Alkane (**a**) and fatty alcohol (**b**) titers, and cell growth (**c**) were demonstrated from each engineered strain after 72 h culture in minimal media. All data represent the mean values and standard deviations from at least triplicate cultures

control strains (OD_{600} of 5.4 at 72 h), which might be attributed to toxicity (Fig. 2c). For this reason, the total amount of alkanes and alkenes produced by the CwADO strain (0.53 mg/L) is lower than with the SeADO strain (0.76 mg/L) (Fig. 2a). Improving cell growth of the CwADO expressing strain could therefore potentially further increase alkane production.

Enhancement of alkane production

In order to relieve the toxicity of expressing CwADO in the cell, we dynamically expressed CwADO by using the *GAL1* promoter (GAL1p) in combination with the *GAL80* deletion. It has been found that the *GAL1* promoter has very low expression in the glucose phase due to Migl repression, but is strongly expressed after glucose consumption in a *GAL80* deletion strain [25]. Hereby CwADO expression could be separated from cell growth, as has been previously applied for improving isoprenoid production by yeast [26]. To further increase alkane biosynthesis, we co-expressed SeADO under the control of the *GAL10* promoter (Gal10p). The resulting strain CSADO had significantly higher specific alkane production of 0.20 mg/L/OD_{600}, (Fig. 3a) which was 35 and 45% higher compared with the CwADO and SeADO strains, respectively (Fig. 3a). We even detected undecane (C11) in the CSADO strain (Additional file 1: Figure S3a). Furthermore, CSADO had 62% higher biomass (OD_{600} of 5.7 at 72 h, Fig. 3c) than the strain CwADO expressed under the TDH3 promoter (Fig. 3c), which indicated that the dynamic control strategy relieved the toxicity of CwADO expression. As a benefit to improved cell growth, the alkane titer reached 1.14 mg/L, which is higher than with our previous strain A6 that had systematic pathway optimization [18]. This suggests that functional AD screening with dynamic expression could be an efficient strategy for enhancing alkane production in yeast.

Discussion

In this study, long-chain alkane biosynthesis has been constructed via decarbonylation of fatty aldehydes by AD enzymes in *S. cerevisiae* [12, 16–18]. However, efficient incorporating of heterologous metabolic pathways into *S. cerevisiae* is challenging and strong endogenous ALRs/ADHs compete with the intermediate fatty aldehydes [18]. Indeed, low catalytic efficiency of ADs has been referred to as a critical bottleneck in alkane biosynthesis in engineered *S. cerevisiae* strains [17–19]. Therefore, it is worthwhile to screen efficient AD enzymes to provide a rationale enzyme for the improvement of alkane biosynthesis in microbial cell factories. To meet this goal, we carried out functional screening of ADs from different origins by comparing alkane and fatty alcohol accumulation in the cells. ADs were introduced using episomal plasmids and expressed in an engineered yeast strain, YJZ60, which provides fatty aldehydes as substrates for alkane biosynthesis. Of all the strains we tested, cyanobacteria ADOs (SeADO and CwADO) synthesized alkanes more efficiently than the CER1 and CYP4G enzymes (Fig. 2a). Even though very-long-chain (VLC) alkane production by CER1 and CYP4G1 have been reported in yeast strains [12, 16], we only found a reduction of fatty alcohol accumulation, but no detectable amounts of alkanes were produced in our yeast strains. We assume substrate preferences of plant and insect ADs might explain this. In fact, plants and insects synthesize

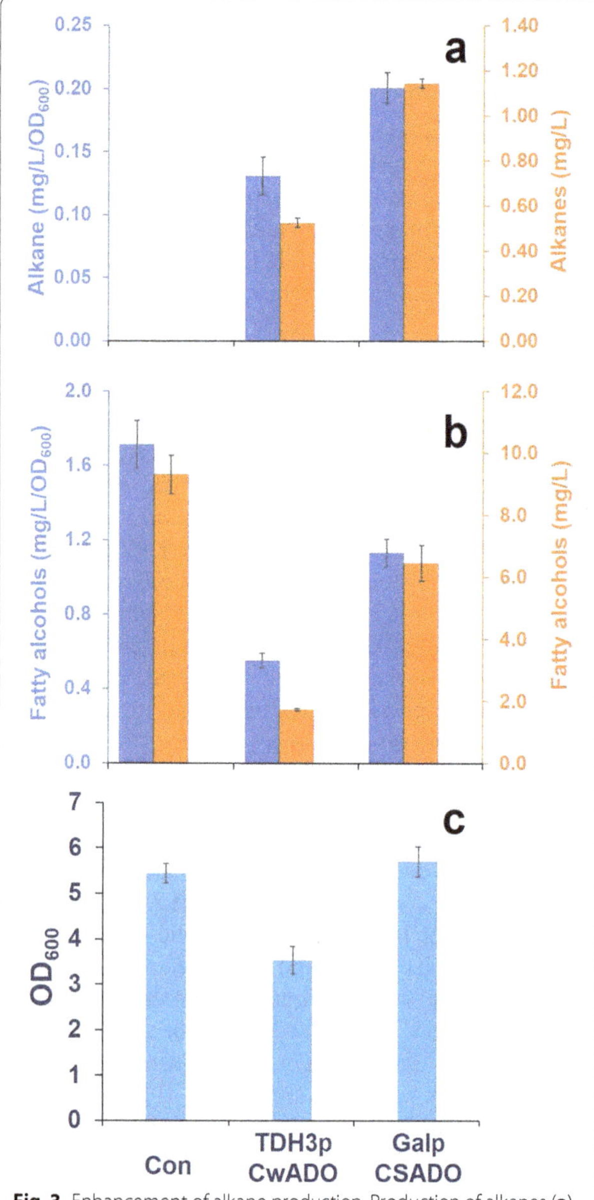

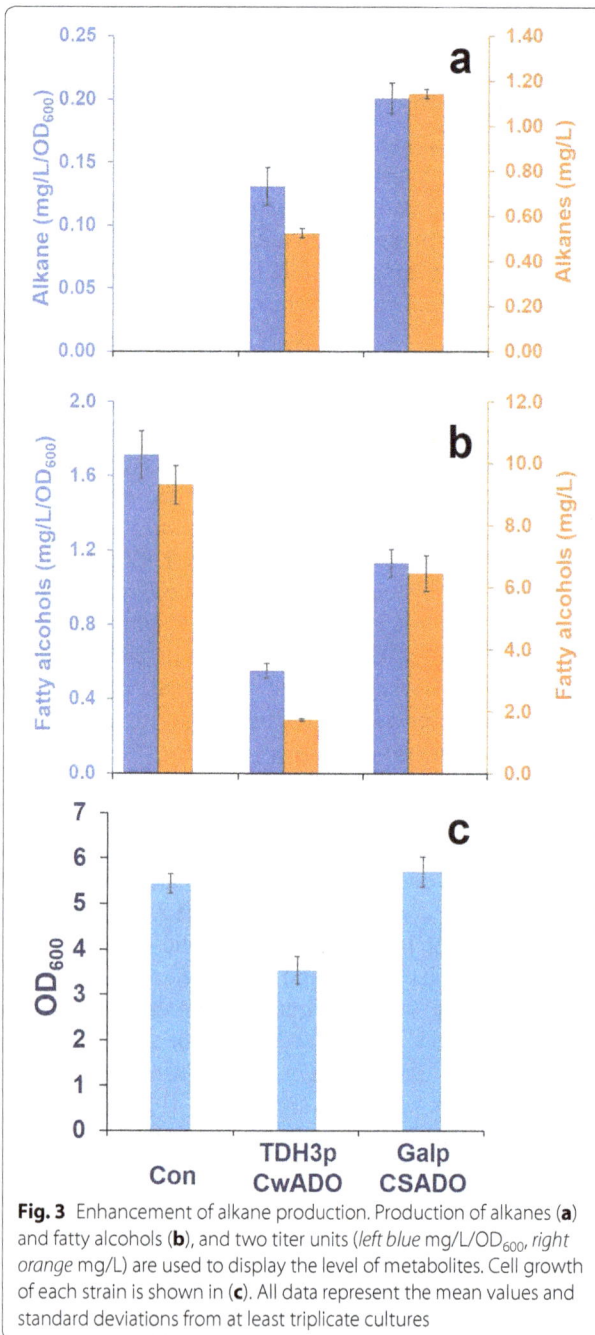

Fig. 3 Enhancement of alkane production. Production of alkanes (**a**) and fatty alcohols (**b**), and two titer units (*left blue* mg/L/OD_{600}, *right orange* mg/L) are used to display the level of metabolites. Cell growth of each strain is shown in (**c**). All data represent the mean values and standard deviations from at least triplicate cultures

VLC alkanes to form a wax layer and cuticular hydrocarbons, respectively, for environmental protection [27, 28]. Arabidopsis (CER1) synthesized VLC alkanes with the range of chain lengths being C27–C31 [16], and insect CYP4G family produces C23–C33 chain lengths VLC alkanes [12, 29, 30]. Through the distribution of fatty alcohol chain lengths (Additional file 1: Figure S2b), we predicted accumulation of fatty aldehydes with the even number of chain lengths, C8–C18 in our background strain, which might be unfavorable substrates for CER1 and CYP4Gs. Meanwhile, major alkane products synthesized by cyanobacteria ADOs are pentadecane (C15) and heptadecane (C17) [13], and both compounds were also major metabolites in our SeADO and CwADO yeast strains (Additional file 1: Figure S2a). Another presumable reason might be the environmental condition for proper function of the AD enzymes. The yeast cytosol may not be an optimized compartment for function of CER1 and CYP4Gs. To date, cyanobacteria ADO is the only group of AD enzymes, which have been demonstrated to have in vitro enzyme activity [31–33]. Plant origin CER1 is an endoplasmic reticulum membrane bound protein and CYP4G1 is localized in oenocytes [12]. The membrane protein expression often causes cell stresses and lower the biomass and expression. In addition, folding and solubility of eukaryotic membrane proteins is generally causing difficulties for performing kinetic studies [34], so no enzyme activity studies have been successfully conducted. Likewise, the membrane association in plant cells may cause problems for proper function of the enzyme in the yeast cytosol. Moreover, the relatively larger size of CER1 and CYP4Gs may cause problems with folding and expression. Moreover, alkane peaks in a GC–MS chromatogram cannot be detected if the AD enzyme has low and slow activity. Because inefficient aldehyde conversion to alkanes leads to high fatty alcohol formation, and the fatty alcohol peaks cover the alkane detection area further causing difficulties in detecting alkanes.

The CwADO enzyme was revealed as a better enzyme compared with the SeADO, but expression of CwADO caused poor growth and negatively affected the final titer of alkanes. Thus, we replaced the *TDH3* promoter with a *GAL1* promoter to control the gene expression, and we placed additional SeADO right after the *GAL10* promoter to co-express CwADO and SeADO in an episomal plasmid (Additional file 1: Figure S1). In our previous work, expression of additional ADO from *Nostoc punctiforme* (SeADO-NpADO) resulted in a 5% increase in alkane titer (0.82 mg/L) compared with only expressing SeADO (0.78 mg/L) [18]. In the case of the CSADO strain (CwADO-SeADO), co-expression of CwADO achieved a significant improvement in alkane titer by 33% (SeADO: 0.76 mg/L, CSADO: 1.14 mg/L) (Figs. 2a, 3a) and surprisingly no fatty alcohols with even number chain lengths C8–C14 were detected (Additional file 1: Figure S3b). In addition, the chain lengths of alkanes were extended from C11 to C17 (Additional file 1: Figure S3a) and growth was greatly improved in the CSADO strain (OD_{600} of 3.5 at 72 h, Fig. 3c) compared with the CwADO strain (OD_{600} of 5.7 at 72 h, Fig. 3c). Even though the CSADO strain lead to increase in alkane production, it was still far from the industrial requirements and even below the alkane titer in engineered *E. coli* (580.8 mg/L) and cyanobacteria (300 mg/L) [13, 35]. Unlike *E. coli* platforms, even the same enzymes involved in alkane biosynthesis produced much smaller quantities of alkanes in *S. cerevisiae* strains. Expression of CER1 enzyme in *E. coli* achieved the highest alkane titer [35], and the ADO enzymes from *T. elongatus*, and *Cyanothece* sp. also produced high amount of alkanes in *E. coli* [13, 22]. However, even trace alkanes were not observed in our CER1, TeADO, and CyADO yeast strains for uncertain reasons. Similar to the AD enzymes, expression of OleT decarboxylase, a terminal alkene producing enzyme, resulted in much higher terminal alkene production in *E. coli* (97.6 mg/L) than in *S. cerevisiae* (3.5 mg/L) [14]. To explain the big differences in alkane titer between *E. coli* and *S. cerevisiae*, other facts should be considered beyond the poor catalytic efficiencies of alkane producing enzymes.

Conclusion

In this study, we examined the functional performance of ADs in engineered yeast strains. Based on the metabolite profiles of our engineered strains, we proposed advisable ADs and their applications to enhance alkane production in *S. cerevisiae*. Our study further provides a platform strain that can be used for screening ADs to be used for alkane production in yeast with the objective to develop a yeast cell factory that can be used for bio-based production of alkanes.

Methods

Construction of plasmids and yeast strains

Plasmids and strains used in this study are shown in Table 1. Plasmid construction was performed by the modular pathway engineering procedure as described by Zhou et al. [36]. DNA fragments for module construction were prepared by PCR amplification and each module was constructed by fusion PCR. PrimeSTAR was used for all the PCR processes, and primers used in this work were listed in Additional file 1: Table S1. Yeast transformation was conducted by LiAc/SS carrier DNA/PEG method [37], and constructed modules and linearized pYX212 plasmid backbone were used as DNA templates. To make yeast competent cells, the YJZ60 yeast strain was cultured at 30 °C and 200 rpm in YPD media, and transformants were selected on synthetic defined (SD) agar plates, which contained 6.9 g/L yeast nitrogen base without amino acids (Formedium, Hunstanton, UK), 0.77 g/L synthetic complete supplement mixture without uracil (Formedium), 20 g/L glucose (Merck Millipore) and 20 g/L agar (Merck Millipore). After the colony selection, yeast plasmids were extracted and introduced into *E. coli* DH5α competent cells to confirm the final plasmid constructs. *E. coli* colonies were selected on Lysogeny Broth (LB) agar plate containing 100 μg/mL ampicillin, and they were confirmed by DNA sequencing.

Alkane biosynthesis and extraction

To produce alkanes, engineered *S. cerevisiae* strains were grown in 100 mL shake flasks containing 15 mL mineral media [38] plus 40 mg/L histidine and 30 g/L glucose at 30 °C and 200 rpm for 72 h. After the cultivation, 10 mL of cell cultures were harvested by centrifugation at 2000*g* for 10 min, and then cell pellets were dried for 48 h in a freeze-dryer. The dried cells were extracted by the method described by Khoomrung [39] by using 4 mL chloroform: methanol (v/v 2:1) solution containing hexadecane (0.5 μg/mL) and pentadecanol (0.01 mg/mL) as internal standards. After centrifugal vacuum concentration, the final dried samples were dissolved in 200 μL hexane.

Metabolite analysis and quantification

Alkanes and alkenes were analyzed by gas chromatography (Focus GC, ThermoFisher Scientific) equipped with a Zebron ZB-5MS GUARDIAN capillary column (30 m × 0.25 mm × 0.25 mm, Phenomenex, Torrance, CA, USA) and a DSQII mass spectrometer (Thermo Fisher Scientific, Waltham, MA, USA). The GC program for alkanes and alkenes was as follows: initial temperature of 50 °C, hold for 5 min; then ramp to 140 °C at a rate of 10 °C per min and hold for 10 min; ramp to 310 °C at a rate of 15 °C per min and hold for 7 min. Fatty alcohols were quantitatively analyzed by GC-FID (Thermo Fisher Scientific, Waltham, MA, USA) equipped with a ZB-5MS GUARDIAN capillary column, and helium was used as carrier gas at a flow rate of 1 mL/min. GC program for fatty alcohol quantification was as follows: initial temperature of 45 °C hold for 2 min; then ramp to 220 °C at a rate of 20 °C per min and hold for 2 min; ramp to 300 °C at a rate of 20 °C per min and hold for 5 min.

Additional file

Additional file 1: Figure S1. Scheme of plasmid constructs for alkane biosynthesis. pYX212 vector was used as a backbone to express ADs in engineered *S. cerevisiae* strains. **Figure S2.** Comparison of alkane and fatty alcohol production by different AD expression in engineered *S. cerevisiae* strains. Alkane (a) and fatty alcohol (b) titers were displayed with the information of chain-length distribution of each engineered strain. All data represent the mean values and standard deviations from at least triplicate cultures. **Figure S3.** Production of alkanes (a) and fatty alcohols (b) with the information of chain-length distribution in the CSADO strain. All data represent the mean values and standard deviations from at least triplicate cultures. **Table S1.** Primers used in this study.

Abbreviations

AD: aldehyde decarbonylase; ADH: alcohol dehydrogenase; ADO: aldehyde deformylating oxygenase; ALR: aldehyde reductase; CAR: carboxylic acid reductase; PDH: pyruvate dehydrogenase; VLC: very-long-chain.

Authors' contributions

The experiments were designed by YJZ and MK. MK and YJZ carried out the strain construction. NA provided preliminary data for enzyme selection. MK performed the cultivation experiments, analyzed the results and drafted the manuscript. YJZ, NA and MK revised the manuscript. JN supervised the design, revised the manuscript and coordinated the study. All authors read and approved the final manuscript.

Author details

[1] Department of Biology and Biological Engineering, Chalmers University of Technology, Kemivägen 10, 412 96 Gothenburg, Sweden. [2] Novo Nordisk Foundation Center for Biosustainability, Chalmers University of Technology, 412 96 Gothenburg, Sweden. [3] Novo Nordisk Foundation Center for Biosustainability, Technical University of Denmark, Kogle allé, 2970 Hørsholm, Denmark. [4] Science for Life Laboratory, Royal Institute of Technology, 17121 Solna, Sweden. [5] Present Address: Division of Biotechnology, Dalian Institute of Chemical Physics, Chinese Academy of Sciences, Dalian 116023, China. [6] Present Address: Evolva Biotech, Lersø Parkalle, 40-42, 2100 Copenhagen, Denmark.

Acknowledgements

This work was financed by the Novo Nordisk Foundation, Vetenskapsrådet, FORMAS and the Knut and Alice Wallenberg Foundation.

Competing interests

The authors declare that they have no competing interests.

References

1. Peralta-Yahya PP, Zhang F, Del Cardayre SB, Keasling JD. Microbial engineering for the production of advanced biofuels. Nature. 2012;488:320–8.
2. Pfleger BF, Gossing M, Nielsen J. Metabolic engineering strategies for microbial synthesis of oleochemicals. Metab Eng. 2015;29:1–11.
3. Chen Y, Nielsen J. Advances in metabolic pathway and strain engineering paving the way for sustainable production of chemical building blocks. Curr Opin Biotechnol. 2013;24:965–72.
4. Nielsen J, Keasling JD. Engineering cellular metabolism. Cell. 2016;164:1185–97.
5. Gustavsson M, Lee SY. Prospects of microbial cell factories developed through systems metabolic engineering. Microb Biotechnol. 2016;9:610–7.
6. Borodina I, Nielsen J. Advances in metabolic engineering of yeast *Saccharomyces cerevisiae* for production of chemicals. Biotechnol J. 2014;9:609–20.
7. Buijs NA, Siewers V, Nielsen J. Advanced biofuel production by the yeast *Saccharomyces cerevisiae*. Curr Opin Chem Biol. 2013;17:480–8.
8. Nielsen J, Larsson C, van Maris A, Pronk J. Metabolic engineering of yeast for production of fuels and chemicals. Curr Opin Biotechnol. 2013;24:398–404.
9. Zhou YJ, Buijs NA, Siewers V, Nielsen J. Fatty acid-derived biofuels and chemicals production in *Saccharomyces cerevisiae*. Front Bioeng Biotechnol. 2014;2:32.
10. Wade L Jr. Organic chemistry. 6th ed. New Jersey: Pearson Prentice Hall; 2006.
11. Post-Beittenmiller D. Biochemistry and molecular biology of wax production in plants. Annu Rev Plant Physiol Plant Mol Biol. 1996;47:405–30.
12. Qiu Y, Tittiger C, Wicker-Thomas C, Le Goff G, Young S, Wajnberg E, Fricaux T, Taquet N, Blomquist GJ, Feyereisen R. An insect-specific P450 oxidative decarbonylase for cuticular hydrocarbon biosynthesis. Proc Natl Acad Sci. 2012;109:14858–63.
13. Schirmer A, Rude MA, Li X, Popova E, del Cardayre SB. Microbial biosynthesis of alkanes. Science. 2010;329(5991):559–62.
14. Kang MK, Nielsen J. Biobased production of alkanes and alkenes through metabolic engineering of microorganisms. J Ind Microbiol Biotechnol. 2016;1–10. doi:10.1007/s10295-016-1814-y.
15. Herman NA, Zhang W. Enzymes for fatty acid-based hydrocarbon biosynthesis. Curr Opin Chem Biol. 2016;35:22–8.
16. Bernard A, Domergue F, Pascal S, Jetter R, Renne C, Faure J-D, Haslam RP, Napier JA, Lessire R, Joubès J. Reconstitution of plant alkane biosynthesis in yeast demonstrates that *Arabidopsis* ECERIFERUM1 and ECERIFERUM3 are core components of a very-long-chain alkane synthesis complex. Plant Cell. 2012;24:3106–18.
17. Buijs NA, Zhou YJ, Siewers V, Nielsen J. Long-chain alkane production by the yeast *Saccharomyces cerevisiae*. Biotechnol Bioeng. 2015;112:1275–9.
18. Zhou YJ, Buijs NA, Zhu Z, Qin J, Siewers V, Nielsen J. Production of fatty acid-derived oleochemicals and biofuels by synthetic yeast cell factories. Nat Commun. 2016;7:11709.
19. Marsh ENG, Waugh MW. Aldehyde decarbonylases: enigmatic enzymes of hydrocarbon biosynthesis. ACS Catal. 2013;3:2515–21.
20. Zhou YJ, Buijs NA, Zhu Z, Gomez DO, Boonsombuti A, Siewers V, Nielsen J. Harnessing yeast peroxisomes for biosynthesis of fatty-acid-derived biofuels and chemicals with relieved side-pathway competition. J Am Chem Soc. 2016;138:15368–77.
21. Bourdenx B, Bernard A, Domergue F, Pascal S, Léger A, Roby D, Pervent M, Vile D, Haslam RP, Napier JA. Overexpression of Arabidopsis *ECERIFERUM1* promotes wax very-long-chain alkane biosynthesis and influences plant response to biotic and abiotic stresses. Plant Physiol. 2011;156:29–45.
22. Coursolle D, Lian J, Shanklin J, Zhao H. Production of long chain alcohols and alkanes upon coexpression of an acyl-ACP reductase and aldehyde-deformylating oxygenase with a bacterial type-I fatty acid synthase in *E. coli*. Mol BioSyst. 2015;11:2464–72.
23. Blazeck J, Garg R, Reed B, Alper HS. Controlling promoter strength and regulation in *Saccharomyces cerevisiae* using synthetic hybrid promoters. Biotechnol Bioeng. 2012;109:2884–95.
24. Hauf J, Zimmermann FK, Müller S. Simultaneous genomic overexpression of seven glycolytic enzymes in the yeast *Saccharomyces cerevisiae*. Enzyme Microb Technol. 2000;26:688–98.
25. Peng B, Plan MR, Carpenter A, Nielsen LK, Vickers CE. Coupling gene regulatory patterns to bioprocess conditions to optimize synthetic metabolic modules for improved sesquiterpene production in yeast. Biotechnol Biofuels. 2017;10:43.
26. Westfall PJ, Pitera DJ, Lenihan JR, Eng D, Woolard FX, Regentin R, Horning T, Tsuruta H, Melis DJ, Owens A, et al. Production of amorphadiene in yeast, and its conversion to dihydroartemisinic acid, precursor to the antimalarial agent artemisinin. Proc Natl Acad Sci. 2012;109:E111–8.
27. Howard RW, Blomquist GJ. Ecological, behavioral, and biochemical aspects of insect hydrocarbons. Annu Rev Entomol. 2005;50:371–93.
28. Samuels L, Kunst L, Jetter R. Sealing plant surfaces: cuticular wax formation by epidermal cells. Annu Rev Plant Biol. 2008;59:683–707.
29. Yu Z, Zhang X, Wang Y, Moussian B, Zhu KY, Li S, Ma E, Zhang J. LmCYP4G102: an oenocyte-specific cytochrome P450 gene required for cuticular waterproofing in the migratory locust, *Locusta migratoria*. Sci Rep. 2016;6:29980.
30. Chen N, Fan Y-L, Bai Y, Li X-D, Zhang Z-F, Liu T-X. Cytochrome P450 gene, CYP4G51, modulates hydrocarbon production in the pea aphid, *Acyrthosiphon* pisum. Insect Biochem Mol Biol. 2016;76:84–94.
31. Eser BE, Das D, Han J, Jones PR, Marsh EN. Oxygen-independent alkane formation by non-heme iron-dependent cyanobacterial aldehyde decarbonylase: investigation of kinetics and requirement for an external electron donor. Biochemistry. 2011;50:10743–50.
32. Andre C, Kim SW, Yu XH, Shanklin J. Fusing catalase to an alkane-producing enzyme maintains enzymatic activity by converting the inhibitory byproduct H_2O_2 to the cosubstrate O_2. Proc Natl Acad Sci. 2013;110:3191–6.

33. Khara B, Menon N, Levy C, Mansell D, Das D, Marsh ENG, Leys D, Scrutton NS. Production of propane and other short-chain alkanes by structure-based engineering of ligand specificity in aldehyde-deformylating oxygenase. Chembiochem. 2013;14:1204–8.
34. Smith SM. Strategies for the purification of membrane proteins. Methods Mol Biol. 2011;681:485–96.
35. Choi YJ, Lee SY. Microbial production of short-chain alkanes. Nature. 2013;502:571–4.
36. Zhou YJ, Gao W, Rong Q, Jin G, Chu H, Liu W, Yang W, Zhu Z, Li G, Zhu G, et al. Modular pathway engineering of diterpenoid synthases and the mevalonic acid pathway for miltiradiene production. J Am Chem Soc. 2012;134:3234–41.
37. Agatep R, Kirkpatrick RD, Parchaliuk DL, Woods RA, Gietz RD. Transformation of *Saccharomyces cerevisiae* by the lithium acetate/single-stranded carrier DNA/polyethylene glycol protocol. Tech Tips Online. 1998;3:133–7.
38. Verduyn C, Postma E, Scheffers WA, Van Dijken JP. Effect of benzoic acid on metabolic fluxes in yeasts: a continuous-culture study on the regulation of respiration and alcoholic fermentation. Yeast. 1992;8:501–17.
39. Khoomrung S, Chumnanpuen P, Jansa-Ard S, Stahlman M, Nookaew I, Boren J, Nielsen J. Rapid quantification of yeast lipid using microwave-assisted total lipid extraction and HPLC-CAD. Anal Chem. 2013;85:4912–9.

13

Functional differentiation of 3-ketosteroid Δ^1-dehydrogenase isozymes in *Rhodococcus ruber* strain Chol-4

Govinda Guevara[1], Laura Fernández de las Heras[2], Julián Perera[1] and Juana María Navarro Llorens[1*]

Abstract

Background: The *Rhodococcus ruber* strain Chol-4 genome contains at least three putative 3-ketosteroid Δ^1-dehydrogenase ORFs (*kstD1*, *kstD2* and *kstD3*) that code for flavoenzymes involved in the steroid ring degradation. The aim of this work is the functional characterization of these enzymes prior to the developing of different biotechnological applications.

Results: The three *R. ruber* KstD enzymes have different substrate profiles. KstD1 shows preference for 9OHAD and testosterone, followed by progesterone, deoxy corticosterone AD and, finally, 4-BNC, corticosterone and 19OHAD. KstD2 shows maximum preference for progesterone followed by 5α-Tes, DOC, AD testosterone, 4-BNC and lastly 19OHAD, corticosterone and 9OHAD. KstD3 preference is for saturated steroid substrates (5α-Tes) followed by progesterone and DOC. A preliminary attempt to model the catalytic pocket of the KstD proteins revealed some structural differences probably related to their catalytic differences. The expression of *kstD* genes has been studied by RT-PCR and RT-qPCR. All the *kstD* genes are transcribed under all the conditions assayed, although an additional induction in cholesterol and AD could be observed for *kstD1* and in cholesterol for *kstD3*. Co-transcription of some correlative genes could be stated. The transcription initiation signals have been searched, both in silico and in vivo. Putative promoters in the intergenic regions upstream the *kstD1*, *kstD2* and *kstD3* genes were identified and probed in an apramycin-promoter-test vector, leading to the functional evidence of those *R. ruber kstD* promoters.

Conclusions: At least three putative 3-ketosteroid Δ^1-dehydrogenase ORFs (*kstD1*, *kstD2* and *kstD3*) have been identified and functionally confirmed in *R. ruber* strain Chol-4. KstD1 and KstD2 display a wide range of substrate preferences regarding to well-known intermediaries of the cholesterol degradation pathway (9OHAD and AD) and other steroid compounds. KstD3 shows a narrower substrate range with a preference for saturated substrates. KstDs differences in their catalytic properties was somehow related to structural differences revealed by a preliminary structural modelling. Transcription of *R. ruber kstD* genes is driven from specific promoters. The three genes are constitutively transcribed, although an additional induction is observed in *kstD1* and *kstD3*. These enzymes have a wide versatility and allow a fine tuning-up of the KstD cellular activity.

Keywords: *Rhodococcus ruber*, 3-Ketosteroid-Δ^1-dehydrogenase, Promoters, Expression, Steroids

Background

Rhodococci are aerobic Gram-positive soil bacteria belonging to the Actinomycetes group. They show a broad catabolic diversity over different substrates, from pollutants to many aromatic compounds, including steroids and sterols [1–3]. Steroids are a source of contamination of soil and waters and their presence has been detected even in drinking water, threatening many ways of life and public health [4–6]. Rhodococci can be useful in this biodegradation field due to their metabolic versatility and steroids degradation capability. On the

*Correspondence: joana@bio.ucm.es
[1] Department of Biochemistry and Molecular Biology I, Universidad Complutense de Madrid, 28040 Madrid, Spain
Full list of author information is available at the end of the article

other hand *Rhodococcus* spp. are potential biotechnological tools [3, 7] as they can provide with key enzymes essential for certain reactions that yield industrial needed intermediaries such as 4-androstene-3,17-dione (AD) and 1,4-androstadiene-3,17-dione (ADD) [8].

But before exploiting all the advantages the different rhodococci offer, it is essential to know how these bacteria degrade steroids and which enzymes are involved in this process.

Steroids are molecules with a carbon skeleton of 4 fused rings (A to D) and a side chain up to 10 carbons. During the last years, the increasing number of studies concerning steroid degradation, and more concretely the degradation of cholesterol in bacteria, have clarified some of the catabolic steps (e.g. initiation of the ring degradation by either a NAD^+-dependent 3β-hydroxysteroid dehydrogenase or a cholesterol oxidase) although other steps still remain unclear (e.g. the processing of the C and D rings of the steroid structure or the relative order in which the different steps of the degradation of ring and chain occurs) [3, 9–12].

In the general scheme of steroid degradation, there are two key enzymes that initiate the opening of the steroid ring: the 3-ketosteroid-Δ^1-dehydrogenase [4-ene-3-oxosteroid: (acceptor)-1-ene-oxoreductase; EC 1.3.99.4)], also known as KstD and the 3-ketosteroid 9α-hydroxylase [Androsta-1,4-diene-3,17-dione; EC 1.14.13.142], also known as Ksh [13]. KstD is a flavoenzyme involved in the Δ^1-dehydrogenation of the steroid molecule leading to the initiation of the breakdown of the steroid nucleus by introducing a double bond into the A-ring of 3-ketosteroids [14, 15]. This flavoprotein converts 4-ene-3-oxosteroids (e.g. AD) to 1,4-diene-3-oxosteroids (e.g. ADD) by trans-axial elimination of the C-1(α) and C-2(β) hydrogen atoms [16]. KstD homologs have been identified in 100 different bacterial species (78 actinobacteria, 20 proteobacteria and 2 firmicutes) and at least in one fungus, *Aspergillus fumigatus* CICC 40167 [17, 18]. Most of these KstD-containing bacteria occur in soil, marine or river sediments and are also able to degrade polycyclic aromatic hydrocarbons [19]. Phylogenetic analysis leads to classify the KstD-like enzymes in at least 4 different groups, in which KstD1, KstD2, KstD3 of *Rhodococcus erythropolis* SQ1 are representatives of three of them [20]. The crystal structure of the enzyme KstD1 of *R. erythropolis* SQ1 has been elucidated [21] confirming the presence of the two domains previously described, namely a N-terminal flavin adenine dinucleotide (FAD) binding motif and a substrate-binding domain [14, 20, 22, 23].

The substrate range of different KstD proteins has been studied in *R. erythropolis* SQ1, being 3-ketosteroids with a saturated A-ring (e.g. 5α-androstane-3,17-dione and 5α-testosterone) the preferred substrates for KstD3 and (9α-hydroxy-)4-androstene-3,17-dione the favourite one for both KstD1 and KstD2 [20]. It should be mentioned that, apart from their role in steroids degradation, KstD proteins could have specific roles depending of their origin; for instance, the KstD of *A. fumigatus* CICC 40167 is involved in fusidane antibiotic biosynthesis [17].

We have previously reported the occurrence of three KstD enzymes in *R. ruber* (NCBI::AFH57399 for KstD1; NCBI::AFH57395 for KstD2 and NCBI::ACS73883 for KstD3) [24]. Growth experiments with single, double or triple *kstD* mutants proved that KstD2 is a key enzyme in the transformation of both AD to ADD and 9α-hydroxy-4-androstene-3,17-dione (9OHAD) to 9α-hydroxy-1,4-androstadiene-3,17-dione (9OHADD) while both KstD2 and KstD3 are involved in the cholesterol catabolism in *R. ruber*. On the other hand, the role of KstD1 on the steroids catabolism remains unclear as *kstD1* mutation did not affect growing of this strain in steroids [24]. In this study, we cloned the three *kstD* ORFs and heterologously expressed them in *R. erythropolis* CECT3014, in order to initiate the biochemical characterization of the encoded enzymes, as the basis for further studies on their applications. The results revealed that KstD3 uses more actively substrates with a saturated ring in contrast to KstD1 and KstD2. Additionally, we located and functionally defined the promoters of the three *kstD* ORFS in order to provide a basis for future research on the regulation of these genes.

Methods

Bacterial strains, plasmids and growth conditions

Rhodococcus ruber strain Chol-4 (CECT 7469; DSM 45280) was isolated from a sewage sludge sample [25]. This strain was routinely grown in Luria-Bertani (LB) or minimal medium (M457 of the DSMZ, Braunschweig, Germany) containing the desired carbon and energy source under aerobic conditions at 30 °C in a rotary shaker (250 rpm) for 1–3 days. For the steroids growth experiments, a LB pre-grown culture was washed two times with minimal medium prior to inoculation. Cholesterol or AD (Sigma), were added directly to the minimal medium culture for growing and/or induction at 0.6 and 0.44 g/L, respectively. *Escherichia coli* strains were grown in LB broth at 37 °C, 250 rpm. For the promoter growth experiments, cells were plated in minimal medium M457 plates containing the desired carbon source and incubated at 30 °C for 3 days. Cholesterol and AD were previously dissolved in methyl-β-cyclodextrin (CD) [26] and prepared as described [27]. Plasmids and bacterial strains used are listed in Additional file 1. Competent cells of *E. coli* DH5αF' and BL21 (DE3) were prepared and transformed by standard protocols [28].

Cloning of *kstD1*, *kstD2* and *kstD3* of *R. ruber* strain Chol-4 and heterologous expression in *Rhodococcus erythropolis* CECT3014 cells

Chromosomal DNA extraction of *R. ruber* grown in a LB agar plate was performed using the hexadecyltrimethylammonium bromide (CTAB) procedure [29] with the following modifications. Bacterial cells were collected, suspended in 400 μL Tris–EDTA buffer (10 mM Tris/HCl, pH 8.1 mM EDTA) and incubated at 80 °C for 20 min. Afterwards, a lysozyme treatment (50 μL of 10 mg/mL stock) was carried out at 37 °C for 1–12 h, and then 75 μL of SDS containing proteinase K (70 μL SDS 10% wt/vol plus 5 μL proteinase K 10 mg/mL) was added and incubated for 10 min at 65 °C. Proteins were precipitated with 100 μL of 5 M NaOH and 100 μL CTAB (0.1 g/mL suspended in 0.7 M NaOH) for 10 min at 65 °C. DNA was purified by extraction with chloroform-isoamyl alcohol (24:1) and phenol-chloroform-isoamyl alcohol (25:24:1) and precipitated with 0.6 vol of isopropanol at room temperature for 30 min. After centrifugation, DNA was washed with 70% ethanol and suspended in distilled water.

The *kstD* ORFs were previously identified using the Bioedit program [25] and they were PCR amplified, from start to stop codon, using primers from Additional file 2. PCR was performed under standard conditions using High Fidelity PCR Enzyme Mix (Fermentas) with a specific high GC buffer (Roche) at 30 cycles of 1 min at 95 °C, 1 min at the desired Tm and 0.5–3 min at 72 °C (unless stated otherwise).

NdeI-*BamHI* (*kstD3*), *NdeI*-*BglII* (*kstD2* and *kstD1*) restricted PCR products were first cloned into pGEM-T Easy Vector (Promega) and then moved to the shuttle vector pTip-QC1 [30]. Expression plasmids either with or without a *kstD* ORF were used to electroporate *Rhodococcus erythropolis* CECT3014. Selection was made in LB with 34 μg/mL Cm. Cultures of *R. erythropolis* harbouring expression plasmids (pTip-KsTD1, pTip-KstD2, pTip-KstD3 or pTip-QC1) were grown (30 °C, 200 rpm) in 50 mL LB broth supplemented with Cm until an OD_{600nm} of 0.6–0.8. After that, expression was induced by adding 1 μg/mL thiostrepton (Sigma) to the culture. Cells were kept growing for 24 h more and they were collected at 5000 rpm for 10 min, washed twice in 50 mM phosphate buffer pH 7.0, concentrated to 5 mL and sonicated. The resulting cell extracts were used for analysis of KstD activity with a range of steroid substrates. Total protein content was measured by Bradford assay [31]. Samples were analysed by PAGE-SDS in 12.5% wt/vol gels and using 10 μg of total protein per lane.

KstD enzymatic assay

The kinetic of the enzymatic extracts was determined as previously stated [20, 32]. The cell-free extracts were incubated with AD (Sigma), 9OHAD (Organon Biosciences), 17β-hydroxy-5α-androstan-3-one or 5α-Tes, DOC (deoxycorticosterone), testosterone, corticosterone, 19OHAD, progesterone (Sigma), 4-BNC (4-pregnen-3-one-20β-carboxylic acid) (Steraloids), 5β-androstane-3,17-dione (Steraloids), 4-cholestene-3-one (Sigma) or 5α-cholestan-3-one (Acros Organics). Structures of steroids used in this study are shown in Additional file 3. Enzyme activities were measured spectrophotometrically at 30 °C using 2,6-dichlorophenol-indophenol (DCPIP, Sigma) as an artificial electron acceptor. The reaction mixture (1 mL) consisted of 50 mM Tris (pH 7), 80 μM DCPIP, cell-free extract and 200 μM steroid in ethanol (methanol in the case of 5α-cholestane-3-one). Four replicates were analysed. Activities are expressed as mean values ± SD in units per milligram of protein; one unit is defined as the amount of enzyme which causes the reduction of 1 μmol of DCPIP/min ($\varepsilon_{600} = 21\ mM^{-1}\ cm^{-1}$) after taking into account the value of the activity of control (cells harbouring the empty pTip-QC1 vector) for a certain steroid. Total protein concentration (mg/mL) was measured by Bradford assay [31]. The kinetics of the KstD enzymes were determined by incubating the cell-free extracts with varying concentrations of steroid substrates. The kinetics parameters were analysed by nonlinear regression curve fitting of the data to the Michaelis- Menten equation using Hyper32 1.0 software (Informer Technologies, http://hyper32.software.informer.com/).

Expression analysis by RT-PCR and RT-qPCR

RNA samples for RT-PCR experiments were obtained from mid-log exponential phase cultures (OD_{600nm} 0.7–0.8). Total RNA was prepared with the RNeasy Mini Kit (Qiagen) following the manufacturer's indications with the following modification: 50 mg of acid-washed glass beads (150 μm diameter) were added in the first step and each sample was shaken at maximum speed in a Bullet Blender for 5 min. The cell debris was removed by centrifugation. The supernatant was subjected to the RNeasy Mini Kit (Qiagen) protocol. The total RNA obtained (0.5–1 μg) was treated once with 5 U of Turbo DNase RNase-Free (Ambion) in a 700 μL volume for 2 h at 37 °C. RNA samples were extracted with 1 volume of acid phenol (Sigma), vigorously shaken and incubated at room temperature for 15 min. After 15 min centrifugation, the upper phase was precipitated by addition of 0.12 volumes of 5 M NH_4Ac, 0.02 volumes of glycogen (5 mg/mL) and 1 volume of isopropanol, washed twice in 70% ethanol and dissolved in water. Samples were treated with DNase until no DNA was detected by PCR to avoid DNA contamination. The RNA concentration was then evaluated using a NanoDrop Spectrophotometer ND-1000.

For the RT-PCR the cDNA was synthesized using SuperScrip III Reverse Transcriptase (Invitrogen) following the manufacturer's indications. cDNA was used as template (25 ng) for PCR reactions (20 μL final volume). Controls without reverse transcriptase (RT-) were used to detect any contamination of undigested DNA in the RNA preparations. PCR products were analysed in 0.8% agarose gels.

To quantify the expression of the three KstD genes, a RT-qPCR analysis was performed using RNA from wild-type strains cultured in M457 minimal medium containing the desired carbon and energy source (2 g/L sodium acetate, 0.44 g/L AD or 0.6 g/L cholesterol). The RNA quality was assessed by using Bioanalyzer 2100 (Agilent). cDNA was synthetized using 1 μg of RNA with the high capacity RNA to cDNA Kit (Applied Biosystems). The RT-qPCR analysis of cDNA was performed on Applied Biosystems QuantStudio 12K Flex Real-Time PCR Systems. The reaction conditions were 10 min at 95 °C followed by 40 cycles of 15 s at 95 °C and 1 min at 60 °C for extension. The temperature of the melting curve was from 60 to 95 °C. The FAD-binding dehydrogenase D092_14375 gene was used as an internal control to normalize messenger RNA levels. All reactions were performed in triplicate. The RT-qPCR experiment and the analysis of the relative fold difference of each gene using the $2^{-\Delta\Delta Ct}$ algorithm was performed in the Genomic unit of Universidad Complutense de Madrid.

The sodium acetate grown culture was used as the reference medium. Therefore, the relative expression indicates how many times the expression level of a certain gene is detected respect to the levels detected when growing on sodium acetate.

In silico analyses

DNASTAR (Lasergene) programs were used to analyse sequences and to design primers. The *R. ruber* strain Chol-4 genomic DNA has been previously sequenced [33]. BioEdit program was used to perform local-blast alignments within the genome data (NCIB::ANGC01000000). Putative signal peptides were predicted by SignalP 4.1 server using a model trained on Gram-positive bacteria [34]. Sigma 70 putative promoters predictions were performed using the BPROM server, a bacterial sigma 70 promoter recognition program with about 80% accuracy and specificity [35], the Neural Network Promoter Prediction (NNPP) based on prokaryotes [36] in all cases with a score value of ≥80%, or the webserver PePPER for prediction of prokaryote promoter elements and regulons [37].

For the protein modelling we employed different software. I-Tasser (http://zhanglab.ccmb.med.umich.edu/I-TASSER) was used for an approach to protein structure and function prediction [38] and PredictProtein (www.predictprotein.org/home) was used for the secondary structure, solvent accessibility and transmembrane helix prediction [39]. COBALT was used as a multiple sequence alignment tool to find similarities among the catalytic residues (www.ncbi.nlm.nih.gov/tools/cobalt/re_cobalt.cgi) [40] and PyMOL (www.pymol.org/) as a molecular visualization system (PyMOL Molecular Graphics System, Version 0.99, Schrödinger).

Promoter cloning and characterization

The *Nhe*I-*Pci*I 0.4 Kb multiple cloning site (mcs) from pSEVA351 [41, 42] was cloned into pNV119 vector [43], from now on named as pNVS (Additional file 4).

The putative *kstD* promoter sequences were amplified by PCR, from the end of the upstream flanking gene to the end of the first six amino acid codifying sequence. The *Xba*I, *Pst*I-flanked *kstD* promoter regions (*kstD1p*, *kstD2p* and *kstD3p*) and the *Kpn*I, *Pst*I-flanked *kstD3^{b}p* minimal promoter were cloned into pNVS. The resulting vectors were designated pNVSP1, pNVSP2, pNVSP3 and pNVSP3^b, respectively.

Apramycin resistance gene (Am^r) was amplified by PCR from plasmid pIJ773 [44], from start to stop codon. The *Nru*I/*Hin*dII digested Am^r fragment was cloned in each one of the previous *kstD* constructions. The resulting plasmids were checked by sequencing (Secugen) and named pNVSP1-A, pNVSP2-A, pNVSP3-A and pNVSP3^b-A, respectively.

All the primers used are listed in Additional file 2. PCR was performed under standard conditions using High Fidelity PCR (Roche) with glycerol 5% and a basic program unless stated otherwise.

As a control, a plasmid without any promoter but carrying the Am^r gene (pNVSA) was made by digesting pNVSP1-A with *Nru*I-*Xba*I; the resulting 4.2 Kb fragment was cut from a 1% agarose gel and purified with GENECLEAN Turbo Kit. Blunt ends suitable for ligation with T4 DNA ligase (Takara) were generated using the End repair kit (DNA terminator, Lucigen). The final ligated product was used to transform *E. coli* DH5αF′. Deletion was checked by sequencing (Secugen).

Every one of the plasmid set was introduced in *R. ruber* strain Chol-4 by electroporation (200 or 400 μL cells with 1 μg DNA at 400 Ω, 25 mA, 2.5 μF; 10–11 ms), the resulting cells were suspended in 800 μL of LB and kept for 6 min at 46 °C, and then for 5 h at 30 °C without shaking. Finally, they were plated on LB Agar with 200 μg/mL kanamycin and kept at 30 °C. To verify the presence of plasmids, two colonies of each plate were picked and grown in 3 mL of LB-200 μg/mL kanamycin. Plasmid were extracted using the method described in Hopwood et al. [45] and used to transform *E. coli* strain DH5αF' [28]. Plasmids obtained from *E. coli* colonies grown at 37 °C in 50 μg/mL kanamycin were verified by sequencing (Secugen). Finally, those colonies of *R. ruber* strain Chol-4 harbouring the right recombinant plasmids were picked and grown in agar minimal media at 30 °C with different carbon source with or without apramycin (300 μg/mL) or kanamycin (200 μg/mL).

To define the transcriptional start sites (TSS) of *kstD* promoters, a transcription start point protocol (ARF-TSS) [46] was used on *R. ruber* cells with the following modifications.

RNA from *R. ruber* cells growing in different carbons sources cultures was isolated. The culture media were: LB for *kstD1* TSS, minimal medium supplemented with AD for *kstD2* TSS and minimal medium supplemented with cholesterol for *kstD3* TSS. Total RNA was isolated as described previously [47]. It was qualified by electrophoresis and quantified by Nanodrop 1000 (NanoDrop Technologies). 20 μg of RNA were reverse transcribed with a gene specific phosphorylated 5′-end primer (R1 for each *kstD*) using SuperScrip III Reverse Transcriptase from Invitrogen. As a result, cDNA fragments with the TSS as 3′-end were generated and purified. After removing of RNA with 0.5 μg/μL RNase A (37 °C for 30 min), cDNA was purified using the GENECLEAN turbo Kit (MPI) and final products were treated with T4 RNA ligase from Fermentas (10U, 37 °C for 30 min). This T4 RNA ligase catalyses the ATP-dependent intra- and intermolecular formation of phosphodiester bonds between 5′-phosphate and 3′-hydroxyl termini of oligonucleotides, single-stranded RNA and DNA. It was used in this study to circularize the cDNA. The circularized cDNAs were then used as template for PCR amplification with Expand High Fidelity (Roche) using R2 and F3 primers specific for each *kstD* ORF. PCR products were purified using the GENECLEAN turbo Kit and sequenced (Secugen). The nucleotide upstream the 5′-end of the R1 primer is the transcription initiation site.

Results and discussion

As we have published earlier, the *R. ruber* strain Chol-4 genome contains three putative 3-ketosteroid Δ^1-dehydrogenase ORFs (*kstD1, kstD2* and *kstD3*) that code for flavoenzymes involved in the steroid ring degradation [24]. Growth experiments with *kstD* mutants proved that KstD2 is the main enzyme involved in the transformation of AD to ADD. *R. ruber kstD2* mutants accumulate 9OHAD from AD due to the action of a 3-ketosteroid-9α-hydroxylase (KshAB). On the other hand, only the strains lacking both KstD2 and KstD3 were unable to grow in minimal medium with cholesterol as the only carbon source. In order to know more about these *R. ruber* enzymes, we have performed transcriptional studies and followed their heterologous expression in *R. erythropolis* to detect their activities on a set of different substrates.

In silico analyses of *kstD* promoter regions of *R. ruber* strain Chol-4

A scheme of the three *R. ruber kstD* ORFs and their genomic surroundings is depicted in Fig. 1 showing also the in silico predicted pP1, pP2, pP3, pP4 and pP5 promoters. The available programs (see "Methods") yielded no putative promoter region just upstream either *kstD1* or *kstD2* ORFs, although it should be noted here that promoter prediction programs are not specific for Gram-positive species.

However, these programs detected putative promoters (pP1, pP2 and pP4) upstream the *cyp450* gene, lying in the intergenic region of the *fadA5-hyp* and *cyp450-kstD1-MFSt* opposite clusters. Flanking the pP1 putative promoter there are two palindromic sequences (TagAACagGTTgtc and TagAACgtGTTccA) (Fig. 2), one of them rather similar and the second one identical to the consensus binding region reported for the KstR regulatory protein of *Mycobacterium* (TnnAACnnGTTnnA) [48]. KstR and KstR2, two TetR family repressor regulators, have been found to control most of the steroid pathways in actinobacteria [48–50]. KstR is a highly conserved TetR

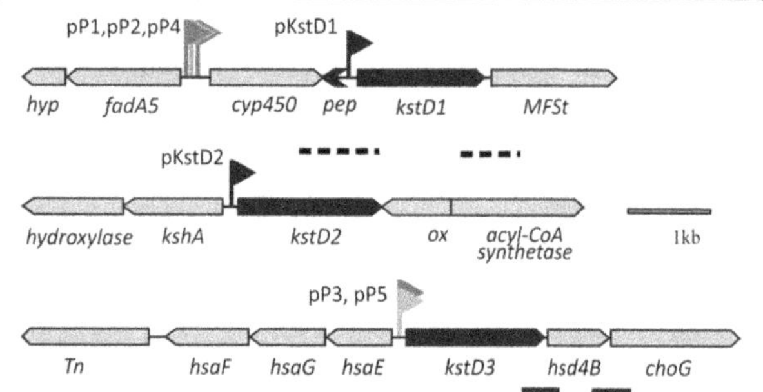

Fig. 1 Schematic representation of *R. ruber* strain Chol-4 DNA *kstD* regions. Putative promoters (pP) predicted using the BPROM (pP1: TTCCTT^{-35}..TGCTTGAAT^{-10}), PePPER (pP2: TTGAATGCTTTTAGAACGTGTTCCACATCgcgaC^{+1} and pP3: TGGACTCACCGCGCCATCATTCTATAACgtgtT^{+1}) or NNPP programs (pP4: GGTTGTCGTGGCGGACAAGGTGTGGTCCGAATGATCGGGAC^{+1}TTGGCGATT and pP5: GAAGGGATGGACTCACCGCGCCATCATTC^{+1}TATAACGTG) are shown in grey flags. TSS derived promoters (pKstD1, pKstD2) appear in black flags. Positive (*solid line*) and negative (*dotted line*) results of the amplification of co-transcribed products are depicted. *hyp* hypothetical protein, *fadA5* acetyl-CoA acyltransferase, *cyp450* cytochrome P450, *pep* hypothetical peptide, *MFSt* major facilitator superfamily transporter, *kshA* 3-ketosteroid 9α-hydroxylase, *ox* oxidoreductase, *Tn transposase*, *hsaF* 4-hydroxy-2-oxovalerate aldolase, *hsaG* acetaldehyde dehydrogenase, *hsaE* 2-hydroxypenta-2,4-dienoate hydratase, *hsd4B* 2-enoyl acyl-CoA hydratase, *choG* cholesterol oxidase

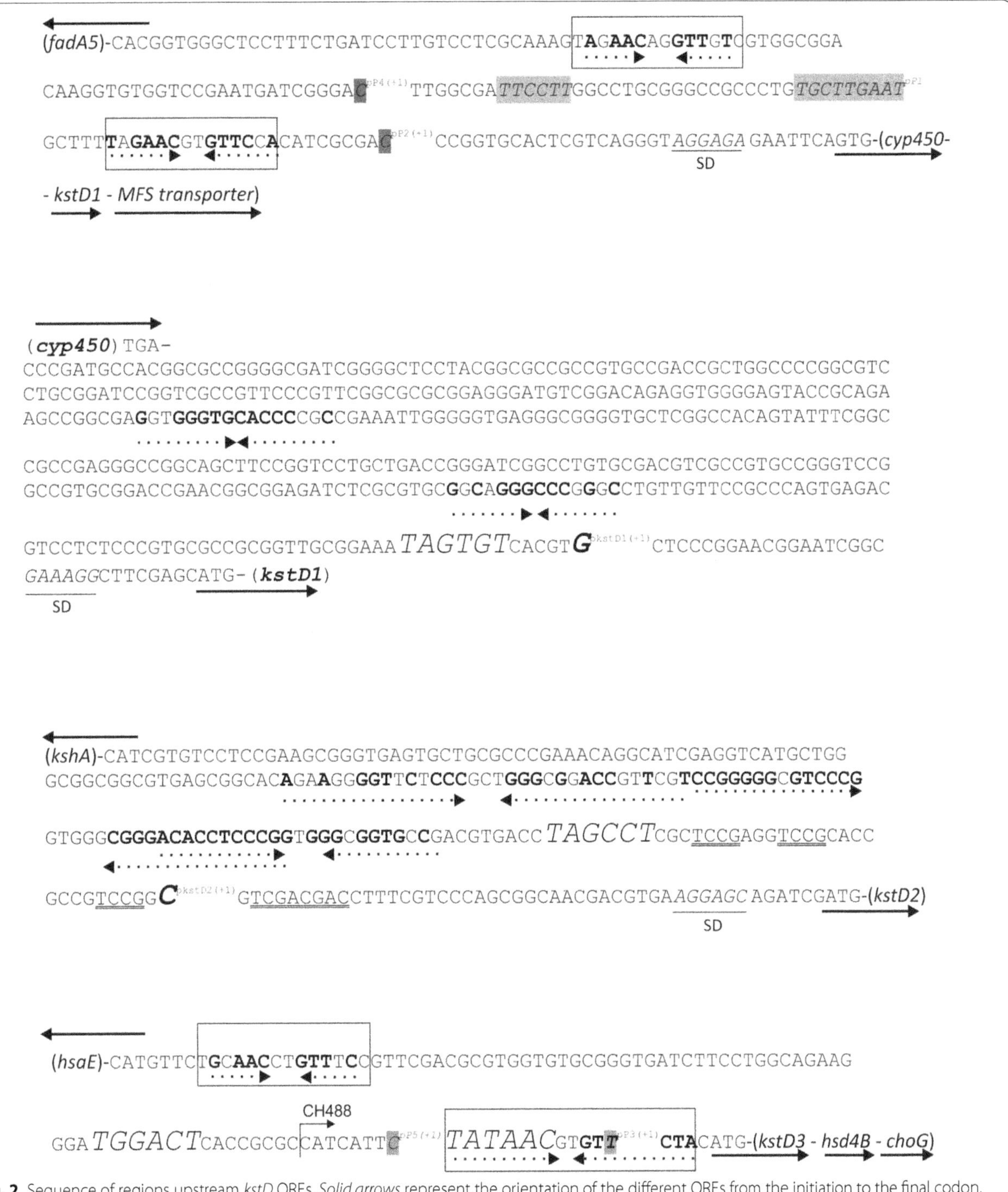

Fig. 2 Sequence of regions upstream *kstD* ORFs. *Solid arrows* represent the orientation of the different ORFs from the initiation to the final codon. Sequences similar to *Mycobacterium* KstR binding sites (TnnAACnnGTTnnA) are within a *square*. Palindromic sequences appear in *bold* characters and *dotted underlined*. Shine Dalgarno (SD) sequences are in *italics* and underlined. The −10, −35 *boxes* are marked in *grey*. The in silico transcription initiation point of the putative pP1, pP2, pP3 and pP4 promoters are in italics and marked as pP(+1). The transcription start sites of *kstD1* and *kstD2* ORF obtained by the ARF-TSS method are shown in *bold italics* and marked as pKstD(+1). The TCCG repeats and the *Sal* box upstream *kstD2* are *double underlined*. Promoter signals similar to others described (e.g. *M. tuberculosis*) [55] appear in bigger size. Primer CH488 indicating the beginning of the minimum *kstD3* promoter is also shown

family repressor that regulates the transcription of genes related to the upper and central pathway of cholesterol catabolism, namely the membrane transport of cholesterol, the degradation of the steroid side chain and the opening of the A and B rings [12, 48]. Upon binding to a 3-oxo-4-cholestenoic acid or to a CoA thioester cholesterol metabolite, KstR releases the DNA and allows transcription to begin [51, 52]. The KstR binding motif has been proved to be conserved within actinobacteria [48, 53].

The possible co-transcription of the *cyp450-kstD1-MSFt* ORF cluster to a polycistronic mRNA would come into contradiction with the transcription of a non-yet described short putative ORF (*pep*, Fig. 1) located in opposite sense in the 425 bp *cyp450-kstD1* intergenic region. This short ORF might code for a 35 amino acid peptide which shows a 98% amino acid identity with part of the hypothetical 78 aa protein RHRU231_750039 (*R. ruber*, 78 aa). Therefore, *kstD1* might be independently transcribed while pP1/pP2/pP4 putative promoters might be only involved in *cyp450* transcription. Moreover, a putative ribosome binding site (GAAAGG) was found 9 bp upstream the *kstD1* initiation codon (Fig. 2) that is identical to the proposed one for *sigA* of *R. ruber* TH [54].

In the case of the *kstD2* ORF, none of the online programs recognized a promoter consensus, although this region contains some quasi-palindromic sequences and a Shine-Dalgarno-like motif (AGGAGC) (Fig. 2).

There are two putative promoters for the *kstD3* ORF (pP3 and pP5, see Figs. 1 and 2) that lie in the intergenic region between *hsaE* and *kstD3* ORFs. The putative promoter pP3 contains the sequence TATAAC similar to the −10 consensus motif described for *M. smegmatis* promoters ($T_{100\%}A_{93\%}T_{50\%}A_{57\%}A_{43\%}T_{71\%}$) [55], and a −35 region (**TGGAC**T) that resembles the *E. coli* promoter consensus motif TTGACA. In this region, there are also a tandem of two putative KstR binding sequences around this promoter (TgcAACctGTTtcc and TatAACgt-GTTctA), one quite similar and the other identical to the KstR binding consensus, in a similar way to the *cyp450* pP1/pP2 putative promoter (Fig. 2). The arrangement of these promoter and regulatory sequences, lying between opposite cluster genes, also occurs in the *R. ruber kstD1* region and it is similar to that found in *Mycobacterium* and *Rhodococcus jostii* RHA1 genomes [48, 56]. On the other hand, Shell et al. have recently described that the abundance of leaderless transcripts (that lack a 5′ UTR and a Shine-Dalgarno sequence and that begin with ATG or GTG) is a major feature of mycobacterial that accounts for around one-quarter of the transcripts [57]. *kstD3* could be a leaderless ORF: there is no evidence of a Shine-Dalgarno sequence in its 5′ region and the putative promoter is quite near to the ATG initiation codon. Moreover, as it will be stated later, the promoter of this intergenic region is functional in *Rhodococcus* but not in *E. coli*, a fact that has also been confirmed in the mycobacterial leaderless messenger translation [57].

Promoter cloning and characterization

To go further in the characterization of the promoter regions of the *R. ruber kstD* ORFs, a promoter-test vector suitable for *R. ruber* strain Chol-4 was constructed. *R. ruber* is sensitive to apramycin so we chose the expression of a gene encoding this resistance as a proof of the promoter activity.

pNV119 [43], a *Nocardian* shuttle vector shown to replicate in *R. ruber* [24], was modified by adding the mcs of pSEVA351 that contains a transcriptional terminator in each extreme [41, 42]. The resulting pNVS plasmid (see Additional file 4) was used to study the activity of the Chol-4 putative promoter regions. The three *kstD* intergenic regions (Fig. 2) plus the first 21 bases of each *kstD* ORF were PCR amplified, transcriptionally fused to the apramycin resistance gene obtained from pIJ773 and cloned into the mcs of pNVS. The recombinant plasmids were introduced into *R. ruber* by electroporation and kanamycin resistant clones were selected. *R. ruber* clones, harbouring the plasmids pNVSP1-A, pNVSP2-A or pNVSP3-A, were then plated in minimal medium supplemented with either 1.5 mM cholesterol, 1.5 mM AD or 10 mM sodium acetate, and in the presence of either 200 μg/mL kanamycin or 300 μg/mL apramycin. The apramycin resistance gene (Am^r) without any upstream promoter region was cloned in pNVS mcs generating the vector pNVSA that was used as a negative control. As a second control, a set of pNVSP vectors that contain every promoter region but do not carry the apramycin resistance gene was used. Figure 3 shows that only the cells harbouring the double system formed by a putative promoter and the apramycin resistance gene were able to grow on apramycin and kanamycin while cells harbouring the pNVSA or the pNVSPs vector were only able to grow in kanamycin plates. These results unambiguously confirm that all the three checked DNA regions contain *R. ruber* promoter sequences functionally active in the conditions used. On the other hand, *E. coli* harbouring the plasmids pNVSP1-A or pNVSP2-A were also able to grow in apramycin, in contrast with those harbouring pNVSP3-A (data not shown). Other actinobacteria promoters (e.g. some of *Mycobacterium* and *Streptomyces* spp.; [55, 58] are also not functional in *E. coli* strains; this fact could related to the occurrence of leaderless genes [57].

Although none of the online programs recognized a promoter consensus for the *kstD2* ORF, the promoter-less

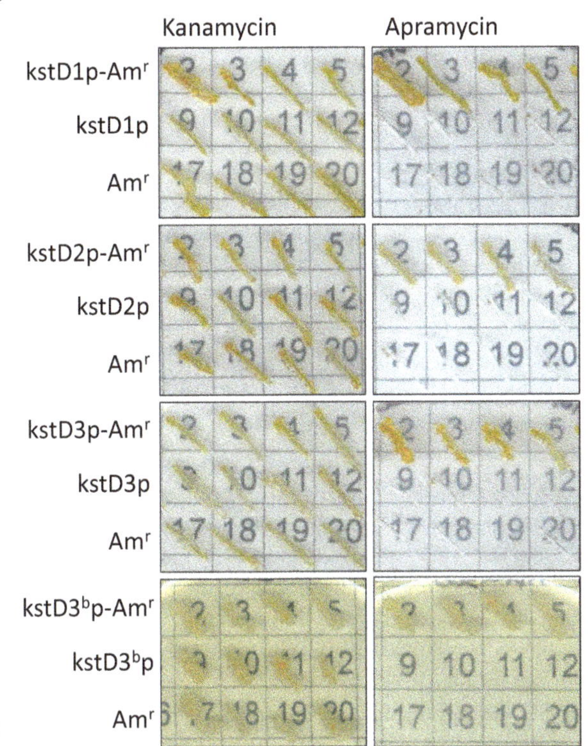

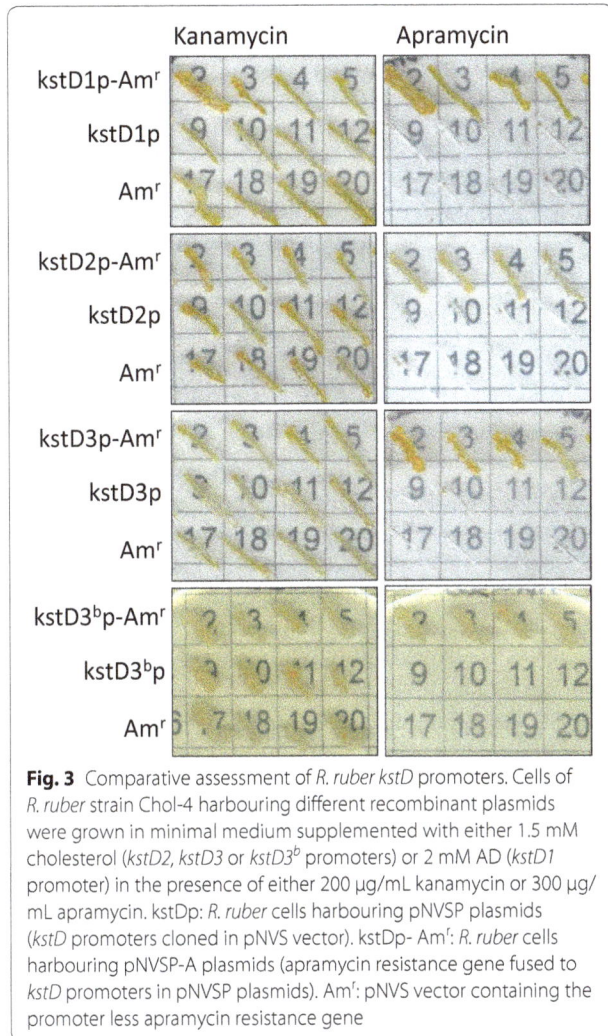

Fig. 3 Comparative assessment of *R. ruber kstD* promoters. Cells of *R. ruber* strain Chol-4 harbouring different recombinant plasmids were grown in minimal medium supplemented with either 1.5 mM cholesterol (*kstD2, kstD3* or *kstD3*^{b} promoters) or 2 mM AD (*kstD1* promoter) in the presence of either 200 μg/mL kanamycin or 300 μg/mL apramycin. kstDp: *R. ruber* cells harbouring pNVSP plasmids (*kstD* promoters cloned in pNVS vector). kstDp- Am^r: *R. ruber* cells harbouring pNVSP-A plasmids (apramycin resistance gene fused to *kstD* promoters in pNVSP plasmids). Am^r: pNVS vector containing the promoter less apramycin resistance gene

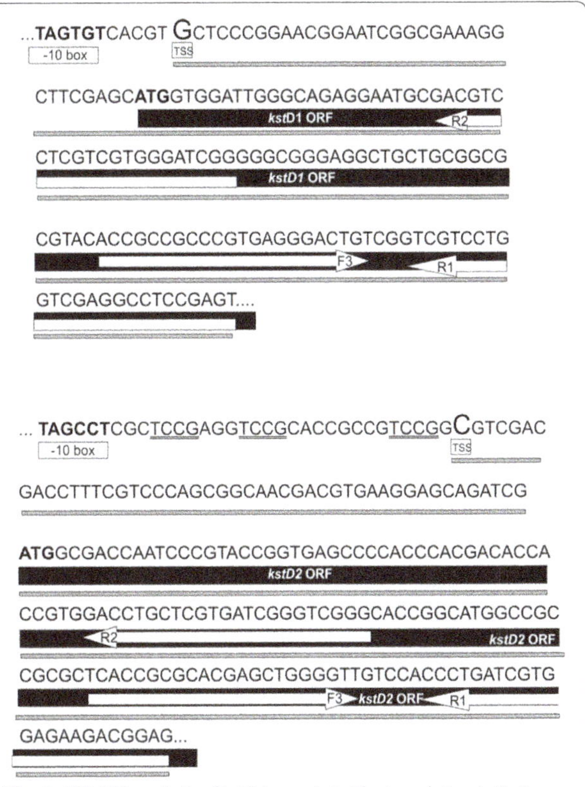

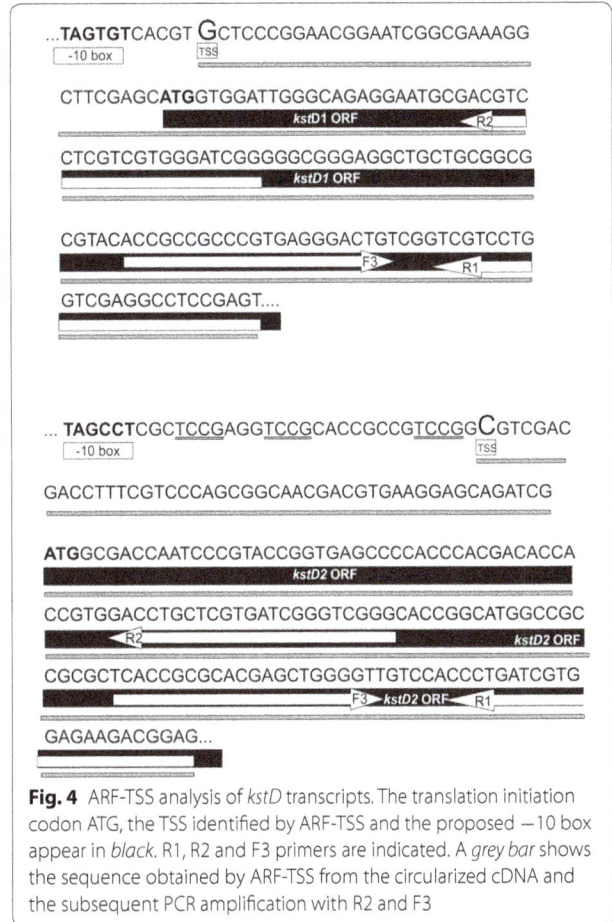

Fig. 4 ARF-TSS analysis of *kstD* transcripts. The translation initiation codon ATG, the TSS identified by ARF-TSS and the proposed −10 box appear in *black*. R1, R2 and F3 primers are indicated. A *grey bar* shows the sequence obtained by ARF-TSS from the circularized cDNA and the subsequent PCR amplification with R2 and F3

vector pNVS has enabled us to check the promoter activity of the intergenic regions. The construction of an improved version of this plasmid that allows a quantitative analysis of promoter strength is under work.

To define the transcription start sites (TSS) of the *kstD* genes, the transcription start point protocol (ARF-TSS) indicated in "Methods" section was followed in *R. ruber* cells. We could conclude that the 5′ terminal base in *kstD1* messenger RNA is a G resulting from the transcription that starts 34 bp upstream the *kstD1* initiation codon. Similarly, the transcription start site of the *kstD2* gene is a C 48 bp upstream the *kstD2* initiation codon (Figs. 2, 4). However, this approach did not yield any result in the case of the *kstD3* gene. In order to better define the limits of the *kstD3* promoter, progressively shorter sections of the intergenic region, keeping the first 21 bases of *kstD3* ORF, were PCR amplified and transcriptionally fused to the apramycin resistance gene. As it can be seen in Figs. 2 and 3, just a minimum region of only 23 pb upstream the ATG of *kstD3* ORF (pNVSP3^b-A vector) was enough to act as a promoter as the *Rhodococcus* cells harbouring this vector were able to grow on medium with either apramycin or kanamycin. This promoter region is partially similar to the putative pP3 and pP5 promoters mentioned before, containing the TATAAC sequence similar to the −10 consensus motif described for *Mycobacterium smegmatis* promoters (Fig. 1).

Putative −35 and −10 hexamers identification was based on the sequence of other actinobacteria promoters. The −10 region motif (TAGTGT) found 40 bases upstream the *kstD1* initiation codon is similar to the −10 consensus region described for *Mycobacterium tuberculosis* promoters ($T_{80\%}A_{90\%}Y_{60\%}G_{40\%}A_{60\%}T_{100\%}$) and identical to the T101 promoter described by Bashyam et al. in the same bacteria [55]. No sequence similar to any −35 motif was found in the *kstD1* region. The absence of −35 motifs seems to be a characteristic of actinobacterial genes [55, 58].

There is a putative −10 sequence (TAGCCT) in the intergenic region between *kshA* and *kstD2* that resembles the T6 promoter of *M. tuberculosis* (TAGGCT) [55]. However, it is a bit far away from the *kstD2* TSS. There is a repetitive motif TCCG in this region (Fig. 2) that also appears in the human *Sal* box element present in the 3′ terminal spacer of rDNA and that constitutes a termination signal for RNA polymerase I (**TCCGCACGGGTCGACCAG**) [59]. In this *kstD2* upstream region we found something similar: TCCGAGGTCCGCACCGCCGTCCGGCGTCGACGAC, where black-underlined letters marked the resemblance between them. Such promoter proximal terminators can appear also upstream of the transcription start site and in this case, they are described to positively affect transcription initiation and to prevent transcriptional interference by reading through of polymerases from the spacer that separates each rDNA repetition [60–62]. The biological significance of this motif in this intergenic *kstD2* region should be further determined.

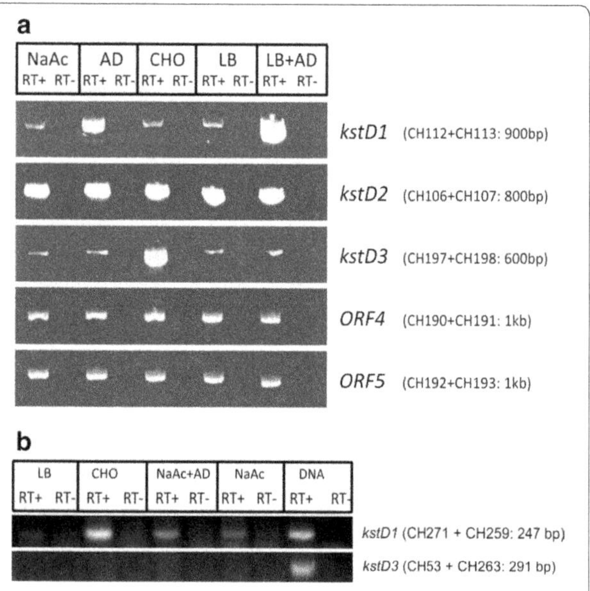

Fig. 5 **a** Analysis of transcription products of *kstD* genes in *R. ruber* strain Chol-4 by specific RT-PCR amplification and agarose gel electrophoresis. RNA samples were isolated from cultures grown in: M457 minimal medium supplemented with sodium acetate (lane NaAc), AD (lane AD) or cholesterol (lane CHO) and LB medium supplemented with (lane LB + AD) or without AD (lane LB). RT—refers to negative controls (not incubated with retrotranscriptase) to exclude DNA contamination in RNA samples. Primers used in each PCR reaction and amplified fragment size are shown in *brackets*. As non-induced expression controls, the transcription of two *R. ruber* Chol-4 enzyme genes was followed in the same conditions: ORF4, that codes for a FAD-binding dehydrogenase D092_14375 (99% identity with the protein NCIB::WP_010594120.1 of *Rhodococcus* sp. P14) and ORF5 (D092_13875) that codes for a fumarate reductase (99% identity with NCIB::WP_010594021.1 of *Rhodococcus* sp. P14; see sequence in Additional file 6). the FAD-binding dehydrogenase D092_14375 and the fumarate reductase D092_13875. **b** RT-PCR assays of *R. ruber* Chol-4 *kstD1* and *kstD3* transcription carried out in *kstD2 R. ruber* mutant growing in: LB (lane LB) or M457 plus CHO (lane CHO), NaAc and AD (lane NaAc + AD) or NaAc (lane NaAc). Lane DNA shows the result of the control experiment: PCR made on *R. ruber* DNA

Transcriptional analysis of *kstD* genes in *R. ruber* strain Chol-4

The transcription of the three *kstD* genes found in *R. ruber* strain Chol-4 was analysed by RT-PCR of RNA samples prepared from cultures grown in either M457 mineral medium or LB medium supplemented with either AD or cholesterol as possible inductors.

Control cultures were grown in either M457 supplemented with sodium acetate (2 g/L NaAc) or LB, both in the absence of any steroid. Using the specific primer pairs designed to search for the transcript of each different ORF (Additional file 2), we could show that all the *kstD* genes are transcribed in all the conditions used in our assays (Fig. 5) and that transcription of some of them is induced by cholesterol or AD. There are also other *Rhodococcus* metabolic genes reported to show a low-level constitutive transcription that can be strongly induced under the presence of a determinate substrate. For instance, phenol degradation genes in *R. erythropolis* are constitutively transcribed and also highly induced by phenol [63].

cDNA templates concentrations were adjusted to the same value in RT-PCR experiments, so we can consider the differences observed in the thickness of some amplification bands to be meaningful (Fig. 5). Specific amplification of *kstD1* cDNA was higher in AD induced than in non-induced cultures, while amplification of *kstD3* cDNA was higher in cultures induced with cholesterol as compared to the other conditions assayed. So we can conclude that these two genes, although constitutively transcribed, they can also be additionally induced by the presence of AD (*kstD1*) and cholesterol (*kstD3*). The induction of *kstD* genes by cholesterol or AD has also been reported in other microorganisms: a 3.3 expression ratio (cholesterol/pyruvate) was reported for one *kstD* in *R. jostii* RHA1 (*ro04532*) [64, 65], while other putative *kstD* genes of the same strain (e.g. *ro02483*, *ro05798* and *ro05813*) were up-regulated in 7-ketocholesterol but not in cholesterol [64]; the main *kstD* in *M. smegmatis* (*MSMEG_5941*) was 13-fold up-regulated in cholesterol respect to glycerol [12]; a 1.8, 4.1 or 1.2 -fold up-regulation (AD/glycerol) was found for *kstD1*, *kstD3* and *kstD2* genes respectively in *Mycobacterium neoaurum* [66]. These differences in regulation among the different *kstD* genes within the same strain highlight the view that KstD proteins may be acting in different metabolic steps and/or pathways, each one having a particular catalytic role.

The amplification of *R. ruber* strain Chol-4 *kstD2* cDNA yielded thick amplification bands in all cases

(Fig. 5), which clearly leads to propose a constitutive expression of *kstD2*. In a similar way, two putative *kstD* genes in *R. jostii* RHA1 genome (*ro90203* and *ro09040*, belonging to the KstD2-branch of the KstD phylogenetic tree) [20] were expressed but not up-regulated neither in cholesterol nor in 7-ketocholesterol, when compared to pyruvate [64].

The expression profile of *R. ruber kstD1*, *kstD2* and *kstD3* genes was determined by real-time PCR. Taking as 1 the expression levels on sodium acetate (unexposed steroid culture), the values obtained for the expression of the three genes were: 7.6, 2.0 and 240.5-fold for *kstD1*, *kstD2* and *kstD3*, respectively, in cultures grown in cholesterol; and 13.6, 0.7 and 0.6-fold for *kstD1*, *kstD2* and *kstD3*, respectively, in cultures grown in AD.

The particular organization of *R. ruber* Chol-4 *kstD1* and *kstD3* genes (Fig. 1) opens the possibility that polycistronic *kstD* mRNAs could be synthesized by the co-transcription of the *cyp450-kstD1-MFS transporter* and *kstD3-hsd4B-choG* gene clusters. The results of the RT-PCR experiments did not show the occurrence in AD culture medium of either *cyp450-kstD1* or *kstD1-MFS transporter* RNA sequences, indicating that *kstD1* gene is independently transcribed.

In contrast, co-transcripts from both *kstD3-hsd4B* ORFs and *hsd4B-choG* ORFs could be amplified from cultures grown in the presence of cholesterol, strongly suggesting that *kstD3-hsd4B-choG* ORFs are co-transcribed into a polycistronic mRNA (Fig. 1). This group of three genes are also described to be co-transcribed in *R. erythropolis* [27].

The adjacent location of *kstD3-hsd4B* ORFs is highly conserved among rhodococci. The *hsd4B* ORF encodes a 2-enoyl acyl-CoA hydratase involved in the β-oxidative cycle of the C-17 cholesterol side chain [65]. The *choG* ORF encodes an extracellular cholesterol oxidase that it is involved in the first step of cholesterol catabolism that implies its conversion to 4-cholesten-3-one [27, 47, 67–69]. All this suggests that *kstD3*, *hsd4B* and *choG* genes are mainly involved in the steroid catabolism.

In this work we showed that *kstD2* is constitutively transcribed, while *kstD1* and *kstD3* are also constitutively but faintly transcribed, although they can be highly induced by AD or CHO. In a previous work [24], we reported the construction of a *kstD2* deletion mutant of *R. ruber* that is unable to grow in minimal medium supplemented with AD. The question is why KstD1 and/or KstD3 cannot substitute KstD2 allowing the *kstD2* deletion mutant to grow on AD. To partially advance in this subject, RT-PCR experiments were also performed on RNA from the *kstD2 R. ruber* mutant to check the expression of the other *kstD* genes (Fig. 5b).

The results showed that the transcription pattern of the *kstD* genes of the mutant is far different than that of the same genes in the wild type. Namely, *kstD1* gene in the mutant was constitutively and slightly transcribed and its expression is induced in CHO. A more noticeable change affects to the *kstD3* gene, which is not transcribed at all in any of the conditions used. These data reveal a complex relationship among the KstD enzymes and their expression control mechanisms. Modification of the *kstD* transcription levels of genes remaining in the cell have also been described in a *M. neoaurum kstD* mutant: the transcription ratio of the *kstD1* ORF (similar to the *kstD3* ORF from *R. ruber*) in AD induced cultures respect to glycerol cultures increases from 1.8 (in the wild type strain) to 2.7-fold (in the *kstD3 M. neoaurum* mutant strain) [66].

An appealing conclusion of that complex situation is that the three KstD proteins of *R. ruber* strain Chol-4 may be differentially involved in distinct pathways of steroid degradation, and that their expression could also be differentially and specifically controlled.

Heterologous expression of KstD1, KstD2 and KstD3 of *R. ruber* strain Chol-4

The three *R. ruber kstD* genes (*kstD1*, *kstD2* and *kstD3*) were cloned into the pTip-QC1 expression vector. *R. erythropolis* CECT3014 cells were electroporated with these constructions and clones harbouring each of those recombinant plasmids were isolated. Expression of the KstD proteins from these vectors in the CECT3014 transformed cells was followed by SDS-PAGE analysis. Molecular weights were 54.8, 60.8 and 61.8 kDa for KstD1, KstD2 and KstD3 respectively (Additional file 5). Cell-free extracts of cultures grown from these clones were used for the analysis of KstDs activities, and the kinetic parameters of the heterologously expressed KstDs from *R. ruber* were followed for different substrates (Table 1; see also Additional file 3 for substrates structure). Control cell extract from the *R. erythropolis* culture harbouring an empty pTip-QC1 vector yielded none or very low basal levels when acting on all the substrates used in the assay and were taken into account for final activities.

The substrate profile of KstD1 showed a clear preference to 9OHAD and testosterone, followed by progesterone, Deoxy corticosterone (DOC) and AD (Table 1). All these compounds display a keto group at C3, a C4-C5 double bond and an electronegative side-chain at C17 (Additional file 3). When comparing to the substrate preference order of *R. erythropolis* SQ1 KstD1 (Prog > 9OHAD, AD > 5α-Tes > BNC > 11βCort), some details highlights: (i) $KstD1_{SQ1}$ has a relative catalytic efficiency (RCE) on progesterone 3.4 times higher than on

Table 1 Substrate profiles of *R. ruber* KstDs expressed an analyzed in cell-free extracts of *R. erythropolis*

Substrate	KstD1				KstD2				KstD3			
	Rel. act %	Km (μM)	RCE	RCE/RCE_{AD}	Rel. act %	Km (μM)	RCE	RCE/RCE_{AD}	Rel. act %	Km (μM)	RCE	RCE/RCE_{AD}
AD	100.0 ± 11.6	34.2 ± 3.8	2.92	1.00	100.0 ± 11.8	40.1 ± 8.7	2.60	1.00	nd	nd		
9OHAD	107.4 ± 13.4	22.1 ± 7.0	4.84	1.66	29.8 ± 6.5	543.2 ± 77.8	0.05	0.02	nd	nd		
4BNC	52.3 ± 7.4	76.1 ± 15.7	0.69	0.24	30.8 ± 7.2	38.2 ± 6.6	0.81	0.31	nd	nd		
Prog	89.7 ± 7.9	27.9 ± 9.5	3.22	1.10	182.0 ± 7.0	33.8 ± 4.9	5.38	2.07	18.6 ± 3.7	43.6 ± 2.9	0.42	0.78
Cort	20.3 ± 5.8	161.6 ± 7.6	0.13	0.04	19.0 ± 3.0	374.3 ± 74.5	0.05	0.02	nd	nd		
Tes	134.6 ± 22.6	28.8 ± 7.4	4.67	1.60	233.3 ± 23.3	107.9 ± 18.5	2.16	0.83	nd	nd		
19OHAD	39.0 ± 10.6	368.8 ± 102.4	0.11	0.04	24.6 ± 7.8	347.4 ± 41.7	0.07	0.03	nd	nd		
DOC	67.0 ± 7.4	21.6 ± 4.5	3.10	1.06	124.2 ± 17.2	42.5 ± 5.8	2.92	1.12	21.6 ± 8.8	111.3 ± 4.0	0.19	0.36
5α-Tes	nd	nd			75.2 ± 4.8	24.57 ± 7.8	3.06	1.18	100.0 ± 21.9	181.1 ± 42.6	0.55	1.00

Rel. act: relative activity values. Enzyme activities are expressed as percentage of activity of AD (for KstD1 and KstD2 with 3.2 U/mg and 1.4 U/mg respectively) or 5α-tes (for KstD3 with 0.3 U/mg) that were set as 100%

nd enzyme activity was not detected for this substrate, *RCE* relative catalytic efficiency given by the ratio Rel. act/km, *Prog* progesterone, *Cort* corticosterone, *Tes* testosterone, *5α-tes* 5-alpha-testosterone, *19OHAD* 19-hydroxy-4-androstene-3,17-dione, *DOC* deoxycorticosterone, *4-BNC* 4-pregnen-3-one-20β-carboxylic acid

AD, a very big difference to the ratio 9OHAD/AD (1.6) showed by *R. ruber* KstD1 (Table 1); (ii) *R. ruber* KstD1 is not active on 5α-Tes in contrast to $KstD1_{SQ1}$.

The order of substrate preference of *R. ruber* KstD2 placed progesterone in the first position, followed by 5α-Tes, DOC, AD and testosterone (Table 1), displaying a substrate profile similar to $KstD2_{SQ}1$ enzyme (Prog > AD > 5α-Tes > 9OHAD, BNC > 11βCort). It is noteworthy the similarity of both KstD profiles, having in mind that they have been expressed in different cellular context (a *Rhodococcus* strain, and *E. coli*). *R. ruber* KstD2 has a broader range of substrates than *R. ruber* KstD1 as it can act on all the KstD1 substrates and also on 5α-Tes that contains a saturated A ring (Table 1).

Rhodococcus ruber KstD3 did not show any activity at all when acting on AD or 9OHAD, which are considered the natural substrates for KstD enzymes, but it showed the highest activity when using 5α-Tes as substrate followed by progesterone and lastly DOC (Table 1). *R. ruber* KstD3 has a very narrow substrate range similarly to the *R. erythropolis* SQ1 and *M. tuberculosis* H37Rv isoforms, being the A-ring saturated 5α-Tes the preferred substrate for all KstD3 enzymes (Table 1, [20]). However, KstD3 affinity for 5α-Tes differs from 33–36 μM in the cases of $KstD3_{H37RV}$ and $KstD3_{SQ1}$ respectively, to 181 μM in *R. ruber* KstD3 (Table 1). Despite this low affinity for 5α-Tes, this is the best substrate for the *R. ruber* enzyme among those that were assayed (Table 1).

On the other hand, although it has been proposed that only steroids carrying a small or no aliphatic side chain at C-17 are suitable substrates for KstD3 [20], a minor activity, not very different to that obtained with progesterone, has been observed in the assays of *R. ruber* KstD3 enzyme on DOC (Additional file 3; Table 1). *R. ruber* KstD3 seems to be more related to the cholesterol metabolism than to the AD metabolism [24] and then it could be acting on some not yet neatly defined intermediaries of the bacterial cholesterol metabolism.

None of the three *R. ruber* KstD proteins displayed detectable activity on 4-cholestene-3-one, 5α-cholestane-3-one or 5β-androstane-3,17-dione (5β-AD), ADD, cholesterol, cholestenone, cholic acid, DHEA, ergosterol, stigmasterol, β-estradiol, sodium deoxycholate or 5-pregen-3-β-nolone. Therefore, these KstDs catalyse preferentially 4-ene-3-oxosteroids.

Molecular modelling of KstD1 of *R. ruber* strain Chol-4

The sequence of the three KstDs of *R. ruber* were described previously [24]. In an attempt to go deeper inside the reasons of the catalytic and kinetic differences among some of these enzymes, we performed protein sequence analysis and modelling studies on the *R. ruber* KstDs using different approaches. The published three-dimensional structure and the catalytic mechanism proposed for KstD1 from *R. erythropolis* SQ1 has been used as a suitable model [21, 70]. The I-Tasser and PredictProtein programmes predicted that none of the three *R. ruber* KstDs contain sulphur bridges or transmembrane segments.

The catalytic mechanism of *R. erythropolis* SQ1 KstD1 is based in the keto-enol tautomerization of the substrate caused by Tyr487 and Gly491 residues that increases the acidity of the C2 hydrogen atoms of the substrate. Then Tyr119 and Tyr318 capture the axial β-hydrogen from C2 as a proton whereas the FAD molecule accepts the axial α-hydrogen from the C1 atom of the substrate as a hydride ion [21, 70]. Tables 2 and 3 collects the residues involved in the active site of $KstD1_{SQ1}$ described for Rohman et al. [21] and the homologue residues of the three KstDs from *R. ruber* found by COBALT

Table 2 Homologous residues of KstDs implicated in the catalytic pocket attending to Rohman et al. [21]

KstD1-SQ1	KstD1-ruber	KstD2-ruber	KstD3-ruber
S52	G53	A59	G50
F116	W118	Y125	Y116
Y119	*Y121*	*Y128*	*Y119*
F294	F295	P337	F327
V296	L297	V339	S330
Y318	*Y319*	*Y366*	*Y354*
I352	T356		I395
I354	V358	F405	P397
L447	L447	L499	L490
Y487	*Y487*	*Y539*	*Y530*
P490	G490	A542	P533
G491	*G491*	*G543*	*G534*
V492	N492	A544	A535
P493	P493	T545	T537

Italic files contain the four key residues of the KstD1 active site reported for *R. erythropolis* SQ1: Tyr119, Tyr318, Tyr487 and Gly491

Table 3 PredictProtein prediction of the secondary structure and solvent accessibility in % of the three KstDs from *R. ruber*

Structure	KstD1	KstD2	KstD3
Strand	12.72	11.88	64.74
Loop	63.99	64.01	14.38
Helix	23.29	24.11	20.88
Accessibility	**KstD1**	**KstD2**	**KstD3**
Exposed	24.1	21.81	22.11
Buried	67.7	71.63	70.5
Intermediate	8.22	6.56	7.54

programme. The four key residues: Tyr119, Tyr318, Tyr487 and Gly491 of the $KstD1_{SQ1}$ active site have a counterpart residue in the *R. ruber* KstDs. However, the I-TASSER model prediction of these *R. ruber* enzymes shows that the orientations of the side chain of the tyrosine residues differ within the catalytic pocket site being specific of each enzyme. The variation in orientation of the key residues inside the catalytic poked is shown in Fig. 6. Particularly, the orientation of KstD3 Y354 and Y530 residues is almost opposite to that of its homologous KstD1 and KstD2 residues. Moreover, amino acid Y128 from KstD2 has a position highly separated compared to its homologous from KstD1 (Y121) and KstD3 (Y119). These differences could justify the different affinity and catalytic properties among the three KstDs. The current shortage of crystalline structures of these proteins greatly limit more detailed conclusions.

A redundancy of KstD and Ksh enzymes have been described in the actinobacteria genomes [20, 24, 64, 70]. This redundancy could provide the cell with a bigger metabolic versatility and a fine-tuned response to the challenging environment. In the case of KstD enzymes, three homologues have been found in *R. erythropolis* strain SQ1 and in *M. neoaurum* that displayed different substrate preferences and that could be involved in different metabolic steps in a strain-dependent way [20, 32, 66].

Even more, it has been shown recently that mutations in the KstDs provokes a different ADD/AD molar ratio [71] and that environmental factors such as an increase of temperature can inhibit the KstD/Ksh action [72] on phytosterol in *Mycobacterium* sp. These multiplicity and versatility give to these enzymes a substantial role in the catalytic dehydrogenation of several related steroid molecules and pose many difficulties to clarify the particular role and way of acting of every single KstD.

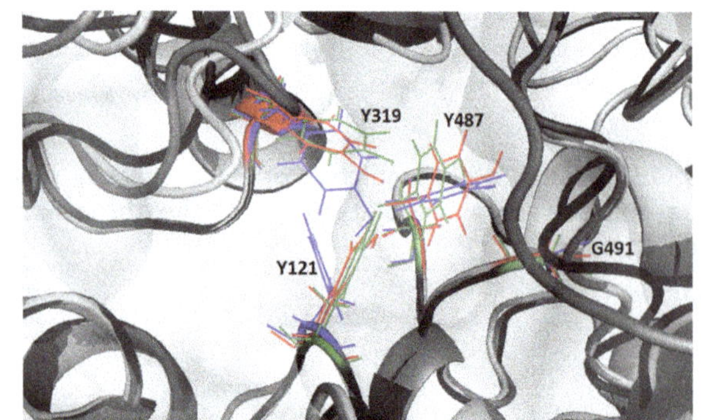

Fig. 6 Modelling of the active site of KstDs. The orientation of the four key residues in the KstD binding pocket is shown. They are superposed and depicted in different colours: *green* for KstD1, *blue* for KstD2 and *red* for KstD3. Only KsD1 residues are named, their counterpart homologues residues of KstD2 and KstD3 being listed in Tables 2 and 3

Our results suggest that both KstD1 and KstD2 of *R. ruber* could act in the conversion of AD to ADD being KstD1 mainly involved in the 9OHAD to 9OHADD conversion, in a similar way to what has been described in *R. erythropolis* SQ1 [32]. However, there are differences between these two strains, as the necessity of a double *kstD1* and *kstD2* mutation to prevent the growth in AD in the case of *R. erythropolis* SQ1 [32] or *R. rhodochrous* DSM43269 [73], while the same effect is obtained by the single *kstD2* deletion in *R. ruber* strain Chol-4 [24], suggesting that this last mutation affects in some way the activity of the KstD1 protein.

kstD3 ORF occurs in the *R. ruber* genome in a quite conserved location within *Rhodococcus* species, clustered with *hsd4B* (which encodes a 2-enoyl acyl-CoA hydratase proposed to be involved in cholesterol side-chain shortening) [65] and *choG* (coding for a cholesterol oxidase that converts cholesterol into cholestenone) [27, 47, 68]. Recently it has been proposed that sterols can be catabolised in *R. equi* USA-18 by two partially different pathways, namely via AD or via Δ1,4-BNC, that converge in the intermediary 9OHADD [74]. Given its substrate preference, the fact that the growth in AD is independent of the KstD3 activity while the growth in cholesterol needs the presence of either KstD2 or KstD3 in *R. ruber* [24], lead us to suggest that KstD3 may be involved in an alternative AD-independent cholesterol catabolic pathway.

Conclusions

To sum up, this study provides biochemical and genetic insights into the three KstD proteins found in *R. ruber*. The kinetic differences between the three KstDs suggest that each enzyme could act on different steps of the steroid catabolic routes. Both KstD1 and KstD2 could be involved in the AD catabolism while KstD1 would have a preference for 9OHAD and KstD2 for progesterone. KstD2 seems to be a more versatile enzyme than KstD1 in *R. ruber* as it can act also on saturated steroid substrates such as 5α-Tes. On the other hand, the narrower range of substrates for KstD3 and its preference for saturated steroid made this enzyme different to KstD1 and KstD2 and suggest that it may be involved in AD-independent steroid catabolism. The differences found in the orientation of catalytic residues of each KstD within the binding pocket site could explain the substrates preferences of each enzyme.

The promoter regions that support transcription of *kstD* genes have been cloned and functionally identified. The three promoter boxes contain different expression patterns, from the TCCG motif found in the *kshA-kstD2* ORF intergenic region to the KstR boxes in the *hsaE-kstD3* intergenic region. Moreover, *kstD3* ORF was transcribed as a polycistronic *kstD3-hsd4B-choG* mRNA and was induced in cholesterol growing media, reinforcing the role of KstD3 in the cholesterol metabolism. Potential functions of *R. ruber* strain Chol-4 KstDs in other steroid pathways remain to be elucidated.

Additional files

Additional file 1. Bacterial strains and plasmids used in this work.

Additional file 2. Primers used in this work. Restriction sites are marked in bold.

Additional file 3. Structure of the steroids used in this work.

Additional file 4. Scheme of the pNVSP vectors construction. The shuttle vector pNV119 was modified to include a mcs from pSEVA351 (pNVS) and after that coupled with the apramycin resistance ORF. Intergenic regions containing the putative *kstD* promoters were cloned in the mcs to obtain pNVSP vectors.

Additional file 5. Expression of KstDs from ptip-QC1 vectors in induced *R. erythropolis* CECT3014 cells. SDS-PAGE analysis on a 12.5% gel was performed using 10 μg of the cell-free extracts. The band corresponding to KstD overexpression is marked with a rectangle. Precision plus protein standard from Bio-Rad was used as size marker.

Additional file 6. *R. ruber* strain Chol-4 ORF4 and ORF5 protein sequences.

Abbreviations

AD: 4-androstene-3,17-dione; ADD: 1,4-androstadiene-3,17-dione; 5β-AD: 5β-androstane-3,17-dione; 9OHAD: 9α-hydroxy-4-androstene-3,17-dione; 9OHADD: 9α-hydroxy-1,4-androstadiene-3,17-dione; 4-BNC: 4-pregnen-3-one-20β-carboxylic acid; 5α-Tes: 17β-hydroxy-5α-androstan-3-one; DOC: deoxy corticosterone; ORF: open reading frame; DCPIP: 2,6-dichlorophenol-indophenol; KstD: 3-ketosteroid-Δ^1-dehydrogenase; LB: Luria-Bertani; mcs: multiple cloning site; TSS: transcriptional start site.

Authors' contributions

GG made the heterologous expression in *R. erythropolis*, the coexpression study, the RT-qPCR and the promoter characterization. JMNLL and GG performed the bioinformatic analyses. LFH carried out the RT-PCR experiment and collaborated with GG in the heterologous expression experiments. JP and JMNLL drafted the manuscript. JP, JMNLL and GG participated in the design and coordination of the study. All authors read and approved the final manuscript.

Author details

[1] Department of Biochemistry and Molecular Biology I, Universidad Complutense de Madrid, 28040 Madrid, Spain. [2] Faculty of Science and Engineering, Microbial Physiology-Gron Inst Biomolecular Sciences & Biotechnology, Nijenborgh 7, 9747 AG Groningen, The Netherlands.

Acknowledgements

We are in debt to R. van der Geize (Netherland) and to the group of J. L. García (CIB, CSIC) for their scientific support and invaluable help. We thank Rodrigo Velasco for his help on molecular modelling and Sara Montero for her help on kinetics.

Competing interests

The authors declare that they have no competing interests.

Funding

This work was funded by projects BFU2009-11545-C03-02, BIO2012-39695-CO2-01 and RTC-2014-2249-1 (GG is under contract of this last project) from the Spanish Ministry of Education and Science.

References

1. Iwabuchi N, Sunairi M, Urai M, Itoh C, Anzai H, Nakajima M, Harayama S. Extracellular polysaccharides of *Rhodococcus rhodochrous* S-2 stimulate the degradation of aromatic components in crude oil by indigenous marine bacteria. Appl Environ Microbiol. 2002;68:2337–43.
2. McLeod MP, Warren RL, Hsiao WW, Araki N, Myhre M, Fernandes C, Miyazawa D, Wong W, Lillquist AL, Wang D, et al. The complete genome of *Rhodococcus* sp. RHA1 provides insights into a catabolic powerhouse. Proc Natl Acad Sci USA. 2006;103:15582–7.
3. Yam KC, Okamoto S. Adventures in *Rhodococcus*—from steroids to explosives. Can J Microbiol. 2011;57:155–68.
4. Barel-Cohen K, Shore LS, Shemesh M, Wenzel A, Mueller J, Kronfeld-Schor N. Monitoring of natural and synthetic hormones in a polluted river. J Environ Manag. 2006;78:16–23.
5. Gracia T, Jones PD, Higley EB, Hilscherova K, Newsted JL, Murphy MB, Chan AK, Zhang X, Hecker M, Lam PK, et al. Modulation of steroidogenesis by coastal waters and sewage effluents of Hong Kong, China, using the H295R assay. Environ Sci Pollut Res Int. 2008;15:332–43.
6. Soto AM, Calabro JM, Prechtl NV, Yau AY, Orlando EF, Daxenberger A, Kolok AS, Guillette LJ Jr, le Bizec B, Lange IG, et al. Androgenic and estrogenic activity in water bodies receiving cattle feedlot effluent in Eastern Nebraska, USA. Environ Health Perspect. 2004;112:346–52.
7. van der Geize R, Dijkhuizen L. Harnessing the catabolic diversity of rhodococci for environmental and biotechnological applications. Curr Opin Microbiol. 2004;7:255–61.
8. Malaviya A, Gomes J. Androstenedione production by biotransformation of phytosterols. Bioresour Technol. 2008;99:6725–37.
9. García JL, Uhía I, Galan B. Catabolism and biotechnological applications of cholesterol degrading bacteria. Microb Biotechnol. 2012;5:679–99.
10. Horinouchi M, Hayashi T, Kudo T. Steroid degradation in *Comamonas testosteroni*. J Steroid Biochem Mol Biol. 2012;129:4–14.
11. Philipp B. Bacterial degradation of bile salts. Appl Microbiol Biotechnol. 2011;89:903–15.
12. Uhía I, Galán B, Kendall SL, Stoker NG, García JL. Cholesterol metabolism in *Mycobacterium smegmatis*. Environ Microbiol Rep. 2012;4:168–82.
13. Petrusma M, van der Geize R, Dijkhuizen L. 3-Ketosteroid 9alpha-hydroxylase enzymes: rieske non-heme monooxygenases essential for bacterial steroid degradation. Antonie Van Leeuwenhoek. 2014;106:157–72.
14. Florin C, Kohler T, Grandguillot M, Plesiat P. *Comamonas testosteroni* 3-ketosteroid-delta 4(5 alpha)-dehydrogenase: gene and protein characterization. J Bacteriol. 1996;178:3322–30.
15. Itagaki E, Hatta T, Wakabayashi T, Suzuki K. Spectral properties of 3-ketosteroid-delta 1-dehydrogenase from *Nocardia corallina*. Biochim Biophys Acta. 1990;1040:281–6.
16. Itagaki E, Wakabayashi T, Hatta T. Purification and characterization of 3-ketosteroid-delta 1-dehydrogenase from *Nocardia corallina*. Biochim Biophys Acta. 1990;1038:60–7.
17. Chen MM, Wang FQ, Lin LC, Yao K, Wei DZ. Characterization and application of fusidane antibiotic biosynthesis enzyme 3-ketosteroid-1-dehydrogenase in steroid transformation. Appl Microbiol Biotechnol. 2012;96:133–42.
18. Kisiela M, Skarka A, Ebert B, Maser E. Hydroxysteroid dehydrogenases (HSDs) in bacteria—a bioinformatic perspective. J Steroid Biochem Mol Biol. 2012;129:31–46.
19. Hilyard EJ, Jones-Meehan JM, Spargo BJ, Hill RT. Enrichment, isolation, and phylogenetic identification of polycyclic aromatic hydrocarbon-degrading bacteria from Elizabeth River sediments. Appl Environ Microbiol. 2008;74:1176–82.
20. Knol J, Bodewits K, Hessels GI, Dijkhuizen L, van der Geize R. 3-Keto-5alpha-steroid Delta(1)-dehydrogenase from *Rhodococcus erythropolis* SQ1 and its orthologue in *Mycobacterium tuberculosis* H37Rv are highly specific enzymes that function in cholesterol catabolism. Biochem J. 2008;410:339–46.

21. Rohman A, van Oosterwijk N, Thunnissen AM, Dijkstra BW. Crystal structure and site-directed mutagenesis of 3-ketosteroid Delta1-dehydrogenase from *Rhodococcus erythropolis* SQ1 explain its catalytic mechanism. J Biol Chem. 2013;288:35559–68.
22. Molnár I, Choi KP, Yamashita M, Murooka Y. Molecular cloning, expression in Streptomyces lividans, and analysis of a gene cluster from *Arthrobacter simplex* encoding 3-ketosteroid-delta 1-dehydrogenase, 3-ketosteroid-delta 5-isomerase and a hypothetical regulatory protein. Mol Microbiol. 1995;15:895–905.
23. Wierenga RK, Terpstra P, Hol WG. Prediction of the occurrence of the ADP-binding beta alpha beta-fold in proteins, using an amino acid sequence fingerprint. J Mol Biol. 1986;187:101–7.
24. de las Heras LF, van der Geize R, Drzyzga O, Perera J, Navarro Llorens JM. Molecular characterization of three 3-ketosteroid-Δ^1-dehydrogenase isoenzymes of *Rhodococcus ruber* strain Chol-4. J Steroid Biochem Mol Biol. 2012;132:271–81.
25. Fernández de las Heras L, García Fernández E, Navarro Llorens JM, Perera J, Drzyzga O. Morphological, physiological, and molecular characterization of a newly isolated steroid-degrading actinomycete, identified as *Rhodococcus ruber* strain Chol-4. Curr Microbiol. 2009;59:548–53.
26. Klein U, Gimpl G, Fahrenholz F. Alteration of the myometrial plasma membrane cholesterol content with b-cyclodextrin modulates the binding affinity of the oxytocin receptor. Biochemistry. 1995;34:13784–93.
27. Fernández de las Heras L, Mascaraque V, García Fernández E, Navarro-Llorens JM, Perera J, Drzyzga O. ChoG is the main inducible extracellular cholesterol oxidase of *Rhodococcus* sp. strain CECT3014. Microbiol Res. 2011;166:403–18.
28. Sambrook J, Fritsch EF, Maniatis T. Molecular cloning: a laboratory manual. New York: Cold Spring Harbor Laboratory Press; 1989.
29. Hosek J, Svastova P, Moravkova M, Pavlik I, Bartos M. Methods of mycobacterial DNA isolation from different biological material: a review. Vet Res. 2006;524:180–92.
30. Nakashima N, Tamura T. Isolation and characterization of a rolling-circle-type plasmid from *Rhodococcus erythropolis* and application of the plasmid to multiple-recombinant-protein expression. Appl Environ Microbiol. 2004;70:5557–68.
31. Bradford MM. A rapid and sensitive method for the quantitation of microgram quantities of protein utilizing the principle of protein-dye binding. Anal Biochem. 1976;72:248–54.
32. van der Geize R, Hessels GI, Dijkhuizen L. Molecular and functional characterization of the kstD2 gene of *Rhodococcus erythropolis* SQ1 encoding a second 3-ketosteroid Δ^1-dehydrogenase isoenzyme. Microbiology. 2002;148:3285–92.
33. de las Heras LF, Alonso S, de Leon AD, Xavier D, Perera J, Navarro Llorens JM. Draft genome sequence of the steroid degrader *Rhodococcus ruber* strain Chol-4. Genome Announc. 2013. doi:10.1128/genomeA.00215-13.
34. Petersen TN, Brunak S, von Heijne G, von Nielsen H. SignalP 4.0: discriminating signal peptides from transmembrane regions. Nat Methods. 2011;8:785–6.
35. Solovyev V, Salamov A. Automatic annotation of microbial genomes and metagenomic sequences. In: Li RW, editor. Metagenomics and its applications in agriculture, biomedicine and environmental studies. New York: Nova Science Publishers; 2011. p. 61–78.
36. Reese MG. Application of a time-delay neural network to promoter annotation in the *Drosophila melanogaster* genome. Comput Chem. 2001;26:51–6.
37. de Jong A, Pietersma H, Cordes M, Kuipers OP, Kok J. PePPER: a webserver for prediction of prokaryote promoter elements and regulons. BMC Genom. 2012;13:299.
38. Yang J, Yan R, Roy A, Xu D, Poisson J, Zhang Y. The I-TASSER Suite: protein structure and function prediction. Nat Methods. 2015;12:7–8.
39. Yachdav G, Kloppmann E, Kajan L, Hecht M, Goldberg T, Hamp T, Honigschmid P, Schafferhans A, Roos M, Bernhofer M, et al. PredictProtein-an open resource for online prediction of protein structural and functional features. Nucl Acids Res. 2014;42:W337–43.
40. Papadopoulos JS, Agarwala R. COBALT: constraint-based alignment tool for multiple protein sequences. Bioinformatics. 2007;23:1073–9.
41. Durante-Rodriguez G, de Lorenzo V, Martinez-Garcia E. The Standard European Vector Architecture (SEVA) plasmid toolkit. Methods Mol Biol. 2014;1149:469–78.
42. Silva-Rocha R, Martínez-García E, Calles B, Chavarría M, Arce-Rodríguez A, de Las Heras A, Paez-Espino AD, Durante-Rodríguez G, Kim J, Nikel PI, et al. The Standard European Vector Architecture (SEVA): a coherent platform for the analysis and deployment of complex prokaryotic phenotypes. Nucl Acids Res. 2013;41:D666–75.
43. Chiba K, Hoshino Y, Ishino K, Kogure T, Mikami Y, Uehara Y, Ishikawa J. Construction of a pair of practical *Nocardia-Escherichia coli* shuttle vectors. Jpn J Infect Dis. 2007;60:45–7.
44. Gust B, Challis GL, Fowler K, Kieser T, Chater KF. PCR-targeted *Streptomyces* gene replacement identifies a protein domain needed for biosynthesis of the sesquiterpene soil odor geosmin. Proc Natl Acad Sci USA. 2003;100:1541–6.
45. Hopwood DA, Bibb MJ, Chater KF, Kieser T, Bruton CJ, Kieser HM, Lydiate DJ, Smith CP, Ward JM, Schrempf H. Genetic manipulation of *Streptomyces*. Norwich: The John Innes Foundation, Cold Spring Harbour Laboratory; 1985.
46. Wang C, Lee J, Deng Y, Tao F, Zhang LH. ARF-TSS: an alternative method for identification of transcription start site in bacteria. Biotechniques. 2012. doi:10.2144/000113858.
47. de las Heras LF, Perera J, Navarro Llorens JM. Cholesterol to cholestenone oxidation by ChoG, the main extracellular cholesterol oxidase of *Rhodococcus ruber* strain Chol-4. J Steroid Biochem Mol Biol. 2014;139:33–44.
48. Kendall SL, Withers M, Soffair CN, Moreland NJ, Gurcha S, Sidders B, Frita R, Ten Bokum A, Besra GS, Lott JS, et al. A highly conserved transcriptional repressor controls a large regulon involved in lipid degradation in *Mycobacterium smegmatis* and *Mycobacterium tuberculosis*. Mol Microbiol. 2007;65:684–99.
49. Crowe AM, Stogios PJ, Casabon I, Evdokimova E, Savchenko A, Eltis LD. Structural and functional characterization of a ketosteroid transcriptional regulator of *Mycobacterium tuberculosis*. J Biol Chem. 2015;290:872–82.
50. Uhía I, Galán B, Medrano FJ, García JL. Characterization of the KstR-dependent promoter of the first step of cholesterol degradative pathway in *Mycobacterium smegmatis*. Microbiology. 2011;157:2670–80.
51. García-Fernández E, Medrano FJ, Galán B, García JL. Deciphering the transcriptional regulation of cholesterol catabolic pathway in mycobacteria: identification of the inducer of KstR repressor. J Biol Chem. 2014;289:17576–88.
52. Ho NA, Dawes SS, Crowe AM, Casabon I, Gao C, Kendall SL, Baker EN, Eltis LD, Lott JS. The structure of the transcriptional repressor KstR in complex with CoA thioester cholesterol metabolites sheds light on the regulation of cholesterol catabolism in *Mycobacterium tuberculosis*. J Biol Chem. 2016;291:7256–66.
53. Shtratnikova VY, Schelkunov MI, Fokina VV, Pekov YA, Ivashina T, Donova MV. Genome-wide bioinformatics analysis of steroid metabolism-associated genes in *Nocardioides simplex* VKM Ac-2033D. Curr Genet. 2016;62:643–56.
54. Ma Y, Yu H. Engineering of *Rhodococcus* cell catalysts for tolerance improvement by sigma factor mutation and active plasmid partition. J Ind Microbiol Biotechnol. 2012;39:1421–30.
55. Bashyam MD, Kaushal D, Dasgupta SK, Tyagi AK. A study of mycobacterial transcriptional apparatus: identification of novel features in promoter elements. J Bacteriol. 1996;178:4847–53.
56. Kendall SL, Burgess P, Balhana R, Withers M, Ten Bokum A, Lott JS, Gao C, Uhia-Castro I, Stoker NG. Cholesterol utilization in mycobacteria is controlled by two TetR-type transcriptional regulators: kstR and kstR2. Microbiology. 2010;156:1362–71.
57. Shell SS, Wang J, Lapierre P, Mir M, Chase MR, Pyle MM, Gawande R, Ahmad R, Sarracino DA, Ioerger TR, et al. Leaderless transcripts and small proteins are common features of the mycobacterial translational landscape. PLoS Genet. 2015;11:e1005641.
58. Strohl WR. Compilation and analysis of DNA sequences associated with apparent *streptomycete* promoters. Nucleic Acids Res. 1992;20:961–74.
59. Pfleiderer C, Smid A, Bartsch I, Grummt I. An undecamer DNA sequence directs termination of human ribosomal gene transcription. Nucleic Acids Res. 1990;18:4727–36.
60. Grummt I, Kuhn A, Bartsch I, Rosenbauer H. A transcription terminator located upstream of the mouse rDNA initiation site affects rRNA synthesis. Cell. 1986;47:901–11.
61. Henderson S, Sollner-Webb B. A transcriptional terminator is a novel element of the promoter of the mouse ribosomal RNA gene. Cell. 1986;47:891–900.

62. Henderson SL, Ryan K, Sollner-Webb B. The promoter-proximal rDNA terminator augments initiation by preventing disruption of the stable transcription complex caused by polymerase read-in. Genes Dev. 1989;3:212–23.
63. Szoköl J, Rucká L, Simciková M, Halada P, Nesvera J, Pátek M. Induction and carbon catabolite repression of phenol degradation genes in *Rhodococcus erythropolis* and *Rhodococcus jostii*. Appl Microbiol Biotechnol. 2014;98:8267–79.
64. Mathieu JM, Mohn WW, Eltis LD, LeBlanc JC, Stewart GR, Dresen C, Okamoto K, Alvarez PJ. 7-ketocholesterol catabolism by *Rhodococcus jostii* RHA1. Appl Environ Microbiol. 2010;76:352–5.
65. van der Geize R, Yam K, Heuser T, Wilbrink MH, Hara H, Anderton MC, Sim E, Dijkhuizen L, Davies JE, Mohn WW, et al. A gene cluster encoding cholesterol catabolism in a soil actinomycete provides insight into *Mycobacterium tuberculosis* survival in macrophages. Proc Natl Acad Sci USA. 2007;104:1947–52.
66. Yao K, Xu LQ, Wang FQ, Wei DZ. Characterization and engineering of 3-ketosteroid- big up tri, open1-dehydrogenase and 3-ketosteroid-9alpha-hydroxylase in *Mycobacterium neoaurum* ATCC 25795 to produce 9alpha-hydroxy-4-androstene-3,17-dione through the catabolism of sterols. Metab Eng. 2014;24:181–91.
67. Kreit J, Sampson NS. Cholesterol oxidase: physiological functions. FEBS J. 2009;276:6844–56.
68. Pollegioni L, Piubelli L, Molla G. Cholesterol oxidase: biotechnological applications. FEBS J. 2009;276:6857–70.
69. Vrielink A, Ghisla S. Cholesterol oxidase: biochemistry and structural features. FEBS J. 2009;276:6826–43.
70. Petrusma M, Hessels G, Dijkhuizen L, van der Geize R. Multiplicity of 3-ketosteroid-9α-hydroxylase enzymes in *Rhodococcus rhodochrous* DSM43269 for specific degradation of different classes of steroids. J Bacteriol. 2011;193:3931–40.
71. Shao M, Zhang X, Rao Z, Xu M, Yang T, Li H, Xu Z, Yang S. A mutant form of 3-ketosteroid-Δ^1-dehydrogenase gives altered androst-1,4-diene-3, 17-dione/androst-4-ene-3,17-dione molar ratios in steroid biotransformations by *Mycobacterium neoaurum* ST-095. J Ind Microbiol Biotechnol. 2016;43:691–701.
72. Xu XW, Gao XQ, Feng JX, Wang XD, Wei DZ. Influence of temperature on nucleus degradation of 4-androstene-3, 17-dione in phytosterol biotransformation by *Mycobacterium* sp. Lett Appl Microbiol. 2015;61:63–8.
73. Liu Y, Shen Y, Qiao Y, Su L, Li C, Wang M. The effect of 3-ketosteroid-Δ^1-dehydrogenase isoenzymes on the transformation of AD to 9alpha-OH-AD by *Rhodococcus rhodochrous* DSM43269. J Ind Microbiol Biotechnol. 2016;43:1303–11.
74. Yeh CH, Kuo YS, Chang CM, Liu WH, Sheu ML, Meng M. Deletion of the gene encoding the reductase component of 3-ketosteroid 9α-hydroxylase in *Rhodococcus equi* USA-18 disrupts sterol catabolism, leading to the accumulation of 3-oxo-23,24-bisnorchola-1,4-dien-22-oic acid and 1,4-androstadiene-3,17-dione. Microb Cell Fact. 2014;13:130.

14

Improving *Escherichia coli* membrane integrity and fatty acid production by expression tuning of FadL and OmpF

Zaigao Tan[1], William Black[1,2], Jong Moon Yoon[1], Jacqueline V. Shanks[1] and Laura R. Jarboe[1*]

Abstract

Background: Construction of microbial biocatalysts for the production of biorenewables at economically viable yields and titers is frequently hampered by product toxicity. Membrane damage is often deemed as the principal mechanism of this toxicity, particularly in regards to decreased membrane integrity. Previous studies have attempted to engineer the membrane with the goal of increasing membrane integrity. However, most of these works focused on engineering of phospholipids and efforts to identify membrane proteins that can be targeted to improve fatty acid production have been unsuccessful.

Results: Here we show that deletion of outer membrane protein *ompF* significantly increased membrane integrity, fatty acid tolerance and fatty acid production, possibly due to prevention of re-entry of short chain fatty acids. In contrast, deletion of *fadL* resulted in significantly decreased membrane integrity and fatty acid production. Consistently, increased expression of *fadL* remarkably increased membrane integrity and fatty acid tolerance while also increasing the final fatty acid titer. This 34% increase in the final fatty acid titer was possibly due to increased membrane lipid biosynthesis. Tuning of *fadL* expression showed that there is a positive relationship between *fadL* abundance and fatty acid production. Combinatorial deletion of *ompF* and increased expression of *fadL* were found to have an additive role in increasing membrane integrity, and was associated with a 53% increase the fatty acid titer, to 2.3 g/L.

Conclusions: These results emphasize the importance of membrane proteins for maintaining membrane integrity and production of biorenewables, such as fatty acids, which expands the targets for membrane engineering.

Keywords: Membrane engineering, Membrane integrity, Outer membrane protein, Tolerance, Fatty acid production

Background

Construction of microbial cell factories for production of biorenewable fuels and chemicals is a promising alternative to current petroleum-driven industries [1, 2]. A variety of microorganisms have been engineered for production of bulk chemicals, biofuels and high-value, fine chemicals [3–7]. However, performance of some biocatalysts can be restricted by various detrimental effects, including toxicity of the product or components of the feedstock [8, 9]. A variety of adverse effects could be the cause of this toxicity, e.g. intracellular acidification; DNA, RNA, protein and membrane damage [10]. Among these, membrane damage has been recognized as a common problem [11–15].

Membrane damage can be compared to a reaction vessel that is vulnerable to corrosion by its contents. In this scenario, a typical response would be to change the composition of the reaction vessel in order to increase resistance to corrosion. For microbial biocatalysts, the composition, function and physical properties of the membrane can be altered through targeted, rational genetic manipulation. Such genetic manipulation is consistent with Cameron and Tong's fifth application of cellular and metabolic engineering, "modification of cell properties" [16]. When enzymes, transporters and regulators are involved in this membrane engineering, it is

*Correspondence: ljarboe@iastate.edu
[1] 4134 Biorenewables Research Laboratory, Department of Chemical and Biological Engineering, Iowa State University, Ames, IA 50011, USA
Full list of author information is available at the end of the article

also consistent with Bailey's 1991 definition of metabolic engineering as "the improvement of cellular activities by manipulation of enzymatic, transport, and regulatory function of the cell with the use of recombinant DNA technology" [17].

This work focuses on membrane engineering to improve production of fatty acids, an attractive class of biorenewable chemicals which can be catalyzed to a variety of products with a large potential market, e.g. alkanes, olefins, esters, fatty aldehydes, and fatty alcohols [18–22]. Unfortunately, these fatty acids have been reported to cause a decrease in membrane integrity of *E. coli* during both exogenous challenge and endogenous production [14]. Engineering of membrane phospholipids has proven as a powerful tool in addressing membrane integrity. Decreasing incorporation of medium-chain fatty acids into the membrane increased the average membrane lipid length, decreased the toxicity of fatty acids and increased fatty acid (C12–C14) production in rich medium from 0.60 to 1.36 g/L [23]. Expression of a thioesterase from *Geobacillus* sp. Y412MC10 that prevents medium-chain unsaturated acyl-ACPs from being incorporated into the phospholipids was shown to increase membrane integrity during fatty acid production, but there was no increase in fatty acid (C8–C14) production after 24 h in rich medium, with titers of 0.65 g/L observed with and without expression of the secondary thioesterase [24]. Both of these works demonstrate the feasibility of engineering the membrane lipid composition in order to increase membrane integrity and possibly enhance fatty acid tolerance and production [23, 24].

As efforts continue to increase the membrane integrity during production of membrane-damaging compounds, it becomes increasingly important to provide a sufficient route of product export. Several studies have shown that increasing the expression of transporters can increase production of inhibitory compounds, such as valine [25] and limonene [8]. With the goal of using this strategy to improve fatty acid production, sixteen possible fatty acid transporters were characterized for their role in fatty acid tolerance and production [26]. This previous study identified several transporters that increased fatty acid tolerance when their expression was increased, but did not identify any such transporters that increased fatty acid production.

The transporters OmpF and FadL were part of the previous study. The OmpF protein exists as a trimer in the outer membrane and participates in the transport of sugars, ions, antibiotics and proteins across the outer membrane [27, 28]. FadL is an outer membrane ligand gated channel that functions in the uptake of exogenous long-chain fatty acids (LCFA), [29, 30], especially palmitic acid (C16:0) and oleic acid (C18:1), yet shows no binding to short-chain fatty acids (SCFA, <C10) [31]. Even though the previous characterization observed that deletion of *ompF* and *fadL* had no impact on fatty acid production [26], several other reports related to these two transporters (Table 1) motivated the further exploration of their role in fatty acid tolerance and production described here.

Two 2015 publications directly implicated OmpF in tolerance of exogenously supplied inhibitors, though in one case OmpF played a protective role and in the other it played a damaging role. Most relevant to our goal of improving fatty acid production is the demonstration that deletion of *ompF* dampened octanoic acid toxicity, with evidence that this deletion of *ompF* reduced SCFA entry into cells [32] (Table 1). This reduced entry of SCFA into cells was assessed by measuring the decrease in intracellular pH during challenge with exogenously supplied octanoic acid. Contrastingly, OmpF was found to be directly related to tolerance of three exogenously provided phenylpropanoids: rutin, naringenin and resveratrol [33]. Specifically, strains with increased expression of OmpF showed increased tolerance to these compounds and strains with decreased expression of OmpF showed decreased tolerance, leading to the proposition that OmpF participates in the removal of phenylpropanoids from the cell interior. Thus, OmpF showed a negative role in SCFA tolerance and a positive role in phenylpropanoid tolerance.

There are also reports of FadL being involved in fatty acid production and tolerance to some inhibitors (Table 1). Increased expression of *fadL* resulted in increased conversion of exogenously supplied palmitic acid to ω-hydroxy palmitic acid [34]. This improved organism performance was attributed to increased uptake of palmitic acid, as data indicated that FadL was not involved in export of the hydroxylated product. Similarly, FadL seemed to play a crucial role in the import of octane for production of octanol, octanal and octanoic acid [35]. Specifically, production of these compounds from exogenously supplied octane was abolished when *fadL* was deleted. However, it was noted that this deletion of *fadL* increased survival during challenge with hexane, with the conclusion that FadL was the main route of hexane entry into the cell [35]. The phenylpropanoid studies described above also noted that FadL abundance was directly related to tolerance of exogenously supplied rutin, naringenin and resveratrol, the same trend was observed for OmpF, with the interpretation that FadL was involved in repairing membrane damage caused by these compounds [33]. However, even though phenol toxicity is often attributed to membrane damage [36], deletion of *fadL* had no impact on survival during phenol challenge [37]. Thus, FadL appears to be important to

Table 1 Previous reports of the role of OmpF and FadL in tolerance of membrane-damaging compounds

Compound	Condition	Result	Reference
OmpF, outer membrane porin F			
C_8–C_{14} fatty acids	Production of ~1 g/L fatty acids during growth in LB with glycerol, 37 °C	Deletion of *ompF* from a derivative of MG1655 had no impact on cell viability or membrane integrity	[26]
Octanoic acid (C8)	Challenge with up to 20 mM C8 in minimal media with glucose, tryptone and yeast extract at pH 7.0 and 37 °C	Deletion of *ompF* from BW25113 decreased sensitivity to C8, and increased expression of *ompF* increased sensitivity to C8. Sensitivity was assessed via the maximum OD. Deletion of *ompF* decreased the magnitude of intracellular acidification	[32]
Phenylpropanoids	Challenge with 1 g/L rutin, naringenin or resveratrol in M9 medium with casamino acids and glucose at 30 °C	Increased expression of *ompF* in BL21 increased the maximum specific growth rate during challenge. Decreased growth rate during challenge was observed when *ompF* expression was decreased	[33]
FadL, long-chain fatty acid outer membrane porin			
C_8–C_{14} fatty acids	Production of ~1 g/L fatty acids during growth in LB with glycerol at 37 °C	Deletion of *fadL* from a derivative of MG1655 had no impact on cell viability or membrane integrity	[26]
Palmitic and ω-hydroxy palmitic acids	Addition of 1 mM palmitic acid in potassium phosphate buffer with glucose or glycerol, 30 °C	Increased expression of *fadL* increased conversion of palmitic acid to ω-hydroxy palmitic acid. The increase was smaller in the presence of glycerol than glucose	[34]
Phenol	Challenge with phenol at 50–75% of the MIC in LB at 37 °C	Deletion of *fadL* from BW25113 had no impact on survival	[37]
Octane	Addition of ~20 vol% octane in LB at 37 °C	Deletion of *fadL* from a BW25113 derivative abolished conversion of octane to octanol, octanal and octanoic acid	[35]
Hexane	Challenge with 10 vol% hexane in LB at 37 °C	Deletion of *fadL* from BW25113 increased survival, as assessed by OD	[35]
Phenylpropanoids	Challenge with 1 g/L rutin, naringenin or resveratrol in minimal medium with casamino acids and glucose at 30 °C	Increased expression of *fadL* in BL21 increased the maximum specific growth rate during challenge. Decreased growth rate during challenge was observed when *fadL* expression was decreased	[33]

the uptake of some fatty acids and alkanes, provides protection from the inhibitory effects of phenylpropanoids, provides entry to some harmful alkanes and yet possibly plays no role in repairing the membrane damage caused by phenol.

Here we have taken another look at the role of OmpF and FadL in fatty acid tolerance and production, with the conclusion that OmpF and FadL have opposite effects. Specifically, fatty acid tolerance, fatty acid production and membrane integrity were all increased when *ompF* was deleted or when expression of *fadL* was increased. Concurrent utilization of these two engineering strategies enabled a roughly 50% increase in production of fatty acids (primarily C14, C16:1 and C16), resulting in a final titer of 2.3 g/L. Although we employed a thioesterase specific for LCFA (C14–C16), some SCFAs (e.g. C8 and C10) were also produced. We propose that deletion of *ompF* prevents re-entry of the SCFA and their corresponding toxic effects. Contrastingly, it seems that FadL may enable the recapture of some of the LCFA for use in membrane biosynthesis and repair.

Methods

Strains and plasmids

All plasmids and strains used in this study are listed in Table 2. One-step recombination method (FLP-FRT) was employed to perform genetic modifications [38]. *E. coli* K-12 MG1655 was employed as the host strain. For modulating expression of *fadL*, the FRT-*cat*-FRT selection marker linked with four different promoters (M1-12, M1-37, M1-46, M1-93) [6, 39, 40] with varying strengths was employed to regulate expression of the original *fadL* gene, yielding engineered strains M1-12-*fadL*, M1-37-*fadL*, M1-46-*fadL* and M1-93-*fadL*, respectively.

For increasing expression of *fadL*, two different strategies were employed. First, the low-copy plasmid pACYC184-Kan-*fadL*, which harbors the native promoter, open reading frame (ORF), and terminator of *fadL* was transformed to MG1655, resulting in Pla-*fadL*. MG1655 with empty pACYC184-Kan served as the corresponding control (Pla-empty). Second, for increased expression of *fadL* from the chromosome, a second copy of the *fadL* gene was inserted into the MG1655 genome at the *ldhA* site, resulting in Gen-*fadL*. The *ldhA* gene was also deleted from MG1655 to generate strain Gen-empty, which serves as a control for strain Gen-*fadL*. Selection of *ldhA* as the integration site was motivated by previous reports [41].

The pXZ18Z plasmid [42] harboring a thioesterase from *Ricinus communis* and the *E. coli* 3-hydroxyacyl-ACP dehydratase (*fabZ*) was used for long-chain fatty acid (LCFA) production. When necessary, ampicillin, kanamycin and chloramphenicol were used at final concentrations of 100, 50 and 34 mg/L, respectively.

Strain tolerance characterization

Octanoic acid tolerance was characterized in 50 mL MOPS defined minimal medium with 2.0% (wt/v)

Table 2 Strains and plasmids used in this study

Strains/plasmids	Genetic characteristics	Source
Strains		
MG1655	Wild type *E. coli* K-12 strain	Lab collection
Δ*ompF*	MG1655, Δ*ompF*	This study
Δ*fadD*	MG1655, Δ*fadD*	This study
Δ*fadL*	MG1655, Δ*fadL*	This study
Pla-empty	MG1655, pACYC184-Kan	This study
Pla-*fadL*	MG1655, pACYC184-Kan-*fadL*	This study
Gen-empty	MG1655, *ldhA*::FRT-*cat*-FRT	This study
Gen-*fadL*	MG1655, *ldhA*::FRT-*cat*-FRT, *fadL*	This study
M1-12-*fadL*	MG1655, FRT-*cat*-FRT, M1-12-*fadL*	This study
M1-37-*fadL*	MG1655, FRT-*cat*-FRT, M1-37-*fadL*	This study
M1-46-*fadL*	MG1655, FRT-*cat*-FRT, M1-46-*fadL*	This study
M1-93-*fadL*	MG1655, FRT-*cat*-FRT, M1-93-*fadL*	This study
Δ*ompF* + Pla-empty	MG1655, Δ*ompF*, pACYC184-Kan	This study
Δ*ompF* + Pla-*fadL*	MG1655, Δ*ompF*, pACYC184-Kan-*fadL*	This study
Plasmids		
pACYC184-Kan	p15A, pACYC184, Kan^r	This study
pACYC184-Kan-*fadL*	pACYC184-Kan harboring *fadL*, Kan^r	This study
pXZ18Z (TE)	pTrc99a-*Ricinus communis* thioesterase-*fabZ*, Amp^r	[42]

dextrose and 10 mM octanoic acid (1.44 g/L) in 250 mL baffled flasks at 220 rpm and initial pH at 7.0, 30 °C. MOPS media contains the following: 8.37 g/L 3-(*N*-morpholino)propanesulfonic acid (MOPS), 0.72 g/L tricine, 2.92 g/L NaCl, 0.51 g/L NH_4Cl, 1.6 g/L KOH, 50 mg/L $MgCl_2$, 48 mg/L K_2SO_4, 348 mg/L K_2HPO_4, 0.215 mg/L Na_2SeO_3, 0.303 mg/L $Na_2MoO_4{\cdot}2H_2O$, 0.17 mg/L $ZnCl_2$, 2.5 μg/L $FeCl_2{\cdot}4H_2O$, 0.092 μg/L $CaCl_2{\cdot}2H_2O$, 0.031 μg/L H_3BO_3, 0.020 μg/L $MnCl_2{\cdot}4H_2O$, 0.0090 μg/L $CoCl_2{\cdot}4H_2O$, and 0.0020 μg/L $CuCl_2{\cdot}4H_2O$ [43, 44]. Specific growth rate μ (h^{-1}) was calculated by fitting the equation $OD = OD_0e^{\mu t}$ over the duration of the exponential growth phase. OD was measured at 550 nm and all estimated μ values had an R^2 of at least 0.95 [45]. Dry cell weight (DCW) was calculated from the optical density at 550 nm (1 $OD_{550} = 0.333$ g DCW/L).

Membrane integrity characterization

Cells were centrifuged, washed twice, and then resuspended in PBS buffer (pH 7.0) at a final OD_{550} of ~1. One hundred microliter (100 μL) of this suspension was mixed with 900 μL of PBS buffer and SYTOX Green (Invitrogen) was added to a final concentration of 5.0 μM. After resting at room temperature for 15 min, cells were analyzed by a BD Biosciences FACSCanto II flow cytometer equipped with standard factory-installed 488 nm excitation laser, signal collection optics, and fluorescence emission filter configuration. Instrument sheath fluid was filtered (0.22 μm) PBS buffer. Green fluorescence from stained cells was collected in the FL1 channel (525/50 nm). Forward scatter (FSC), side scatter (SSC), and FL1 (Green) parameters were collected as logarithmic signals. All data collections were performed at low flow rate setting (~12 μL/min) and cell concentrations were such that the event rate was below 5000 events/s. All samples were analyzed immediately after staining. Background noise and small debris was eliminated from data collection via a side scatter signal threshold that was established by examining samples containing only SYTOX Green staining buffer. Bacteria in SYTOX Green-stained samples were readily identified on the basis of FSC and SSC signals and an appropriate "Cell" gate was drawn to limit FL1 analysis to bacteria and exclude non-cell events. A minimum of 20,000 cell-gated events were collected for each sample. Green fluorescence data for these "cell" events were plotted as histograms showing the signal distribution of bacteria in the sample [14]. Flow cytometry data for this work is available via Flow Repository (https://flowrepository.org) FR-FCM-ZY2B.

Membrane lipid composition characterization

The membrane lipids were extracted by using the Bligh and Dyer method with minor modifications [14, 46]. Cells were centrifuged, washed twice with cold double-distilled water (ddH_2O), resuspended in 1.4 mL methanol and transferred to a new glass tube. Ten μicroliter of 1 μg/μL pentadecanoic acid (C15:0) dissolved in ethanol was added as internal standard. Then, samples were sonicated, incubated at 70 °C for 15 min and centrifuged at 5000×*g* for 5 min. The supernatant was transferred to a new glass tube and the cell pellet was resuspended in 0.75 mL of chloroform, shaken at 37 °C, 150 rpm for 5 min. Transferred supernatant and pellet suspension were combined, vortexed for 1 min and centrifuged at 5000×*g* for 2 min. The bottom phase was transferred to a new glass tube and dried under nitrogen gas. Two milliliter of methanol:sulfuric acid (98:2 v/v) mixture was added and the mixture was vortexed and incubated at 80 °C for 30 min. Finally, 2 mL of 0.9% (wt/v) sodium chloride (NaCl) and 1 mL of hexane were added, vortexed and centrifuged at 2000×*g* for 2 min. The top hexane layer was then analyzed by gas chromatography–mass spectrometry (GC–MS). The temperature for GC–MS analysis was initially held at 50 °C for 2 min, ramped to 200 °C at 25 °C/min, held for 1 min, then raised to 315 °C at 25 °C/min, held for 2 min. Helium was used as a carrier gas and the flow rate was 1 mL/min through a DB-5MS separation column (30 m, 0.25 mm ID, 0.25 μm, Agilent). Methods for calculating average membrane lipid length and lipid saturated:unsaturated ratio can be found in [14].

Membrane lipid content measurement

Thirty milliliters of mid-log *E. coli* cells were centrifuged, washed by ddH_2O and adjusted to OD_{550} ~10. Then, 1.8 mL of cell suspension was centrifuged at 14,000×*g* for 5 min and the resulting cell pellets were resuspended in 1.4 mL methanol. As described in "Membrane lipid composition characterization" section, the total membrane bound fatty acid was measured. Given that membrane-bound fatty acids account for 71% (w/w) of lipid mass [47], we use the following formula to calculate the membrane lipid content: total membrane lipid (mg/g DCW) = membrane fatty acids (mg)/0.71 × g DCW.

Real-time quantitative PCR

Bacterial cultures were grown and collected by centrifugation at 10,000×*g* for 2 min. Total RNA was extracted by using the RNeasy Mini Kit (Qiagen), and the residual DNA was removed by TURBO DNA-free™ Kit (Life Technology). Superscript III First-Strand Synthesis SuperMix (Invitrogen) was employed for the cDNA synthesis, then the cDNA was diluted 100-fold and used as template for quantitative real-time PCR (qRT-PCR) analysis with SYBR Green ER™ qPCR SuperMix (Invitrogen). The *E. coli* 16S *rrsA* gene was employed as the housekeeping gene for *fadL* mRNA abundance analysis.

Sequences of *fadL* primers for qRT-PCR are CTGAAATGTGGGAAGTGTC/GAAGGTCCAGTTATCATCGT, Primers for *rrsA* are TGGCTCAGATTGAACGC/ATCCGATGGCAAGAGGC. The qRT-PCR was performed with the StepOnePlus™ Real-Time PCR System (Thermo Fisher Scientific). The PCR mixture was held at 95 °C for 10 min and then subjected to 40 cycles of incubation at 95 °C for 15 s, then 60 °C for 1 min.

Fermentation for fatty acid production

Individual colonies were selected from Luria Broth (LB) plates with ampicillin and inoculated into 3 mL of LB liquid medium with ampicillin for 4 h. Then, 0.5 mL of culture was added to 20 mL LB with ampicillin at 30 °C, 220 rpm overnight for seed culture preparation. Seed culture was collected, resuspended in MOPS 2.0% (wt/v) dextrose medium, and transferred into 50 mL MOPS 2.0% (wt/v) dextrose containing ampicillin and 1 mM of isopropyl-β-D-thiogalactopyranoside (IPTG) in 250 mL baffled flasks. The target initial cell density was OD_{550} ~0.1. Cultures were grown in 250 mL baffled flasks with initial pH 7.0 at 30 °C, 220 rpm for 72 h.

Determination of carboxylic acid titers

Carboxylic acid production was quantified by an Agilent 7890 gas chromatograph equipped with an Agilent 5975 mass spectroscope using flame ionization detector and mass spectrometer (GC–MS) after carboxylic acid extraction. Briefly, 100 μL of whole liquid media sample was taken and 10 μL of 1 μg/μL C7:0/C11:0/C17:0 was added as internal standards. Two milliliter of ethanol: sulfuric acid (98:2 v/v) mixture was added, mixed and incubated at 65 °C for 30 min. Then, 2 mL of 0.9% (wt/v) NaCl solution and 1 mL of hexane were added, vortexed and centrifuged at 2000×*g* for 2 min. The top hexane layer was then analyzed by GC–MS, as described in "Strain tolerance characterization" section.

Statistical analysis

The two-tailed t test method was employed to analyze the statistical significance of all data in this study and P value <0.05 is deemed statistically significant.

Results

Effects of *ompF* or *fadL* deletion on tolerance and production of fatty acids

It was previously reported that OmpF facilitates transport of SCFA, such as octanoic acid (C8), into *E. coli*, and that deletion of *ompF* in *E. coli* BW25113 decreased the impact of C8 on biomass production [32]. To evaluate the effect of OmpF on C8 tolerance in MG1655, we also constructed an *ompF* deletion strain (Δ*ompF*) and confirmed that this engineering strategy improved tolerance to C8. In the absence of C8, the specific growth rates (μ) of both strains were approximately 0.39 h^{-1}. During C8 challenge, the specific growth rate of the Δ*ompF* mutant was 0.33 h^{-1}, which is 7% higher than that of MG1655 (0.31 h^{-1}) (Fig. 1a), which is consistent with the previous report [32].

Decreased membrane integrity has been previously described as a primary cause of C8 toxicity, where decreased membrane integrity is evidenced by leakage of metabolites and ions, such as Mg^{2+}, out of the cell or the entry of membrane-impermeable molecules, such as SYTOX, into the cell [14, 24, 48]. We next characterized the membrane integrity changes after disruption of *ompF*. Consistent with the growth results, deletion of *ompF* dampened the impact of C8 on membrane integrity. Specifically, the percentage of cells with intact membranes, i.e. SYTOX impermeable, during challenge with exogenously provided 10 mM C8, increased by 18% compared with the wild-type control strain ($P < 0.05$) (Fig. 1b).

Given that increased tolerance might lead to increased production of bio-products, we next applied the *ompF* deletion strategy to fatty acid production. The plasmid pXZ18Z (TE) harboring the heterologous thioesterase from *R. communis* [42], which primarily releases tetradecanoic acid (C14:0), palmitoleic acid (C16:1) and hexadecanoic acid (C16:0), was transformed into the Δ*ompF* strain and the corresponding control for fatty acid production in minimal MOPS 2.0% (wt/v) dextrose medium. We observed that deletion of *ompF* increased fatty acid production (Fig. 1c): in the Δ*ompF* + TE mutant, the titer of C14:0 was increased by 10% ($P = 0.03$) to 875 mg/L, C16:1 was increased by 17% ($P = 0.24$) to 71 mg/L and C16:0 was increased by 11% ($P = 0.01$) to 711 mg/L. All of these increases led to a 10% improvement of total fatty acids produced by the Δ*ompF* + TE mutant compared to MG1655 + TE strain, with titers of 1500 ± 20 and 1660 ± 40 mg/L, respectively ($P = 0.005$). It should be noted that previous studies concluded that deletion of *ompF* from *E. coli* strain TY05 did not significantly increase fatty acid (C8–C14) production [26]. The difference from this previous report and the findings presented here may be due to the use of different thioesterases (from *U. californica* vs. from *R. communis*), growth media (nutrient-rich LB + 0.4% (v/v) glycerol vs. minimal MOPS + 2% (wt/v) glucose) and temperature (37 vs. 30 °C).

While OmpF has been previously characterized in terms of SCFA transport, FadL predominantly functions in the uptake of LCFA [29, 30]. To investigate the effect of FadL on fatty acid tolerance and production, a *fadL* deletion mutant (Δ*fadL*) was constructed. Interestingly, the Δ*fadL* mutant showed decreased tolerance to C8.

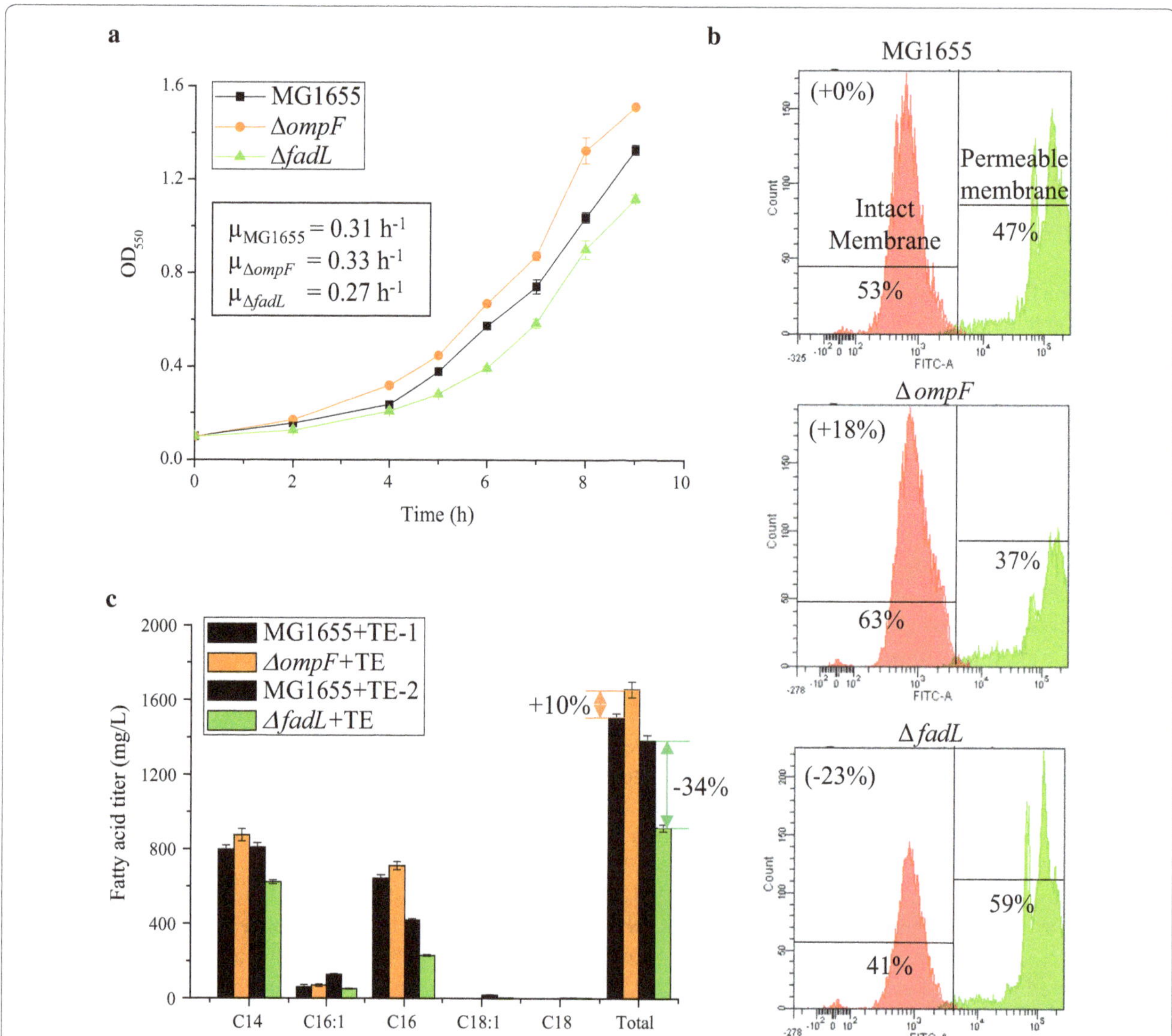

Fig. 1 Effects of *ompF* or *fadL* deletion on membrane integrity during short-chain fatty acid challenge, short-chain fatty acid tolerance and production of C12 and C14 fatty acids. **a** Deletion of *ompF* or *fadL* impact the specific growth rate relative to the wild type MG1655 during challenge with 10 mM C8. *Inset values* are the specific growth rate, h^{-1}. **b** Deletion of *ompF* or *fadL* alters the percentage of cells with intact membranes (membrane integrity), assessed using SYTOX Green, during challenge with 10 mM C8. **c** Deletion of *ompF* increased fatty acid production and deletion of *fadL* decreased fatty acid production. MG1655 + TE-1 and MG1655 + TE-2 indicates experiments performed with the same strain, but on different days. For **a** and **b**, experiments were performed in MOPS + 2% (wt/v) dextrose shake flasks at 220 rpm 30 °C with an initial pH of 7.0, 10 mM octanoic acid (C8). For **c**, strains carry the pXZ18Z plasmid (TE) for LCFA (C14–C16) production. Fermentations were performed in MOPS + 2% (wt/v) dextrose shake flasks at 220 rpm 30 °C with an initial pH of 7.0, 1.0 mM IPTG. Values are the average of at least three biological replicates with *error bars* indicating one standard deviation. Percent increase values are shown only for differences that were deemed statistically significant ($P < 0.05$)

For example, the specific growth rate of the Δ*fadL* strain was 12% lower than that of MG1655 (0.27 vs. 0.31 h^{-1}) ($P < 0.05$) (Fig. 1a). Further membrane characterization showed that the percentage of cells with intact membranes was 23% lower for the Δ*fadL* strain than MG1655 ($P < 0.05$) (Fig. 1b). When this *fadL* deletion strategy was applied to fatty acid production (+TE), titers of C14:0 decreased by 23% to 623 mg/L, C16:1 decreased by 60% to 51 mg/L and C16:0 decreased by 45% to 230 mg/L. Each of these changes had a P value less than 0.05. Together, these changes led to a 34% reduction of total fatty acids in the Δ*fadL* + TE mutant compared with MG1655 + TE strain (from 1390 ± 30 to 920 ± 20 mg/L) ($P < 0.05$) (Fig. 1c). It should be noted that the fatty acid titer of MG1655 + TE here (1390 ± 30 mg/L) is slightly lower than the 1500 ± 20 mg/L described above for the *ompF* results, due to differences between batches, similar to the results described elsewhere [26]. As with deletion of *ompF*, our results differ from previous reports of the effect of *fadL* deletion on fatty acid production. This previous characterization employed *E. coli* strain TY05 in rich medium with glycerol and found no significant change in production of C8–C14 fatty acids upon deletion of *fadL* [26]. However, our observation that deletion of *fadL* can increase sensitivity to membrane-damaging short-chain fatty acids is consistent with observations made for phenylpropanoid tolerance [33].

Increased expression of *fadL* increased fatty acid tolerance and production

Given that the deletion of *fadL* decreased fatty acid tolerance and production, it is reasonable to expect that increased expression of *fadL* might improve fatty acid tolerance and production. To this end, two different strategies were employed in *E. coli* MG1655 for increased expression of *fadL*: plasmid expression (Pla-*fadL*) and genomic integration of a second copy of *fadL* (Gen-*fadL*). Consistent with our hypothesis, both of these increased expression strategies significantly improved C8 tolerance. Specifically, the specific growth rate of Pla-*fadL* (0.33 h^{-1}) and Gen-*fadL* (0.33 h^{-1}) were 8 and 7% higher than Pla-empty (0.31 h^{-1}) and Gen-empty (0.31 h^{-1}) ($P < 0.05$) (Fig. 2a). Membrane damage, as evidenced by entry of the SYTOX nucleic acid dye into the cell, was decreased in the two strains engineered for increased *fadL* expression. Specifically, Pla-*fadL* showed a 25% increase in membrane integrity and Gen-*fadL* had a 14% increase in membrane integrity ($P < 0.05$) (Fig. 2b).

Further characterization showed that both of the strains with increased *fadL* expression also had increased fatty acid production capability. This significantly ($P < 0.05$) increased fatty acid titer was observed for C14:0 and the total fatty acid pool, though the increase was slightly higher for C16:1 and C16:0 in both cases (Fig. 2c). Specifically, the plasmid-based strain produced 1150 mg/L of C14:0, 556 mg/L of C16:0 and 1800 mg/L of total fatty acid, which was 57, 10 and 34% higher than the corresponding control encoding the thioesterase and an empty plasmid. This control strain produced 728 mg/L C14:0, 505 mg/L C16:0 and 1340 mg/L total fatty acids. A similar trend was also observed for genome-based *fadL* expression tuning: 872 mg/L of C14:0, 531 mg/L of C16:0 and 1580 mg/L of total fatty acids were produced by the engineered Gen-*fadL* + TE strain, which was 23, 6 and 18% higher than in the 710, 500 and 1340 mg/L produced by the corresponding Gen-empty + TE control. These results demonstrate the effectiveness of increasing *fadL* expression for increasing fatty acid production.

In order to further characterize the relationship between the expression level of *fadL* and fatty acid production, additional strains were constructed (+TE) and characterized. Specifically, different promoters (M1-12, M1-37, M1-46, M-93) with varied strengths [6, 39, 40] were employed to regulate the expression of the native *fadL* (Fig. 2d). A positive relationship between mRNA relative abundance of *fadL* and fatty acid titers was observed (Fig. 2d). For instance, mRNA relative abundance of *fadL* increased nearly 120-fold in M1-93-*fadL* strain relative to M1-12-*fadL* (of which *fadL* expression level was deemed as 1), and it also produced 1250 mg/L of fatty acid, which is 37% higher than the 915 mg/L produced by M1-12-*fadL*. It should be noted that expression level of *fadL* under all artificial promoters used here is lower than the native promoter, which suggests that expression of *fadL* is held at a relatively high level in *E. coli* MG1655.

Deletion of *ompF* and increased expression of *fadL* have an additive effect in increasing fatty acid production

Given that deletion of *ompF* and increased expression of *fadL* were each found to increase tolerance and production of fatty acids, we proposed that combinatorial utilization of both engineering strategies would further increase performance. To this end, the plasmid-based expression of *fadL* was selected as the strategy for increasing expression of *fadL*, due to its substantial increase in tolerance and production of fatty acid.

Consistent with our hypothesis, combinatorial utilization of the *ompF* deletion and increased expression of *fadL* was found to have an additive effect for improving tolerance to C8 (Fig. 3a). The specific growth rate of Δ*ompF* + Pla-*fadL* strain reached up to 0.36 h^{-1} in the presence of 10 mM C8, which exceeds that of Pla-empty ($\mu = 0.31\ h^{-1}$) by 18%, and is also 10% higher than individual deletion of *ompF* (Δ*ompF* + Pla-empty, $\mu = 0.33\ h^{-1}$) and 12% higher than individual increased expression of *fadL* (Pla-*fadL*, $\mu = 0.32\ h^{-1}$) ($P < 0.05$) (Fig. 3a). Besides increased tolerance, membrane integrity was significantly increased in the Δ*ompF* + Pla-*fadL* strain during challenge with C8. Compared with Pla-empty, the percentage of Δ*ompF* + Pla-*fadL* cells with intact membranes increased by 37% ($P < 0.05$) (Fig. 3b).

Combination of *ompF* deletion and increased expression of *fadL* also increased the specific growth rate during fatty acid production (data not shown), and final fatty acid titers (Fig. 3c). Specifically, the combination of these engineering strategies in the Δ*ompF* + Pla-*fadL* + TE strain resulted in a specific growth rate of 0.25 h^{-1} in the first 12 h of fermentation, where this exceeds that of Pla-empty ($\mu = 0.16\ h^{-1}$) by 53% ($P < 0.05$). Correspondingly, the Δ*ompF* + Pla-*fadL* + TE strain produced 1310 mg/L

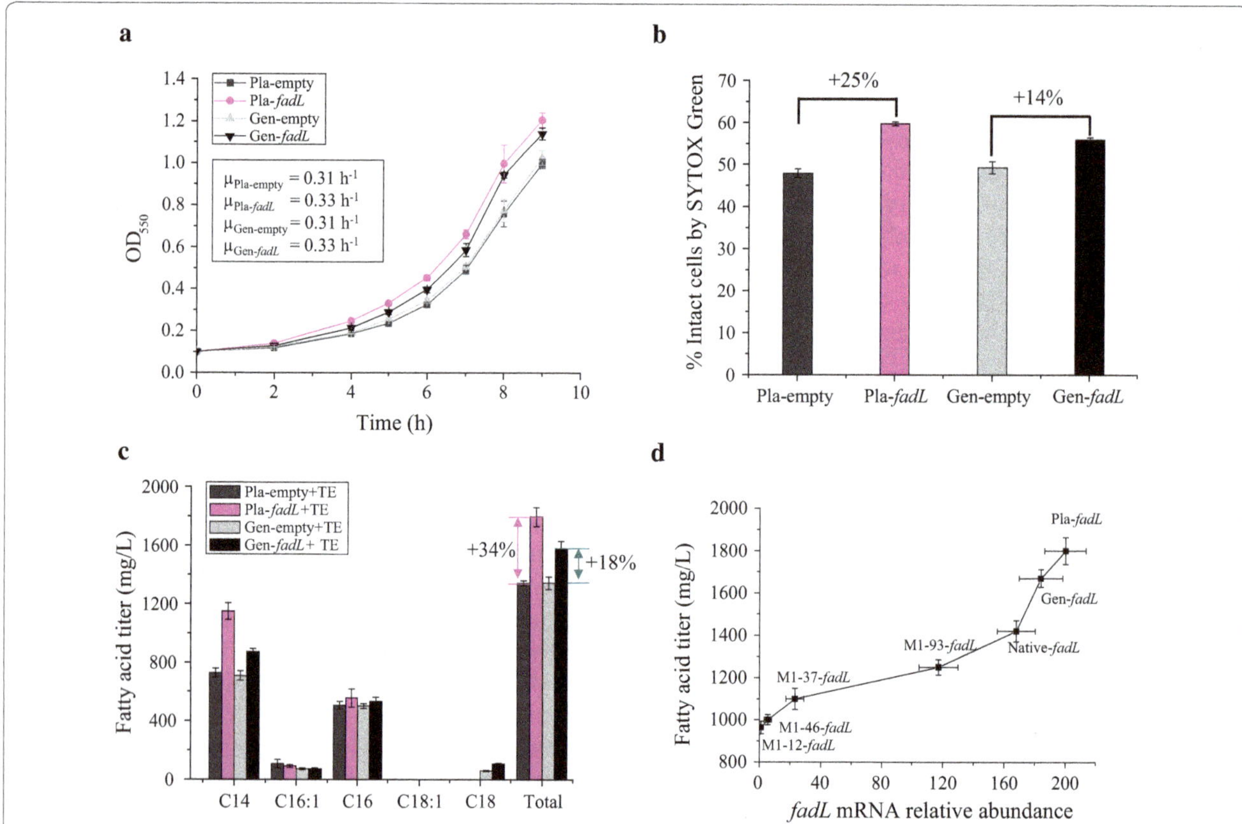

Fig. 2 Increased expression of *fadL* increases membrane integrity, fatty acid tolerance and production. **a** Increased expression of *fadL* from a plasmid (Pla-*fadL*) or a genomic insertion (Gen-*fadL*) both increase the specific growth rate relative to the corresponding controls (Pla-empty, Gen-empty) during challenge with 10 mM C8. *Inset values* are the specific growth rate, h^{-1}. **b** Percentage of cells with intact membrane (membrane integrity), assessed using SYTOX Green. Strains with increased expression of *fadL*, Pla-*fadL* and Gen-*fadL*, have improved membrane integrity relative to their corresponding controls, Pla-empty and Gen-empty, during challenge with 10 mM C8. **c** Strains with increased expression of *fadL*, Pla-*fadL* and Gen-*fadL*, produce more fatty acid than the corresponding controls, Pla-empty and Gen-empty. **d** The *fadL* mRNA relative abundance at 48 h has a positive relationship with the fatty acids titer after 72 h. Four different promoters (M1-12, M1-37, M1-46 and M1-93) were used to replace the native promoter of *fadL*. The mRNA abundance of *fadL* in M1-12-*fadL* strain was set as 1. The 16S *rrsA* gene was used as normalizing factor. For **a** and **b**, experiments were performed in shake flasks containing MOPS + 2% (wt/v) dextrose with 10 mM octanoic acid (C8) at an initial pH of 7.0, shaken at 220 rpm, and maintained at 30 °C. For **c** and **d**, all strains carry the pXZ18Z plasmid (TE, *fabZ*) for LCFA (C14–C16) production. Fermentations were performed in MOPS + 2% (wt/v) dextrose shake flasks at 220 rpm 30 °C with an initial pH of 7.0, 1.0 mM IPTG. Values are the average of at least three biological replicates with *error bars* indicating one standard deviation. Percent increase values are shown only for differences that were deemed statistically significant ($P < 0.05$). Pla-empty: MG1655 + pACYC184-Kan; Pla-*fadL*: MG1655 + pACYC184-Kan-*fadL*; Gen-empty: MG1655 *ldhA*::FRT-*cat*-FRT; Gen-*fadL*: MG1655 *ldhA*::FRT-*cat*-FRT, *fadL*. TE: pXZ18Z plasmid

of C14:0, 90 mg/L of C16:1, 930 mg/L of C16:0 and 2330 mg/L of total fatty acids after 72 h fermentation. These titers are 47, 25, 29 and 38% higher than the strain in which only the *ompF* deletion was implemented (Δ*ompF* + Pla-empty + TE, 885 mg/L of C14:0, 72 mg/L of C16:1, 722 mg/L of C16:0 and 1680 mg/L of total fatty acid) and 25, 10, 18 and 20% higher than the strain in which only the *fadL* overexpression was implemented (Pla-*fadL* + TE, 1040 mg/L of C14:0, 83 mg/L of C16:1, 786 mg/L of C16:0 and 1930 mg/L of total fatty acid). Note that all of these comparisons have $P < 0.05$, except for C16:1. The combined strain has an approximately 50% improvement in fatty acid titers relative to the corresponding un-engineered control, Pla-empty + TE, which produced 801 mg/L of C14:0, 65 mg/L of C16:1, 653 mg/L of C16:0 and 1520 mg/L of total fatty acid (Fig. 3c). These results again demonstrate the effectiveness of concurrent utilization of *ompF* deletion and increased expression of *fadL* for increasing fatty acid production.

Functional mechanism of OmpF and FadL on increased membrane integrity

In this study, engineering the abundance of the membrane proteins OmpF and FadL increased membrane integrity, fatty acid tolerance and fatty acid production.

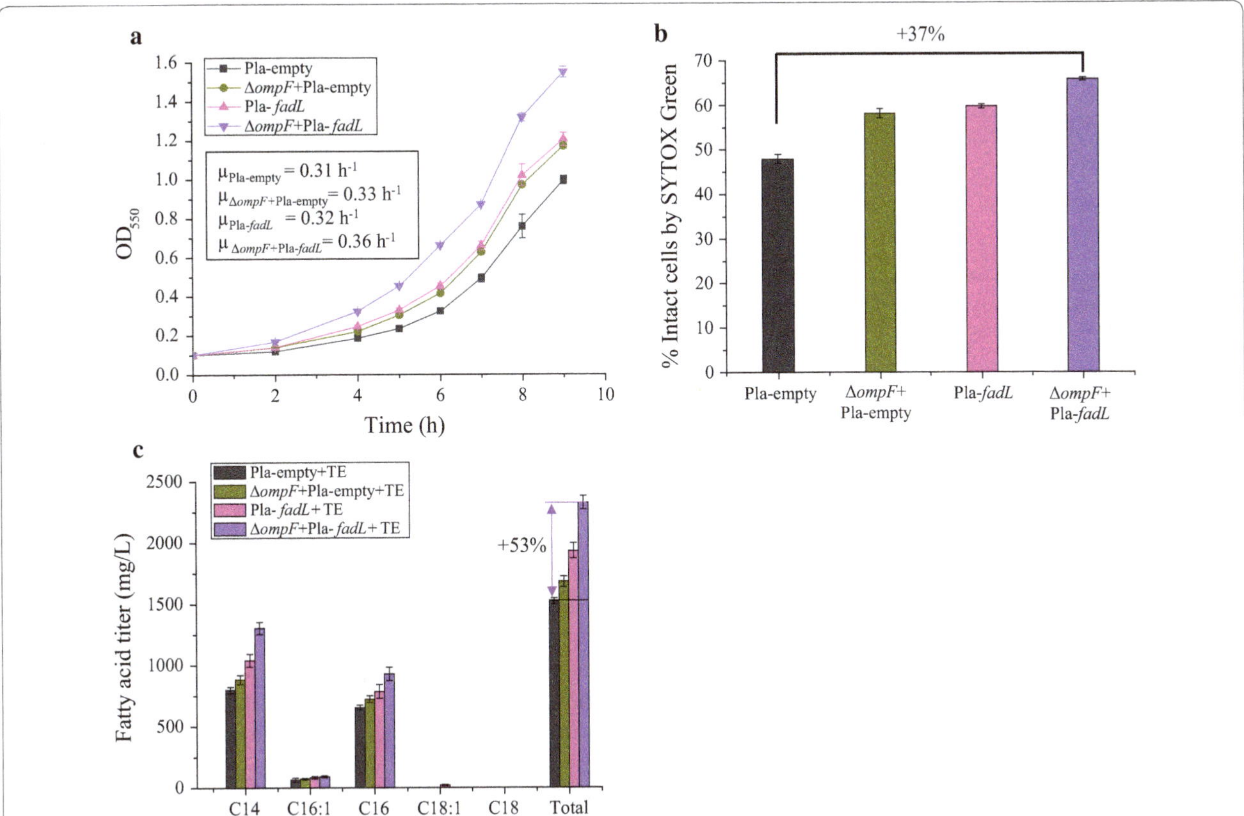

Fig. 3 Deletion of *ompF* and increased expression of *fadL* have an additive effect on increasing membrane integrity, fatty acid tolerance and production. **a** Combinatorial deletion of *ompF* (Δ*ompF*) and increased expression of *fadL* (Pla-*fadL*) increases the specific growth rate during challenge with 10 mM C8 relative to the starting strain (Pla-empty), individual *ompF* deletion strain (Δ*ompF* + Pla-empty), and individual overexpression of *fadL* (Pla-*fadL*). *Inset values* are the specific growth rate, h^{-1} **b** Percentage of cells with intact membrane (membrane integrity), assessed using SYTOX Green. Combinatorial deletion of *ompF* and increased expression of *fadL* improves membrane integrity during challenge with 10 mM C8 relative to Pla-empty, Δ*ompF* + Pla-empty and Pla-*fadL* strains. **c** The combined implementation of *ompF* deletion and increased expression of *fadL* supports increased fatty acid titers relative to each engineering strategy implemented individually. For **a** and **b**, experiments were performed in MOPS + 2% (wt/v) dextrose shake flasks at 220 rpm 30 °C with an initial pH of 7.0, 10 mM octanoic acid (C8). For **c**, all strains carry the pXZ18Z plasmid (TE, *fabZ*) for LCFA (C14–C16) production. Fermentations were performed in MOPS + 2% (wt/v) dextrose shake flasks at 220 rpm 30 °C with an initial pH of 7.0, 1.0 mM IPTG. Values are the average of at least three biological replicates with *error bars* indicating one standard deviation. Percent increase values are shown only for differences that were deemed statistically significant ($P < 0.05$). Pla-empty: MG1655 + pACYC184-Kan; Δ*ompF* + Pla-empty: MG1655, Δ*ompF* + pACYC184-Kan; Pla-*fadL*: MG1655 + pACYC184-Kan-*fadL*; Δ*ompF* + Pla-*fadL*: MG1655, Δ*ompF* + pACYC184-Kan-*fadL*

Prior studies showed that increasing the average length or the saturated:unsaturated (S/U) ratio of *E. coli* membrane lipids can alleviate the decreased membrane integrity caused by fatty acids [23, 24]. In order to determine whether the increased membrane integrity here could be attributed to such changes in the phospholipid tail distribution, we measured the membrane lipid composition in the wild-type MG1655, Δ*ompF*, Δ*fadL* and Pla-*fadL* strains (Table 3). However, no significant changes in membrane composition were observed. Similarly, the average lipid length in wild-type MG1655 was 16.4 ± 0.2, which is comparable to the value observed for the Δ*ompF*, Δ*fadL* and Pla-*fadL* strains (Table 3). Additionally, the membrane lipid S/U ratio in the wild-type MG1655 was 1.06 ± 0.02, which is similar to the ratios for the Δ*ompF*, Δ*fadL* and Pla-*fadL* strains (Table 3). These results indicate that the previously-described membrane engineering mechanisms of increasing the membrane lipid and S/U ratio are not the underlying reason for increased membrane integrity here.

Since the membrane consists of lipids and proteins, altering the abundance of FadL and OmpF might affect the total membrane lipid content. The Δ*ompF* strain had a comparable membrane lipid content to MG1655 (Table 3), which indicates that *ompF* deletion did not significantly impact membrane lipid production. However, unlike *ompF*, altering the abundance of *fadL* remarkably affected membrane lipid content. For example, the

Table 3 Membrane lipid content and composition changes in the wild type MG1655, Δ*ompF*, Δ*fadL*, Pla-*fadL* strains

Strain	Membrane lipid content (mg/g DCW)	Membrane lipid composition (mol %)							Membrane lipid length	Membrane lipid S/U ratio
		C14:0	C16:1	C16:0	C17cyc	C18:1	C18:0	C19cyc		
MG1655	69.4 ± 0.3	1.3 ± 0.1	13.6 ± 0.2	48.5 ± 0.2	14.1 ± 0.1	19.1 ± 0.4	1.70 ± 0.03	1.8 ± 0.1	16.4 ± 0.2	1.06 ± 0.02
Δ*ompF*	71.3 ± 0.5 (+2.7%)	1.1 ± 0.1	13.6 ± 0.1	48.7 ± 0.1	13.6 ± 0.3	19.4 ± 0.1	1.9 ± 0.1	1.7 ± 0.1	16.4 ± 0.1	1.07 ± 0.01
Δ*fadL*	62 ± 3 (−10%)	1.2 ± 0.1	12.7 ± 0.1	48.3 ± 0.4	14.5 ± 0.1	19.4 ± 0.3	1.9 ± 0.1	1.9 ± 0.1	16.4 ± 0.2	1.06 ± 0.01
Pla-*fadL*	78 ± 1 (+13%)	1.3 ± 0.1	13.2 ± 0.2	48.2 ± 0.1	13.4 ± 0.1	20.7 ± 0.1	1.9 ± 0.2	1.2 ± 0.1	16.4 ± 0.1	1.00 ± 0.02

Each value is an average and standard deviation of three biological replicates

All experiments were performed in MOPS + 2% (wt/v) dextrose shake flasks at 220 rpm 30 °C with an initial pH of 7.0, 10 mM octanoic acid (C8). All values are the average of at least three biological replicates with the associated standard deviation indicated. Percent increase values are only shown for differences that were deemed statistically significant (P < 0.05)

DCW dry cell weight, *S/U ratio* membrane saturated: unsaturated lipid ratio

membrane lipid content of Δ*fadL* is only 62 ± 3 mg/g DCW, which is an 11% decrease compared to MG1655 (P < 0.05). Consistently, Pla-*fadL* had a 13% increase in membrane lipid content relative to MG1655 (P < 0.05) (Table 3). This result indicates that, unlike OmpF, FadL might be involved in membrane lipid synthesis, and therefore altering the abundance of *fadL* affects the membrane lipid content and thus membrane integrity. It should be noted that the relative distribution of the lipid tails is not changed in the Pla-*fadL* strain (Table 3).

Discussion

Product toxicity is often an obstacle for cost-effective production of biofuels and chemicals [9, 10]. Therefore, construction of robust production organisms tolerant to these biorenewables is critical for industrial applications and has attracted increasing attention in recent years [12, 45, 49, 50]. Given its importance to overall cell function, membrane integrity has become an attractive engineering target for enhancing robustness [13, 24]. In the case of fatty acids, a variety of engineering efforts have been applied to increasing membrane integrity, with mixed results. Most of these engineering strategies focused on altering the distribution of the membrane lipids of *E. coli*, such as by altering the average lipids length or degree of saturation [23, 24], though there have also been efforts to identify an efflux system that can improve fatty acid production [26].

Here we focused on two membrane proteins, OmpF and FadL, and found that they have distinct effects on maintaining membrane integrity during fatty acid challenge and production. OmpF has been reported to function as the general diffusion porin of *E. coli*, through which a variety of inhibitory molecules, e.g. antibiotics, colicin and SCFA, can enter the cell [32, 51, 52]. Rodriguez-Moya et al. showed that OmpF facilitates transport of C8 into *E. coli*, disrupting intracellular pH and oxidative balance [32]. It has also been suggested that OmpF is involved in the removal of phenylpropanoids from the cell interior [33]. In this study, we further characterized the role of OmpF in maintaining membrane integrity and used the *ompF* deletion strategy to increase fatty acid production. Although we employed the thioesterase specific for release of LCFA (C14–C16), some SCFAs were produced (e.g. C8 and C10) (Additional file 1: Figure S1). These endogenously produced SCFAs can be exported, i.e. by AcrAB-TolC [26], to the extracellular environment. Conversely, they can also re-enter across the outer membrane through *E. coli* porins (e.g. OmpF) (Fig. 4), which can cause severe membrane damage to *E. coli* even at low concentrations [14].

One possible explanation for our observations is that after the endogenously produced fatty acids exit the cell, presumably via AcrAB-TolC [26], some of the SCFA re-enter the cell via OmpF. Deletion of *ompF* blocks this re-entry and thereby increases membrane integrity, which in turn reduces the leakage of important cellular molecules such as Mg^{2+} [14, 53], thereby elevating fatty acid tolerance and production (Fig. 4). The unexpected driving force for such transport may be due to the nature of the AcrAB-TolC transporter. Specifically, this transporter spans the periplasmic space [54–56] and thus the periplasm should be relatively depleted in fatty acids.

Our results demonstrate that, in addition to membrane engineering strategies that alter the distribution of the membrane lipid tails, altering the abundance of membrane protein OmpF can also affect membrane integrity and production of fatty acids, which provides another strategy for future membrane engineering. Increasing the expression of an efflux pump has been shown to improve the production of inhibitory products, such as valine [25] and limonene [8] and these efflux pumps are also an important part of antibiotic resistance [57]. To the best of our knowledge, this is the first demonstration that deletion of a transporter is associated with increased production of a membrane-damaging compound.

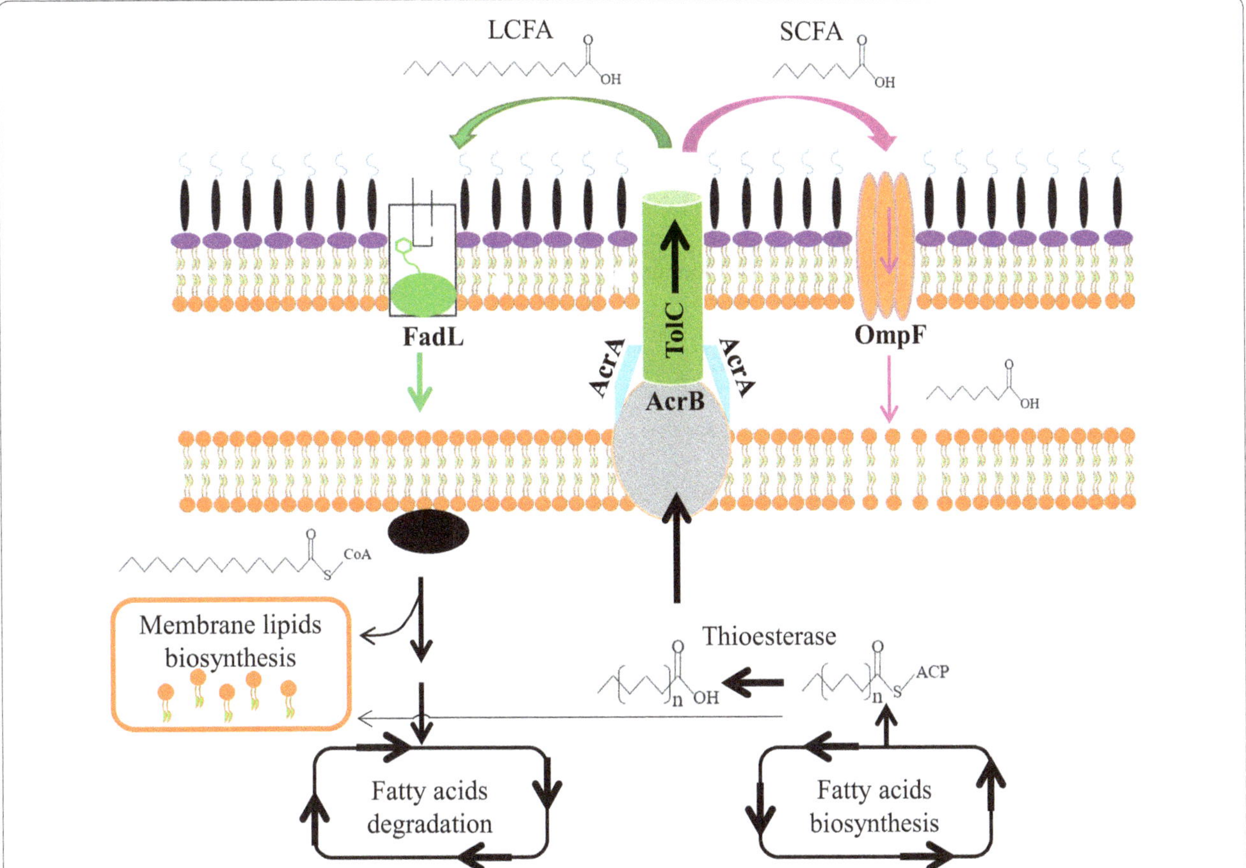

Fig. 4 Schematic of the proposed role of *ompF* and *fadL* in maintenance of membrane integrity during fatty acid production in *E. coli*. The elongated acyl-ACP formed during the fatty acids biosynthesis will have two major destinations. Partial acyl-ACPs are hydrolyzed by thioesterase to release free fatty acids. Residual acyl-ACPs serve as precursor for membrane lipids biosynthesis. Among the produced free fatty acids, LCFA (C14–C16) predominates while there is still some SCFA (<C10). It is proposed that LCFA and SCFA are both transported from the cytoplasm directly to the extracellular medium with the AcrAB-TolC complex [26]. However, the low abundance of these compounds in the periplasmic space relative to the extracellular medium results in a driving force for SCFA entry via OmpF and LCFA entry via FadL. LCFAs imported by FadL can be catalyzed by FadD to acyl-CoA, which then serve as fatty acyl precursors for synthesis of phospholipids or enter the β-oxidation cycle for degradation. SCFAs that enter the cell through OmpF, can damage the inner membrane. Increased expression of *fadL* contributes to import of exogenous LCFA, providing precursors for membrane lipids biosynthesis, thereby increasing membrane integrity and supporting fatty acids production, while deletion of OmpF prevents re-entry of the harmful SCFA. LCFA, long chain fatty acids; SCFA, short-chain fatty acids

In contrast to the *ompF* deletion strategy, deletion of *fadL* was found to decrease membrane integrity, tolerance and production of fatty acid. FadL is the only known outer membrane protein capable of importing exogenous hydrophobic LCFA compounds in *E. coli* [32, 34, 58, 59]. Imported LCFA can be degraded through the β-oxidation pathway as sources of carbon and energy, or serve as precursors for membrane phospholipid biosynthesis [30, 59–61]. Since there was still residual glucose at the end of our experiments (data not shown), it is not likely that the decreased fatty acid tolerance and decreased fatty acid production of the Δ*fadL* mutant was caused by carbon or energy limitations. Membrane lipid biosynthesis in *E. coli* requires acyl chains (C16:0, C16:1 and C18:1), of which there are two sources: (1) endogenous long chain acyl-ACP produced by the fatty acid biosynthesis pathway; and (2) long chain acyl-CoA derived from exogenous LCFA [62, 63]. Upon inactivation of FadL, uptake of exogenous LCFA will be decreased and thus membrane lipid biosynthesis will be impaired (Fig. 4). Our experimental results verify this hypothesis, as membrane lipid content was decreased in the Δ*fadL* strain and increased in the Pla-*fadL* strain. Since lipids are the primary structural component of the membrane, changing the membrane lipid content is likely to alter the membrane integrity. This altered membrane lipid content by Δ*fadL* or Pla-*fadL* does not change the distribution of the different membrane lipid types (Table 3), which suggests that FadL is only responsible for supplying LCFA precursors instead of directly participating in the biosynthesis of phospholipids.

As with OmpF, a driving force for fatty acid uptake via FadL is not expected to exist during fatty acid production. Here, we again refer to the nature of the AcrAB-TolC efflux pump as a possible reason for the existence of this driving force. Since the AcrAB-TolC system spans the periplasmic space [54–56], the periplasm may be depleted of fatty acids relative to the extracellular medium. This direct relationship between *fadL* expression and tolerance of membrane-damaging compounds has been noted elsewhere, specifically in regards to phenylpropanoids [33]. This protective effect of FadL against rutin, naringenin and resveratrol was attributed to FadL's role in repairing membrane damage, though there is no apparent exogenous source of the fatty acids used for this membrane repair [33].

Current membrane engineering strategies focus on altering membrane lipids composition, such as with the goal of increasing membrane lipid length or S/U ratio, to increase membrane integrity. Our results show that increasing the whole membrane lipid content possibly also contributes to increased membrane integrity, tolerance and production of fatty acids, which may serve as a novel strategy for membrane engineering in the future.

Our qRT-PCR results showed that there is a positive relationship between *fadL* mRNA abundance and fatty acid titer, and they also show that the native *fadL* gene is maintained at a high expression level, which indicates the importance of FadL in maintaining normal phospholipids biosynthesis. Concurrent deletion of *ompF* and increased expression of *fadL* synergistically increased fatty acid tolerance and production, accompanied by increased membrane integrity, possibly due to an increase in membrane lipid content and prevention of re-entry of the SCFA.

Bae et al. [34] found that deletion of *fadD* and overexpression of *fadL* in *E. coli* increased hydroxy long-chain fatty acid production. In that study, it was concluded that overexpression of *fadL* contributes to the improvement in the production of ω-hydroxy palmitic acid, primarily due to increased ability to transport exogenously fed palmitic acid (C16). The present work mainly focuses on the effect of *fadL* overexpression on the import of exogenous LCFA for membrane lipid synthesis and thus maintaining membrane integrity during the production of or challenge with membrane-damaging fatty acids. Prior research showed that deletion of *ompF* or *fadL* in *E. coli* did not affect fatty acid production [26], which is different from our results. There are two possible reasons for this difference: (A) the use of different thioesterases; and (B) the use of different growth conditions. The previous studies used a C8–C14-producing thioesterase enzyme from *U. californica*, while here we used a C14–C16-producing thioesterase from *R. communis*. This previous study also used nutrient-rich LB with 0.4% (v/v) glycerol at 37 °C, while we used the nutrient-poor minimal MOPS with 2% (wt/v) dextrose at 30 °C. It is interesting to note that the studies that identified a positive relationship between OmpF abundance, FadL abundance and phenylpropanoid tolerance were also performed at 30 °C [33]. The use of glycerol in the previous fatty acid production studies may also be a complicating factor. The increase in hydroxy-palmitic acid production upon overexpression of FadL was smaller in the presence of glycerol relative to glucose [34] and the presence of glycerol has previously been reported to alter the phospholipid composition of microbial cell membranes [64–66]. Under different growth conditions, the membrane composition and associated amount of membrane damage caused by the fatty acids is expected to vary, and therefore the roles of OmpF and FadL may differ.

This engineering method appears to increase fatty acid production as a direct function of increased abundance of the microbial biocatalyst. Thus, it differs from a previously described membrane engineering method that increased fatty acid titers by 50% without impacting the final culture OD [23] and evolutionary strain development that improved fatty acid production fivefold while only increasing growth during fatty acid production threefold [50]. The strategy described here also differs from provision of valine-producing *E. coli* with a valine exporter, which increased valine titers by 50% without changing the final OD [25]. Thus, additional strain engineering would be needed in order for this strategy to be effective in improving fatty acid production in fed-batch or continuous culture systems. However, this work clearly demonstrates that these two membrane proteins are two viable engineering targets for improving fatty acid production.

Conclusions

Membrane damage of the microbial biocatalyst is a widespread problem in the problem of biorenewable fuels and chemicals. Here we have demonstrated two strategies for dealing with membrane damage in our condition. The first is to increase the abundance of FadL, which we propose increases the ability of the organism to repair the membrane damage incurred by fatty acids. The second method is to delete OmpF, which we propose prevents re-entry of the inhibitory product.

Additional file

Additional file 1: Figure S1. Fatty acids profile of *E. coli* MG1655 harboring pXZ18Z plasmid which carries thioesterase gene from *Ricinus communis* and *fabZ* gene from *E. coli*. Some short chain fatty acids (e.g. butanedioic acid, octanoic acid and decanoic acid) were found in the fermentation broth.

Abbreviations

OmpF: outer membrane porin F; FadL: long-chain fatty acid outer membrane porin; SCFA: short-chain fatty acids; LCFA: long chain fatty acids; C7: heptanoic acid; C8: octanoic acid; C11: undecanoic acid; C14: tetradecanoic acid; C15: pentadecanoic acid; C16:1: palmitoleic acid; C16:0: hexadecanoic acid; C17: heptadecanoic acid; TE: pXZ18Z; DCW: dry cell weight; IPTG: isopropyl-β-D-thiogalactopyranoside; ORF: open reading frame; *ldhA*: lactate dehydrogenase gene.

Authors' contributions

ZT and LRJ designed research. ZT, WB, and JMY performed research. ZT, JVS and LRJ analyzed data, and LRJ wrote the paper. All authors read and approved the final manuscript.

Author details

[1] 4134 Biorenewables Research Laboratory, Department of Chemical and Biological Engineering, Iowa State University, Ames, IA 50011, USA. [2] Present Address: Department of Chemical Engineering and Materials Sciences, University of California, 916 Engineering Tower Irvine, Irvine, CA 92697-2575, USA.

Acknowledgements

We thank ISU Flow Cytometry Facility for help with SYTOX Green cells analysis, ISU DNA Facility for help with real-time quantitative PCR analysis and ISU W.M. Keck Metabolomics Research Laboratory for help with membrane fluidity analysis and GC–MS analysis. We would also like to thank Edward Yu and Thomas Mansell for helpful discussion of these results.

Competing interests

This work will be included in patent applications by Iowa State University.

Funding

This work was supported by the NSF Engineering Research Center for Biorenewable Chemicals (CBiRC), NSF Award number EEC-0813570. The funders had no role in study design, data collection and analysis, decision to publish, or preparation of the manuscript.

References

1. Gallezot P. Process options for converting renewable feedstocks to bioproducts. Green Chem. 2007;9:295–302.
2. Larson ED. A review of life-cycle analysis studies on liquid biofuel systems for the transport sector. Energy Sustain Dev. 2006;10:109–26.
3. Thakker C, Martinez I, San KY, Bennett GN. Succinate production in *Escherichia coli*. Biotechnol J. 2012;7:213–24.
4. Park J, Rodriguez-Moya M, Li M, Pichersky E, San KY, Gonzalez R. Synthesis of methyl ketones by metabolically engineered *Escherichia coli*. J Ind Microbiol Biotechnol. 2012;39:1703–12.
5. McKenna R, Nielsen DR. Styrene biosynthesis from glucose by engineered *E. coli*. Metab Eng. 2011;13:544–54.
6. Zhu X, Tan Z, Xu H, Chen J, Tang J, Zhang X. Metabolic evolution of two reducing equivalent-conserving pathways for high-yield succinate production in *Escherichia coli*. Metab Eng. 2014;24C:87–96.
7. Atsumi S, Cann AF, Connor MR, Shen CR, Smith KM, Brynildsen MP, Chou KJ, Hanai T, Liao JC. Metabolic engineering of *Escherichia coli* for 1-butanol production. Metab Eng. 2008;10:305–11.
8. Dunlop MJ, Dossani ZY, Szmidt HL, Chu HC, Lee TS, Keasling JD, Hadi MZ, Mukhopadhyay A. Engineering microbial biofuel tolerance and export using efflux pumps. Mol Syst Biol. 2011;7:487.
9. Jarboe LR, Liu P, Royce LA. Engineering inhibitor tolerance for the production of biorenewable fuels and chemicals. Curr Opin Chem Eng. 2011;1:38–42.
10. Nicolaou SA, Gaida SM, Papoutsakis ET. A comparative view of metabolite and substrate stress and tolerance in microbial bioprocessing: from biofuels and chemicals, to biocatalysis and bioremediation. Metab Eng. 2010;12:307–31.
11. Huffer S, Clark ME, Ning JC, Blanch HW, Clark DS. Role of alcohols in growth, lipid composition, and membrane fluidity of yeasts, bacteria, and archaea. Appl Environ Microbiol. 2011;77:6400–8.
12. Lennen RM, Kruziki MA, Kumar K, Zinkel RA, Burnum KE, Lipton MS, Hoover SW, Ranatunga DR, Wittkopp TM, Marner WD 2nd, Pfleger BF. Membrane stresses induced by overproduction of free fatty acids in *Escherichia coli*. Appl Environ Microbiol. 2011;77:8114–28.
13. Liu P, Chernyshov A, Najdi T, Fu Y, Dickerson J, Sandmeyer S, Jarboe L. Membrane stress caused by octanoic acid in *Saccharomyces cerevisiae*. Appl Microbiol Biotechnol. 2013;97:3239–51.
14. Royce LA, Liu P, Stebbins MJ, Hanson BC, Jarboe LR. The damaging effects of short chain fatty acids on *Escherichia coli* membranes. Appl Microbiol Biotechnol. 2013;97:8317–27.
15. Zaldivar J, Ingram LO. Effect of organic acids on the growth and fermentation of ethanologenic *Escherichia coli* LY01. Biotechnol Bioeng. 1999;66:203–10.
16. Cameron DC, Tong IT. Cellular and metabolic engineering—an overview. Appl Biochem Biotechnol. 1993;38:105–40.
17. Bailey JE. Toward a science of metabolic engineering. Science. 1991;252:1668–75.
18. Korstanje TJ, van der Vlugt JI, Elsevier CJ, de Bruin B. Hydrogenation of carboxylic acids with a homogeneous cobalt catalyst. Science. 2015;350:298–302.
19. Lennen RM, Braden DJ, West RM, Dumesic JA, Pfleger BF. A process for microbial hydrocarbon synthesis: overproduction of fatty acids in *Escherichia coli* and catalytic conversion to alkanes. Biotechnol Bioeng. 2010;106:193–202.
20. Lopez-Ruiz JA, Davis RJ. Decarbonylation of heptanoic acid over carbon-supported platinum nanoparticles. Green Chem. 2014;16:683–94.
21. Kim S, Cheong S, Chou A, Gonzalez R. Engineered fatty acid catabolism for fuel and chemical production. Curr Opin Biotechnol. 2016;42:206–15.
22. Sanchez MA, Torres GC, Mazzieri VA, Pieck CL. Selective hydrogenation of fatty acids and methyl esters of fatty acids to obtain fatty alcohols—a review. J Chem Technol Biotechnol. 2017;92:27–42.
23. Sherkhanov S, Korman TP, Bowie JU. Improving the tolerance of *Escherichia coli* to medium-chain fatty acid production. Metab Eng. 2014;25:1–7.
24. Lennen RM, Pfleger BF. Modulating membrane composition alters free fatty acid tolerance in *Escherichia coli*. PLoS ONE. 2013;8:54031.
25. Park JH, Lee KH, Kim TY, Lee SY. Metabolic engineering of *Escherichia coli* for the production of L-valine based on transcriptome analysis and in silico gene knockout simulation. Proc Natl Acad Sci USA. 2007;104:7797–802.
26. Lennen RM, Politz MG, Kruziki MA, Pfleger BF. Identification of transport proteins involved in free fatty acid efflux in *Escherichia coli*. J Bacteriol. 2013;195:135–44.
27. Cowan SW, Schirmer T, Rummel G, Steiert M, Ghosh R, Pauptit RA, Jansonius JN, Rosenbusch JP. Crystal structures explain functional properties of two *E. coli* porins. Nature. 1992;358:727–33.
28. Nikaido H. Outer-membrane barrier as a mechanism of antimicrobial resistance. Antimicrob Agents Chemother. 1989;33:1831–6.
29. Lepore BW, Indic M, Pham H, Hearn EM, Patel DR, van den Berg B. Ligand-gated diffusion across the bacterial outer membrane. Proc Natl Acad Sci USA. 2011;108:10121–6.
30. van den Berg B, Black PN, Clemons WM Jr, Rapoport TA. Crystal structure of the long-chain fatty acid transporter FadL. Science. 2004;304:1506–9.
31. Black PN. Characterization of FadL-specific fatty-acid binding in *Escherichia coli*. Biochim Biophys Acta. 1990;1046:97–105.
32. Rodriguez-Moya M, Gonzalez R. Proteomic analysis of the response of *Escherichia coli* to short-chain fatty acids. J Proteom. 2015;122:86–99.
33. Zhou JW, Wang K, Xu S, Wu JJ, Liu PR, Du GC, Li JH, Chen J. Identification of membrane proteins associated with phenylpropanoid tolerance and transport in *Escherichia coli* BL21. J Proteom. 2015;113:15–28.
34. Bae JH, Park BG, Jung E, Lee PG, Kim BG. *fadD* deletion and *fadL* overexpression in *Escherichia coli* increase hydroxy long-chain fatty acid productivity. Appl Microbiol Biotechnol. 2014;98:8917–25.

35. Call TP, Akhtar MK, Baganz F, Grant C. Modulating the import of medium-chain alkanes in *E. coli* through tuned expression of FadL. J Biol Eng. 2016;10:5.
36. Heipieper HJ, Keweloh H, Rehm HJ. Influence of phenols on growth and membrane-permeability of free and immobilized *Escherichia coli*. Appl Environ Microbiol. 1991;57:1213–7.
37. Zhang DF, Li H, Lin XM, Wang SY, Peng XX. Characterization of outer membrane proteins of *Escherichia coli* in response to phenol stress. Curr Microbiol. 2011;62:777–83.
38. Datsenko KA, Wanner BL. One-step inactivation of chromosomal genes in *Escherichia coli* K-12 using PCR products. Proc Natl Acad Sci USA. 2000;97:6640–5.
39. Tang J, Zhu X, Lu J, Liu P, Xu H, Tan Z, Zhang X. Recruiting alternative glucose utilization pathways for improving succinate production. Appl Microbiol Biotechnol. 2013;97:2513–20.
40. Tan Z, Zhu X, Chen J, Li Q, Zhang X. Activating phosphoenolpyruvate carboxylase and phosphoenolpyruvate carboxykinase in combination for improvement of succinate production. Appl Environ Microbiol. 2013;79:4838–44.
41. Zhang X, Jantama K, Moore JC, Shanmugam KT, Ingram LO. Production of L-alanine by metabolically engineered *Escherichia coli*. Appl Microbiol Biotechnol. 2007;77:355–66.
42. San K-Y, Li M, Zhang X. Bacteria and method for synthesizing fatty acids. Google Patents; 2011.
43. Neidhard FC, Bloch PL, Smith DF. Culture medium for Enterobacteria. J Bacteriol. 1974;119:736–47.
44. Wanner BL. Methods in molecular genetics. New York: Academic; 1994.
45. Tan Z, Yoon JM, Nielsen DR, Shanks JV, Jarboe LR. Membrane engineering via trans unsaturated fatty acids production improves *Escherichia coli* robustness and production of biorenewables. Metab Eng. 2016;35:105–13.
46. Bligh EG, Dyer WJ. A rapid method of total lipid extraction and purification. Can J Biochem Physiol. 1959;37:911–7.
47. Torella JP, Ford TJ, Kim SN, Chen AM, Way JC, Silver PA. Tailored fatty acid synthesis via dynamic control of fatty acid elongation. Proc Natl Acad Sci USA. 2013;110:11290–5.
48. Lian J, McKenna R, Rover MR, Nielsen DR, Wen Z, Jarboe LR. Production of biorenewable styrene: utilization of biomass-derived sugars and insights into toxicity. J Ind Microbiol Biotechnol. 2016;43:595–604.
49. Chubukov V, Mingardon F, Schackwitz W, Baidoo EEK, Alonso-Gutierrez J, Hu QJ, Lee TS, Keasling JD, Mukhopadhyay A. Acute limonene toxicity in *Escherichia coli* is caused by limonene hydroperoxide and alleviated by a point mutation in alkyl hydroperoxidase AhpC. Appl Environ Microbiol. 2015;81:4690–6.
50. Royce LA, Yoon JM, Chen Y, Rickenbach E, Shanks JV, Jarboe LR. Evolution for exogenous octanoic acid tolerance improves carboxylic acid production and membrane integrity. Metab Eng. 2015;29:180–8.
51. Kim YC, Tarr AW, Penfold CN. Colicin import into *E. coli* cells: a model system for insights into the import mechanisms of bacteriocins. Biochim Biophys Acta. 2014;1843:1717–31.
52. Ziervogel BK, Roux B. The binding of antibiotics in OmpF porin. Structure. 2013;21:76–87.
53. Jarboe LR, Royce LA, Liu P. Understanding biocatalyst inhibition by carboxylic acids. Front Microbiol. 2013;4:272.
54. Du DJ, Wang Z, James NR, Voss JE, Klimont E, Ohene-Agyei T, Venter H, Chiu W, Luisi BF. Structure of the AcrAB-TolC multidrug efflux pump. Nature. 2014;509:512–5.
55. Tikhonova EB, Zgurskaya HI. AcrA, AcrB, and TolC of *Escherichia coli* form a stable intermembrane multidrug efflux complex. J Biol Chem. 2004;279:32116–24.
56. Touze T, Eswaran J, Bokma E, Koronakis E, Hughes C, Koronakis V. Interactions underlying assembly of the *Escherichia coli* AcrAB-TolC multidrug efflux system. Mol Microbiol. 2004;53:697–706.
57. Blair JMA, Richmond GE, Piddock LJV. Multidrug efflux pumps in Gram-negative bacteria and their role in antibiotic resistance. Future Microbiol. 2014;9:1165–77.
58. Black PN, DiRusso CC. Transmembrane movement of exogenous long-chain fatty acids: proteins, enzymes, and vectorial esterification. Microbiol Mol Biol Rev. 2003;67:454–72 **(table of contents)**.
59. Hearn EM, Patel DR, Lepore BW, Indic M, van den Berg B. Transmembrane passage of hydrophobic compounds through a protein channel wall. Nature. 2009;458:367–70.
60. Hearn EM, Patel DR, van den Berg B. Outer-membrane transport of aromatic hydrocarbons as a first step in biodegradation. Proc Natl Acad Sci USA. 2008;105:8601–6.
61. Fujita Y, Matsuoka H, Hirooka K. Regulation of fatty acid metabolism in bacteria. Mol Microbiol. 2007;66:829–39.
62. Rock CO. Fatty acid and phospholipid metabolism in prokaryotes. In: Biochemistry of lipids, lipoproteins and membranes. 5th ed. 2008; p. 59–96.
63. Rock CO, Jackowski S. Pathways for the incorporation of exogenous fatty-acids into phosphatidylethanolamine in *Escherichia coli*. J Biol Chem. 1985;260:2720–4.
64. Du GC, Yang G, Qu YB, Chen J, Lun SY. Effects of glycerol on the production of poly(gamma-glutamic acid) by *Bacillus licheniformis*. Process Biochem. 2005;40:2143–7.
65. Kautharapu KB, Rathmacher J, Jarboe LR. Growth condition optimization for docosahexaenoic acid (DHA) production by *Moritella marina* MP-1. Appl Microbiol Biotechnol. 2013;97:2859–66.
66. Pramanik J, Keasling JD. Effect of *Escherichia coli* biomass composition on central metabolic fluxes predicted by a stoichiometric model. Biotechnol Bioeng. 1998;60:230–8.

15

Engineered microbial biosensors based on bacterial two-component systems as synthetic biotechnology platforms in bioremediation and biorefinery

Sambandam Ravikumar[1†], Mary Grace Baylon[2†], Si Jae Park[2*] and Jong-il Choi[1*]

Abstract

Two-component regulatory systems (TCRSs) mediate cellular response by coupling sensing and regulatory mechanisms. TCRSs are comprised of a histidine kinase (HK), which serves as a sensor, and a response regulator, which regulates expression of the effector gene after being phosphorylated by HK. Using these attributes, bacterial TCRSs can be engineered to design microbial systems for different applications. This review focuses on the current advances in TCRS-based biosensors and on the design of microbial systems for bioremediation and their potential application in biorefinery.

Keywords: Two-component regulatory system, Biosensor, Bioremediation, Genetic circuit, Biorefinery

Background

Toxic chemicals have currently been released into the environment by accidental spills and the improper management of chemical industries. These toxic chemicals include inorganic products such as heavy metals and organic products such as benzene, toluene, ethylbenzene, biphenyl, and styrene, accidental release of which into environment are a significant threat to the environment. Heavy metals and oil products are difficult to remove from the environment and cannot be easily degraded. Thus, they are ultimately indestructible and constitute a global environmental hazard. As a result, soil and groundwater contamination has become a major problem at these polluted sites and requires urgent remediation technology to protect the environment.

*Correspondence: parksj93@ewha.ac.kr; choiji01@jnu.ac.kr
†Sambandam Ravikumar and Mary Grace Baylon contributed equally to this work
[1] Biomolecules Engineering Lab, Department of Biotechnology and Bioengineering, Chonnam National University, 77 Yongbong-ro, Gwangju 61186, Republic of Korea
[2] Division of Chemical Engineering and Materials Science, Ewha Womans University, 52 Ewhayeodae-gil, Seodaemun-gu, Seoul 03760, Republic of Korea

Over the past few decades, several technologies based on novel analytical methods have been developed to remove certain metals and organic pollutants from the environment [1]. Unfortunately, many conventional techniques have been found to be ineffective and/or expensive due to low permeability, different subsurface conditions, and contaminant mixtures. Owing to the limitations of traditional methods, researchers have focused on in situ bioremediation, which uses microorganisms to degrade petroleum products or immobilize heavy metal contaminants. Bioremediation strategies have been proposed as potential alternatives for the removal of organic and inorganic pollutants due to their safety, speed, low cost, and high efficiency in removing pollutants from the environment.

The central principle of bioremediation is that microorganisms are able to produce energy they need to grow and reproduce by degrading hazardous contaminants. In some cases, bioremediation occurs spontaneously because the essential materials required for bacterial growth are naturally present at the contaminated sites. More often, bioremediation requires an engineered bacterial system to accelerate the tailor-made biodegradation of organic compounds or bio-adsorption of inorganic elements as we desired [2, 3]. It is also needed to further optimize the environmental conditions, in which the microorganisms carry out the detoxification reactions by employing several engineered microorganism systems such as cell surface display- and secretion-based strategies to remediate the contaminated environment. Cell surface display technologies have widely been used in both pharmaceutical and bioremediation applications such as live vaccine development, antibody production, peptide library screening, biosensors, bio-adsorption of organic and inorganic pollutants, and whole-cell biocatalysis (Fig. 1) [4–7].

Heavy metals are common pollutants that are byproducts of various industrial activities. Microorganisms usually mobilize metals from one location and scavenge metals from another. Recently, recombinant bacterial systems displaying chimeric proteins on the cell surface have been developed for use in the bio-adsorption of specific heavy metals. To address organic products, microorganisms have been engineered to produce extracellular enzymes or display enzymes as outer membrane proteins, and they act as a whole-cell catalyst to break down petroleum hydrocarbons and their derivatives. However, all of these constructs require expensive inducers, or the constitutive expression of a membrane protein on the cell surface may affect the growth of the host system. Additionally, none of these engineered bacteria can sense the particular bio-component to be degraded. Therefore, engineered bacteria should be designed to monitor the environmental pollutants, and the design should also include a well-defined removal system. The engineered bacterial system should behave normally until it senses the target in the environment. Once the target is detected, the system should modulate bacterial genes in response. In this way, the genes needed to remove the target are only transcribed and expressed when required. Therefore, it is essential to construct an inexpensive system that can efficiently examine and remove hazardous materials present in the environment.

Nature has provided an excellent solution to this problem. Interestingly, cells have evolved many intricate sensory apparatuses to control cellular growth and behavior. Thus, some cells not only sense light, temperature, oxygen, and pH, but also detect the toxic status of the external environment. An essential requirement for a biosensor or bioremediation process is promoting contact between the contaminants and microbes. As a result of this contact, the microbes adapt their cellular functions in response to the surrounding environmental conditions and then express the relevant genes when needed. If the aim is to monitor and remove an individual toxic compound from the environment, then a synthetic biological strategy will be more feasible because the necessary genetic circuits can be assembled to sense and reduce the level of the exogenous toxin. These synthetic genetic circuits can be assembled using a two-component regulatory system (TCRS) in bacteria [8].

Two-component regulatory systems are widely found in prokaryotes, but only a few have been identified in eukaryotic organisms that can be coupled to environmental stimuli for an appropriate cellular response. This system senses environmental changes and regulates cellular metabolism in response to these changes thereby allowing bacteria to grow, thrive and adapt in different environments. A prototypical TCRS has two components: a histidine kinase (HK) and a response regulator (RR). The HK sensor is a homodimeric integral membrane protein that contains a sensor domain as an extracellular loop located between two membrane-spanning segments (TM1 and TM2) and a transmitter domain located in the last transmembrane segment confined to the cytoplasm. All HK domains contain two highly

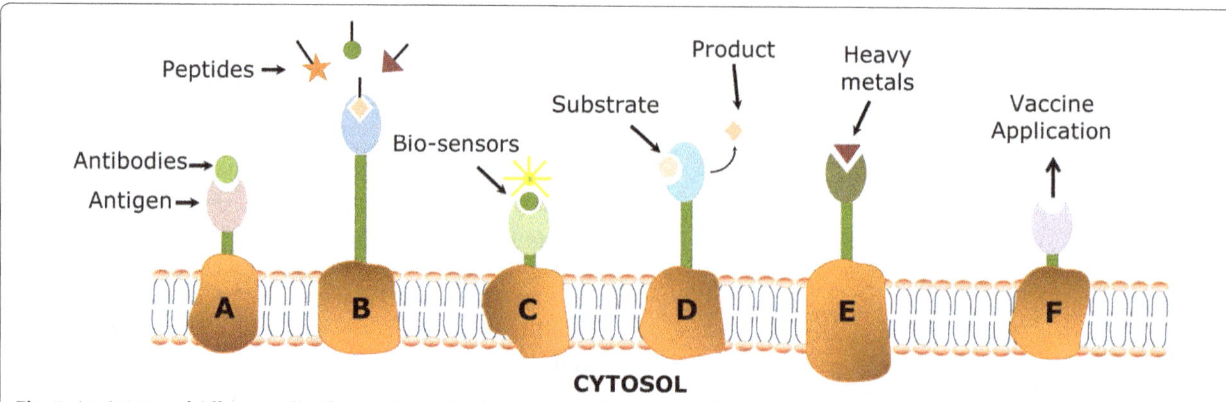

Fig. 1 Application of different cell surface display technologies in *A* antibody production, *B* peptide library screening, *C* biosensors, *D* biocatalysts, *E* bio-adsorption, and *F* vaccine development

conserved domains: dimerization and histidine phosphotransfer domain (DHp) and catalytic ATP-binding domain (CA). The periplasmic or extracellular region serves mostly as the signal recognition domain. The DHp and CA domains are responsible for the molecular recognition of the cognate RR as well as the hydrolysis of ATP. The transmitter domain, which serves as a signal transmitter linking the periplasmic and cytoplasmic regions, contains three domains that are named after the proteins where they were first discovered: PAS (Periodic circadian proteins, Aryl hydrocarbon nuclear translocator proteins and Single-minded proteins), HAMP (HKs, Adenylate cyclases, Methyltransferases, and Phosphodiesterases), and GAF (cGMP-specific phosphodiesterases, adenylyl cyclases, and formate hydrogenases). These domains can either transmit signals from the periplasmic region or directly recognize the cytoplasmic signals. Therefore, the HK senses stimuli from the external environment and autophosphorylates conserved histidine residues in the kinase itself. The RR is regulated by the HK, which phosphorylates aspartate residues on the RR. The phosphorylated RR generates output by binding to promoters and thus activates or represses gene expression [8].

Aside from the application of TCRSs in the development of engineered microorganisms for coupled detection and degradation of environmental pollutants, recently, the potential application of TCRSs to metabolically engineered microorganisms has also been extensively examined for different biotechnological purposes. Thus, the recent advances in TCRS-based biosensors designed for cell-mediated bioremediation in response to different environmental pollutants are discussed along with the potential application of TCRSs for the development of engineered host microorganisms in biorefinery process to produce bio-based chemicals.

TCRS sensing of heavy metals and organic pollutants

Two-component regulatory systems can detect a broad range of environmental signals, such us light, oxygen, pH, temperature, and even some heavy metals and organic contaminants [9]. Many types of TCRS-based environmental biosensors have been reported, but only a few heavy metal- and organic pollutant-based sensors have been developed to date (Fig. 2). Bacteria use several TCRSs to sense specific heavy metals. Because heavy metals are cations that are both toxic and essential, bacterial cells use TCRSs to regulate the homeostasis of these metal cations. A HydHG TCRS (also known as ZraSR) was identified in *Escherichia coli* that senses and controls the expression of *zraP* gene encoding zinc efflux protein under high concentrations of Zn^{2+} and Pb^{2+} in aerobic condition [10]. HydH protein is tightly bound

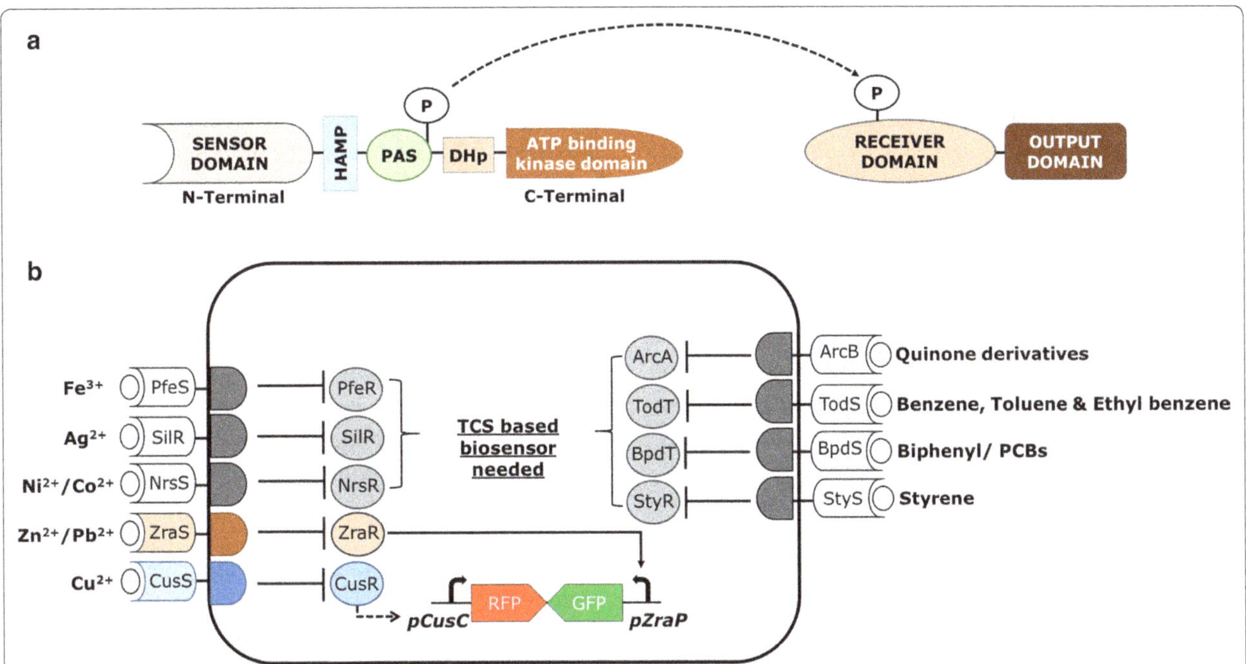

Fig. 2 **a** Domain structure of bacterial two-component regulatory systems (TCRS). Typical two-component phosphotransfer systems contain a sensor domain and a cytoplasmic response regulator (RRs). **b** A multi-component phosphorelay system containing the HAMP, PAS, and phosphotransfer domains. The periplasmic metal-sensing receptors sense heavy metals and phosphorylate the HK domain and activate the corresponding RR. The RR activates the synthetic genetic circuit of the TCRS resulting in the expression of the reporter protein. The genetic circuit shown in *gray* can be developed as a biosensor

to the cell membrane and is assumed to be responsible for sensing high periplasmic Zn^{2+} and Pb^{2+} concentration. Then, in the presence of a phosphoryl donor, HydG binds to the intergenic region within *zraP-hydHG* resulting in the upregulated expression of ZraP [10]. Likewise, the CusRS (*ylcA, ybcZ*) TCRS found in *E. coli* K-12 is responsive to Cu^{2+} ions and is required for the inducible expression of *pcoE*, belonging to the plasmid-borne *pco* operon, the induction of the genes in this operon activates the copper efflux system thereby allowing the excess Cu^{2+} to exit the cell [11]. Some TCRS can regulate the expression of several specific genes in an operon or a whole operon. The SilRS TCRS increases the resistance of *Salmonella enterica* to silver cations through the coupled sensing and activation expression of the periplasmic silver-specific binding protein, SilE encoded by *silE* gene and two parallel efflux pumps, SilP and SilCBA [12]. This is also in the case of NrsSR TCRS identified in *Synechocystis* sp. PCC6803. NrsSR senses Ni^{2+} and Co^{2+} ions and regulates the expression of the *nrsBACD* operon that encodes proteins involved in Ni^{2+} resistance [13]. In another study, a PfeS/R TCRS senses ferric enterobactin and induces the production of the enterobactin receptor PfeA in *Pseudomonas aeruginosa* [14].

Aromatic compounds are the most abundant organic contaminants. However, utilizing these compounds is disruptive to most bacteria. Due to the genetic and metabolic flexibility of bacteria, some microorganisms can use organic contaminants as their sole carbon source. Several TCRSs have been identified to be involved in catabolizing aromatic compounds by inducing and activating the aromatic metabolism pathways. The TodST TCRS of *Pseudomonas putida* can be induced by different aromatic substrates such as toluene, xylene, benzene, and ethylbenzene. This TCRS modulates the expression of the *tod* genes, which encode enzymes for the catabolism of these aromatic compounds [15]. The StySR TCRS identified in *Pseudomonas* sp. strain Y2 activates the expression of the *styABCD* genes in response to changes in styrene concentration in the environment [16]. Another TCRS, BpdST, potentially controls biphenyl or polychlorobiphenyl degradation in *Rhodococcus* sp. [15].

TCRS-based heavy metal bio-adsorption coupled with a biosensor

One of the best approaches to a biosensor-based method is to use a genetically modified microorganism that emits a clear signal when the microbes encounter a target molecule [17, 18]. To date, many metal-specific and a few petroleum product-based bacterial sensors have been developed [19–23]. Based on the nature of the cells used, a variety of TCRS-based environmental contaminant sensors has been constructed by several research groups. However, to remediate environmental pollutants, new synthetic genetic circuits are needed so that the bacterial system can have both sensor and remediation activities. Future research on the application of biosensors in bioremediation should focus on the development of such TCRSs. Some of the TCRS-based heavy metal biosensors for use in bioremediation applications have been developed and are reviewed below.

A zinc adsorption system was developed by using the ZraSR TCRS and chimera Zinc binding OmpC. In normal microbial system, ZraSR detects and induces the membrane protein ZraP, which is responsible for the efflux of Zn^{2+} ions. Engineered zinc adsorption system was based on normal ZraSR TCRS, in which ZraS is used for detecting Zn^{2+} ions, but the ZraR activates the *ompC*-Zinc binding peptide chimeric gene under the ZraP promoter instead of native ZraP. The zinc binding peptides displayed in the cell surface can adsorb exogenous Zinc. This system is sensitive to zinc even at low concentrations (0.001 mM) [24].

In the same manner, simultaneous detection and removal of copper ions in the bacterial surface was achieved through the combined application of CuSR TCRS and cell surface displayed copper binding peptides (CBP) fused to the membrane protein OmpC. In this system, CuSR induces the expression of the chimera OmpC-CBP upon sensing Cu^{2+} ions. Then, the chimera proteins expressed in bacterial cell surface can adsorb the copper ions [25].

An interesting feature of these adsorption systems is that the expression of the chimeric OmpC with the metal binding site is induced by heavy metals (Table 1). Hence, the construction of a heavy metal biosensor in combination with a bio-adsorption system would complement analytical heavy metal detection methodologies and enable the rapid monitoring and removal of toxic levels of bioavailable metal contaminants in industrial settings. The above biosensor combined with bio-adsorption was able to absorb heavy metals efficiently without any induction system. Following this scheme, this synthetic bacterial system is an excellent paradigm for developing multifunctional synthetic systems that can be applied both in the efficient removal and recovery of the target compound.

Engineering chimeric TCRSs for detecting novel compounds

The successful design and construction of TCRS provide a better understanding of the system to obtain a chimeric TCRS customized for achieving a desired input/output. The HK domain, which has a variety of signal recognition capabilities, may be used to couple or shuffle a broad range of input signals to the appropriate output responses

Table 1 Two-component regulatory systems based on microbial biosensors coupled with bio-adsorption

Field of application	TCRS	Function	Host chassis	Promoter-reporter	Chemical target	Detection range (mM)	References
Bioremediation	ZraSR (also known as HydHG)	Biosensor	*E. coli* XL1-blue	*zraP-gfp*-HydG	Zinc	0.01–1	[66]
	CuSR	Biosensor	*E. coli* XL1-blue	*cusC-gfp*-CusR	Copper	0.004–1	[25]
	ZraSR and CusSR	Biosensor coupled with bio-adsorption	*E. coli* XL1-blue	*zraP-gfp, cusC-gfp*	Zinc and Copper	0.05–1	[67]
	ZraSR	Biosensor coupled with bio-adsorption	*E. coli* TOP10	*zraP-gfp-ompC*	Lead	0.3–1	[68]
	ZraSR	Biosensor coupled with bio-adsorption	*E. coli* XL1-blue	*zraP-gfp*	Zinc	0.1–1	[24]
Biorefinery	DcuSZ (Chimeric)	Biosensor	*E. coli* BL21 (DE3)	*ompC-gfp*	Fumarate	0.1–10	[55]
	MalKZ (Chimeric)	Biosensor	*E. coli* BL21 (DE3)	*ompC-gfp*	Malate	0.1–10	[56]
	AauSZ (Chimeric)	Biosensor	*E. coli* BL21 (DE3)	*ompC-gfp*	Acidic amino acid	0.05–10	[57]
	Tazl (Chimeric)	Biosensor	*E. coli* RU1012	*ompC-lacZ*	Aspartate	0.2–1	[28]

through a conserved phosphotransfer process. This shuffling can be achieved by cross-linking the domains of evolutionarily distinct TCRSs, and a chimeric TCRS with the desired sensing ability can be obtained. Most of the domain shuffling required for rational design of chimeric proteins is between HKs and rarely between RRs. At present, several research groups have successfully constructed a chimeric two-component sensor protein by fusing the HK domain to the sensory domain of another kinase or a completely unrelated protein. These studies improve our understanding of the molecular events that occur during signal transduction across membranes in these organisms.

Engineering receptor kinases mainly involve a domain swapping or shuffling strategy in which a receptor protein or another HK contributes their functional module. The domain swapping in HKs implies that these proteins are flexible, allowing the construction of new kinases using a rational design strategy. The domain swapping strategy has been used to produce chimeric TCRSs that include chemotaxis proteins. There are several periplasmic chemotactic receptors, such as Tsr, Tar, Trg, and Aer, that recognize specific chemicals, and they can be coupled with the cytoplasmic domain of EnvZ to allow signal transduction [26]. EnvZ is the most studied HK protein that regulates the phosphorylation state of OmpR in response to osmolarity changes. OmpR is an RR protein responsible for the controlled expression of *ompF* and *ompC* genes encoding for the membrane porin proteins OmpF and OmpC, respectively. Aside from OmpR, EnvZ can also regulate the phosphotransfer of 11 different RRs found in *E. coli* [27]. Because the EnvZ–ompR complex in *E. coli* is a well-studied TCRS that is widely-distributed in bacteria, the DHp and CA domains of EnvZ are commonly used for the domain swapping strategy. A good example of this is the hybridization of Tar, a chemoreceptor transmembrane protein that can detect aspartate and EnvZ. By replacing the cytoplasmic signaling domain of Tar protein with the cytoplasmic kinase/phosphatase domain of EnvZ, the hybridized proteins were able to carry out both the sensing capability of Tar for aspartate and the regulation capability of EnvZ towards OmpR thereby consequently activating *ompC* [28]. This strategy also worked in the hybridization of Trg protein and EnvZ, allowing the recognition of ribose-binding peptides and activation expression of *ompC* [29]. In addition to functioning as chemotactic receptors, HK domains are also involved in light sensing, and kinases that sense C_4-dicarboxylate, sugar, aspartate, and acidic amino acids have been engineered with the EnvZ cytoplasmic domain to provide a better sensing ability for the desired substance (Table 1). This approach to engineering novel two-component sensor proteins not only acts as a high throughput screening system but also provides knowledge of the newly identified two-component signaling pathways.

Chimeric TCRS-based screening and regulation of microbial chemical production

In line with the depletion of fossil fuels, renewable biomass is being exploited as a sustainable substitute for petroleum. Among the renewable biomass resources, lignocellulosic biomass is one of the most promising due to its abundance. Lignocellulosic biomass undergoes different pretreatment methods that result in a hydrolysate containing mixed sugars and inhibitors that can be detrimental to the growth of microbial cells during fermentation [30].

Metabolic engineering strategies have been developed in systems level for the development of metabolically engineered microorganisms as host strains in biorefinery processes to produce bio-based fuels [31–35], chemicals [36–41] and polymers [42–47] from renewable resources. Also, engineered strains that have high levels of growth and tolerance in the presence of high concentrations of sugars and inhibitors are extensively being developed to utilize biomass-derived renewable resources [48–53]. Therefore, it is important to develop a high-throughput screening method to identify the high-producing strains. High-producing strains can be screened using a riboselector, which is composed of a riboswitch that can detect the target compound and a selection module such as *tetA*, which will enable favorable growth of a lysine-accumulating cell in the presence of selection pressure ($NiCl_2$) [54]. Likewise, chimeric TCRS can be potentially used in screening for high-producing strains (Fig. 3). DcuSZ is an EnvZ/OmpR-based chimeric TCRS that was constructed by fusing the DcuS HK sensory domain with the cytoplasmic domain of EnvZ. The chimeric DcuSZ is highly specific to fumarate in such a way that the expression of the *gfp* gene under the control of the *ompR*-regulated *ompC* promoter is proportional to different fumarate concentrations in the medium [55]. Other chimeric TCRSs based on EnvZ/OmpR were constructed by fusing the HK sensory domain of MalK and AauS to the EnvZ catalytic domain to detect high malate- and aspartate-producing strains, respectively [56, 57].

Two-component regulatory systems may also be used to develop tightly regulated gene expression systems. Tightly regulated gene expression is important in engineering metabolic pathways to avoid leaky expression that may cause a metabolic burden to the microbial cell. Typical induction strategies include the use of

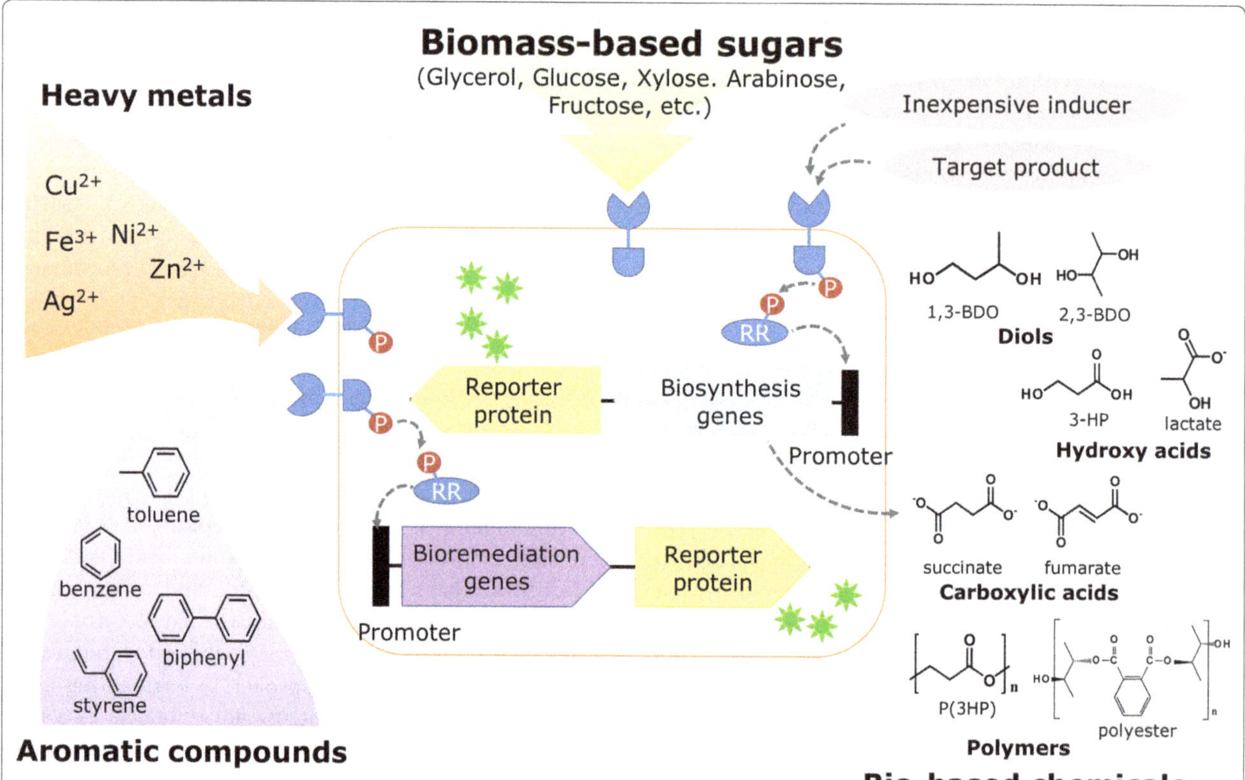

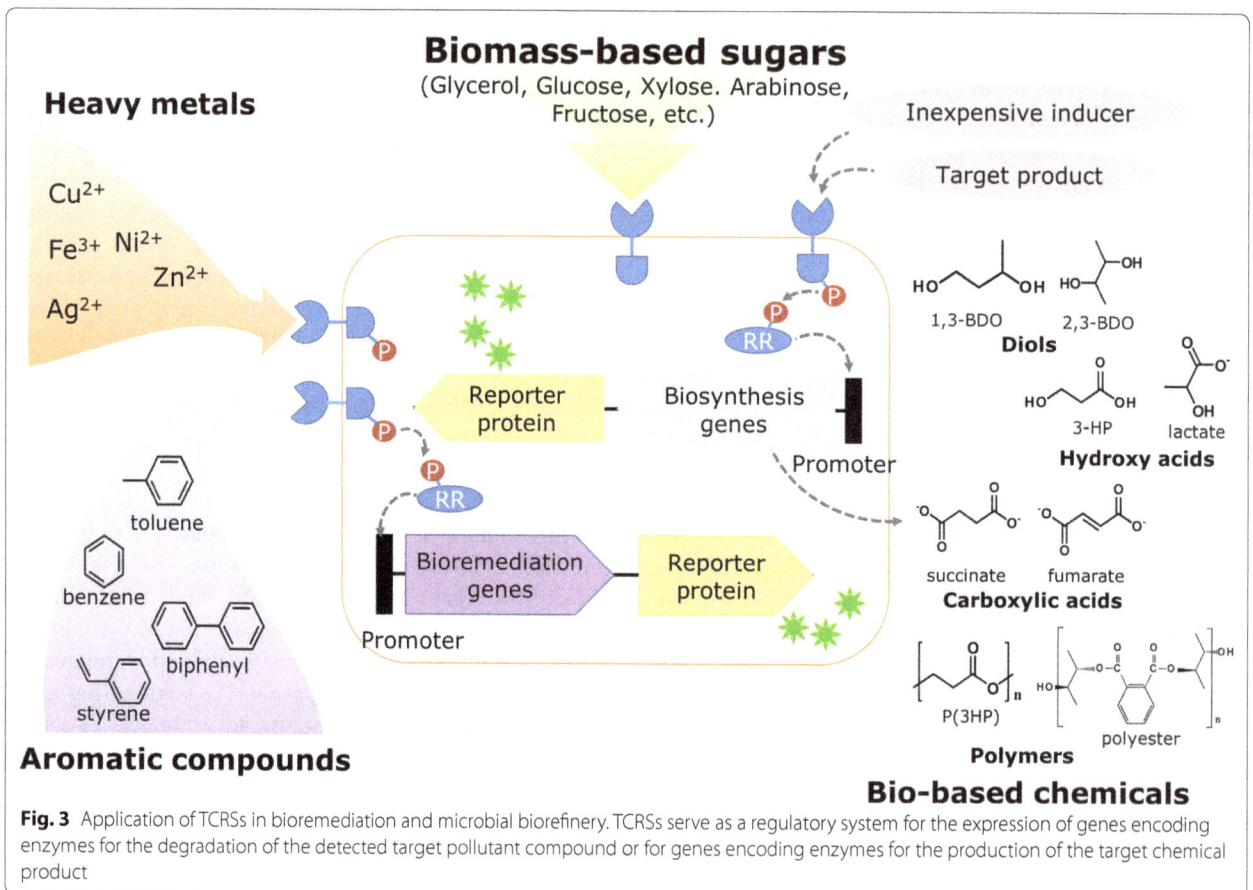

Fig. 3 Application of TCRSs in bioremediation and microbial biorefinery. TCRSs serve as a regulatory system for the expression of genes encoding enzymes for the degradation of the detected target pollutant compound or for genes encoding enzymes for the production of the target chemical product

isopropyl-β-D-thiogalactopyranoside (IPTG). However, IPTG is expensive and can be toxic to cells at high concentrations. An example of tightly regulated gene expression induced by an inexpensive substrate is the invertible promoter system. In this system, the promoter is active or 'ON' when the target substrate that serves as an inducer is present and 'OFF' (inverted orientation) when absent [58]. Based on this invertible promoter system's mechanism, the coupled sensing and regulating activities of TCRSs can be modified to achieve tightly regulated gene expression.

Summary and perspectives

To date, some TCRSs have been identified that sense organic compounds (benzene, toluene, ethylbenzene, biphenyl, styrene, fumarate, and malate) and regulate the gene expression of proteins involved in catabolic pathways. These compounds can be metabolized and used as a carbon source for most groups of microorganisms [9]. In TCRSs, the signal recognized by the sensor kinase domain catalyzes the ATP-dependent phosphorylation of a conserved histidine residue in the protein. The phosphoryl group is then transferred from the histidine to an aspartate residue located in the RR. The phosphorylated RR binds to specific promoter sequences to either activate or repress transcription. At present, a wide range of synthetic genetic circuits has been developed that can couple a sensor output to a desired biological activity [59]. In addition, numerous genetic switches are also available to turn on gene expression once a target molecule has reached its activation threshold. A switch can be assembled using transcriptional repressors or activators, which allows the connection between the sensor output and regulation of the biological response [58]. Several switch types have been developed to control the cellular response: inverter switches that produce a reciprocal response [60]; biphasic switches that use both negative and positive regulation and respond to small amounts of input [61]; toggle switches that use two repressors that cross-regulate each other's promoters [62]; and riboswitches that regulate gene expression by inhibiting protein synthesis [63]. Likewise, many logic gate types have been developed for biological circuits, including 'NAND', 'NOT IF' and 'NOR' [64].

Integrated approaches provide a better perspective for developing a specific biosensor designed to catalyze the production and/or degradation of the desired compound. To achieve this, it is necessary to rewire the genetic circuits of bacteria using the above synthetic devices. Design of the engineered system should be based on strategies for building sensory regulation components that incorporate a target substrate-responsive TCRS in any desired host (Fig. 4). Introducing a sensory regulation device in a host cell enables it to sense the target compound and trigger the genetic circuit, achieving real-time monitoring of the compound present and upregulation of the effector protein's gene expression. Use of engineered TCRSs in bacteria would prevent the production of

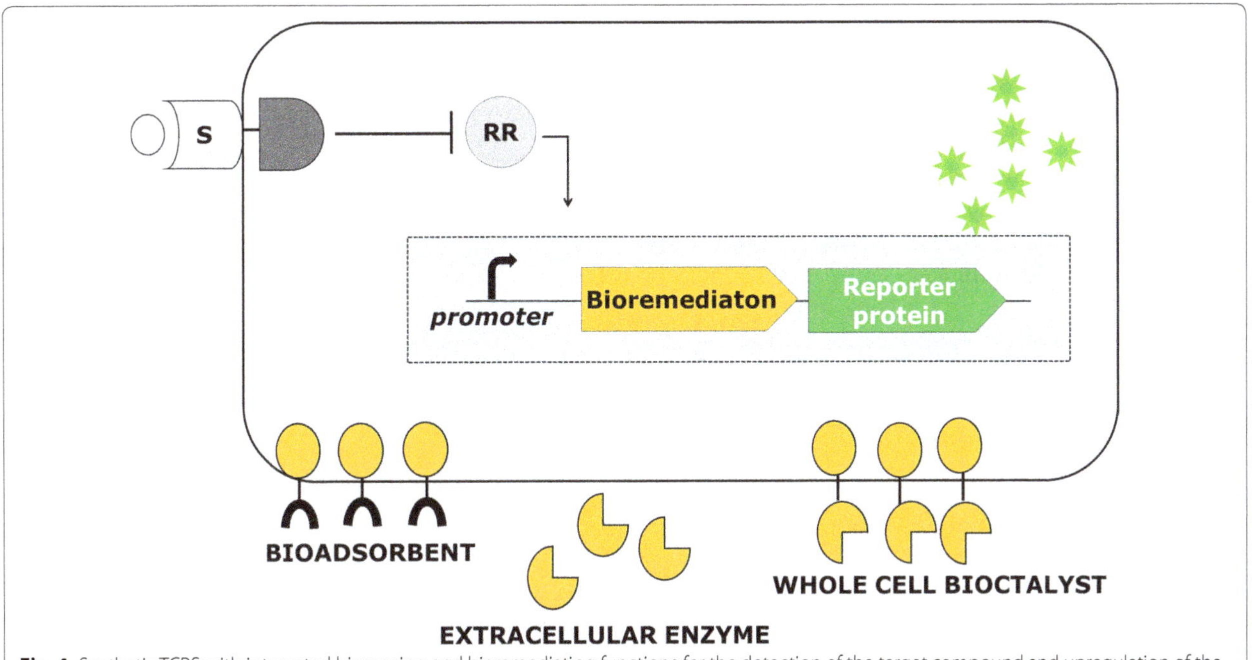

Fig. 4 Synthetic TCRS with integrated biosensing and bioremediating functions for the detection of the target compound and upregulation of the effector protein that allow real-time detection of controlled gene expression

redundant proteins at the initial growth phase and avoid the use of toxic and costly chemical inducers.

Although a large number of accessible sensor parts are available for TCRSs, employing these sensors in a domain shuffling strategy can be challenging. To attach the sensor domain to the HK domain of the protein, structural and functional information on both proteins is needed [65]. When designing chimeric TCRS-based biosensors, great care is required in domain swapping to maximize the kinase activity of the chimeric protein. In the majority of the chimeric TCRS-based biosensors, monitoring of the extracellular targets and the response to these targets is achieved by producing a reporter protein [55–57]. Moreover, biosensors have also been modified with other synthetic biology tools such as the bio-absorption of heavy metals with a cell surface display system and expression of an extracellular enzyme to degrade aromatic compounds. Therefore, such a synthetic genetic circuit can be switched on when a signal is detected to remove certain pollutants, and after the input signal disappears, the microbes behave like normal bacteria.

Conclusions

In this review, we have discussed numerous TCRSs engineered in different prokaryotic species that can sense inorganic and organic pollutants, and examined the recent developments in cellular biosensors coupled with bioremediation. The TCRS-based biosensor coupled with bioremediation approach has the potential to advance even further using the recent developments in bioengineering in strain development. However, only a few studies on TCRS-based biosensors have been reported, and much effort is needed to obtain a complete picture of the TCRS-based control of downstream catabolic pathways. To achieve these goals, a thorough understanding of TCRS mechanisms is essential to engineer strains for use in efficient biosensor systems coupled with bio-degradation or bio-adsorption functionality. Moreover, more studies are required to extend its use in food, pharmaceutical and industrial biotechnology applications.

Abbreviations

TCRS: two-component regulatory system; HK: histidine kinase; RR: response regulator; DHp: histidine phosphotransfer domain; CA: catalytic ATP-binding domain; ATP: adenosine triphosphate.

Authors' contributions

JC and SJP conceived the project. SR, MGB, SJP, and JC wrote the manuscript. All authors read and approved the final manuscript.

Acknowledgements

This work was supported by a grant by the National Research Foundation of Korea (NRF) funded by the Ministry of Science, ICT, and Future Planning (MSIP) (NRF-2015R1A2A2A01004733), a Golden Seed Project Grant (213008-05-1-SB910) funded by Ministry of Oceans and Fisheries, and the Mid-career Researcher Program from MSIP through NRF of Korea (NRF-2016R1A2B4008707).

Competing interests

The authors declare that they have no competing interests.

Funding

Funding sources are declared in the acknowledgement section.

References

1. Wanekaya AK, Chen W, Mulchandani A. Recent biosensing developments in environmental security. J Environ Monit. 2008;10:703–12.
2. Liu X, Germaine KJ, Ryan D, Dowling DN. Development of a GFP-based biosensor for detecting the bioavailability and biodegradation of polychlorinated biphenyls (PCBs). J Environ Eng Landsc. 2007;15:261–8.
3. Lovley DR, Coates JD. Bioremediation of metal contamination. Curr Opin Biotechnol. 1997;8:285–9.
4. Bae W, Wu CH, Kostal J, Mulchandani A, Chen W. Enhanced mercury biosorption by bacterial cells with surface-displayed MerR. Appl Environ Microbiol. 2003;69:3176–80.
5. Harvey BR, Georgiou G, Hayhurst A, Jeong KJ, Iverson BL, Rogers GK. Anchored periplasmic expression, a versatile technology for the isolation of high-affinity antibodies from *Escherichia coli*-expressed libraries. Proc Natl Acad Sci U S A. 2004;101:9193–8.
6. Lee JS, Shin KS, Pan JG, Kim CJ. Surface-displayed viral antigens on *Salmonella* carrier vaccine. Nat Biotechnol. 2000;18:645–8.
7. Taschner S, Meinke A, Gabain AV, Boyd AP. Selection of peptide entry motifs by bacterial surface display. Biochem J. 2002;367:393–402.
8. Casino P, Rubio V, Marina A. The mechanism of signal transduction by two-component systems. Curr Opin Struct Biol. 2010;20:763–71.
9. Tropel D, Van Der Meer JR. Bacterial transcriptional regulators for degradation pathways of aromatic compounds. Microbiol Mol Biol Rev. 2004;68:474–500.
10. Leonhartsberger S, Huber A, Lottspeich F, Böck A. The *hydH/G* genes from *Escherichia coli* code for a zinc and lead responsive two-component regulatory system. J Mol Biol. 2001;307:93–105.
11. Munson GP, Lam DL, Outten FW, O'Halloran TV. Identification of a copper-responsive two-component system on the chromosome of *Escherichia coli* K-12. J Bacteriol. 2000;182:5864–71.
12. Gupta A, Matsui K, Lo JF, Silver S. Molecular basis for resistance to silver cations in *Salmonella*. Nat Med. 1999;5:183–8.
13. López-Maury L, García-Domínguez M, Florencio FJ, Reyes JC. A two-component signal transduction system involved in nickel sensing in the cyanobacterium *Synechocystis* sp. PCC 6803. Mol Microbiol. 2002;43:247–56.
14. Dean CR, Poole K. Expression of the ferric enterobactin receptor (PfeA) of *Pseudomonas aeruginosa*: involvement of a two-component regulatory system. Mol Microbiol. 1993;8:1095–103.
15. Díaz E, Prieto MA. Bacterial promoters triggering biodegradation of aromatic pollutants. Curr Opin Biotechnol. 2000;11:467–75.
16. Velasco A, Alonso S, Garcia JL, Perera J, Díaz E. Genetic and functional analysis of the styrene catabolic cluster of *Pseudomonas* sp. strain Y2. J Bacteriol. 1998;180:1063–71.
17. Sagi E, Hever N, Rosen R, Bartolome AJ, Premkumar JR, Ulber R, Lev O, Scheper T, Belkin S. Fluorescence and bioluminescence reporter functions in genetically modified bacterial sensor strains. Sens Actuators B Chem. 2003;90:2–8.
18. Yong YC, Zhong JJ. A genetically engineered whole-cell pigment-based bacterial biosensing system for quantification of N-butyryl homoserine lactone quorum sensing signal. Biosens Bioelectron. 2009;25:41–7.
19. Biran I, Babai R, Levcov K, Rishpon J, Ron EZ. Online and in situ monitoring of environmental pollutants: electrochemical biosensing of cadmium. Environ Microbiol. 2000;2:285–90.
20. Corbisier P, Ji G, Nuyts G, Mergeay M, Silver S. *luxAB* gene fusions with the arsenic and cadmium resistance operons of *Staphylococcus aureus* plasmid pI258. FEMS Microbiol Lett. 1993;110:231–8.
21. Ivask A, Hakkila K, Virta M. Detection of organomercurials with sensor bacteria. Anal Chem. 2001;73:5168–71.
22. Selifonova O, Burlage R, Barkay T. Bioluminescent sensors for detection of bioavailable Hg(II) in the environment. Appl Environ Microbiol. 1993;59:3083–90.

23. Tauriainen S, Karp M, Chang W, Virta M. Luminescent bacterial sensor for cadmium and lead. Biosens Bioelectron. 1998;13:931–8.
24. Ravikumar S, Yoo IK, Lee SY, Hong SH. A study on the dynamics of the *zraP* gene expression profile and its application to the construction of zinc adsorption bacteria. Bioprocess Biosyst Eng. 2011;34:1119–26.
25. Ravikumar S, Yoo IK, Lee SY, Hong SH. Construction of copper removing bacteria through the integration of two-component system and cell surface display. Appl Biochem Biotechnol. 2011;165:1674–81.
26. Grebe TW, Stock J. Bacterial chemotaxis: the five sensors of a bacterium. Curr Biol. 1998;8:R154–7.
27. Skerker JM, Prasol MS, Perchuk BS, Biondi EG, Laub MT. Two-component signal transduction pathways regulating growth and cell cycle progression in a bacterium: a system-level analysis. PLoS Biol. 2005;3:e334.
28. Utsumi R, Brissette RE, Rampersaud A, Forst SA, Oosawa K, Inouye M. Activation of bacterial porin gene expression by a chimeric signal transducer in response to aspartate. Science. 1989;245:1246–9.
29. Baumgartner JW, Kim C, Brissette RE, Inouye M, Park C, Hazelbauer GL. Transmembrane signalling by a hybrid protein: communication from the domain of chemoreceptor Trg that recognizes sugar-binding proteins to the kinase/phosphatase domain of osmosensor EnvZ. J Bacteriol. 1994;176:1157–63.
30. Oh YH, Eom IY, Joo JC, Yu JH, Song BK, Lee SH, Hong SH, Park SJ. Recent advances in development of biomass pretreatment technologies used in biorefinery for the production of bio-based fuels, chemicals and polymers. Korean J Chem Eng. 2015;32:1945–59.
31. Oh YH, Eom GT, Kang KH, Joo JC, Jang YA, Choi JW, Song BK, Lee SH, Park SJ. Construction of heterologous gene expression cassettes for the development of recombinant *Clostridium beijerinckii*. Bioprocess Biosyst Eng. 2016;39:555–63.
32. Kim S, Jang YS, Ha SC, Ahn JW, Kim EJ, Lim JH, Cho C, Ryu YS, Lee SK, Lee SY, Kim KJ. Redox-switch regulatory mechanism of thiolase from *Clostridium acetobutylicum*. Nat Commun. 2015;6:8410.
33. Jang YS, Malaviya A, Cho C, Lee J, Lee SY. Butanol production from renewable biomass by *Clostridia*. Bioresour Technol. 2012;123:653–63.
34. Jojima T, Noburyu R, Sasaki M, Tajima T, Suda M, Yukawa H, Inui M. Metabolic engineering for improved production of ethanol by *Corynebacterium glutamicum*. Appl Microbiol Biotechnol. 2015;99:1165–72.
35. Gaida SM, Liedtke A, Jentges AH, Engels B, Jennewein S. Metabolic engineering of *Clostridium cellulolyticum* for the production of n-butanol from crystalline cellulose. Microb Cell Fact. 2016;15:6.
36. Shin JH, Park SH, Oh YH, Choi JW, Lee MH, Cho JS, Jeong KJ, Joo JC, Yu J, Park SJ, Lee SY. Metabolic engineering of *Corynebacterium glutamicum* for enhanced production of 5-aminovaleric acid. Microb Cell Fact. 2016;15:174.
37. Oh YH, Choi JW, Kim EY, Song BK, Jeong KJ, Park K, Kim IK, Woo HM, Lee SH, Park SJ. Construction of synthetic promoter-based expression cassettes for the production of cadaverine in recombinant *Corynebacterium glutamicum*. Appl Biochem Biotechnol. 2015;176:2065–75.
38. Choi JW, Yim SS, Lee SH, Kang TJ, Park SJ, Jeong KJ. Enhanced production of gamma-aminobutyrate (GABA) in recombinant *Corynebacterium glutamicum* by expressing glutamate decarboxylase active in expanded pH range. Microb Cell Fact. 2015;14:21.
39. Park SJ, Kim EY, Noh W, Park HM, Oh YH, Lee SH, Song BK, Jegal J, Lee SY. Metabolic engineering of *Escherichia coli* for the production of 5-aminovalerate and glutarate as C5 platform chemicals. Metab Eng. 2013;16:42–7.
40. Abdel-Rahman MA, Xiao Y, Tashiro Y, Wang Y, Zendo T, Sakai K, Sonomoto K. Fed-batch fermentation for enhanced lactic acid production from glucose/xylose mixture without carbon catabolite repression. J Biosci Bioeng. 2015;119:153–8.
41. Li Y, Wang X, Ge X, Tian P. High production of 3-hydroxypropionic acid in *Klebsiella pneumoniae* by systematic optimization of glycerol metabolism. Sci Rep. 2016;6:26932.
42. Choi SY, Park SJ, Kim WJ, Yang JE, Lee H, Shin J, Lee SY. One-step fermentative production of poly(lactate-*co*-glycolate) from carbohydrates in *Escherichia coli*. Nat Biotechnol. 2016;34:435–40.
43. Yang JE, Kim JW, Oh YH, Choi SY, Lee H, Park AR, Shin J, Park SJ, Lee SY. Biosynthesis of poly(2-hydroxyisovalerate-*co*-lactate) by metabolically engineered *Escherichia coli*. Biotechnol J. 2016;11:1572–85.
44. Chae CG, Kim YJ, Lee SJ, Oh YH, Yang JE, Joo JC, Kang KH, Jang YA, Lee H, Park AR, Song BK, Lee SY, Park SJ. Biosynthesis of poly(2-hydroxybutyrate-*co*-lactate) in metabolically engineered *Escherichia coli*. Biotechnol Bioproc Eng. 2016;21:169–74.
45. Park SJ, Jang YA, Noh W, Oh YH, Lee H, David Y, Baylon MG, Shin J, Yang JE, Choi SY, Lee SH, Lee SY. Metabolic engineering of *Ralstonia eutropha* for the production of polyhydroxyalkanoates from sucrose. Biotechnol Bioeng. 2015;112:638–43.
46. Meng DC, Wang Y, Wu LP, Shen R, Chen JC, Wu Q, Chen GQ. Production of poly(3-hydroxypropionate) and poly(3-hydroxybutyrate-*co*-3-hydroxypropionate) from glucose by engineering *Escherichia coli*. Metab Eng. 2015;29:189–95.
47. Zhang S, Liu Y, Bryant DA. Metabolic engineering of *Synechococcus* sp. PCC 7002 to produce poly-3-hydroxybutyrate and poly-3-hydroxybutyrate-*co*-4-hydroxybutyrate. Metab Eng. 2015;32:174–83.
48. Kim HS, Oh YH, Jang YA, Kang KH, David Y, Yu JH, Song BK, Choi JI, Chang YK, Joo JC, Park SJ. Recombinant *Ralstonia eutropha* engineered to utilize xylose and its use for the production of poly(3-hydroxybutyrate) from sunflower stalk hydrolysate solution. Microb Cell Fact. 2016;15:95.
49. Oh YH, Lee SH, Jang YA, Choi JW, Hong KS, Yu JH, Shin J, Song BK, Mastan SG, David Y, Baylon MG, Lee SY, Park SJ. Development of rice bran treatment process and its use for the synthesis of polyhydroxyalkanoates from rice bran hydrolysate solution. Bioresour Technol. 2015;181:283–90.
50. Park SJ, Park JP, Lee SY. Production of poly(3-hydroxybutyrate) from whey by fed-batch culture of recombinant *Escherichia coli* in a pilot-scale fermenter. Biotechnol Lett. 2002;24:185–9.
51. Kim DY, Yim SC, Lee PC, Lee WG, Lee SY, Chang HN. Batch and continuous fermentation of succinic acid from wood hydrolysate by *Mannheimia succiniciproducens* MBEL55E. Enzyme Microb Technol. 2004;35:648–53.
52. Saha BC, Nichols NN, Qureshi N, Kennedy GJ, Iten LB, Cotta MA. Pilot scale conversion of wheat straw to ethanol via simultaneous saccharification and fermentation. Bioresour Technol. 2015;175:17–22.
53. Zhang Y, Kumar A, Hardwidge PR, Tanaka T, Kondo A, Vadlani PV. D-lactic acid production from renewable lignocellulosic biomass via genetically modified *Lactobacillus plantarum*. Biotechnol Prog. 2016;32:271–8.
54. Yang J, Seo SW, Jang S, Shin SI, Lim CH, Roh TY, Jung GY. Synthetic RNA devices to expedite the evolution of metabolite-producing microbes. Nat Commun. 2013;4:1413.
55. Ganesh I, Ravikumar S, Lee SH, Park SJ, Hong SH. Engineered fumarate sensing *Escherichia coli* based on novel chimeric two-component system. J Biotechnol. 2013;168:560–6.
56. Ganesh I, Ravikumar S, Yoo IK, Hong SH. Construction of malate-sensing *Escherichia coli* by introduction of a novel chimeric two-component system. Bioprocess Biosyst Eng. 2015;38:797–804.
57. Ravikumar S, Ganesh I, Maruthamuthu MK, Hong SH. Engineering *Escherichia coli* to sense acidic amino acids by introduction of a chimeric two-component system. Korean J Chem Eng. 2015;32:2073–7.
58. Ham TS, Lee SK, Keasling JD, Arkin AP. A tightly regulated inducible expression system utilizing the fim inversion recombination switch. Biotechnol Bioeng. 2006;94:1–4.
59. Voigt CA. Genetic parts to program bacteria. Curr Opin Biotechnol. 2006;17:548–57.
60. Yokobayashi Y, Weiss R, Arnold FH. Directed evolution of a genetic circuit. Proc Natl Acad Sci U S A. 2002;99:16587–91.
61. Isaacs FJ, Hasty J, Cantor CR, Collins JJ. Prediction and measurement of an autoregulatory genetic module. Proc Natl Acad Sci USA. 2003;100:7714–9.
62. Gardner TS, Cantor CR, Collins JJ. Construction of a genetic toggle switch in *Escherichia coli*. Nature. 2000;403:339–42.
63. Mandal M, Breaker RR. Gene regulation by riboswitches. Nat Rev Mol Cell Biol. 2004;5:451–63.
64. Guet CC, Elowitz MB, Hsing W, Leibler S. Combinatorial synthesis of genetic networks. Science. 2002;296:1466–70.
65. Zhang F, Keasling J. Biosensors and their applications in microbial metabolic engineering. Trends Microbiol. 2011;19:323–9.
66. Pham VD, Ravikumar S, Lee SH, Hong SH, Yoo IK. Modification of response behavior of zinc sensing HydHG two-component system using a self-activation loop and genomic integration. Bioprocess Biosyst Eng. 2013;36:1185–90.
67. Ravikumar S, Ganesh I, Yoo IK, Hong SH. Construction of a bacterial biosensor for zinc and copper and its application to the development of multifunctional heavy metal adsorption bacteria. Process Biochem. 2012;47:758–65.

16

Nutrient starvation leading to triglyceride accumulation activates the Entner-Doudoroff pathway in *Rhodococcus jostii* RHA1

Antonio Juarez[1,2], Juan A. Villa[3], Val F. Lanza[3], Beatriz Lázaro[3], Fernando de la Cruz[3], Héctor M. Alvarez[4] and Gabriel Moncalián[3*]

Abstract

Background: *Rhodococcus jostii* RHA1 and other actinobacteria accumulate triglycerides (TAG) under nutrient starvation. This property has an important biotechnological potential in the production of sustainable oils.

Results: To gain insight into the metabolic pathways involved in TAG accumulation, we analysed the transcriptome of *R jostii* RHA1 under nutrient-limiting conditions. We correlate these physiological conditions with significant changes in cell physiology. The main consequence was a global switch from catabolic to anabolic pathways. Interestingly, the Entner-Doudoroff (ED) pathway was upregulated in detriment of the glycolysis or pentose phosphate pathways. ED induction was independent of the carbon source (either gluconate or glucose). Some of the diacylglycerol acyltransferase genes involved in the last step of the Kennedy pathway were also upregulated. A common feature of the promoter region of most upregulated genes was the presence of a consensus binding sequence for the cAMP-dependent CRP regulator.

Conclusion: This is the first experimental observation of an ED shift under nutrient starvation conditions. Knowledge of this switch could help in the design of metabolomic approaches to optimize carbon derivation for single cell oil production.

Keywords: *Rhodococcus*, Triacylglycerol, Nutrient starvation, RNA-Seq, Entner-Doudoroff pathway, CRP

Background

Microbial triglycerides, called single cell oils (SCO), have biotechnological potential in the production of sustainable oils for their use either as biodiesel or as commodity oils. Biodiesel is produced by transesterification of triacylglycerides with short-chain alcohols (mainly methanol). Vegetable oils and animal fats such as soybean oil, rapeseed oil, palm oil or waste cooking oils are used as feedstocks for biodiesel production [1]. However, this strategy has been criticized for being a non-sustainable process since it leads to a reduction in edible oil feedstocks [2]. Production of biodiesel using SCO is considered as a promising alternative solution [3]. SCO produce high quality biodiesel esters according to currently existing standards [4, 5]. SCO are appropriate for their use as a biodiesel source since the producing microorganisms can grow using a variety of substrates, show rapid life cycles and can be easily modified by genetic engineering.

Several microorganisms, including bacteria, yeasts, molds and microalgae, can be considered as oleaginous microorganisms [6]. Regarding bacteria, the accumulation of the neutral lipids triacylglycerols (TAGs), wax esters (WEs) and polyhydroxyalkanoates (PHAs) has been reported. The main purpose of this accumulation is to store carbon and energy under growth-limiting

*Correspondence: moncalig@unican.es
[3] Departamento de Biología Molecular (Universidad de Cantabria) and Instituto de Biomedicina y Biotecnología de Cantabria IBBTEC (CSIC-UC), C/Albert Einstein 22, 39011 Santander, Spain
Full list of author information is available at the end of the article

conditions. While PHAs are synthesized in a wide variety of bacteria [7], the accumulation of triacylglycerols (TAGs) has only been described for a few bacteria belonging to the proteobacteria and actinobacteria groups (for a review see [8]). *Acinetobacter* [9] *Mycobacterium* [10], *Streptomyces* [11] or *Rhodococcus* [12] are such examples. Accumulation of TAGs is remarkably high in the actinobacteria *Rhodococcus* and *Gordonia*, which accumulate up to 80% of the cellular dry weight in the form of neutral lipids with maximal TAG production of 88.9 and 57.8 mg/l, respectively [13].

Rhodococcus are aerobic, non-sporulating soil bacteria, with unique enzymatic activities used for several environmental and biotechnological processes [14]. *Rhodococcus* strains are industrially used for large-scale production of acrylamide and acrylic acid as well as for the production of bioactive steroid compounds and fossil fuel biodesulfurization [15]. Moreover, *Rhodococcus* are able to degrade contaminant hydrophobic natural compounds and xenobiotics. *R. jostii* RHA1 has been shown to convert lignocellulose into different phenolic compounds [16] while it also has the potential to use this waste material for the production of valuable oils [17].

Due to its capability for degrading hydrocarbons, *R. jostii* RHA1 is one of the best studied *Rhodococcus* species in the terms of biotechnological applications [18–20]. Moreover, high TAG accumulating capability has been reported [21] and its genomic sequence is available [22].

In this article we decipher the metabolic changes associated to nutrient starvation conditions that influence TAG accumulation.

Methods

Bacterial strain and growth conditions

Rhodococcus jostii strain RHA1 was grown aerobically at 30 °C in *Streptomyces* medium, Fluka (Rich Medium, RM, 4.0 g/l glucose, 4.0 g/l Yeast extract, and 10.0 g/l Malt extract). After 48 h, 25 ml of *R. jostii* cells in RM were collected by centrifugation, washed with mineral salts medium M9 (Minimal Medium, MM, [23], 95 mM Na_2HPO_4, 44 mM KH_2PO_4, 17 mM NaCl, 0.1 mM $CaCl_2$ and 2 mM $MgSO_4$) containing 20% w/v sodium gluconate (MMGln) or 20% w/v glucose (MMGls) as the sole carbon sources and transfer into 25 ml of MMGln or MMGls. The concentration of ammonium chloride in MM was reduced to 10 mM to enhance lipid accumulation.

Extraction and analysis of lipids

Pelleted cells were extracted with hexane/isopropanol (3:1 v/v). An aliquot of the whole cell extract was analyzed by thin layer chromatography (TLC) on silica gel plates (Merck) applying n-hexane/diethyl ether/acetic acid (80:20:1, v/v/v) as a solvent system. Lipid fractions were revealed using iodine vapour. Trioleine and oleic acid (Merck) were used as standards.

RNA extraction

RNA was extracted from RM and MM-grown cells originally harvested from 3 ml of culture. Total RNA isolation involved vortexing of the pellet with 6 ml of RNA Protect (QIAGEN) followed by centrifugation. The pellet was thereafter lysed using 280 μl of lysis buffer (10% Zwittergent (Calbiochem), 15 mg/ml Lysozime (Sigma) and 20 mg/ml Proteinase K (Roche) in TE buffer). Total RNA was purified with RNeasy mini kit (QIAGEN, Valencia, CA) combined with DNase I (QIAGEN) according to the manufacturer's instructions. The quantity and quality of RNA were assessed using a NanoDrop ND-1000 spectrophotometer (NanoDrop Technology, Rockland, DE) and Experion Automated Electrophoresis using the RNA StdSens Analysis Kit (Bio Rad).

mRNA enrichment

Removal of 16S and 23S rRNA from total RNA was performed using MicrobExpress™ Bacterial mRNA Purification Kit (Ambion) according to the manufacturer's protocol with the exception that no more than 5 μg total RNA was treated per enrichment reaction. Each RNA sample was divided into multiple aliquots of ≤5 μg RNA and separate enrichment reactions were performed for each sample. Enriched mRNA samples were pooled and run on the 2100 Bioanalzyer (Agilent) to confirm reduction of 16S and 23S rRNA prior to preparation of cDNA fragment libraries.

Preparation of cDNA fragment libraries

Ambion RNA fragmentation reagents were used to generate 60–200 nucleotide RNA fragments with an input of 100 ng of mRNA. Following precipitation of fragmented RNA, first strand cDNA synthesis was performed using random N6 primers and Superscript II Reverse Transcriptase, followed by second strand cDNA synthesis using RNaseH and DNA pol I (Invitrogen, CA). Double stranded cDNA was purified using Qiaquick PCR spin columns according to the manufacturer's protocol (Qiagen).

RNA-Seq using the Illumina genome analyzer

The Illumina Genomic DNA Sample Prep kit (Illumina, Inc., San Diego, CA) was used according to the manufacturer's protocol to process double-stranded cDNA for RNA-Seq. This process included end repair, A-tailing, adapter ligation, size selection, and pre-amplification. Amplified material was loaded onto independent flow cells. Sequencing was carried out by running 36 cycles on the Illumina Genome Analyzer IIx. The quality of the RNA-Seq reads was analyzed by assessing the relationship between the quality score and error probability. These analyses were performed on Illumina RNA-Seq quality scores that were converted to phred format (http://www.phrap.com/phred/).

Computational methods

To filter genes with low signal/noise ratio we built 3 subsets of each condition taking randomly 70% of the total sequenced reads for each subset. The alignment was performed by Bowtie [24] against the *R. jostii* RHA1 reference genomes of the chromosome and three endogenous plasmids (Genome Reviews CP000431-4_GR). Gene expression was determined by Samtools [25], Artemis [26] and home-made perl scripts. We represent gene expression as reads per kilobase (RPK) and the data was normalized by quantiles according to [27]. Statistical analysis was performed by DESeq package [28] and R software.

Quantitative real-time RT-PCR (qRT-PCR)

cDNA was generated from 1.5 µg of total RNA using the iScript kit (BioRad) according to manufacturer's instructions. 1 µl of the cDNA template was then used in quantitative real-time PCR reactions using iQ SUYBRE Green Supermix (BioRad) and a iCycler iQ5(BioRad). Primers were designed using Primer3 (http://primer3.sourceforge.net). The cycle of threshold (Ct) was determined for each reaction using the iQ5 Optical System Software 2.0 (BioRad). All qRT-PCR reactions were done in triplicate.

KDPG aldolase activity assay

KDPG aldolase activity was quantified by a lactate dehydrogenase (LDH) coupled assay where the production of pyruvate is related to the NADH consumption, as described in [29]. 2 ml of *R. jostii RHA1* RM or MMGls cultures were harvested and resuspended in 1 ml of buffer TrisHCl 100 mM pH 7.5, NaCl 300 mM, EDTA 1 mM, DTT 1 mM and PMSF 1 mM. The cells were lysed using 0.2 mm silica beads and a Fast Prep-24 system (MP Biomedicals) for 3 cycles of 60 s and centrifuged at 100,000*g* for 25 min at 4 °C. 150 µl aliquots of the resulting RM or MMGls total extracts were then treated with 1 µl of LDH (5 U/µL), 0.70 µl of NADH (50 mM) and 1 µl of KDPG (50 mM). Decrease in NADH absorbance at 340 nm was measured in quartz microcuvettes (150 µl) in a UV-1603 spectrophotometer (Shimadzu) for 5 min. Total protein concentration was determined by Bradford assays using BSA as standard. KDGP activity was calculated as moles of NADH consumed per mg of total protein per second (mol/s/mg).

Results and discussion

Culture conditions for *R. jostii* RHA1, TAGs accumulation and RNA-Seq analysis

R. jostii RHA1 is able to transform a diverse range of organic substrates into large quantities of TAGs [21]. The best conditions for TAG accumulation in *R. opacus* occur when gluconate is used as carbon source in a nitrogen-limited medium [30]. We have checked TAG accumulation over time in *R. jostii* RHA1 cells transferred to M9 medium with 10 mM ammonium chloride and sodium gluconate (20% w/v) as carbon source (MMGln medium, Fig. 1). While TAG accumulation was already detected upon 4 h in MMGln (Fig. 1), no TAG accumulation was observed at any time in a complex rich-nutrient medium (RM). TAGs were also accumulated in an M9 medium with 20 mM ammonium chloride (MMN) and even when MMN was enriched with 0.2% casamino acids (data not shown). Thus, for comparative analysis of the *R. jostii* transcriptome under conditions that lead or do not lead to TAG accumulation, RNA-Seq analyses were performed on two RNA samples collected from *R.jostii* RHA1 strain incubated either 24 h in RM medium (exponential phase) or 4 h in MMGln after 48 h in RM medium. cDNA was generated from mRNA-enriched total RNA preparations from each strain and sequenced using the Illumina Genome Analyzer IIx as described in Methods, to yield a total number of 9,611,145 reads for MMGln and 14,330,620 reads for RM (Table 1).

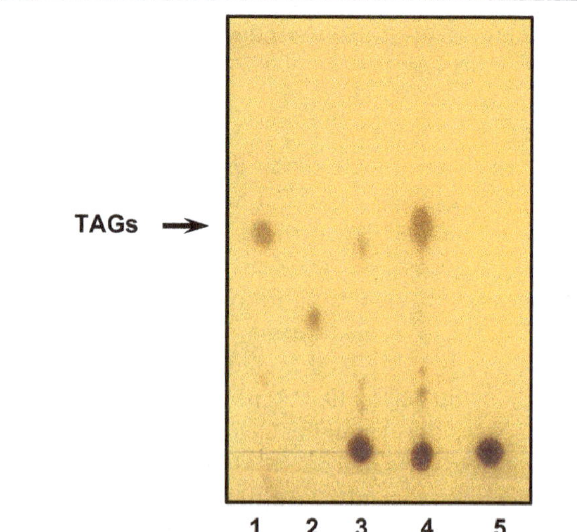

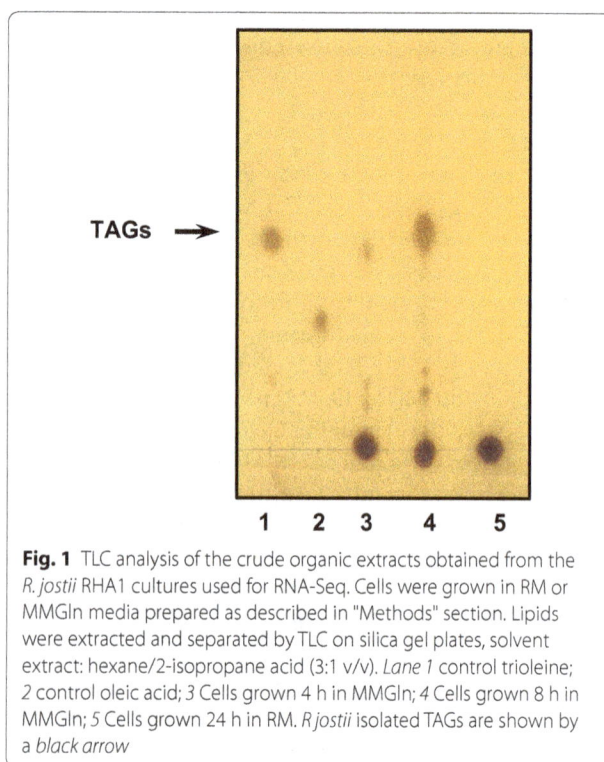

Fig. 1 TLC analysis of the crude organic extracts obtained from the *R. jostii* RHA1 cultures used for RNA-Seq. Cells were grown in RM or MMGln media prepared as described in "Methods" section. Lipids were extracted and separated by TLC on silica gel plates, solvent extract: hexane/2-isopropane acid (3:1 v/v). *Lane 1* control trioleine; *2* control oleic acid; *3* Cells grown 4 h in MMGln; *4* Cells grown 8 h in MMGln; *5* Cells grown 24 h in RM. *R jostii* isolated TAGs are shown by a *black arrow*

For comparative analysis of the *R. jostii* transcriptome under conditions that lead or do not lead to TAG accumulation, reads per kilobase (RPK) were calculated for each of the 9145 annotated *R jostii* genes [22] and normalized for each condition as described in "Methods" section (Additional file 1: Table S1). After data processing, we observed 701 upregulated genes (twofold or greater, MMGln vs RM) and 538 downregulated genes (twofold or greater, MMGln vs RM) (Table 2; Fig. 2a). Whereas the percentage of chromosomal upregulated and downregulated genes was similar (6.3 vs 6.8%), the percentage of plasmid upregulated genes was much higher than the percentage of downregulated genes (13.3 vs. 2.0% in pRHL1, 11.7 vs. 4.4% in pRHL2 and 11.4 vs. 0.9% in pRHL3) (Table 2). Predominant gene upregulation is a common feature of different bacterial stress conditions where a quick response to environmental changes is needed [31]. It is also apparent that, for the whole

Table 1 Summary of the *R. jostii* cDNA samples sequenced using the Illumina genome analyzer

Sequenced sample	Total mapped reads	Total mapped bps ($\times 10^6$)	Mapped mRNA reads	Mapped mRNA bp ($\times 10^6$)	mRNA reads (% of all mapped reads)
MMGln	9,611,145	336.39	2,751,223	96.292	28.6
RM	14,330,620	501.57	1,554,502	54.408	10.8

Table 2 Distribution of the upregulated and downregulated genes in the chromosome and plasmids of *R. jostii* RHA1

	pRHL3	pRHL2	pRLH1	Chrom	Total
Up	38	53	153	457	701
Normal	293	381	970	6262	7906
Down	3	20	23	492	538
All	334	454	1146	7211	9145
	%pRHL3	**%pRHL2**	**%pRHL1**	**%Chrom**	**Total**
Up	11.38	11.67	13.35	6.34	7.66
Normal	87.72	83.92	84.64	86.84	86.45
Down	0.90	4.40	2.00	6.82	5.88
All	100	100	100	100	100

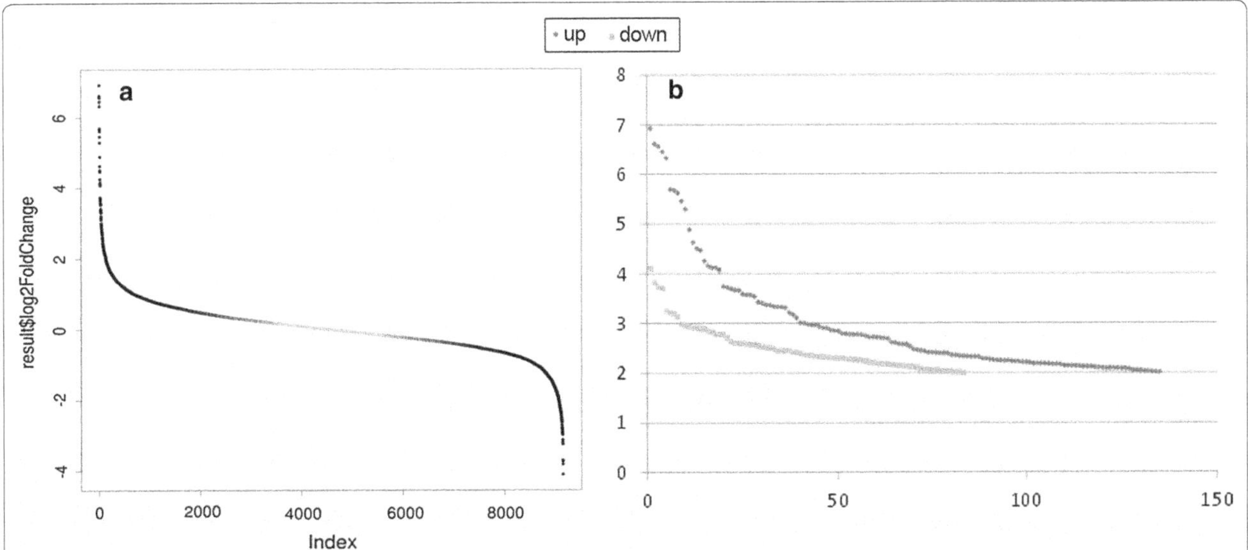

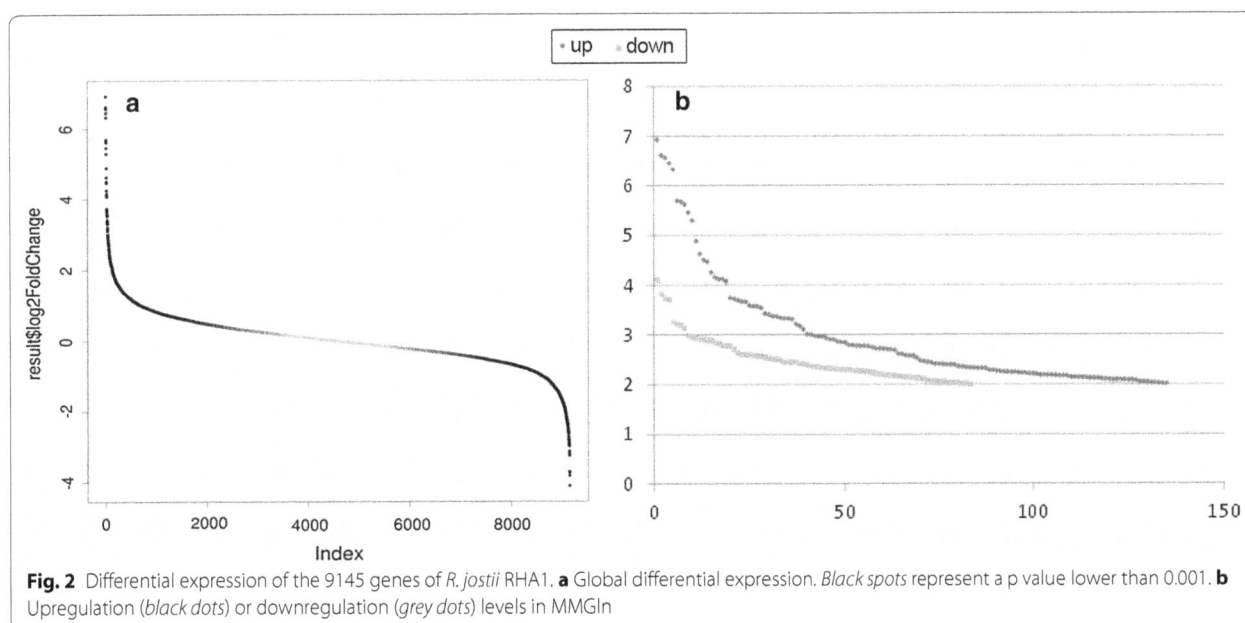

Fig. 2 Differential expression of the 9145 genes of *R. jostii* RHA1. **a** Global differential expression. *Black spots* represent a p value lower than 0.001. **b** Upregulation (*black dots*) or downregulation (*grey dots*) levels in MMGln

genome, genes showing high induction predominate over genes showing high repression (Fig. 2b). 42 genes showed eightfold or higher upregulation, while only 8 genes showed eightfold or higher downregulation (Additional file 1: Table S1).

Comparative analysis of *R. jostii* RHA1 transcriptome under nutrient-rich and nutrient-limiting (TAG accumulating) conditions

For an overview of the metabolic changes that occurred after nutrient deprivation maintaining the carbon source excess, we identified the KEGG pathways [32] corresponding to the up- or downregulated genes. For some functional categories (i.e., oxidative phosphorylation, pentose phosphate, ABC transporters, fatty acid metabolism), upregulated genes predominate (Fig. 3). In contrast, for other categories (i.e., amino acids metabolism and inositol phosphate metabolism), downregulated genes predominate. To better understand the global effects of nutrient deprivation, we looked at specific pathways rather than to functional categories. Downregulation is the rule in several metabolic activities, both catabolic and biosynthetic, as well as in the turnover of macromolecules. Key assimilatory pathways were repressed (Phosphate and sulphate assimilation, synthesis of glutamine synthetase, synthesis of C1-carriers). DNA duplication machinery and several biosynthetic pathways (i.e., pyrimidine, peptidoglycan) were also repressed. With respect to the catabolic pathways, repression occurred in: (i) degradation of several alternative carbon sources and (ii) sugar transport via phosphotransferase system (PTS). Turnover by RNA degradation was also repressed. These downregulated pathways can be interpreted as a result of cells stopping metabolic activities that lead to cell proliferation as a consequence of nutrient starvation.

Other alterations in gene expression can be directly correlated to specific starvation conditions: excess of the carbon source or depletion of the nitrogen source. Hence, significant alterations of metabolic pathways are related to nitrogen starvation: (i) amino acid catabolism is repressed and (ii) reactions that might render free ammonia from organic compounds are induced (i.e., formamidase and ethanolamine ammonia lyase). Finally, a set of metabolic activities are induced as a consequence of the fact that nutrient-starved cells can still incorporate the carbon source leading, for instance, to the synthesis of TAGs. In fact, induction of glycerol-3P-acyltransferase, fatty acid synthesis, acyl-carrier protein and biotin biosynthetic enzymes was observed. The transcriptome analysis of *R. opacus* PD630 under TAG accumulating conditions has been recently reported [33]. 3 h after cells were transferred to a minimal medium (MSM3)

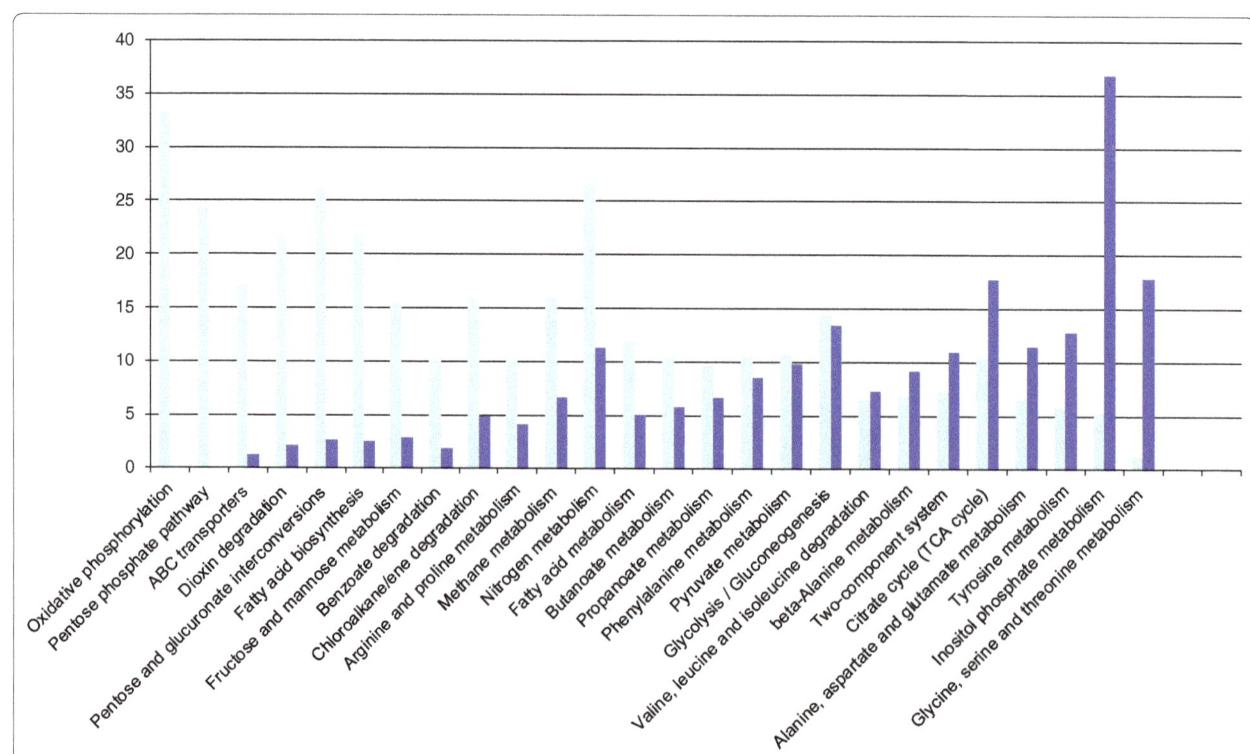

Fig. 3 Number of up- and downregulated MMGln *R. jostii* genes in the corresponding KEGG functional pathways. The *bars* represent the number of genes with upregulation of twofold or greater (*cyan bars*) or a downregulation of twofold or greater (*blue bars*)

similar to our MMGln medium, 21.15% of the genes were upregulated >2-fold and 9.36% downregulated >2-fold. Globally, genes related to biogenesis were upregulated while genes involved in energy production or carbohydrate metabolism were downregulated. 4273 *R. jostii* RHA1 homologous genes have been found in *R. opacus* PD630 chromosome. Most of the upregulated genes in *R. jostii* MMGln are also upregulated in *R. opacus* MSM3 (Additional file 1: Table S3), thus confirming the metabolic shift observed for *R. jostii* under TAG accumulating conditions.

Genes of the Entner-Doudoroff (ED) pathway are highly upregulated

Switching metabolism to the synthesis of TAGs not only requires the upregulation of enzymes specifically involved in the corresponding biosynthetic pathways, but also the upregulation of the corresponding pathways that generate the appropriate building blocks, ATP and reducing power [34]. One of the main functional categories presenting upregulated genes that were activated when *R. jostii* cells were grown in MMGln was the pentose phosphate pathway (Fig. 3). However, a detailed analysis of the specific genes of this functional category that are upregulated showed them to belong to the ED catabolic pathway. The ED pathway is, in addition to the Embden-Meyerhof-Parnas (EMP) and pentose phosphate pathways, one of three pathways that process 6-carbon sugars [35, 36]. The first step in the ED pathway is the formation of gluconate-6-phosphate by oxidation of glucose-6-phosphate or phosphorylation of gluconate. Then, the 6-phosphogluconate dehydratase catalyzes the dehydration of 6-phosphogluconate to produce KDPG. Finally, the cleavage of KDPG catalysed by the KDPG aldolase yields pyruvate and glyceraldehyde-3-phosphate. Electrons drawn in reactions catalysed by the glucose-6P-dehydrogenase are transferred to $NADP^+$. According to the RNA-Seq transcriptomic analysis, every gene coding for the different enzymes of the ED pathway was highly upregulated in the MMGln conditions (Fig. 4; Table 3).

Consistently, genes involved in ED pathway were also found amongst the genes upregulated in the TAG accumulating medium in *R. opacus* PD630 (Additional file 1: Table S3).

For RNA-Seq transcriptomic analysis, we used gluconate as a carbon source in MMGln because gluconate led to the highest level of TAG accumulation in *R. opacus* [30]. Therefore, induction of the ED pathway could be the consequence of the use of gluconate as the sole carbon source and not of a general mechanism for TAG accumulation under nutrient-deprived conditions. To solve this question, we tested whether the presence of glucose in MMGls also induces TAG accumulation and the ED

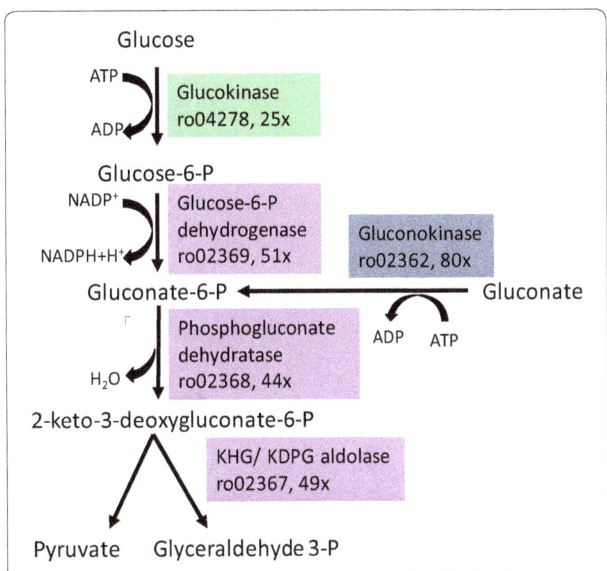

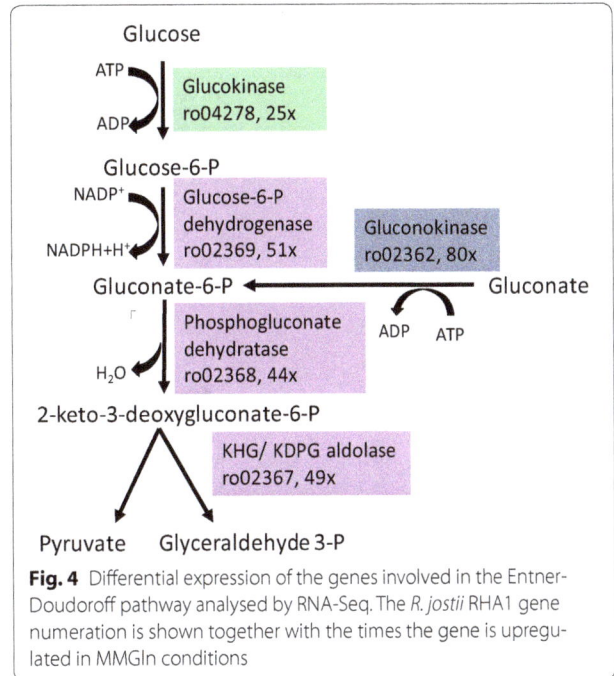

Fig. 4 Differential expression of the genes involved in the Entner-Doudoroff pathway analysed by RNA-Seq. The *R. jostii* RHA1 gene numeration is shown together with the times the gene is upregulated in MMGln conditions

pathway in *R. jostii*. TAG accumulation in MM containing either glucose or gluconate as carbon source was evaluated by fluorescence measurements using red nile and the Victor-3 fluorometer system (Perkin Elmer). We observed that glucose was also able to induce TAG accumulation in *R jostii*, but to a lower extent than gluconate (data not shown). Two likely hypotheses to explain this are: (i) only gluconate is able to induce the ED pathway and glucose is metabolized to TAG by the EM pathway, or (ii) glucose is also metabolized by the ED pathway but with a slightly lower yield, because glucose has to be transformed first to gluconate.

To check if glucose was also able to activate the ED pathway under nutrient-limiting conditions, we used RT-qPCR to measure the expression of the most upregulated genes involved in the ED pathway. The expression of these genes was compared in RM and in MM with gluconate or glucose as carbon source. As shown in Table 4, the three selected genes (ro2369: glucose-6-phosphate 1-dehydrogenase, ro02367: KHG/KDPG aldolase, and ro02362: gluconokinase) were again highly upregulated when gluconate was used as carbon source in the nutrient-limited medium. Interestingly, similar upregulation was observed when the MM contained glucose instead of gluconate. Thus, the ED is also activated with glucose as carbon source supporting that the activation is due to the metabolic stress and not due to the use of gluconate as carbon source. We have selected the gene ro00588 (cold shock protein) as control or housekeeping gene. Expression of this gene led to a 1.008 fold change (MMGln vs RM) in

Table 3 A subset of the *R. jostii* RHA1 most upregulated genes in the MMGln nutrient-deprived medium

Gene ID	RPK RM	RPK MMGln	FoldChange	Protein name
RHA1_ro02363	1190	144,105	121	Gluconate permease family protein
RHA1_ro04139	1035	101,265	98	Metabolite transporter, MFS superfamily
RHA1_ro06058	610	57,816	95	Possible ATP-dependent protease
RHA1_ro06057	1465	128,281	88	Probable 1,3-propanediol dehydrogenase
RHA1_ro02362	2544	203,714	80	Probable gluconokinase
RHA1_ro02369	2036	105,557	52	Glucose-6-phosphate 1-dehydrogenase
RHA1_ro04138	1740	89,076	51	Possible hydratase
RHA1_ro02367	2703	133,001	49	KHG/KDPG aldolase
RHA1_ro02368	2874	126,594	44	Phosphogluconate dehydratase
RHA1_ro06059	569	22,385	39	Hypothetical protein
RHA1_ro04137	702	20,726	30	Reductase
RHA1_ro04278	2020	49,893	25	Glucokinase
RHA1_ro04140	3796	86,218	23	Probable phosphoglycerate dehydrogenase
RHA1_ro02361	2216	49,009	22	Probable lipase
RHA1_ro03288	1117	21,311	19	Probable glutamate dehydrogenase (NAD(P) +)
RHA1_ro04279	2964	52,890	18	Possible transcriptional regulator, WhiB family
RHA1_ro03287	1634	28,443	17	Hypothetical protein
RHA1_ro06083	1756	30,473	17	Probable ethanolamine permease, APC superfamily
RHA1_ro01051	1414	23,886	17	Hypothetical protein

Table 4 qRT-PCR evaluation of the ED pathway gene expression in MM medium containing glucose or gluconate as sole carbon source

Gene	Annotation	Carbon source	Ct[a]	ΔCt[b]	FoldChange
ro02362	Probable gluconokinase	RM	30.254	–	
		MMGln	25.539	5.135	35.14
		MMGls	26.316	4.335	20.18
ro02367	KHG/KDPG aldolase	RM	24.194	–	
		MMGln	20.708	3.378	10.39
		MMGls	21.324	2.762	6.78
ro02369	Glucose-6-P 1-dehydrogenase	RM	24.084	–	
		MMGln	20.857	3.129	8.75
		MMGls	22.018	2.478	5.57
ro00588	Cold shock protein	RM	21.195	–	
		MMGln	21.086	−0.109	0.92
		MMGls	21.596	0.401	1.32

[a] Ct is the cycle threshold or number of cycles requires for the fluorescence signal to cross the threshold. The Cts shown are the mean of three experiments

[b] ΔCt = Ct (MM) − Ct (RM)

RNA-Seq and it was also almost unaffected in any of the three used media in the RT-qPCR experiment (Table 4).

We have also analysed the enzymatic activity of the KHG/KDPG aldolase in crude extracts of *R. jostii RHA1* grown on MMGls or RM as described in Methods. In accordance with the transcriptomic results, KDPG aldolase activity (Additional file 2: Figure S1) was 8.75 times higher in MMGls (3.5 nmol/s/mg) than in RM (0.4 nmol/s/mg).

Catabolism of the carbon source (either glucose or gluconate) by the ED pathway renders two moles of pyruvate per mole of carbon source. One mole of ATP is generated also. However, generation of reduced coenzymes depends on the carbon source. Whereas catabolism of 1 mol of glucose by the ED pathway generates 1 mol NADPH and 1 mol NADH, catabolism of gluconate generates only 1 mol NADH (see below).

Energy and redox metabolism in *R. jostii* RHA1 cells grown in MMGln

More than 30 genes that code for proteins of the oxidative phosphorylation process are upregulated and none of these genes is downregulated (Fig. 3). More specifically, the upregulated genes mainly code for subunits of the complex I or NADH dehydrogenase, while the genes of the F1-ATPase remain unchanged. Hence, respiratory activity may provide part of the ATP required for TAG biosynthesis.

The highest transcriptional repression was observed for the ro03923 gene coding for a NADPH dehydrogenase (Table 5). Oxidation of glucose to pyruvate by the EMP has a net yield of 2 ATP and 2 NADH per molecule of glucose. In contrast, if the ED pathway is used, the net yield is 1 ATP, 1 NADH and 1 NADPH per molecule of glucose. It should be pointed out here that if, instead of glucose, gluconate is oxidized by the ED pathway, the net yield should be 1 ATP and 1 NADH per molecule of gluconate (see Fig. 4). According to [37], the synthesis of fatty acids requires stoichiometric amounts of ATP and acetyl-CoA, NADPH and NADH for each C2 addition. Considering that catabolism of gluconate to pyruvate by the ED pathway renders NADH and not NADPH, there is a requirement for this latter reduced coenzyme for TAG biosynthesis. This may explain the downregulation of the NADPH dehydrogenase (ro03923, 0.06x).

Different metabolic pathways lead to acetyl-CoA generation from pyruvate. Pyruvate dehydrogenase, partially repressed, may account for the conversion of a fraction of the total pyruvate available to acetyl-CoA. Induction of other enzymes, such as acetyl-CoA synthase (8 homologs in RHA1 like ro04332 and ro11190, 6.9× and 5.9× upregulated, respectively) (Additional file 1: Table S1), that can generate acetyl-CoA from acetate without a requirement for NAD^+ suggests that a fraction of the available pyruvate could be converted to acetyl-CoA by enzymes that do not generate NADH.

Induction of the Kennedy pathway for TAG accumulation

The glyceraldehyde-3-phosphate generated by the ED enzyme KDPG aldolase could be used for pyruvate formation, but also for conversion to dihydroxyacetone-phosphate by a reaction catalyzed by the triose-phosphate isomerase enzyme (TpiA). Then, the dihydroxyacetone-phosphate intermediate may be converted into glycerol-3-phosphate by a NAD(P)-dependent glycerol-3-phosphate dehydrogenase enzyme (GpsA). Glycerol-3-phosphate is later sequentially acylated, after removing the phosphate group, to form TAG (Kennedy pathway). Interestingly, the genes *tpiA* (ro07179, 1.76×) and *gpsA* (ro06505, 1.78×) were both upregulated to some extent by cells during cultivation in nutrient starvation conditions. Moreover, genes involved in the de novo fatty acid biosynthesis were also upregulated. An acetyl-CoA carboxylase enzyme (ACC) coded by ro04222 (2.36×) was significantly induced in starved

Table 5 A subset of the *R. jostii* RHA1 most downregulated genes in the MMGln nutrient-deprived medium

Gene ID	RPK RM	RPK MMGln	FoldChange	Protein name
RHA1_ro04379	10,415	1519	0.146	Transcriptional regulator, GntR family
RHA1_ro04433	18,405	2666	0.145	Hypothetical protein
RHA1_ro03412	1295	183	0.142	Hypothetical protein
RHA1_ro02813	65,807	9173	0.139	Probable NADP dependent oxidoreductase
RHA1_ro03320	27,980	3784	0.135	Pyruvate dehydrogenase E1 component beta subunit
RHA1_ro04380	9371	1267	0.135	Probable multidrug resistance transporter, MFS superfamily
RHA1_ro01994	57,602	7661	0.133	Probable succinate-semialdehyde dehydrogenase (NAD(P) +)
RHA1_ro05024	76,619	10,173	0.133	Reductase
RHA1_ro03319	29,880	3927	0.131	Dihydrolipoyllysine-residue acetyltransferase, E2 component of pyruvate dehydrogenase complex
RHA1_ro03811	58,271	7458	0.128	Probable carboxylesterase
RHA1_ro03321	39,387	4982	0.126	Pyruvate dehydrogenase E1 component alpha subunit
RHA1_ro06364	18,591	2126	0.114	Probable cyanate transporter, MFS superfamily
RHA1_ro03916	27,486	2996	0.109	Hypothetical protein
RHA1_ro01380	88,898	9590	0.108	Hypothetical protein
RHA1_ro03318	48,005	5041	0.105	Dihydrolipoyl dehydrogenanse
RHA1_ro03207	21,864	1671	0.076	Hypothetical protein
RHA1_ro03206	48,313	3638	0.075	Dehydrogenase
RHA1_ro03208	38,493	2729	0.071	Polysaccharide deacetylase
RHA1_ro03923	62,548	3639	0.058	NADPH dehydrogenase

cells. ACC catalyzes the formation of malonyl-CoA molecules, which are used for fatty acid biosynthesis by the enzymatic complex known as fatty acid synthase I (FAS-I). FAS-I, a unique, large protein with different catalytic activities, is responsible for fatty acid biosynthesis in rhodococci, which are used for phospholipids and TAG synthesis. FAS-I coded by ro01426 (2.81×) was highly upregulated in cells under nutrient starvation conditions. Although the genes coding for several enzymes of the Kennedy pathway were not significantly upregulated in MMGln, some of the diacylglycerol acyltransferase genes were indeed upregulated (Fig. 5). The acyltransferase enzymes involved in the upper reactions of the Kennedy pathway were slightly upregulated in MMGln, such as ro05648 (GPAT) 1.99×, ro01115 (AGPAT) 1.67×, and ro05647 (AGPAT) 1.70× (Fig. 5 and Additional file 1: Table S1). Wax ester synthase/acyl coenzyme A:diacylglycerol acyltransferases (WS/DGATs) are key bacterial enzymes that catalyze the final step of TAG biosynthesis (acylation of DAG intermediates). Fourteen WS/DGAT genes were identified in *R. jostii* [21]. The WS/DGAT genes ro05356 (Atf8) and ro02966 (Atf7) were upregulated almost sixfold and fourfold, respectively. Indeed, *atf8* transcripts were also the most abundant WS/DGAT transcripts during RHA1 grow on benzoate under nitrogen-limiting conditions, being this enzyme determinant for TAG accumulation [16]. Moreover, the genes ro01601 (Atf6) and ro05649 (Atf9) were expressed 2 times more in MMGln than in RM. These four WS/DGAT enzymes are expected to be specifically involved in the TAG synthesis. Finally, ro02104 (*tadA*), another gene described to be involved in TAG accumulation, was upregulated 3.7 times in MMGln (Additional file 1: Table S1). TadA is a predicted apolipoprotein associated with lipid droplets in *R. jostii RHA1* [38] and *R. opacus PD630* [33]. *TadA* mutant was described to accumulate 30–40% less TAG than the parental *R. opacus PD630* strain [39]. This protein may mediate lipid body formation in TAG-accumulating rhodococcal cells with a similar structural role than apolipoproteins in eukaryotes [39].

Putative CRP binding sites are present in the highly expressed genes

Alternative sigma factors such as sigma54 are widely used in bacteria as a quick response to cope with environmental changes such as nutrient deprivation. To find if these alternative factors are being used for the upregulation of the *R. jostii* genes in MMGln, the program BPROM (http://www.softberry.com/) for the recognition of sigma70 promoters was used with the 150 bp

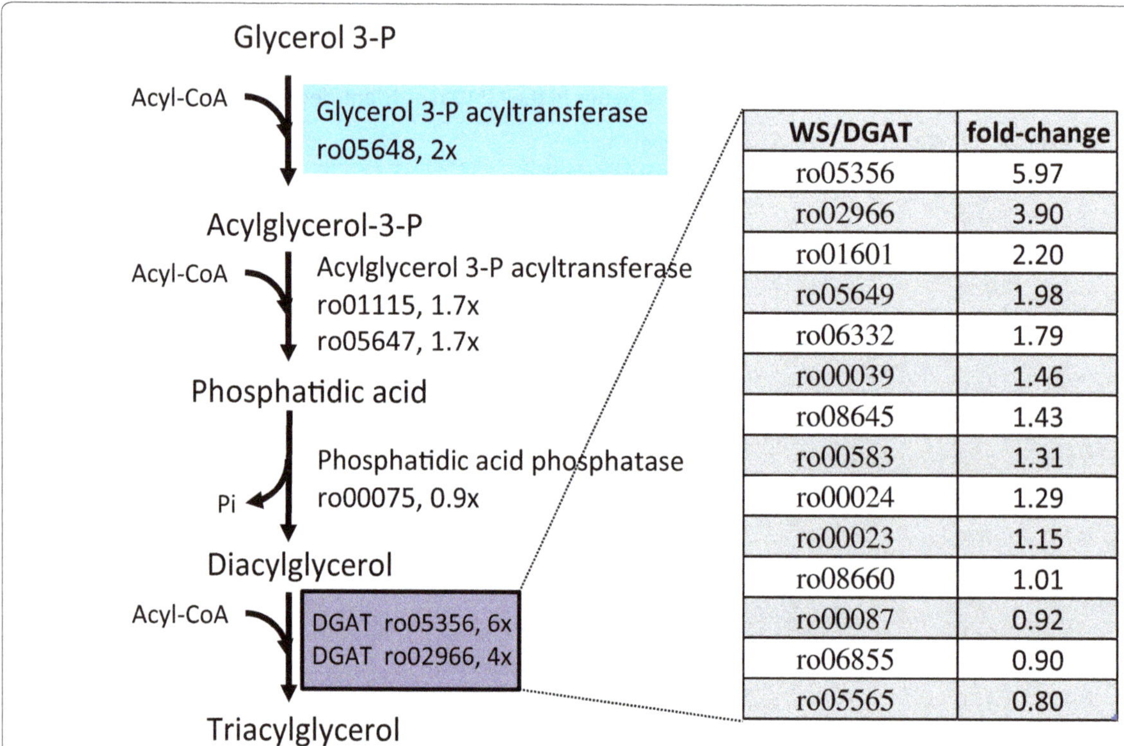

WS/DGAT	fold-change
ro05356	5.97
ro02966	3.90
ro01601	2.20
ro05649	1.98
ro06332	1.79
ro00039	1.46
ro08645	1.43
ro00583	1.31
ro00024	1.29
ro00023	1.15
ro08660	1.01
ro00087	0.92
ro06855	0.90
ro05565	0.80

Fig. 5 Differential expression of the genes involved in the Kennedy pathway for TAG synthesis analysed by RNA-Seq. The expression of the 14 putative *R. jostii* WS/DGAT genes is shown

immediately upstream from each ORF start. A putative sigma70 binding site was found in most upregulated genes. Hence, regulatory element(s) alternative to sigma70 subunit must be responsible for the transcriptional activation of the *R. jostii* genes in MMGln. These element(s) should target conserved binding sites in some of the altered genes.

The identification and localization of conserved sequences within the upstream regions of the upregulated genes was performed by the MEME Suite [40]. The consensus sequence 5′-GTGANNTGNGTCAC-3′ was found in almost every promoter region of the 40 highest upregulated genes, as shown in Additional file 1: Table S2 and Fig. 6a. This conserved sequence is identical to the cAMP Receptor Protein (CRP) consensus binding site found either in *E. coli* (5′-tGTGANNNNNNTCACa-3′, [41]) or *Pseudomonas aeruginosa* (5′-ANWWTGNGAWNYAGWTCACAT-3′ [42]. Moreover, the protein coded by ro04321 is 90% identical (Fig. 6b) to the corresponding CRP protein in *Mycobacterium tuberculosis* [43]. Structural modelling by Phyre 2 [44] of the putative *R. jostii* CRP correctly predicts a CRP fold with 223 residues (92%) modelled at >90% accuracy.

Bacterial CRPs are transcription factors that respond to cAMP by binding at target promoters when cAMP concentration increases. 254 CRP-binding sites have been found in *E. coli*, regulating at least 378 promoters [41]. In *R. jostii*, 371 putative CRP binding sites have been found (Additional file 1: Table S2). Thus, there is a CRP binding site per, approximately, each 25 genes. However, the density increases significantly up to 1 site per 4 genes in the genes that we identified as highly upregulated (eightfold or greater) when *Rhodococcus* cells grow in MMGln. Specifically, in all the promoters controlling genes involved in the ED pathway there is at least one CRP binding site. Most of these promoters are divergent promoters and both of the controlled operons are upregulated. Moreover, CRP binding sites have also been found in the promoter regions of the two main upregulated WS/DGAT genes (ro05356 and ro02966), but not in the promoter regions of the other WS/DGAT genes. Strikingly, the promoter regions of the most upregulated operons in *R. opacus* PD630 also contain a CRP putative binding sequence (Additional file 1: Table S3).

In *E. coli*, gluconate was shown to lower both CRP and cAMP to nearly the same extent as glucose [45]. Hence, it is likely that in *R. jostii*, the predicted cAMP increase, rather than being related to the carbon source, is related to the stress generated by depletion of nutrients.

We also searched for the presence of a CRP binding site in the upstream regulatory region of the orthologs of the 40 *Rhodococcus* genes in other microorganisms using the MEME Suite (Additional file 1: Table S4). According to the results, it seems that the CRP mediated activation of the ED pathway is only conserved in *R. opacus*, also an oleogenic rhodococci. CRP binding sites were also found in the promoter regions of a few genes in the other two *Rhodococcus* genomes analyzed (*R. equi* and *R. erythropolis*). However, no consensus CRP binding sequence was found in the promoter regions of the orthologous genes in *Escherichia coli* or *Pseudomonas putida*. We have also searched without success for CRP binding sites in similar operons of non-oleaginous organisms containing WS/DGAT enzymes, such as *Mycobacterium tuberculosis*, *Acinetobacter baumanii* or *Marinobacter aquaolei*. Thus, it seems the upregulation of these *R. jostii* genes by CRP is related to the TAG accumulation.

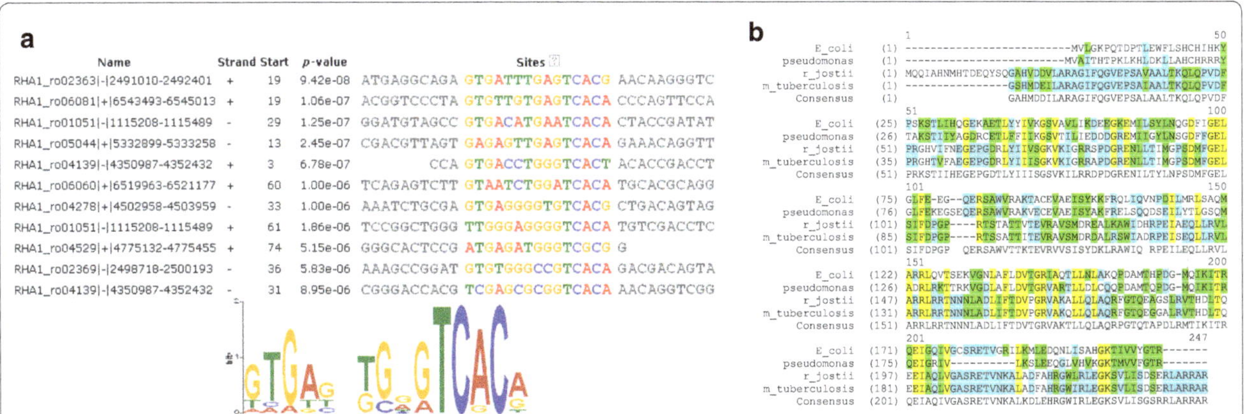

Fig. 6 **a** Conserved sequences found by using the meme program within the 11 most upregulated *R. jostii* promoters in MMGln. The consensus sequence is also shown. **b** Alignment of the *R. jostii* putative CRP sequence (YP_704269) with the CRP sequences of *E. coli* (PDB 1O3Q), *P. Aeruginosa* (PDB 2OZ6) and *M. tuberculosis* (PDB 3D0S)

Conclusions

Different microorganisms are able to accumulate TAGs or other neutral lipids to serve as carbon and energy sources during starvation. One of these microorganisms is *R. jostii* strain RHA1. Transcriptomic analysis of *R. jostii* RHA1 under conditions that lead or do not lead to TAG accumulation allowed us to identify the metabolic pathways that are relevant for oxidation of the carbon source, biosynthesis and TAG accumulation under nutrient-deprivation.

Two interesting results arose from our work. First, under nutrient-deprivation, *Rhodococcus* metabolizes carbohydrates such as glucose or gluconate by the Entner-Doudoroff pathway. Up- or downregulation of other key enzymes (i.e., pyruvate dehydrogenase, acetyl CoA synthetase, NADH oxidase), provides the ATP, reducing equivalents and building blocks for TAG synthesis. Second, the metabolic shift is likely driven by an increase in cAMP concentration that activates the expression of several operons via CRP.

Both observations could help in engineering metabolic modifications to improve TAG yield for biotechnological applications.

Additional files

Additional file 1: Table S1. RPK values of all genes. **Table S2.** CRP binding sequences in *R. jostii*. **Table S3.** *R. jostii* RHA1comparison to *R. opacus* PD630. **Table S4.** CRP binding sequences in rhodoccoci.

Additional file 2: Figure S1. (A) Kinetic determination of KDPG aldolase activity in MMGls and RM. (B). KDPG aldolase activity calculated from the kinetic curves. Error bars show the standard deviation from three independent experiments.

Authors' contributions

AJ analysed data and wrote the article, JAV performed the experiments, VF carried out the bioinformatics analysis, BL performed the experiments, FC wrote the article, HMA analysed data and wrote the article and GM designed research, analysed data and wrote the article. All authors read and approved the final manuscript.

Author details

[1] Institut de Bioenginyeria de Catalunya, Parc Científic de Barcelona, 08028 Barcelona, Spain. [2] Departamento de Microbiología, Facultad de Biología, Universidad de Barcelona, Avda Diagonal, 643., 08028 Barcelona, Spain. [3] Departamento de Biología Molecular (Universidad de Cantabria) and Instituto de Biomedicina y Biotecnología de Cantabria IBBTEC (CSIC-UC), C/Albert Einstein 22, 39011 Santander, Spain. [4] INBIOP (Instituto de Biociencias de la Patago nia), Consejo Nacional de Investigaciones Científicas y Técnicas, Facultad de Ciencias Naturales, Universidad Nacional de la Patagonia San Juan Bosco, Ruta Provincial No 1, Km 4-Ciudad Universitaria 9000, Comodoro Rivadavia, Chubut, Argentina.

Acknowledgements

We are grateful to Dr. Juan Maria Garcia-Lobo and Dr. Maria Cruz Rodriguez for RNA-Seq analysis performed in the massive sequencing service at the IBBTEC. We thank Dr. Lindsay Eltis for the gift of the strain *R. jostii* RHA1.

Competing interests

The authors declare that they have no competing interests.

Funding

This work was financed by Grants BIO2010-14809 from the Spanish Ministry of Science and Innovation and BFU2014-55534-C2-2-P from the Spanish Ministry of Economy and Competitiveness to GM. H.M. Alvarez is a career investigator of the Consejo Nacional de Investigaciones Científicas y Técnicas (CONICET), Argentina.

References

1. Charpe TW, Rathod VK. Biodiesel production using waste frying oil. Waste Manag. 2011;31:85–90.
2. Hawley C. Criticism mounts against biofuels. BusinessWeek: Europe [Internet]. 2008. http://www.businessweek.com/globalbiz/content/jan2008/gb20080124_071995.htm. Accessed 24 Oct 2011.
3. Shi S, Valle-Rodríguez JO, Siewers V, Nielsen J. Prospects for microbial biodiesel production. Biotechnol J. 2011;6:277–85.
4. Ratledge C, Cohen Z. Microbial and algal oils: do they have a future for biodiesel or as commodity oils? Lipid Technol. 2008;20:155–60.
5. Vicente G, Bautista LF, Gutiérrez FJ, Rodríguez R, Martínez V, Rodríguez-Frómeta RA, et al. Direct transformation of fungal biomass from submerged cultures into biodiesel. Energy Fuels. 2010;24:3173–8.
6. Cohen Z, Ratledge C. Single cell oils: microbial and algal oils. 2nd ed. Champaign: AOCS Publishing; 2010.
7. Steinbüchel A, Hustede E, Liebergesell M, Pieper U, Timm A, Valentin H. Molecular basis for biosynthesis and accumulation of polyhydroxyalkanoic acids in bacteria. FEMS Microbiol Lett. 1992;103:217–30.
8. Alvarez HM, Steinbüchel A. Triacylglycerols in prokaryotic microorganisms. Appl Microbiol Biotechnol. 2002;60:367–76.
9. Kalscheuer R, Steinbüchel A. A novel bifunctional wax ester synthase/acyl-CoA:diacylglycerol acyltransferase mediates wax ester and triacylglycerol biosynthesis in *Acinetobacter calcoaceticus* ADP1. J Biol Chem. 2003;278:8075–82.
10. Daniel J, Deb C, Dubey VS, Sirakova TD, Abomoelak B, Morbidoni HR, et al. Induction of a novel class of diacylglycerol acyltransferases and triacylglycerol accumulation in *Mycobacterium tuberculosis* as it goes into a dormancy-like state in culture. J Bacteriol. 2004;186:5017–30.
11. Kaddor C, Biermann K, Kalscheuer R, Steinbüchel A. Analysis of neutral lipid biosynthesis in *Streptomyces avermitilis* MA-4680 and characterization of an acyltransferase involved herein. Appl Microbiol Biotechnol. 2009;84:143–55.
12. Alvarez HM, Kalscheuer R, Steinbüchel A. Accumulation and mobilization of storage lipids by *Rhodococcus opacus* PD630 and *Rhodococcus ruber* NCIMB 40126. Appl Microbiol Biotechnol. 2000;54:218–23.
13. Gouda MK, Omar SH, Aouad LM. Single cell oil production by *Gordonia* sp. DG using agro-industrial wastes. World J Microbiol Biotechnol. 2008;24:1703–11.
14. Bell KS, Philp JC, Aw DWJ, Christofi N. A review: the genus *Rhodococcus*. J Appl Microbiol. 1998;85:195–210.
15. van der Geize R, Dijkhuizen L. Harnessing the catabolic diversity of rhodococci for environmental and biotechnological applications. Curr Opin Microbiol. 2004;7:255–61.
16. Ahmad M, Taylor CR, Pink D, Burton K, Eastwood D, Bending GD, et al. Development of novel assays for lignin degradation: comparative analysis of bacterial and fungal lignin degraders. Mol BioSyst. 2010;6:815–21.
17. Alvarez HM. Biotechnological Production and Significance of Triacylglycerols and Wax Esters. In: Timmis KN, editor. Handbook of Hydrocarbon and Lipid Microbiology [Internet]. Springer Berlin Heidelberg; 2010. p. 2995–3002. http://link.springer.com/referenceworkentry/10.1007/978-3-540-77587-4_222. Accessed 7 Feb 2017.
18. Seto M, Kimbara K, Shimura M, Hatta T, Fukuda M, Yano K. A novel transformation of polychlorinated biphenyls by *Rhodococcus* sp. strain RHA1. Appl Environ Microbiol. 1995;61:3353–8.
19. Navarro-Llorens JM, Patrauchan MA, Stewart GR, Davies JE, Eltis LD, Mohn WW. Phenylacetate catabolism in *Rhodococcus* sp. strain RHA1: a central pathway for degradation of aromatic compounds. J Bacteriol. 2005;187:4497–504.
20. Patrauchan MA, Florizone C, Dosanjh M, Mohn WW, Davies J, Eltis LD. Catabolism of benzoate and phthalate in *Rhodococcus* sp. strain RHA1: redundancies and convergence. J Bacteriol. 2005;187:4050–63.

21. Hernández M, Mohn W, Martínez E, Rost E, Alvarez A, Alvarez H. Biosynthesis of storage compounds by *Rhodococcus jostii* RHA1 and global identification of genes involved in their metabolism. BMC Genom. 2008;9:600.
22. McLeod MP, Warren RL, Hsiao WWL, Araki N, Myhre M, Fernandes C, et al. The complete genome of *Rhodococcus* sp. RHA1 provides insights into a catabolic powerhouse. Proc Natl Acad Sci USA. 2006;103:15582–7.
23. Sambrook J. Molecular cloning: a laboratory manual, Third Edition. 3rd ed. New york: Cold Spring Harbor Laboratory Press; 2001.
24. Langmead B, Trapnell C, Pop M, Salzberg S. Ultrafast and memory-efficient alignment of short DNA sequences to the human genome. Genome Biol. 2009;10:R25.
25. Li H, Handsaker B, Wysoker A, Fennell T, Ruan J, Homer N, et al. The sequence alignment/map format and SAMtools. Bioinformatics. 2009;25:2078–9.
26. Rutherford K, Parkhill J, Crook J, Horsnell T, Rice P, Rajandream MA, et al. Artemis: sequence visualization and annotation. Bioinformatics. 2000;16:944–5.
27. Bullard J, Purdom E, Hansen K, Dudoit S. Evaluation of statistical methods for normalization and differential expression in mRNA-Seq experiments. BMC Bioinform. 2010;11:94.
28. Anders S, Huber W. Differential expression analysis for sequence count data. Genome Biol. 2010;11:R106.
29. Chen X, Schreiber K, Appel J, Makowka A, Fähnrich B, Roettger M, et al. The Entner-Doudoroff pathway is an overlooked glycolytic route in cyanobacteria and plants. Proc Natl Acad Sci USA. 2016;113:5441–6.
30. Alvarez HM, Mayer F, Fabritius D, Steinbüchel A. Formation of intracytoplasmic lipid inclusions by *Rhodococcus opacus* strain PD630. Arch Microbiol. 1996;165:377–86.
31. Wall ME, Hlavacek WS, Savageau MA. Design of gene circuits: lessons from bacteria. Nat Rev Genet. 2004;5:34–42.
32. Kanehisa M, Goto S, Sato Y, Furumichi M, Tanabe M. KEGG for integration and interpretation of large-scale molecular data sets. Nucleic Acids Res. 2011;40:D109–14.
33. Chen Y, Ding Y, Yang L, Yu J, Liu G, Wang X, et al. Integrated omics study delineates the dynamics of lipid droplets in *Rhodococcus opacus* PD630. Nucleic Acids Res. 2014;42:1052–64.
34. Dávila Costa JS, Herrero OM, Alvarez HM, Leichert L. Label-free and redox proteomic analyses of the triacylglycerol-accumulating *Rhodococcus jostii* RHA1. Microbiology (Reading, Engl.). 2015;161:593–610.
35. Conway T. The Entner-Doudoroff pathway: history, physiology and molecular biology. FEMS Microbiol Lett. 1992;103:1–27.
36. Peekhaus N, Conway T. What's for dinner?: Entner-Doudoroff metabolism in *Escherichia coli*. J Bacteriol. 1998;180:3495–502.
37. Alvarez H. Biology of Rhodococcus. Berlin, Heidelberg: Springer; 2010.
38. Ding Y, Yang L, Zhang S, Wang Y, Du Y, Pu J, et al. Identification of the major functional proteins of prokaryotic lipid droplets. J Lipid Res. 2012;53:399–411.
39. MacEachran DP, Prophete ME, Sinskey AJ. The *Rhodococcus opacus* PD630 heparin-binding hemagglutinin homolog TadA mediates lipid body formation. Appl Environ Microbiol. 2010;76:7217–25.
40. Bailey TL, Boden M, Buske FA, Frith M, Grant CE, Clementi L, et al. MEME SUITE: tools for motif discovery and searching. Nucleic Acids Res. 2009;37:W202–8.
41. Shimada T, Fujita N, Yamamoto K, Ishihama A. Novel roles of cAMP receptor protein (CRP) in regulation of transport and metabolism of carbon sources. PLoS ONE. 2011;6:e20081.
42. Kanack KJ, Runyen-Janecky LJ, Ferrell EP, Suh S-J, West SEH. Characterization of DNA-binding specificity and analysis of binding sites of the *Pseudomonas aeruginosa* global regulator, Vfr, a homologue of the *Escherichia coli* cAMP receptor protein. Microbiology (Reading, Engl.). 2006;152:3485–96.
43. Stapleton M, Haq I, Hunt DM, Arnvig KB, Artymiuk PJ, Buxton RS, et al. Mycobacterium tuberculosis cAMP receptor protein (Rv3676) differs from the *Escherichia coli* paradigm in Its cAMP binding and DNA binding properties and transcription activation properties. J Biol Chem. 2010;285:7016–27.
44. Kelley LA, Sternberg MJE. Protein structure prediction on the Web: a case study using the Phyre server. Nat Protoc. 2009;4:363–71.
45. Hogema BM, Arents JC, Inada T, Aiba H, Van Dam K, Postma PW. Catabolite repression by glucose 6-phosphate, gluconate and lactose in *Escherichia coli*. Mol Microbiol. 1997;24:857–67.

17

A novel strategy to improve protein secretion via overexpression of the SppA signal peptide peptidase in *Bacillus licheniformis*

Dongbo Cai[1†], Hao Wang[1†], Penghui He[1], Chengjun Zhu[1], Qin Wang[1], Xuetuan Wei[2], Christopher T. Nomura[1,3] and Shouwen Chen[1*]

Abstract

Background: Signal peptide peptidases play an important role in the removal of remnant signal peptides in the cell membrane, a critical step for extracellular protein production. Although these proteins are likely a central component for extracellular protein production, there has been a lack of research on whether protein secretion could be enhanced via overexpression of signal peptide peptidases.

Results: In this study, both nattokinase and α-amylase were employed as prototypical secreted target proteins to evaluate the function of putative signal peptide peptidases (SppA and TepA) in *Bacillus licheniformis*. We observed dramatic decreases in the concentrations of both target proteins (45 and 49%, respectively) in a *sppA* deficient strain, while the extracellular protein yields of nattokinase and α-amylase were increased by 30 and 67% respectively in a strain overexpressing SppA. In addition, biomass, specific enzyme activities and the relative gene transcriptional levels were also enhanced due to the overexpression of *sppA*, while altering the expression levels of *tepA* had no effect on the concentrations of the secreted target proteins.

Conclusions: Our results confirm that SppA, but not TepA, plays an important functional role for protein secretion in *B. licheniformis*. Our results indicate that the *sppA* overexpression strain, *B. licheniformis* BL10GS, could be used as a promising host strain for the industrial production of heterologous secreted proteins.

Keywords: *Bacillus licheniformis*, Signal peptide peptidase, Protein secretion, *sppA*, *tepA*

Background

Heterologous expression is an effective strategy to improve the production of secreted proteins. Many industrial enzymes (proteases, α-amylase, etc.) have been produced by heterologous expression in bacteria and fungi [e.g. *Escherichia coli*, *Bacillus* species, and *Saccharomyces*] [1–3]. Compared to other expression systems, *Bacillus* species are regarded as promising host strains with numerous advantages including: non-toxicity, convenience for gene modification and high yields of target proteins [4–6]. In particular, *Bacillus licheniformis* has been shown to be an effective host strain for protein production in previous studies [7, 8].

The genetic engineering of host strain to improve production is regarded as an efficient tactic due to its universality and efficiency [9, 10]. A number of examples of this effectiveness are available. For example, *B. licheniformis* BL10 was constructed as a highly efficient host strain for protein production by knocking out ten genes coding for extracellular proteases (Mpr, Vpr, AprX, Epr, Bpr, WprA, AprE, BprA), flagellin and amylase in *B. licheniformis* WX-02. Using this engineered *B. licheniformis* BL10, a previous study has demonstrated that nattokinase activity could be increased by 39% [7]. Also, the translocation elements (SecA, SecDF etc.), post-translocation

*Correspondence: mel212@126.com
†Dongbo Cai and Hao Wang contributed equally to this work
[1] Hubei Collaborative Innovation Center for Green Transformation of Bio-Resources, College of Life Sciences, Hubei University, No. 368 Youyi Avenue, Wuchang District, Wuhan 430062, Hubei, People's Republic of China
Full list of author information is available at the end of the article

chaperone PrsA and signal peptidase I have been engineered for production of recombinant protective antigen, α-amylase, subtilisin and Dsrs [9, 11–14].

In general, secreted proteins are extracellularly transported through three steps: protein synthesis, translocation, and release. Signal peptides play an important role in protein secretion, these peptide sequences are cleaved from the mature protein by signal peptidase I during the late stage of the secretion process. The cleaved remnant signal peptides are left in the cell membrane and could affect protein transportation [15–17]. Signal peptide peptidase (SPP) is a membrane-bound enzyme that uses a serine/lysine catalytic dyad mechanism to cleave the remnant signal peptides in the cellular membrane and aids in protein secretion [18, 19]. TepA and SppA have been previously identified as putative signal peptide peptidases in *B. subtilis*, and the concentrations of secreted target proteins decreased dramatically for a *sppA*-null strain, which suggested that SppA was involved in protein secretion in *B. subtilis* [18]. Despite the central role for SppA in protein secretion, prior to our study, there have been no studies focused on the improvement of protein secretion via overexpression of SPP.

In this study, both nattokinase and α-amylase were selected as prototype target proteins for secretion to evaluate the function of SppA and TepA in *B. licheniformis* BL10, derived from the native strain *B. licheniformis* WX-02 [7]. Our results identified and confirmed that SppA is the main SPP for protein secretion, and the concentrations of total extracellular proteins and target proteins increased significantly for a strain overexpressing *sppA*. As a result of these studies, the host strain *B. licheniformis* BL10GS was successfully constructed as a highly efficient platform for protein secretion.

Methods

Bacterial strains and plasmids

The bacterial strains and plasmids used in this study were listed in Table 1. *B. licheniformis* BL10 was employed as the original strain for gene modification, and *E. coli* DH5α was served as the host strain for plasmid construction. The expression vector pHY300PLK was used for constructing protein expression vectors, and T_2(2)-Ori was applied for constructing the gene knockout vectors and integrating expression vectors in this study.

Media and culture conditions

The LB medium was served as the basic medium for bacterial growth, and the corresponding titer of antibiotic (50 μg/mL ampicillin, 25 μg/mL tetracycline or 20 μg/mL kanamycin) was added into the medium if necessary. The ME medium for cell growth contains 20 g/L glucose, 10 g/L sodium citrate, 7 g/L NH_4Cl, 0.5 g/L $K_2HPO_4 \cdot 3H_2O$, 0.5 g/L $MgSO_4 \cdot 7H_2O$, 0.04 g/L $FeCl_3 \cdot 6H_2O$, 0.104 g/L $MnSO_4 \cdot H_2O$, 0.15 g/L $CaCl_2 \cdot 2H_2O$. The inoculum (3%) was added into 30 mL of ME media in a 250 mL flask, and was incubated at 180 rpm and 37 °C for 24 h. All the fermentation experiments were repeated at least three times.

The medium for α-amylase production was supplied as follow: 5 g/L corn starch, 5 g/L yeast extract, 10 g/L peptone, 12 g/L sodium citrate, 1 g/L $K_2HPO_4 \cdot 3H_2O$, 0.5 g/L $MgSO_4 \cdot 7H_2O$, 0.15 g/L $CaCl_2 \cdot 2H_2O$, pH 7.2; The medium for nattokinase production containing 20 g/L glucose, 10 g/L soy peptone, 10 g/L peptone, 10 g/L corn starch, 15 g/L yeast extract, 10 g/L sodium chloride, 3 g/L $K_2HPO_4 \cdot 3H_2O$, 6 g/L $(NH_4)_2SO_4$, pH 7.2. The inoculum (3%) was added into 30 mL of fermentation media in a 250 mL flask, and was incubated at 180 rpm and 37 °C for 48 h. All the fermentation experiments were repeated at least three times.

Gene knockout in *B. licheniformis*

The genes *sppA* and *tepA* were deleted individually in *B. licheniformis* BL10 according to our previously reported method [20, 21], and the construction procedure of the *sppA* deficient strain was described briefly as following. First, the upstream (a) and downstream (b) regions of *sppA* were amplified respectively by the corresponding primers sppA-KF1/sppA-KR1 and sppA-KF2/sppA-KR2 listed in Additional file 1: Table S1, and fused by splicing overlap extension (SOE)-PCR using the primers sppA-KF1/sppA-KR2. The fused fragments were cloned into T_2(2)-Ori at the restriction sites *XbaI*/*SacI*, diagnostic PCR and DNA sequencing confirmed that the vector, named T_2-sppA, was constructed successfully (Fig. 1).

Then, the vector T_2-sppA was electro-transferred into *B. licheniformis* BL10 and was verified by diagnostic PCR and plasmids extractions [22, 23]. The positive transformants were cultivated in the LB liquid medium with 20 μg/mL kanamycin at 45 °C, and subcultured for three generations, then transferred into kanamycin-free LB medium at 37 °C for another six generations. The cells were then plated on LB agar medium with or without kanamycin and incubated for 20 h, respectively. The primers sppA-KYF/sppA-KYR were used for PCR to verify the double crossover strains, and DNA sequencing confirmed that the *sppA* deficient strain (*B. licheniformis* BL10S) was constructed successfully. The *tepA* deficient strain (*B. licheniformis* BL10T) was constructed using methods similar to those employed for the construction of *B. licheniformis* BL10S.

Construction of the protein expression vector

The nattokinase expression vector used in this study was obtained from our previously reported research [7],

Table 1 The strains and plasmids used in this study

Strains	Relevant properties	Source of reference
Escherichia coli		
DH5α	*sup*E44 Δ*lac*U169 (f 80 *lacZ*ΔM15) *hsd* R17 *recA*1 *gyr*A96 *thi*1 *rel*A1	[7]
Bacillus licheniformis		
WX-02	Polyglutamate productive strain (CCTCC M208065)	CCTCC
BL10	WX-02(Δ*mpr*, Δ*vpr*, Δ*aprX*, Δ*epr*, Δ*bpr*, Δ*wprA*, Δ*aprE*, Δ*bprA*, Δ*hag*, Δ*amyL*)	[7]
BL10/pHY-amyL	BL10 harboring pHY-amyL	This study
BL10/pP43SacCNK	BL10 harboring pP43SacCNK(CCTCC M2014253)	CCTCC
BL10T	BL10(Δ*tepA*)	This study
BL10S	BL10(Δ*sppA*)	This study
BL10T/pHY-amyL	BL10T harboring pHY-amyL	This study
BL10S/pHY-amyL	BL10S harboring pHY-amyL	This study
BL10T/pP43SacCNK	BL10T harboring pP43SacCNK	This study
BL10S/pP43SacCNK	BL10S harboring pP43SacCNK	This study
BL10GT	Overexpression of *tepA* in BL10	This study
BL10GS	Overexpression of *sppA* in BL10	This study
BL10GT/pHY-amyL	BL10GT harboring pHY-amyL	This study
BL10GS/pHY-amyL	BL10GS harboring pHY-amyL	This study
BL10GT/pP43SacCNK	BL10GT pP43SacCNK	This study
BL10GS/pP43SacCNK	BL10GS pP43SacCNK	This study
Plasmids		This study
T_2(2)-ori	*Bacillus* knockout vector; Kan^r	[7]
T_2-sppA	T2(ori)-sppA(A + B); to knock out *sppA*	This study
T_2-tepA	T2(ori)-tepA(A + B); to knock out *tepA*	This study
T_2-GsppA	T2(ori)-sppA(A + B+*sppA*); to over-express *sppA*	This study
T_2-GtepA	T2(ori)-tepA(A + B+*tepA*); to over-express *tepA*	This study
pHY300PLK	*E. coli* and *B. s* shuttle vector; Amp^r, Tet^r	[7]
pP43SacCNK	PHY300PLK + Promotor-P43 (*B. subtilis* 168) + signal peptide of SacC (*B. subtilis* 168) + *aprN*(*B. subtilis* MBS 04-6) + Terminator of *amyL* (*B. licheniformis* WX-02)	[7]
pHY-amyL	PHY300PLK + Promotor-P43 (*B. subtilis* 168) + *amyL* (containing its own signal peptide and terminator, *B. licheniformis* WX-02)	This study

and the α-amylase expression vector was constructed according to the following method. P43 promoter was from *B. subtilis* 168. The gene *amyL* (FJ556804.1) coding for α-amylase (containing its own signal peptide and terminator) was amplified from the genome DNA of *B. licheniformis* WX-02 with the corresponding primers (Additional file 1: Table S1), purified and fused by SOE-PCR, and inserted into the expression vector pHY-300PLK at the restriction sites *Eco*RI and *Xba*I, plasmid extraction and DNA sequencing have confirmed that this new plasmid was constructed successfully, named pHY-amyL. The green fluorescent protein (GFP) expression vector (pHY-GFP) was constructed by using a similar method.

Gene integration in *B. licheniformis*

To develop strains to over express *sppA* and *tepA*, the individual genes *sppA* and *tepA* mediated by P43 promoter were integrated into the genome of *B. licheniformis* by the previous reported method [8]. The resulting strains *B. licheniformis* BL10GS, capable of overexpressing *sppA* and *B. licheniformis* BL10GT, capable of overexpressing *tepA*, were verified by PCR and used for further assays.

Enzyme activity assay

Two proteins, nattokinase and α-amylase, were served as the target proteins to evaluate the efficiency of host strains constructed in this research, and the methods for activity assay were described previously [8].

α-amylase activity assay

The mixture of 20 mL soluble starch solution (2 g/mL) and 5 mL 0.12 M phosphate buffer (pH 6) was added into the 50 mL centrifuge tube, and incubated at 60 °C for 8 min. The fermentation supernatant at a volume of 1 mL was added into the tube and incubated at 60 °C for another 10 min. Then, the 1-mL reaction solution

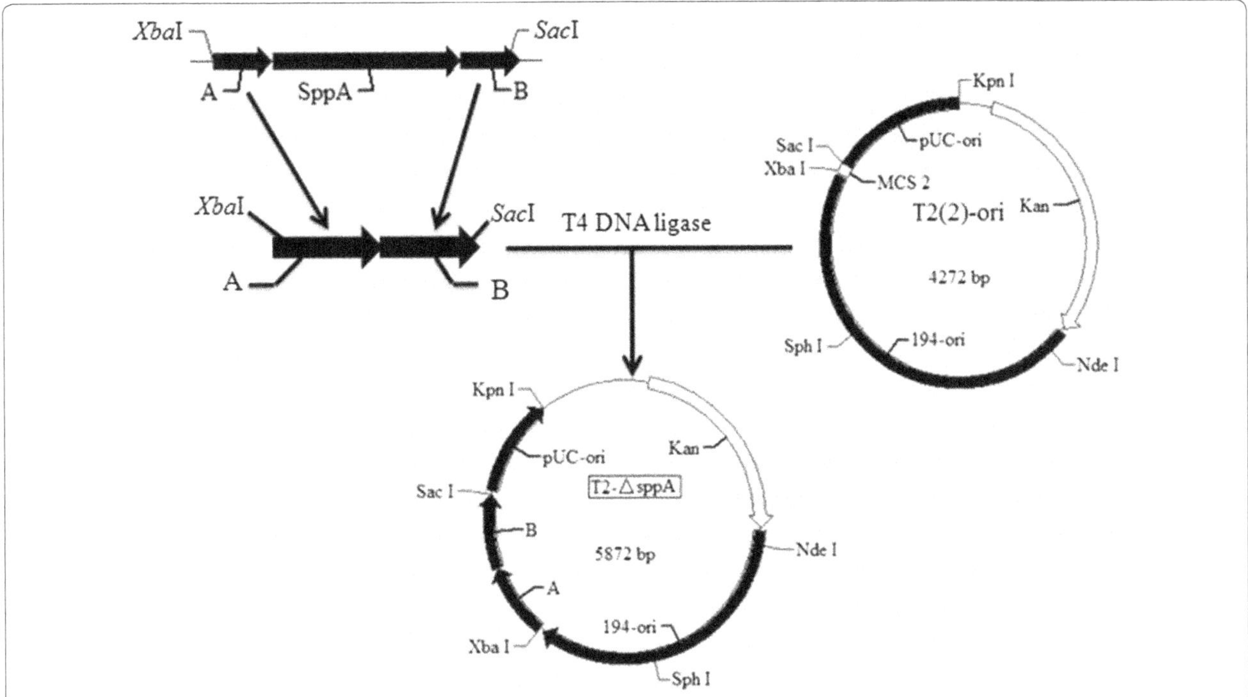

Fig. 1 The construction procedure of gene knockout vector. *A* and *B* represented the *upstream* and *downstream* homologous arms of *sppA*, respectively

was transferred into a new centrifuge tube containing 0.5 mL 0.1 mol/L HCl and 5 mL dilute iodine solution (266 mM KI and 0.35 mM I_2), and the absorbance was detected under 660 nm wavelength. The mixture containing 0.5 mL 0.1 mol/L HCl and 5 mL dilute iodine solution was served as the control. The α-amylase activity was calculated by the standard curve made from the different concentrations of starch solution (0.00, 0.10, 0.20, 0.30, 0.40, 0.50, 0.60, 0.70, 0.80, 0.90, 1.00 mg/mL). One Unit of α-amylase activity (U) was defined as the amount of enzyme for catalyzing 1 mg starch to release reducing sugar (glucose) per minute at 60 °C (pH 6) [24].

Nattokinase activity assay

The nattokinase activity was measured according to our previous reported method [8]: the mixture of 0.4 mL fibrinogen solution (0.72%, w/v) and 1.4 mL Tris–HCl (50 mM, pH 7.8) was incubated in a test tube at 37 °C for 10 min, followed by adding 0.1 mL thrombin solution (20 U/mL) to form the fibrin at 37 °C for 10 min. Then, the diluted sample solution (D) at the volume of 0.1 mL was added into the tube and shaken at 37 °C for 60 min at the interval of every 15 min, followed by adding 2 mL trichloroacetic acid (TCA) solution (0.2 M) to stop the reaction (AT). As for the control group, the mixture of 0.1 mL sample solution and 2 mL TCA solution (0.2 M) was used after incubating at 37 °C for 60 min (AB). Finally, all mixtures were centrifuged at 12000*g* for 10 min, and the absorbance of the supernatant was determined at 275 nm. One unit of nattokinase activity (1 FU) was defined as the amount of enzyme leading to the 0.01 increase for A_{275} in 1 min, and the formula was as follows:

$$\text{Nattokinase activity (FU/mL)} = (\text{AT} - \text{AB}) \times 100 \times \text{D}/6$$

Determination of the total extracellular protein and target protein concentrations

To determine the extracellular protein concentration, the volume of 100 μL fermentation broth was mixed with an equal volume of Coomassie Brilliant Blue G-250 and the absorbance was measured at 595 nm. The concentration of total extracellular protein was calculated by the standard curve made from different concentrations of bovine serum albumin (BSA) [25]. Meanwhile, the cells were washed three times with physiological saline, followed with sonication (pulse: 1 s on; 2 s off; total 4 min) to disrupt cells, and the concentration of intracellular protein was determined with the similar method.

For determining the target protein concentration, the mixture of 900 μL fermentation broth with 100 μL TCA (6.12 M) was maintained at 4 °C overnight, and centrifuged at 13,000*g* for 10 min. The precipitate was washed three times with ethanol and dried, and re-dissolved in a solution containing 2 M thiourea and 8 M urea. The sample was then mixed with equal volume of 2× SDS-PAGE loading buffer, and the volume of 10 μL solution was subjected to SDS-PAGE analysis. BSA was used as the standard. Protein concentration of the sample was calculated by comparing the area and pixel counts of the bands imaged from the gel with that of the standard, using a Bio-Rad GS-800 calibrated densitometer and Quantity One software [26].

Analysis of transcription level

The total RNA of the strain was extracted by the TRIzol® Reagent (Invitrogen, USA) according to our previously reported method [20], and contaminant DNA was digested by the Rnase-free DNase I enzyme (TaKaRa, Japan). The RevertAid First Strand cDNA Synthesis Kit (Thermo, USA) was applied to amplify the first stand of cDNA. The primers in the (seeing the Additional file: Table S2) were used for amplifying the corresponding genes in the 10 μL Real-Time PCR mixture (containing 5 μL SYBR® Select Master Mix, 0.5 μL primers, 1 μL cDNA, 3.5 μL DEPC water), and 16S rRNA from *B. licheniformis* BL10 was used as the reference gene to normalize the data. The Vii7 Real-Time PCR system (ABI, USA) was used for the Real-Time PCR reaction (95 °C for 3 min; 40 cycles of 95 °C for 30 s, 60 °C for 30 s). The transcriptional levels for genes in the recombinant strain were compared with those of the control strain after normalization to the reference gene 16S rRNA, and the experiments were performed in triplicate.

Statistical analyses

All samples were analyzed in triplicate, and the data were presented as the mean ± the standard deviation for each sample point. All data were conducted to analyze the variance at $P < 0.05$ and $P < 0.01$, and a *t* test was applied to compare the mean values using the software package Statistica 6.0 [27].

Results

Establishment of the signal peptide peptidase gene deficient strains

Based on the genome sequence and annotation of *B. licheniformis* WX-02, two genes (*sppA* and *tepA*) were predicted to function as SPPs in *B. licheniformis* WX-02 [28]. In order to investigate the function of these gene products, *sppA* and *tepA* were deleted in the parent strain *B. licheniformis* BL10, respectively. Figure 1 shows the construction procedure of the *sppA* knockout vector, and the bands amplified from mutants had the same length of the homologous arms without that of target genes (Additional file 1: Figure S1), confirming that *sppA* was deleted successfully. The new strain was named *B. licheniformis* BL10S. Similarly, *tepA* was also deleted using a similar method, and the resultant strain was named *B. licheniformis* BL10T.

Effects of signal peptide peptidase deficiency on the total extracellular protein secretion

The *sppA* and *tepA* deficient strains, as well as the parental strain BL10, were cultivated in ME media for 24 h, respectively, and the concentrations of total extracellular proteins were assayed to evaluate the importance of these genes on total extracellular protein secretion. As shown in Fig. 2a, the yield of total extracellular proteins decreased by 35% in the *sppA* deficient strain, compared to BL10. Meanwhile, there was no difference for the concentrations of total extracellular proteins produced by the *tepA* deficient strain and BL10, and these results were positively correlated with the SDS-PAGE results in Fig. 2b, which suggested that *sppA* gene product had the strongest effect on protein secretion in *B. licheniformis*.

Effects of signal peptide peptidase deficiency on the target protein production

In this study, nattokinase and α-amylase were selected as the target proteins to evaluate the function of *sppA* and *tepA*. Nattokinase expression vector pP43SacCNK was made in a previous study [7], and the α-amylase expression vector pHY-amyL harboring P43 promoter (K02174.1), *amyL* (FJ556804.1) coding for α-amylase (containing its own signal peptide and terminator) was verified by DNA sequence. The pP43SacCNK and pHY-amyL plasmids were individually electro-transferred into the parental strain *B. licheniformis* BL10, as well as the SPP gene deficient strains BL10S and BL10T. PCR verification and plasmid extraction confirmed that the recombinant strains were constructed successfully.

Furthermore, the extracellular activities of α-amylase and nattokinase produced by different strains were measured. As shown in Fig. 3, the extracellular activities of α-amylase and nattokinase were decreased by 51 and 48%, and the maximum biomass was decreased by 18 and 15% respectively due to the deficiency of *sppA*. Furthermore, the specific activities of α-amylase and nattokinase in BL10S were 17011.86 U/g_{DCW} and 3436.71 FU/g_{DCW} (1 $OD_{600} = 0.363$ g_{DCW}/L), which were 40 and 38% lower compared with those of BL10 (28238.63 U/g_{DCW} and 5585.39 FU/g_{DCW}), respectively. There were no differences in the yields of the target proteins and specific enzyme activities between the *tepA* deficient strain BL10T and BL10. These results suggest that SppA and not TepA, plays an important role for protein secretion. Additionally, the deficiency of *sppA* might not only affect the target protein yields in *B. licheniformis*, but also the cell biomass was decreased in the *sppA* deficient strains (Fig. 4a, c).

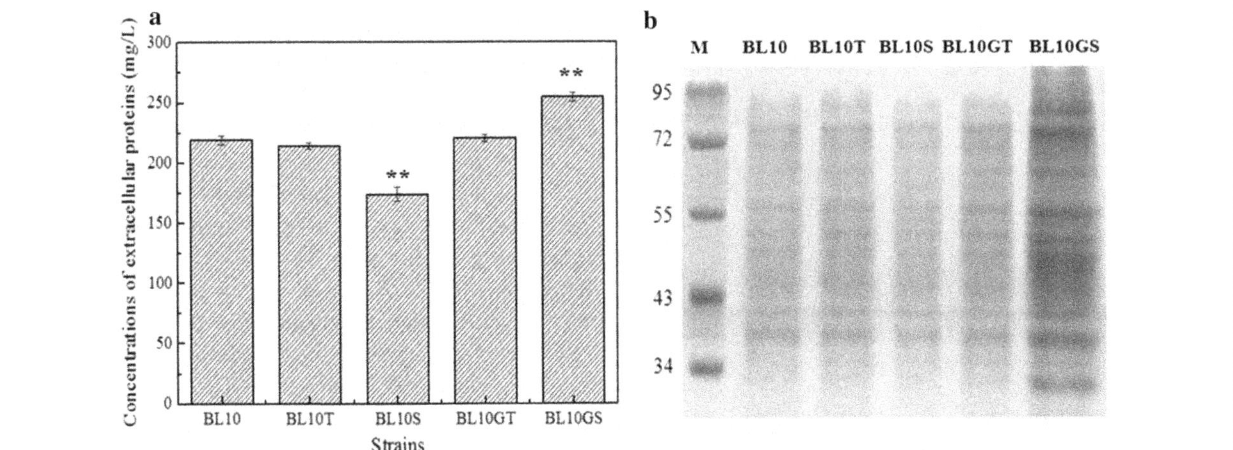

Fig. 2 Effects of deficiency or over-expression of *sppA* and *tepA* on the extracellular secretion. **a** The concentrations of total extracellular proteins of different strains; **b** SDS-PAGE analysis of the extracellular proteins from different host strains. *M* protein marker (95, 72, 55, 43, 34 kDa); *Lane 1 B. licheniformis* BL10; *Lane 2 B. licheniformis* BL10T; *Lane 3 B. licheniformis* BL10S; *Lane 4 B. licheniformis* BL10GT; *Lane 5 B. licheniformis* BL10GS. Data are represented as the means of three replicates and *bars* represent the standard deviations, $^{*}P < 0.05$; and $^{**}P < 0.01$ indicate the significance levels between recombinant strains and control strain

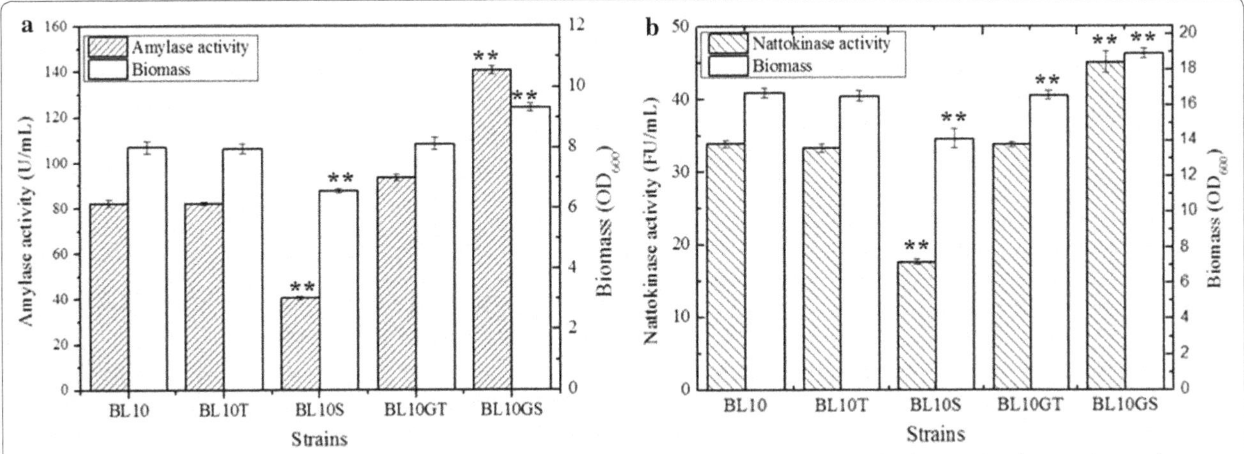

Fig. 3 The activities of target proteins and biomass of different host strains harboring pHY-amyL or pP43SacCNK. **a** α-Amylase activities and biomass of different host strains harboring pHY-amyL; **b** nattokinase activities and biomass of different host strains harboring pP43SacCNK. Data are represented as the means of three replicates and *bars* represent the standard deviations, $^{*}P < 0.05$; and $^{**}P < 0.01$ indicate the significance levels between recombinant strains and control strain

Effects of overexpression of signal peptide peptidase on the protein secretion

To verify the vital role of SPP in extracellular protein production, the genes *sppA* and *tepA* mediated by P43 promoter were further overexpressed in *B. licheniformis* BL10 by inserting the relevant expression cassettes into the chromosome by homologous recombination, respectively [8]. The *sppA* insert was verified by PCR and the bands amplified from the mutant strain exhibited the same length exactly as the length combination of the homologous arms and the *sppA* gene plus P43 promoter. Furthermore, DNA sequencing results confirmed the successful integration of *sppA* into *B. licheniformis* BL10, resulting in the new strain *B. licheniformis* BL10GS. The *tepA* overexpression strain was constructed and verified using the similar methods. The resulting new strain was named *B. licheniformis* BL10GT.

Both SPP overexpression strains BL10GS and BL10GT were cultivated in the ME medium for 24 h. As shown in Fig. 2a, the yield of total extracellular proteins increased by 37% in the *sppA* overexpression strain BL10GS, compared with that of the parental strain BL10. These results positively correlated with the SDS-PAGE results in Fig. 2b. Meanwhile, overexpression of *tepA* had no effect

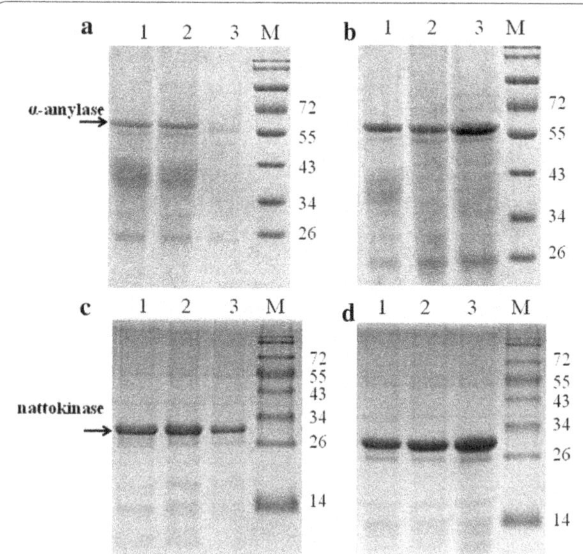

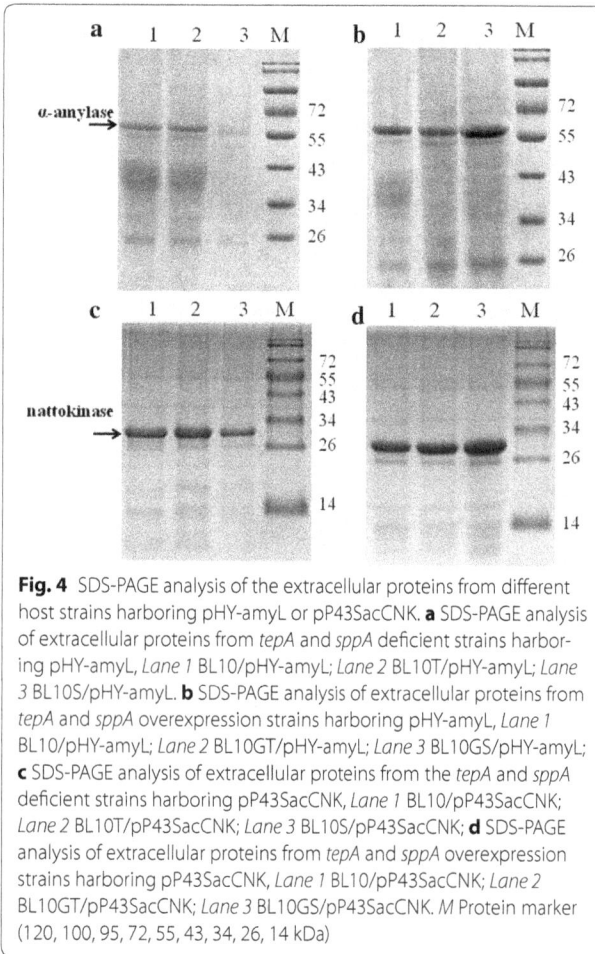

Fig. 4 SDS-PAGE analysis of the extracellular proteins from different host strains harboring pHY-amyL or pP43SacCNK. **a** SDS-PAGE analysis of extracellular proteins from *tepA* and *sppA* deficient strains harboring pHY-amyL, *Lane 1* BL10/pHY-amyL; *Lane 2* BL10T/pHY-amyL; *Lane 3* BL10S/pHY-amyL. **b** SDS-PAGE analysis of extracellular proteins from *tepA* and *sppA* overexpression strains harboring pHY-amyL, *Lane 1* BL10/pHY-amyL; *Lane 2* BL10GT/pHY-amyL; *Lane 3* BL10GS/pHY-amyL; **c** SDS-PAGE analysis of extracellular proteins from the *tepA* and *sppA* deficient strains harboring pP43SacCNK, *Lane 1* BL10/pP43SacCNK; *Lane 2* BL10T/pP43SacCNK; *Lane 3* BL10S/pP43SacCNK; **d** SDS-PAGE analysis of extracellular proteins from *tepA* and *sppA* overexpression strains harboring pP43SacCNK, *Lane 1* BL10/pP43SacCNK; *Lane 2* BL10GT/pP43SacCNK; *Lane 3* BL10GS/pP43SacCNK. *M* Protein marker (120, 100, 95, 72, 55, 43, 34, 26, 14 kDa)

on the concentration of the total extracellular proteins. Furthermore, in order to analyze the influence of *sppA* overexpression on cell lysis, the green fluorescent protein (GFP) was expressed in the parental strain BL10, the *sppA* mutant strain BL10S and the *sppA* overexpression strain BL10GS (Additional file 1: Figure S2). As shown in Additional file 1: Figure S3, GFP was expressed successfully in these strains, and no GFP was detected in ME medium after cultivation for 24 h. These results suggest that overexpression of *sppA* has no effect on the cell lysis. Furthermore, these results indicate that SppA could be used to improve the concentration of extracellular proteins produced by *B. licheniformis*.

The nattokinase and α-amylase expression vectors were electro-transferred into the *sppA* and *tepA* overexpression strains, and the recombinant strains were cultivated in the nattokinase and amylase production media for 48 h to evaluate the effects of overexpression of SPP genes on the target protein production, respectively. As shown in Fig. 3, the activities of target enzymes and cell yields both increased in the *sppA* overexpression strain, which was consistent with the improvement of target protein concentrations (Table 2). Based on our results, the cell biomass of *sppA* deficiency strains were decreased by 16 and 13% during α-amylase and nattokinase production, and the decreases in cell biomass observed were much lower than decreases in protein yields (67 and 30%, respectively). Also, the specific activities of α-amylase and nattokinase were increased by 43 and 18% respectively in the *sppA* overexpression strain, compared with those produced by *B. licheniformis* BL10. These results indicate that overexpression of *sppA* improves both cell biomass and protein yields. Meanwhile, in the *tepA* overexpression strain, only a 12% improvement of specific α-amylase activity was obtained and overexpression of *tepA* had no effect on the concentrations of nattokinase (Fig. 4b, d). These results were consistent with the enzymes activities (Fig. 3). Moreover, the concentrations of intracellular protein were measured in these strains duringα-amylase and nattokinase production, the SPPs deficiency and overexpression have no effects on the concentrations of intracellular protein (Additional file 1: Table S3).

Effect of SppA on the α-amylase secretion and cell growth during the fermentation process

Figure 5 shows the time courses of the cell growth and α-amylase yield of *B. licheniformis* BL10/pHY-amyL, *B. licheniformis* BL10S/pHY-amyL, *B. licheniformis* BL10T/pHY-amyL, *B. licheniformis* BL10GT/pHY-amyL, and *B. licheniformis* BL10GS/pHY-amyL in the α-amylase production medium. The culture was sampled every 4 h, and α-amylase activity and biomass were measured at the defined time. Based on our results, the α-amylase activities of BL10GS/pHY-amyL were higher than those of BL10/pHY-amyL and BL10S/pHY-amyL throughout the whole fermentation process, and the α-amylase activity was the highest in *B. licheniformis* BL10GS/pHY-amyL and the lowest in BL10S/pHY-amyL. Also, the biomass of BL10GS/pHY-amyL was higher than those of other

Table 2 Effects of deletion or overexpression of *sppA* and *tepA* on the concentrations of target proteins

	Strains	BL10	BL10T	BL10S	BL10GT	BL10GS
Concentrations (mg/L)	pHY-amyL	89.04 (± 5.43)	86.93 (± 6.25)	46.78 (± 2.43)	104.29 (± 9.54)	147.95 (± 11.32)
	pP43SacCNK	215.30 (± 10.19)	205.41 (± 21.64)	108.94 (± 8.11)	212.62 (± 14.46)	281.19 (± 12.65)

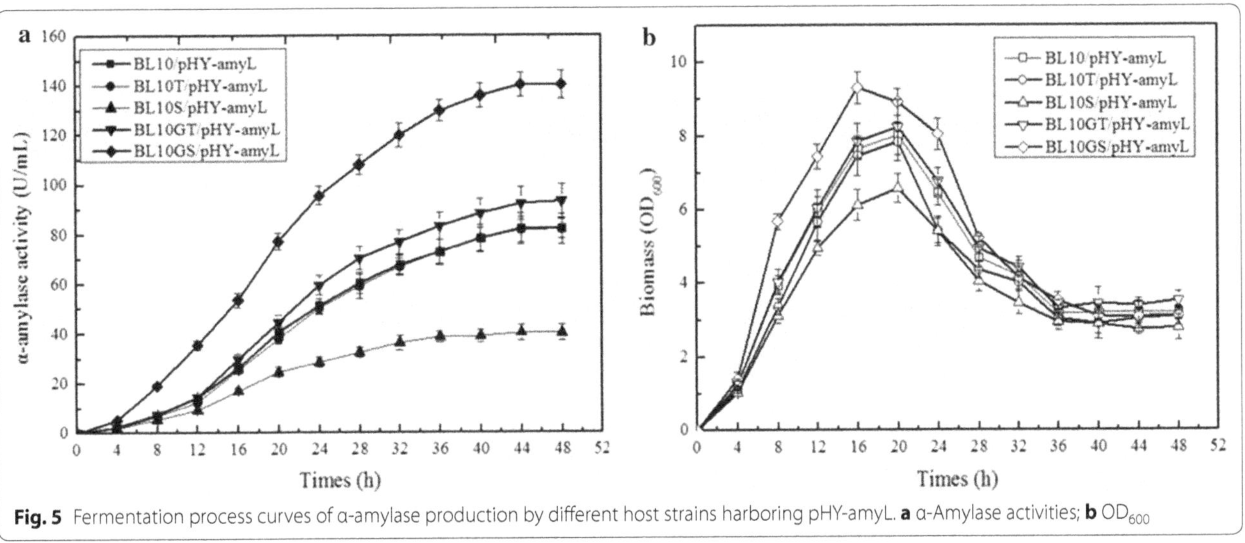

Fig. 5 Fermentation process curves of α-amylase production by different host strains harboring pHY-amyL. **a** α-Amylase activities; **b** OD_{600}

strains throughout the fermentation process. Meanwhile, deletion or overexpression of *tepA* had no effect on the cell growth and target production during the production of α-amylase (Fig. 5). These results indicated that overexpression of *sppA* could not only improve the protein secretion level, but also enhance the biomass yield.

The relative gene expression levels of BL10/pHY-amyL, BL10T/pHY-amyL, BL10S/pHY-amyL, BL10GT/pHY-amyL and BL10GS/pHY-amyL were also evaluated at the mid-log phase during the α-amylase production process (8 h). As shown in Fig. 6, the transcriptional expression levels of *sppA* and *tepA* could not be detected in the *sppA* and *tepA* deficient strains, and the *sppA* and *tepA* transcriptional expression levels increased by 11.43- and 11.47-fold in the BL10GS and BL10GT strains, respectively compared to the parental strain.

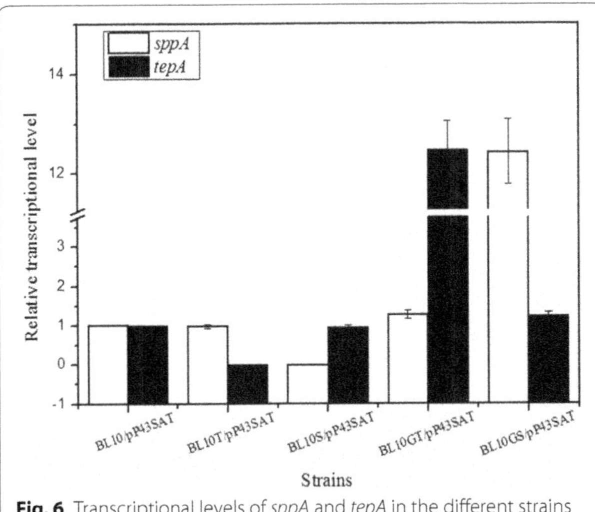

Fig. 6 Transcriptional levels of *sppA* and *tepA* in the different strains during α-amylase production

Discussion

Heterologous expression is an effective strategy to improve the target protein production, and many strategies have been carried out to improve the protein yield [3, 9, 29–31]. It has been previously shown that SPP can play an important role during protein secretion [32], However, prior to the current study, no research has been reported on the improvement of protein secretion by overexpression of SPP. In this study, we have demonstrated that the SppA SPP plays an important role for protein secretion in *B. licheniformis*, and overexpression of *sppA* could improve the yields of the total extracellular proteins and target proteins.

In this study, the yields and specific activities of target proteins were decreased significantly in the *sppA* deficient strain, and they were increased in the *sppA* overexpression strain, compared with those of the parental strain BL10. However, different expression levels of an alternative SPP, TepA, had no effect on the concentrations of heterologously produced extracellular target proteins. These results suggest that SppA is very important for protein secretion in *B. licheniformis*, and are consistent with previously reported results in *B. subtilis* [18, 19]. Previously, several researches have implied that the TepA of *B. subtilis* is a cytoplasmic, ClpP-like germination protease involved in spore outgrowth, and it is expressed almost exclusively during sporulation, and researches proposed that the TepA of *B. subtilis* is not the SPP [33, 34]. Since *B. licheniformis* have high homology with *B. subtilis*, it was suggested that the TepA of *B. licheniformis* BL10 might also not be the SPP by our results. Meanwhile, SPPs are required for degradation of signal peptides that are inhibitory to protein translocation [35, 36], and the deficiency of *sppA* might influence the ability to remove remnant signal peptides in the cellular membrane. It has been proposed that the accumulation of these remnant signal peptides might block the membrane channel for protein translocation and hinder protein secretion [18]. Therefore, we postulated that overexpression of *sppA* would increase the removal of the remnant signal peptides, therefore improving protein secretion. The extracellular activities of α-amylase produced by *B. licheniformis* BL10GS/pHY-amyL were higher than those of BL10/pHY-amyL throughout the whole fermentation process. These results indicate that overexpression of *sppA* could improve the α-amylase yield from the beginning of the fermentation, and SppA is necessary for highly efficient protein secretion. Meanwhile, the concentrations of intracellular protein showed no differences among these recombinant strains, which indicated that the higher levels of extracellular proteins might be due to the increase level of recombinant proteins production. Thus, in-depth research on remnant signal peptides in the cell membrane should be carried out to further understand this mechanism. In addition, our results indicate that overexpression of *sppA* could improve the cell growth (Fig. 3), which partially contributed to the increase of the α-amylase activity. However, since the increase in cell biomass observed was much lower than the increase in the α-amylase activity, we concluded that the enhanced α-amylase activity was mainly due to the increased protein secretion.

In order to improve extracellular protein production, several strategies have been developed for signal peptide processing by manipulating the signal peptidase and signal peptidase [37]. Malten et al. [29] have overexpressed the type I signal peptidase SipM in *Bacillus megaterium* MS941, and the yield of target protein Dsrs, mediated by its own signal peptide, was increased by 3.7-fold. Also, the signal peptidase I SipV was confirmed to play a vital role for nattokinase (mediated by the signal peptide of AprE) production in our previous study [8]. Previous studies have shown that either signal peptidase I (SipS or SipT) was sufficient for protein secretion [38]. Signal peptidases serve as the "scissors" to cut off the signal peptides from the pre-proteins [39], and multiple signal peptidases were responsible for the cleavage of different signal peptides [40]. Therefore, overexpression of a single signal peptidase could not be used as a universal strategy for enhancing protein secretion. In this research, the yield of total extracellular proteins was improved markedly in the *sppA* overexpression strain, and the concentrations of nattokinase and α-amylase mediated by different signal peptides were increased by 30 and 67%, respectively. These results implied that overexpression of *sppA* might act as an efficient and useful strategy for protein production.

In conclusion, this study implied that SppA is the main functional SPP for protein secretion in *B. licheniformis*, and overexpression of *sppA* indeed improved the concentrations of target proteins, biomass and specific activities. In this study, overexpression of *sppA* might act as a useful and efficient strategy to improve protein secretion. The host strain *B. licheniformis* BL10GS demonstrated high efficiency for protein secretion and could be employed as a promising industrial strain for extracellular production of valuable proteins.

Additional file

Additional file 1. All the sequences of the primers and concentrations of total intracellular proteins were listed in the Additional file (Table S1, Table S2 and Table S3). The agarose gel electrophoresis analysis of the SPPs deficient strains and the fluorescence detection of the cell and the fermentation supernatant of the BL10/pHY-GFP, BL10S/pHY-GFP, BL10GS/pHY-GFP were also contained in the Additional file (Fig. S1, Fig. S2 and Fig. S3). This information is available free of charge via the Internet http://microbialcellfactories.biomedcentral.com/.

Authors' contributions

DC, XW and SC designed the study. DC and HW carried out the molecular biology studies and construction of engineering strains. DC, HW and PH carried out the fermentation studies. DC, HW, CZ, QW, CTN and SC analyzed the data and wrote the manuscript. All authors read and approved the final manuscript.

Author details

[1] Hubei Collaborative Innovation Center for Green Transformation of Bio-Resources, College of Life Sciences, Hubei University, No. 368 Youyi Avenue, Wuchang District, Wuhan 430062, Hubei, People's Republic of China. [2] College of Food Science and Technology, Huazhong Agricultural University, Wuhan 430070, China. [3] Department of Chemistry, The State University of New York College of Environmental Science and Forestry (SUNY ESF), Syracuse, NY 13210, USA.

Acknowledgements

Not applicable.

Competing interests

The authors declare that they have no competing interests.

Funding

This work was supported by the National Science & Technology Pillar Program during the Twelfth Five-year Plan Period (2013AA102801-52), the Science and Technology Program of Wuhan (20160201010086).

References

1. Yang H, Liu L, Shin HD, Chen RR, Li J, Du G, Chen J. Comparative analysis of heterologous expression, biochemical characterization optimal production of an alkaline alpha-amylase from alkaliphilic *Alkalimonas amylolytica* in *Escherichia coli* and *Pichia pastoris*. Biotechnol Prog. 2013;29:39–47.
2. Lin S, Zhang M, Liu J, Jones GS. Construction and application of recombinant strain for the production of an alkaline protease from *Bacillus licheniformis*. J Biosci Bioeng. 2015;119:284–8.
3. Wang P, Wang P, Tian J, Yu X, Chang M, Chu X, Wu N. A new strategy to express the extracellular alpha-amylase from *Pyrococcus furiosus* in *Bacillus amyloliquefaciens*. Sci Rep. 2016;6:22229.
4. Westers L, Westers H, Quax WJ. *Bacillus subtilis* as cell factory for pharmaceutical proteins: a biotechnological approach to optimize the host organism. Biochim Biophys Acta. 2004;1694:299–310.
5. van Dijl JM, Hecker M. *Bacillus subtilis*: from soil bacterium to super-secreting cell factory. Microb Cell Fact. 2013;12:3.
6. Pohl S, Harwood CR. Heterologous protein secretion by *Bacillus* species from the cradle to the grave. Adv Appl Microbiol. 2010;73:1–25.
7. Wei X, Zhou Y, Chen J, Cai D, Wang D, Qi G, Chen S. Efficient expression of nattokinase in *Bacillus licheniformis*: host strain construction and signal peptide optimization. J Ind Microbiol Biotechnol. 2015;42:287–95.
8. Cai D, Wei X, Qiu Y, Chen Y, Chen J, Wen Z, Chen S. High-level expression of nattokinase in *Bacillus licheniformis* by manipulating signal peptide and signal peptidase. J Appl Microbiol. 2016;121:704–12.
9. Kang Z, Yang S, Du G, Chen J. Molecular engineering of secretory machinery components for high-level secretion of proteins in *Bacillus* species. J Ind Microbiol Biotechnol. 2014;41:1599–607.
10. Wang Y, Liu Y, Wang Z, Lu F. Influence of promoter and signal peptide on the expression of pullulanase in *Bacillus subtilis*. Biotechnol Lett. 2014;36:1783–9.
11. Bolhuis A, Broekhuizen CP, Sorokin A, van Roosmalen ML, Venema G, Bron S, Quax WJ, van Dijl JM. SecDF of *Bacillus subtilis*, a molecular Siamese twin required for the efficient secretion of proteins. J Biol Chem. 1998;273:21217–24.
12. Hunt JF, Weinkauf S, Henry L, Fak JJ, McNicholas P, Oliver DB, Deisenhofer J. Nucleotide control of interdomain interactions in the conformational reaction cycle of SecA. Science. 2002;297:2018–26.
13. Chen J, Fu G, Gai Y, Zheng P, Zhang D, Wen J. Combinatorial Sec pathway analysis for improved heterologous protein secretion in *Bacillus subtilis*: identification of bottlenecks by systematic gene overexpression. Microb Cell Fact. 2015;14:92.
14. Chen J, Gai Y, Fu G, Zhou W, Zhang D, Wen J. Enhanced extracellular production of alpha-amylase in *Bacillus subtilis* by optimization of regulatory elements and over-expression of PrsA lipoprotein. Biotechnol Lett. 2015;37:899–906.
15. Fu LL, Xu ZR, Li FW, Shuai JB, Lu P, Hu CX. Protein secretion pathways in *Bacillus subtilis*: implication for optimization of heterologous protein secretion. Biotechnol Adv. 2007;25:1–12.
16. Harwood CR, Cranenburgh R. *Bacillus* protein secretion: an unfolding story. Trends Microbiol. 2008;16:73–9.
17. van Roosmalen ML, Geukens N, Jongbloed JD, Tjalsma H, Dubois JY, Bron S, van Dijl JM, Anne J. Type I signal peptidases of Gram-positive bacteria. Biochim Biophys Acta. 2004;1694:279–97.
18. Bolhuis A, Matzen A, Hyyrylainen HL, Kontinen VP, Meima R, Chapuis J, Venema G, Bron S, Freudl R, MaartenvanDijl J. Signal peptide peptidase- and ClpP-like proteins of *Bacillus subtilis* required for efficient translocation and processing of secretory proteins. J Bio Chem. 1999;274:24585–92.
19. Nam SE, Paetzel M. Structure of signal peptide peptidase A with C-termini bound in the active sites: insights into specificity, self-processing, and regulation. Biochemistry. 2013;52:8811–22.
20. Qiu Y, Xiao F, Wei X, Wen Z, Chen S. Improvement of lichenysin production in *Bacillus licheniformis* by replacement of native promoter of lichenysin biosynthesis operon and medium optimization. Appl Microbiol Biotechnol. 2014;98:8895–903.
21. Qi G, Kang Y, Li L, Xiao A, Zhang S, Wen Z, Xu D, Chen S. Deletion of meso-2,3-butanediol dehydrogenase gene *budC* for enhanced D-2,3-butanediol production in *Bacillus licheniformis*. Biotechnol Biofuels. 2014;7:16.
22. Xue GP, Johnson BP, Dalrymple BP. High osmolarity improves the electrotransformation efficiency of the gram-positive bacteria *Bacillus subtilis* and *Bacillus licheniformis*. J Microbiol Meth. 1999;34:183–91.
23. Qiu Y, Zhang J, Li L, Wen Z, Nomura CT, Wu S, Chen S. Engineering *Bacillus licheniformis* for the production of meso-2,3-butanediol. Biotechnol Biofuels. 2016;9:117.
24. Chen J, Zhou Y, Zhao X, Chen S, Wei X. Comparative analysis of different *Bacillus licheniformis* host strains on the secretion expression of α-amylase. Food Sci. 2015;9:275–84.
25. Grintzalis K, Georgiou CD, Schneider YJ. An accurate and sensitive Coomassie Brilliant Blue G-250-based assay for protein determination. Anal Biochem. 2015;480:28–30.
26. Voigt B, Schweder T, Sibbald MJ, Albrecht D, Ehrenreich A, Bernhardt J, Feesche J, Maurer KH, Gottschalk G, van Dijl JM, Hecker M. The extracellular proteome of *Bacillus licheniformis* grown in different media and under different nutrient starvation conditions. Proteomics. 2006;6:268–81.
27. Tian G, Fu J, Wei X, Ji Z, Qi G, Chen S. Enhanced expression of *pgdS*, gene for high production of poly-γ-glutamic aicd with lower molecular weight in *Bacillus licheniformis* WX-02. J Chem Technol Biotechnol. 2013;89:1825–32.
28. Yangtse W, Zhou Y, Lei Y, Qiu Y, Wei X, Ji Z, Qi G, Yong Y, Chen L, Chen S. Genome sequence of *Bacillus licheniformis* WX-02. J Bacteriol. 2012;194:3561–2.
29. Malten M, Nahrstedt H, Meinhardt F, Jahn D. Coexpression of the type I signal peptidase gene *sipM* increases recombinant protein production and export in *Bacillus megaterium* MS941. Biotechnol Bioeng. 2005;91:616–21.
30. Degering C, Eggert T, Puls M, Bongaerts J, Evers S, Maurer KH, Jaeger KE. Optimization of protease secretion in *Bacillus subtilis* and *Bacillus licheniformis* by screening of homologous and heterologous signal peptides. Appl Environ Microbiol. 2010;76:6370–6.
31. Zhang JK, Kang Z, Ling Z, Cao W, Liu L, Wang M, Du G, Chen J. High-level extracellular production of alkaline polygalacturonate lyase in Bacillus subtilis with optimized regulatory elements. Bioresour Technol. 2013;146:543–8.
32. Nam SE, Kim AC, Paetzel M. Crystal structure of *Bacillus subtilis* signal peptide peptidase A. J Mol Biol. 2012;419:347–58.
33. Westers H, Darmon E, Zanen G, Veening JW, Kuipers OP, Bron S, Quax WJ, Van Dijl JM. The *Bacillus* secretion stress response is an indicator for alpha-amylase production levels. Lett Appl Microbiol. 2004;39:65–73.
34. Traag BA, Pugliese A, Setlow B, Setlow P, Losick R. A conserved ClpP-like protease involved in spore outgrowth in *Bacillus subtilis*. Mol Microbiol. 2013;90:160–6.
35. Chen CY, Malchus NS, Hehn B, Stelzer W, Avci D, Langosch D, Lemberg MK. Signal peptide peptidase functions in ERAD to cleave the unfolded protein response regulator XBP1u. EMBO J. 2014;33:2492–506.
36. Voss M, Schroder B, Fluhrer R. Mechanism, specificity, and physiology of signal peptide peptidase (SPP) and SPP-like proteases. Biochim Biophys Acta. 2013;1828:2828–39.
37. Dalbey RE, Wang P, van Dijl JM. Membrane proteases in the bacterial protein secretion and quality control pathway. Microbiol Mol Biol Rev. 2012;76:311–30.
38. Tjalsma H, Bolhuis A, van Roosmalen ML, Wiegert T, Schumann W, Broekhuizen CP, Quax WJ, Venema G, Bron S, van Dijl JM. Functional analysis of the secretory precursor processing machinery of *Bacillus subtilis*: identification of a eubacterial homolog of archaeal and eukaryotic signal peptidases. Genes Dev. 1998;12:2318–31.
39. Auclair SM, Bhanu MK, Kendall DA. Signal peptidase I: cleaving the way to mature proteins. Protein Sci. 2012;21:13–25.
40. Song Y, Nikoloff JM, Zhang D. Improving protein production on the level of regulation of both expression and secretion pathways in *Bacillus subtilis*. J Microbiol Biotechnol. 2015;25:963–77.

PERMISSIONS

All chapters in this book were first published in MCF, by BioMed Central; hereby published with permission under the Creative Commons Attribution License or equivalent. Every chapter published in this book has been scrutinized by our experts. Their significance has been extensively debated. The topics covered herein carry significant findings which will fuel the growth of the discipline. They may even be implemented as practical applications or may be referred to as a beginning point for another development.

The contributors of this book come from diverse backgrounds, making this book a truly international effort. This book will bring forth new frontiers with its revolutionizing research information and detailed analysis of the nascent developments around the world.

We would like to thank all the contributing authors for lending their expertise to make the book truly unique. They have played a crucial role in the development of this book. Without their invaluable contributions this book wouldn't have been possible. They have made vital efforts to compile up to date information on the varied aspects of this subject to make this book a valuable addition to the collection of many professionals and students.

This book was conceptualized with the vision of imparting up-to-date information and advanced data in this field. To ensure the same, a matchless editorial board was set up. Every individual on the board went through rigorous rounds of assessment to prove their worth. After which they invested a large part of their time researching and compiling the most relevant data for our readers.

The editorial board has been involved in producing this book since its inception. They have spent rigorous hours researching and exploring the diverse topics which have resulted in the successful publishing of this book. They have passed on their knowledge of decades through this book. To expedite this challenging task, the publisher supported the team at every step. A small team of assistant editors was also appointed to further simplify the editing procedure and attain best results for the readers.

Apart from the editorial board, the designing team has also invested a significant amount of their time in understanding the subject and creating the most relevant covers. They scrutinized every image to scout for the most suitable representation of the subject and create an appropriate cover for the book.

The publishing team has been an ardent support to the editorial, designing and production team. Their endless efforts to recruit the best for this project, has resulted in the accomplishment of this book. They are a veteran in the field of academics and their pool of knowledge is as vast as their experience in printing. Their expertise and guidance has proved useful at every step. Their uncompromising quality standards have made this book an exceptional effort. Their encouragement from time to time has been an inspiration for everyone.

The publisher and the editorial board hope that this book will prove to be a valuable piece of knowledge for researchers, students, practitioners and scholars across the globe.

LIST OF CONTRIBUTORS

Martina Aulitto, Gabriella Fiorentino, Danila Limauro, Emilia Pedone, Simonetta Bartolucci and Patrizia Contursi
Dipartimento di Biologia, Università degli Studi di Napoli Federico II, Complesso Universitario Monte S. Angelo, Via Cinthia, 80126 Naples, Italy

Salvatore Fusco
Division of Industrial Biotechnology, Department of Biology and Biological Engineering, Chalmers University of Technology, Gothenburg, Sweden

Björn D. Heijstra, Ching Leang and Alex Juminaga
LanzaTech, Inc., 8045 Lamon Ave, Suite 400, Skokie, IL, USA

Justyna Ruchala
Department of Biotechnology and Microbiology, University of Rzeszow, Zelwerowicza 4, 35-601 Rzeszow, Poland

Olena O. Kurylenko and Kostyantyn V. Dmytruk
Department of Molecular Genetics and Biotechnology, Institute of Cell Biology, Drahomanov Str., 14/16, Lviv 79005, Ukraine

Andriy A. Sibirny
Department of Biotechnology and Microbiology, University of Rzeszow, Zelwerowicza 4, 35-601 Rzeszow, Poland
Department of Molecular Genetics and Biotechnology, Institute of Cell Biology, Drahomanov Str., 14/16, Lviv 79005, Ukraine

Nitnipa Soontorngun
King Mongkut Technical University, Thonbury, Thailand

Iain B. H. Wilson and Martin Dragosits
Department of Chemistry, University of Natural Resources and Life Sciences, Muthgasse 11, 1190 Vienna, Austria

Josef W. Moser
Department of Chemistry, University of Natural Resources and Life Sciences, Muthgasse 11, 1190 Vienna, Austria
Austrian Centre of Industrial Biotechnology (ACIB), 1190 Vienna, Austria

Roland Prielhofer and Diethard Mattanovich
Austrian Centre of Industrial Biotechnology (ACIB), 1190 Vienna, Austria
Department of Biotechnology, University of Natural Resources and Life Sciences, Vienna, Austria

Samuel M. Gerner
University of Applied Sciences FH-Campus Wien, Bioengineering, Vienna, Austria

Alexandra B. Graf
Austrian Centre of Industrial Biotechnology (ACIB), 1190 Vienna, Austria
University of Applied Sciences FH-Campus Wien, Bioengineering, Vienna, Austria

Fenghua Chai, Ying Wang, Xueang Mei, Mingdong Yao, Yan Chen, Hong Liu, Wenhai Xiao and Yingjin Yuan
Key Laboratory of Systems Bioengineering (Ministry of Education), Tianjin University, 92, Weijin Road, Nankai District, Tianjin 300072, People's Republic of China
SynBio Research Platform, Collaborative Innovation Center of Chemical Science and Engineering (Tianjin), School of Chemical Engineering and Technology, Tianjin University, Tianjin 300072, People's Republic of China

Rémi Dulermo, François Brunel, Thierry Dulermo, Rodrigo Ledesma-Amaro, Jérémy Vion, Marion Trassaert, Stéphane Thomas, Jean-Marc Nicaud and Christophe Leplat
Micalis Institute, INRA-AgroParisTech, UMR1319, Team BIMLip: Integrative Metabolism of Microbial Lipids, Université Paris-Saclay, domaine de Vilvert, 78350 Jouy-en-Josas, France

Tuulikki Seppänen-Laakso, Sandra Castillo and Brian Gibson
VTT Technical Research Centre of Finland, Tietotie 2, 02044 Espoo, Finland

Kristoffer Krogerus
VTT Technical Research Centre of Finland, Tietotie 2, 02044 Espoo, Finland

Department of Biotechnology and Chemical Technology, Aalto University, School of Chemical Technology, Kemistintie 1, Aalto, 00076 Espoo, Finland

Paulo Gonçalves Teixeira, Raphael Ferreira, Yongjin J. Zhou and Verena Siewers
Department of Biology and Biological Engineering, Chalmers University of Technology, 412 96 Gothenburg, Sweden
Novo Nordisk Foundation Center for Biosustainability, Chalmers University of Technology, 412 96 Gothenburg, Sweden

Francesca Zerbini, Ilaria Zanella, Davide Fraccascia, Enrico König, Carmela Irene, Luca F. Frattini, Michele Tomasi, Laura Fantappiè, Luisa Ganfini, Elena Caproni and Guido Grandi
Synthetic and Structural Vaccinology Unit, CIBIO, University of Trento, Via Sommarive, 9, Povo, 38123 Trento, Italy

Matteo Parri and Alberto Grandi
Toscana Life Sciences Scientific Park, Via Fiorentina, 1, 53100 Siena, Italy

Wei Tao and Li Lv
School of Life Sciences, Tsinghua University, Beijing 100084, China

Guo-Qiang Chen
School of Life Sciences, Tsinghua University, Beijing 100084, China
Center for Synthetic and Systems Biology, Tsinghua University, Beijing 100084, China
Tsinghua-Peking Center for Life Sciences, Tsinghua University, Beijing 100084, China
Center for Nano- and Micro-Mechanics, Tsinghua University, Beijing 100084, China
MOE Key Lab of Industrial Biocatalysis, Department Chemical Engineering, Tsinghua University, Beijing 100084, China

Jianzhi Zhao, Chen Li, Yan Zhang, Yu Shen and Jin Hou
State Key Laboratory of Microbial Technology, School of Life Science, Shandong University, Jinan 250100, China

Xiaoming Bao
State Key Laboratory of Microbial Technology, School of Life Science, Shandong University, Jinan 250100, China
Shandong Provincial Key Laboratory of Microbial Engineering, School of Bioengineering, QiLu University of Technology, Jinan 250353, China

Min-Kyoung Kang
Department of Biology and Biological Engineering, Chalmers University of Technology, Kemivägen 10, 412 96 Gothenburg, Sweden

Yongjin J. Zhou
Department of Biology and Biological Engineering, Chalmers University of Technology, Kemivägen 10, 412 96 Gothenburg, Sweden
Novo Nordisk Foundation Center for Biosustainability, Chalmers University of Technology, 412 96 Gothenburg, Sweden
Division of Biotechnology, Dalian Institute of Chemical Physics, Chinese Academy of Sciences, Dalian 116023, China

Jens Nielsen
Department of Biology and Biological Engineering, Chalmers University of Technology, Kemivägen 10, 412 96 Gothenburg, Sweden
Novo Nordisk Foundation Center for Biosustainability, Chalmers University of Technology, 412 96 Gothenburg, Sweden
Novo Nordisk Foundation Center for Biosustainability, Technical University of Denmark, Kogle allé, 2970 Hørsholm, Denmark
Science for Life Laboratory, Royal Institute of Technology, 17121 Solna, Sweden

Nicolaas A. Buijs
Department of Biology and Biological Engineering, Chalmers University of Technology, Kemivägen 10, 412 96 Gothenburg, Sweden
Evolva Biotech, Lersø Parkalle, 40-42, 2100 Copenhagen, Denmark

Govinda Guevara, Julián Perera and Juana María Navarro Llorens
Department of Biochemistry and Molecular Biology I, Universidad Complutense de Madrid, 28040 Madrid, Spain

Laura Fernández de las Heras
Faculty of Science and Engineering, Microbial Physiology-Gron Inst Biomolecular Sciences & Biotechnology, Nijenborgh 7, 9747 AG Groningen, The Netherlands

Zaigao Tan, Jong Moon Yoon, Jacqueline V. Shanks and Laura R. Jarboe
4134 Biorenewables Research Laboratory, Department of Chemical and Biological Engineering, Iowa State University, Ames, IA 50011, USA

William Black
4134 Biorenewables Research Laboratory, Department of Chemical and Biological Engineering, Iowa State University, Ames, IA 50011, USA
Department of Chemical Engineering and Materials Sciences, University of California, 916 Engineering Tower Irvine, Irvine, CA 92697-2575, USA

Sambandam Ravikumar and Jong-il Choi
Biomolecules Engineering Lab, Department of Biotechnology and Bioengineering, Chonnam National University, 77 Yongbong-ro, Gwangju 61186, Republic of Korea

Mary Grace Baylon and Si Jae Park
Biomolecules Engineering Lab, Department of Biotechnology and Bioengineering, Chonnam National University, 77 Yongbong-ro, Gwangju 61186, Republic of Korea
Division of Chemical Engineering and Materials Science, Ewha Womans University, 52 Ewhayeodae-gil, Seodaemun-gu, Seoul 03760, Republic of Korea

Antonio Juarez
Institut de Bioenginyeria de Catalunya, Parc Científic de Barcelona, 08028 Barcelona, Spain
Departamento de Microbiología, Facultad de Biología, Universidad de Barcelona, Avda Diagonal, 643., 08028 Barcelona, Spain

Juan A. Villa, Val F. Lanza, Beatriz Lázaro, Gabriel Moncalián and Fernando de la Cruz
Departamento de Biología Molecular (Universidad de Cantabria) and Instituto de Biomedicina y Biotecnología de Cantabria IBBTEC (CSIC-UC), C/ Albert Einstein 22, 39011 Santander, Spain

Héctor M. Alvarez
INBIOP (Instituto de Biociencias de la Patagonia), Consejo Nacional de Investigaciones Científicas y Técnicas, Facultad de Ciencias Naturales, Universidad Nacional de la Patagonia San Juan Bosco, Ruta Provincial No 1, Km 4-Ciudad Universitaria 9000, Comodoro Rivadavia, Chubut, Argentina

Dongbo Cai, Hao Wang, Shouwen Chen, Penghui He, Chengjun Zhu and Qin Wang
Hubei Collaborative Innovation Center for Green Transformation of Bio-Resources, College of Life Sciences, Hubei University, No. 368 Youyi Avenue, Wuchang District, Wuhan 430062, Hubei, People's Republic of China

Xuetuan Wei
College of Food Science and Technology, Huazhong Agricultural University, Wuhan 430070, China

Christopher T. Nomura
Hubei Collaborative Innovation Center for Green Transformation of Bio-Resources, College of Life Sciences, Hubei University, No. 368 Youyi Avenue, Wuchang District, Wuhan 430062, Hubei, People's Republic of China
Department of Chemistry, The State University of New York College of Environmental Science and Forestry (SUNY ESF), Syracuse, NY 13210, USA

Index

P

R

S

T

W

X

Y

www.ingramcontent.com/pod-product-compliance
Lightning Source LLC
Chambersburg PA
CBHW082049190326
41458CB00010B/3490
* 9 7 8 1 6 8 2 8 6 7 4 0 2 *